Reliability Engineering

Stefan Bracke

Reliability Engineering

Data analytics, modeling, risk prediction

Stefan Bracke
Lehrstuhl für Zuverlässigkeitstechnik
und Risikoanalytik
Bergische Universität Wuppertal
Wuppertal, Germany

ISBN 978-3-662-67448-2   ISBN 978-3-662-67446-8 (eBook)
https://doi.org/10.1007/978-3-662-67446-8

This book is a translation of the original German edition „Technische Zuverlässigkeit" by Bracke, Stefan, published by Springer-Verlag GmbH, DE in 2022. The translation was done with the help of artificial intelligence (machine translation by the service DeepL.com). A subsequent human revision was done primarily in terms of content, so that the book will read stylistically differently from a conventional translation. Springer Nature works continuously to further the development of tools for the production of books and on the related technologies to support the authors.

© Springer-Verlag GmbH Germany, part of Springer Nature 2024

This work is subject to copyright. All rights are reserved by the Publisher, whether the whole or part of the material is concerned, specifically the rights of translation, reprinting, reuse of illustrations, recitation, broadcasting, reproduction on microfilms or in any other physical way, and transmission or information storage and retrieval, electronic adaptation, computer software, or by similar or dissimilar methodology now known or hereafter developed.
The use of general descriptive names, registered names, trademarks, service marks, etc. in this publication does not imply, even in the absence of a specific statement, that such names are exempt from the relevant protective laws and regulations and therefore free for general use.
The publisher, the authors, and the editors are safe to assume that the advice and information in this book are believed to be true and accurate at the date of publication. Neither the publisher nor the authors or the editors give a warranty, expressed or implied, with respect to the material contained herein or for any errors or omissions that may have been made. The publisher remains neutral with regard to jurisdictional claims in published maps and institutional affiliations.

This Springer imprint is published by the registered company Springer-Verlag GmbH, DE, part of Springer Nature.
The registered company address is: Heidelberger Platz 3, 14197 Berlin, Germany

For
Manmoleja
scientia, agnitio et opus: ars gratia artis

Foreword

A technically complex product or a complex production process fulfils on the one hand the requirements in terms of a usually wide range of functions, and on the other hand also the requirements for technical reliability with regard to the expected product life cycle and maintainability. The development of technical products over the past 100 years shows an enormous increase in terms of product complexity, the constant enlargement of the range of functions and functionalities. In addition, there is the interaction of components from different disciplines such as mechanical engineering, electrical engineering and computer science. This development has led to a focus on technical reliability in addition to the pure fulfilment of requirements with regard to the functional spectrum.

Against this background, technical reliability has established itself as an independent, interdisciplinary discipline, first in military technology and then in all civil technology disciplines and industries, such as power plant technology, aeronautical technology, automotive technology, apparatus/plant engineering, electrical/electronic consumer goods and medical technology. Technical reliability is an essential, elementary contributor with regard to the safe confident realization of function-critical and safety-critical product characteristics in the development, realization and operation of technically complex products. Furthermore, requirements of reliability engineering have also an influence on the manufacturing processes.

The focus of this book is on data analysis for mapping and assessing technical reliability by the use of probabilistics, statistics and modelling. The methods of technical reliability are applied in the elementary phases of the product development process (concept and series development, production) as well as during the field use of technical products.

The book is aimed equally at engineering students and engineers who deal with technical reliability in the context of the development and manufacture of complex technical products and in the context of field data analysis. The presentation of the methods and procedures of technical reliability follows the maxim "theory-guided – practice-oriented", so that this book can be used both as a reference work and as a textbook.

In the preparation of this work I have received support from various persons. Special thanks go to Alicia Puls for the extensive proofreading: The critical review of all texts, the careful examination of the procedures shown, as well as the discussions on technical content with patience and perseverance have contributed significantly to the success of the present work. Furthermore, I would like to thank Brigitte Fricke as well as Renate Rutter for their support in the preparation of numerous graphics and schematic diagrams. Furthermore, I would like to thank Marcin Hinz for the manifold exchange on the topic of accelerated testing, Christoph Rosebrock for the technical discussions on probabilistics and Dominik Brüggemann for the discussions on the topic of significance tests.

I would like to thank the ladies and gentlemen of Springer Verlag, in particular Susanne Schemann and Alexander Grün, for their dedicated advice and patience in the preparation of this book.

A very special, personal thanks goes to Monika Piskala for her constant support in this book project over many years. Finally, I would also like to thank Mr. Grauburgunder for his occasional presence at the one or other philosophical discussion and for his pleasant, cool and dry contributions.

Despite all efforts in the preparation of the texts and the presentation of the methods, errors are not excluded. Comments from readers on improvements or corrections to the technical content are welcome and can be sent to the subsequently following adress:

Univ.-Prof. Dr.-Ing. Stefan Bracke
Chair of Reliability Engineering and Risk Analytics
University of Wuppertal, Germany
e-mail address: bracke@uni-wuppertal.de

Cologne,   Stefan Bracke
May 2023

Contents

1 Introduction 1
   1.1 Structure of the Book 2
   1.2 History of the Discipline Technical Reliability 5

2 Technical Reliability and Product Emergence Process 9
   2.1 Product Emergence Process and Product Life Cycle 9
      2.1.1 Basic Features of the Product Emergence Process 9
      2.1.2 Concept and Series Development Phases 11
      2.1.3 Production and Distribution Phases 13
   2.2 Technical Reliability in the Product Emergence Process 15
      2.2.1 Use of Technical Reliability Methods 15
      2.2.2 Risk Elimination and Risk Prevention 18

3 Elementary Terms 21
   3.1 Statistics and Specification 21
   3.2 Product and Process Characteristics and Their Specifity 24
   3.3 Technical Reliability and Maintenance 28
      3.3.1 Technical Reliability According to Bitter and DIN 28
      3.3.2 Discussion of the Definition of Reliability
            According to Bitter and DIN 30
      3.3.3 Definition of Technical Reliability 31
      3.3.4 Technical Reliability and Failure Behaviour 32
      3.3.5 Data, Damage Data, Product Fleet and Censoring 33
      3.3.6 Key Indicators 34
      3.3.7 Maintenance and Quality 35
   3.4 Measuring and Testing 36
   3.5 Risk and Risk Analysis 39

4 Descriptive Statistics: Forms of Visulization 43
   4.1 Introduction and Case Study Engine Connecting Rod 43
   4.2 Value Pattern 45
   4.3 Bar Diagram 46
   4.4 Histogram 47
   4.5 Cumulative Diagram 48

	4.6	Mathematical Papers: Log Representations................	49
		4.6.1 Logarithmic paper	49
		4.6.2 Weibull Probability Paper	50
	4.7	Boxplot..	52
	4.8	Isochrone Diagram...	54
	4.9	X–Y Plot	58
5	**Probability and Random Experiments**		**61**
	5.1	Probability	61
	5.2	Probabilistics	68
	5.3	Examples for Probabilistics	71
		5.3.1 Sampling Inspection in the Incoming Inspection......	71
		5.3.2 Urn Model: The Ball Drawing...................	72
		5.3.3 The Monty Hall Problem (Goat Problem)	75
6	**Data Collection and First Analysis Steps**		**79**
	6.1	Sample	79
	6.2	Data Classification	80
	6.3	Estimation Methods for Unknown Parameters Mean and Dispersion of a Distribution	83
		6.3.1 Estimators...................................	84
		6.3.2 Estimators and Estimates for Mean M and Variance σ^2.	85
		6.3.3 Estimation Function and Estimation of the Proportion Value p	86
	6.4	Basic Estimators and Characteristics of a Sample............	87
		6.4.1 Case Study Engine Connecting Rod: Characteristic Description, Characteristic Values of a Sample	87
		6.4.2 Estimator for the Mean Value Based on a Sample.....	88
		6.4.3 Estimator for the Dispersion Based on a Sample......	89
		6.4.4 Mean, Dispersion and Skewness of a Measured Value Series	90
7	**Distribution Models and Functions**............................		**93**
	7.1	Fundamentals of a Distribution Model	93
	7.2	Continuous Distribution Models	94
		7.2.1 Basics and Fields of Application	94
		7.2.2 Normal Distribution............................	96
		7.2.3 Standard Normal Distribution and Application	99
		7.2.4 Folded Normal Distribution......................	104
		7.2.5 Rayleigh Distribution...........................	106
		7.2.6 Logarithmic Normal Distribution	107
		7.2.7 Weibull Distribution............................	109
		7.2.8 Exponential Distribution	111
		7.2.9 Uniform Distribution	112
		7.2.10 Gamma Distribution............................	114

Contents

		7.2.11	Erlang Distribution	115
		7.2.12	Extreme Value Distribution Models	117
		7.2.13	Distribution Model and Technical Reliability Analysis	119
	7.3	Test Distributions		121
		7.3.1	t-Distribution According to Gosset (Respectively Student)	121
		7.3.2	χ^2-Distribution	123
		7.3.3	Fisher's F-Distribution	125
	7.4	Discrete Distribution Models		126
		7.4.1	Basics	127
		7.4.2	Binomial Distribution	128
		7.4.3	Poisson Distribution	132
		7.4.4	Hypergeometric Distribution	135
	7.5	Growth Process and Saturation Function		137
		7.5.1	Logistic Function	137
		7.5.2	Application Example Verhulst Model: Corrosion Process	139
	7.6	Mixed Distribution Models		140
		7.6.1	Alternative Models	141
		7.6.2	Competing Models	143
8	**Distribution Models: Parameter Estimation**			**145**
	8.1	Regression Analysis		146
	8.2	Maximum Likelihood Estimation Method (Fisher)		147
	8.3	Special Parameter Estimators with Reference to Specific Distribution Models		150
		8.3.1	Dubey	150
		8.3.2	Gumbel	152
		8.3.3	Tintner and Rhodes	153
9	**Confidence Intervals**			**155**
	9.1	Introduction Confidence Interval		155
	9.2	Confidence Intervals for Estimators with Normally Distributed Population		157
		9.2.1	Confidence Interval for Unknown Mean with Known Variance	157
		9.2.2	Confidence Interval for Unknown Mean with Unknown Variance	160
		9.2.3	Confidence Interval for Unknown Variance	161
		9.2.4	Application Example: Confidence Interval for Mean and Variance	163
	9.3	Confidence Interval for Proportion Value with Binomially Distributed Population		167
		9.3.1	Approximation via Normal Distribution	167

		9.3.2	Application Example: Confidence Interval for a Proportion Value p..........................	170
		9.3.3	Approaches Based on Clopper-Pearson and Fisher.....................................	172
	9.4	Weibull Distribution Model: Confidence Intervals of Parameters.................................		175
		9.4.1	Shape Parameter.................................	175
		9.4.2	Characteristic Life Span..........................	175
		9.4.3	Threshold (Threshold, Failure-Free Time)..........	176
	9.5	Confidence Interval for a Function (Distribution Model)......		177

10 Correlation and Regression 181
   10.1 Basics of Correlation and Regression 181
   10.2 Correlation Analysis..................................... 182
       10.2.1 Correlation According to Bravais-Pearson........... 185
       10.2.2 Correlation According to Spearman................ 185
   10.3 Regression Analysis..................................... 186
   10.4 Case Studies on Correlation and Regression Analysis......... 190
   10.5 Spurious Causality and Correlation....................... 196

11 System Analysis: Function, Fault Tree and Failure Mode and Effects 201
   11.1 Basics of Function and Fault Tree Analysis................. 201
       11.1.1 Fault tree: Development and Visualisation.......... 202
       11.1.2 Boolean Algebra................................. 205
       11.1.3 Illustration of System Function and Failure Analysis............................ 208
       11.1.4 Importance Parameters........................... 212
   11.2 Application Examples: Function and Fault Tree Analysis...... 213
       11.2.1 Braking System of a Motor Vehicle................ 213
       11.2.2 Bridge Circuit................................... 216
       11.2.3 Double Bridge Configuration...................... 219
   11.3 Failure Mode and Effects Analysis (FMEA)................. 225

12 Analysis of the Failure Behaviour of Components, Assemblies, Systems 231
   12.1 Reliability Analysis Based on Damage Data................. 231
       12.1.1 Base of Operations and Overview 232
       12.1.2 Data Review.................................... 232
       12.1.3 Mapping of the Failure Behaviour................. 233
       12.1.4 Key Indicators for the Characterisation of the Failure Behaviour.......................... 235
       12.1.5 Confidence Intervals for Parameters, Key Indicators and Model 236
       12.1.6 Survival Probability and Failure Rate 237

		12.1.7	Interpretation of Key Indicators and Model Parameters	238
		12.1.8	Candidate Prognosis	242
	12.2	Case Study: Analysis of Failure Behaviour of a Coolant Pump		243
	12.3	Candidate Prognosis in Testing and Field		251
		12.3.1	Sudden Death Test and Field Data Analytics	252
		12.3.2	Johnson Ranking	253
		12.3.3	Procedure According to Nelson	254
		12.3.4	Product Use Profile: Empirical Distribution Function	255
		12.3.5	Kaplan and Meier Method	257
		12.3.6	Eckel Method	257
	12.4	RAPP – Analysis of Serial Damages		258
		12.4.1	Risk Analysis and Prognosis of complex Products (RAPP)	259
		12.4.2	Case Study RAPP: Damage Focus in a Fleet	264
	12.5	Case Study Sensor: Failure Behaviour; Saturation Model Versus Weibull Model		269
13	**Significance Tests**			275
	13.1	Introduction and Application Example		275
	13.2	Basics of Significance Tests		278
		13.2.1	Performance of Significance Tests, p-Value	278
		13.2.2	Error 1st Kind, Error 2nd Kind and Test Efficiency	280
		13.2.3	Dependent and Independent Samples	282
		13.2.4	Parametric and Parameter-Free Significance Tests	283
		13.2.5	Overview: Statistical Tests for the One, Two, Multiple Sample Case	284
	13.3	Single Sample Case		286
		13.3.1	Wallis-Moore Test for Randomness	286
		13.3.2	Kolmogorov–Smirnov Test: Test of Goodness of Fit	291
		13.3.3	Sign Test: Test for Location Parameters	296
		13.3.4	Test for Outliers Within an Observation Series	302
		13.3.5	t-Test: Comparison of Sample Mean Against Reference Value	306
		13.3.6	Cox and Stuart Test: Trend Analysis	307
	13.4	Two Sample Case		311
		13.4.1	Mann–Whitney U Test: Comparison of Two Centroids	311
		13.4.2	Siegel-Tukey Test: Dispersion Comparison	317
		13.4.3	t-Test: Comparison of Two Mean Values	324
		13.4.4	F-Test: Comparison of Two Variances	331

	13.5	Multi-Sample Case.	333
		13.5.1 Kruskal–Wallis H Test and Conover Post hoc Analysis: Centroids	334
		13.5.2 Bartlett-Test: Comparison of Variances	337
		13.5.3 Meyer-Bahlburg Test: Comparison of Variances	339
14	**Prototype Testing and Accelerated Testing**		341
	14.1	Accelerated Testing	341
	14.2	Quantitative Accelerated Testing: Models	344
		14.2.1 Arrhenius Model	344
		14.2.2 Wöhler Test and Damage Accumulation Hypotheses	348
		14.2.3 Inverse Power Law: Introduction	356
		14.2.4 Coffin-Manson Model	357
		14.2.5 Inverse Power Weibull Model	357
		14.2.6 Lundberg-Palmgren Equation	358
		14.2.7 Taylor Equation	359
		14.2.8 Eyring Model	360
		14.2.9 Power Acceleration Factor	360
	14.3	Minimum Test Scope in a Prototype Test	360
		14.3.1 Test Scope on the Basis of Survival Probability	361
		14.3.2 Test Scope Based on Failure Behaviour and Test Duration	362
	14.4	Case Study: Accelerated Testing and Weibull Distribution Model	366
		14.4.1 Introduction	366
		14.4.2 Accelerated Testing: Example Paper Clip Bending	367
	14.5	Qualitative Accelerated Testing: HALT and HASS	372
15	**Test Process Suitability**		377
	15.1	Basics of the Analysis of Measuring and Testing Processes	377
	15.2	Test Equipment Capability and Test Process Suitability	381
	15.3	Short-Term Test Equipment Capability: C_g-/C_{gk}-Study	384
		15.3.1 Basics of the C_g-/C_{gk}-Study	384
		15.3.2 Application Example C_g-/C_{gk}-Study: Cylinder Head Testing	386
	15.4	Repeatability and Reproducibility: %GRR Study	387
		15.4.1 Basics of the %GRR Study	387
		15.4.2 Application Example %GRR Study: Engine Piston Rod Testing	389
	15.5	Test Process Suitability: Analysis of Measurement Uncertainty (Q_{MS}-/Q_{MP}-Study)	392
		15.5.1 Analysis of the measurement uncertainty	392
		15.5.2 Notes on the Analysis of Measurement Uncertainty	396

Contents

    15.5.3 Application Example Q_{MS}-/Q_{MP}-Study:
       Measurement of Body-in-White 399
  15.6 Notes and References: Procedure X 402

16 Statistical Process Control (SPC) 403
  16.1 Basics and Definitions 404
    16.1.1 Process suitability analyses 406
    16.1.2 Inspection Strategy: Full Inspection versus
       Sampling Inspection Procedure 408
  16.2 Planning and Realization of SPC 409
    16.2.1 Procedure 409
    16.2.2 Summary: Scope of Analysis, Indices and Notes 412
  16.3 Machine and Process Analysis: Capability Indices 413
    16.3.1 Machine Capability and Preliminary
       Process Capability 413
    16.3.2 Process Capability in the Current Series 414
    16.3.3 Capability Analysis with Arbitrary
       Distribution Model: Quantile Approach 417
    16.3.4 Capability Analysis and Interpretation
       of Process Images 420
    16.3.5 Case Study: Capability Analysis
       of a Machining Process 423
  16.4 Multivariate Process Evaluation 424
    16.4.1 Multivariate Process Capability Index 426
    16.4.2 Reject Probability Regarding Characteristic Sets 428
  16.5 Process Visualization and Control Technology:
    Control Charts 431
    16.5.1 Basics: Structure of Control Charts 432
    16.5.2 Control Charts for Continuous Characteristics 435
    16.5.3 Case Study Engine Piston Rod: Acceptance
       and Shewhart Control Charts 440
    16.5.4 Control Charts for Discrete Characteristics 441
    16.5.5 Case Study Grey Cast Iron Component
       Production: Design of a Control Chart 445
    16.5.6 Analysis of Typical Process Scenarios
       by Means of Control Charts 446

Appendix 449

References 493

Index 501

Introduction 1

A technical product is characterised by a high level of functionality, good usability and an attractive design, as well as a high level of technical reliability. The technical reliability of a product is defined as follows:

"Reliability is the probability that a unit will not fail with respect to its function- and safety–critical characteristics during a defined time interval or interval related to a time-correlated lifetime variable under given functional and environmental conditions"; cf. in detail Sect. 3.3.

The development and manufacture of reliable products, components and systems in the various specialist disciplines and industries – such as power plant engineering, automotive engineering, aerospace engineering, toolmaking and mechanical engineering, plant engineering, apparatus/plant engineering, consumer goods industry – takes place under challenges and conditions that have been subject to dynamic development for decades:

- Increasing product complexity, number of derivatives and variants,
- Continuous expansion of the requirement profile for the range of functions and functionality,
- Interdependencies between mechanical and electronic components,
- Integration of complex product and process controls: High data volume, complex data compression and focused information appropriation,
- Different product life cycles of mechanical and electronic components,
- Interaction of mechanical components, hardware and software,
- Interaction through networking of products and processes,
- Complex value networks: in-house production versus outsourcing,
- Reduction of product development time due to competitive situation,
- Legislation regarding product safety and product liability.

The technical reliability of a product can be analysed in every phase of the product development process (concept, series development and production) as well as in the field on the basis of data, represented in models and also assessed and influenced on the basis of this.

Therefore, the focus of this book is on data analytics with the aid of probabilistics, statistics, and modeling for mapping and assessing the technical reliability of a product or a production process. The presented methods of technical reliability, which are mostly located in statistics, can be used for data analysis in all phases of the product development process and in the field.

Both the theories of the methods of technical reliability are presented and the application of the same is illustrated by numerous case studies. The application substantiates the methods of technical reliability. Special emphasis is given to the manual comprehensibility of the methods shown (with the exception of those approaches that require a numerical solution), in order to provide the reader with a basic approach as well as an in-depth understanding.

1.1 Structure of the Book

In the introduction to this work, the structure of the book is explained and an outline of the historical development of the discipline of technical reliability is given (Chap. 1).

Chapter 2 is dedicated to the product emergence process respectively to the product life cycle and the associated elementary phases and activities with a focus on technical reliability with the aid of examples from automotive engineering. Furthermore, the possibilities of influencing the technical reliability of a product and the decision-making with regard to the activities in the context of risk prevention in development, production and field use are shown.

The basic terms of technical reliability and related disciplines are outlined in Chap. 3. The presentation of terms from the subject areas of statistics, product and process characteristics, technical reliability and maintenance, measurement and testing, and risk and risk analysis are supplemented selectively with technical perspectives to facilitate understanding in the engineering application context.

Chapter 4 focuses on basic options for visualizing observed values and measurement or test data. The forms of data representation shown are used for illustration and comparability in the data analysis of a technical issue and are explained using the engine piston rod as a case study.

The basics of probability and probability theory (probabilistics) in the context of single- and multi-stage random experiments as well as various application examples for substantiation – sampling in incoming goods inspection, the urn model as well as the Monty Hall problem – form the focus of Chap. 5.

Chapter 6 contains simple principles of data collection as well as first steps of data analysis (data summarization). Sampling, data classification and basic estimation methods for unknown parameters of a distribution are presented. The chapter concludes with the application of different estimators for the unknown parameters mean (centroid) and dispersion on the basis of available samples.

Chapter 7 is devoted to important distribution models that are used in reliability and risk analysis. First, an introduction to the general principles of a distribution model is given. This is followed by an explanation of continuous distribution models and test distributions. Furtermore, an outline of some discrete distribution models is given. An insight into the topic of growth or saturation functions, which can also be used in the analysis of complex damage events, is given. The chapter concludes with an overview of mixed distribution models with a view to mapping complex failure behaviour.

The focus of Chap. 8 is the estimation of parameters with respect to fitting a distribution model to available measured values. The focus is on two methods: First, the robust, and in many cases manually easy to perform, regression analysis is outlined. On the other hand, Fisher's maximum likelihood estimator (MLE) is explained, which is characterized by its universal applicability and is therefore a frequently used method. Furthermore, special parameter estimation methods with reference to distribution models are listed, which are used in technical reliability analysis.

Chapter 9 deals with the determination of confidence intervals for different parameter estimators as well as for adjustment models. After an introduction to the basics, procedures for the determination of confidence intervals of the parameter estimators for mean and dispersion in the case of a normally distributed population as well as for the proportion value in the case of a binomially distributed population are presented. Furthermore, possibilities for the determination of confidence intervals for estimators of the Weibull distribution model parameters are presented. The chapter concludes with the procedure for determining the confidence interval for an adjustment function or for a distribution model fit.

Correlation analysis and the regression analysis that often goes hand in hand with it can be used to depict the correlation between two or more variables – e.g. of function- or safety – critical characteristics. The fundamentals and fields of application of correlation and regression analysis are shown in Chap. 10. Bivariate correlation analysis and regression analysis are illustrated with the aid of various case studies. The chapter concludes with a discussion of spurious causality and correlation.

Chapter 11 focuses on the analysis of the function and failure of technical systems. First, the fundamentals of function and fault tree analysis (FTA), Boolean algebra and importance parameters are presented. After illustration by some application examples, the Failure Mode and Effects Analysis (FMEA) is briefly explained in conclusion: The FMEA is also one of the important methods for risk prevention and, with its inductive approach, represents the counterpart to the Fault Tree Analysis (FTA) with its deductive character.

Chapter 12 focuses on outlining procedures for performing a reliability analysis based on damage data. The damage data refer to products, components or systems and may have been collected during testing (development phase), during end-of-line testing (production) or on the basis of a product fleet in field use. After explaining a general procedure for reliability analysis based on damage data, the application follows in the context of a case study from automotive engineering.

Furthermore, procedures for candidate prediction – candidates for potential failure – in testing and in the field are explained. The presentation of the procedure "Risk Analysis and Prognosis of complex Products (RAPP) for the risk analysis of serial damages within a product fleet in the field forms the completition of Chap. 12.

The main area of Chap. 13 is data analysis on the basis of significance tests (hypothesis tests). After an introduction to the topic of significance testing, possible objectives for the use of significance tests are discussed in the context of an application example. The basics of significance testing as well as the associated key variables and terminology form the starting point for all the significance tests presented below for analyses in the one-sample case, two-sample case and multiple-sample case. The focus of the investigations is on observations such as trends, periodicities or outliers in a series of measured values. Furthermore, differences in centroids and dispersion can be investigated, for example to examine different characteristics of different prototypes in the development phase or different production batches. Furthermore, different failure behaviors of products in the field based on failure data can be the subject of data analysis. In order to enable analyses with explicitly small data sets (e.g. in prototype tests or at an early stage of an growing amount of field damage events), special emphasis is placed on the explanation of parameter-free significance analyses.

In the context of series development of a product development process, prototype testing plays an important role in analyzing the technical reliability of a product at a certain stage of development. Chapter 14 is dedicated to the basics of accelerated testing, followed by an explanation of important models that can be used in the planning and implementation of accelerated testing activities. Furthermore, approaches to determine a minimum scope of prototypes to be tested within the testing are shown. A case study is used to illustrate accelerated testing using the Weibull distribution model and the corresponding detailed data analysis. The chapter concludes with an explanation of the qualitative procedures Highly Accelerated Life Test (HALT) and Highly Accelerated Stress Screening (HASS).

The starting point of a reliability or risk analysis is based on collected measurement or test data. The requirement for the collection of measurement and test data is the analysis and determination of test process suitability. The measurement system analysis is the focus of Chap. 15. First, an introduction to the basics of the analysis of measurement and test processes as well as a comparison to the known methods of test process suitability analysis are given. Then the analysis of the short-term test equipment capability (C_g-, C_{gk}-study), the repeatability and reproducibility (%GRR-study) as well as the comprehensive analysis of the test process suitability (Q_{MS}-/Q_{MP}-study) with determination of the measurement uncertainty U are explained.

Chapter 16 focuses on the control and quality capability of production processes for the manufacture of technically complex products. The term "reliability" is used here in deviation from the above definition and refers to process control and quality capability with respect to functional critical product characteristics. Statistical Process Control (SPC) has established itself in industrial practice as a method for ensuring controlled and quality-capable processes. After presenting the basics of SPC, the univariate process analysis by means of process capability indices as well as process visualizations within a statistical process control in the context of the different phases of the product manufacturing process (preproduction and series production) are shown. Furthermore, approaches to multivariate process analysis using process capability indices are outlined. The final part is the control engineering with regard to production processes with its components process visualization, control chart design and analysis.

The contents of this book are based on numerous references. The bibliography (appendix) lists all the literature sources used. The tables used are also reproduced in the appendix.

1.2 History of the Discipline Technical Reliability

In this chapter, a brief insight into the history of the discipline of technical reliability in Germany is given and the development of reliability methods is outlined using the example of the Fault Tree Analysis (FTA) and Failure Mode and Effects Analysis (FMEA) methods in Germany and the USA.

The discipline of technical reliability developed, like many technical disciplines, initially in the field of military technology. After the formulation and publication of the first standards and methods for reliability analysis in the 1950s and 1960s by the US Army (e.g.: United States Military Procedure (MIL-P 1949) or later cumulated in (MIL 1998)), technical reliability established itself as an independent, interdisciplinary discipline in the engineering sciences. Civil engineering disciplines have adopted and steadily expanded the subject content of technical reliability. These engineering disciplines from the civil sector were (and still are) mostly related to product and process safety, such as power plant engineering, aeronautical engineering, automotive engineering, process control engineering and medical engineering. Furthermore, technical reliability plays an important role in the development of products and processes, when it is not only a matter of the assessment of safety–critical features but also of the assessment of function-critical features. Due to this development, technical reliability has established itself in almost all engineering disciplines and industries, such as apparatus and plant engineering, plant construction, automotive engineering, household technology, and electrical/electronic consumer goods.

The following explanations trace the historical development of the discipline of technical reliability at the institutional level in Germany and at the methodological level using the example of probabilistics and the Fault Tree Analysis (FTA) method in the USA.

The development of the first thrust-propelled missiles, explicitly the Fieseler Fi 103 cruise missile, was carried out at the Fieseler Werke in the 1940s by engineers Robert Lusser and Willy Fiedler; cf. NRS Tait (1995). It was found that the overall system had low reliability due to unforeseen component failures: the higher the system complexity and the more components the system had, the lower the overall reliability of the system. Lusser formulated as a law that the system survival probability R(t) is determined by the product of the survival probabilities of the system components (given a series structure). Thus, the survival probability R(t) of a system is always lower than the survival probability of the weakest component. The law has been called Lusser's Law (NRS Tait 1995). In the end, this is the application of the multiplication law (for independent events) from probabilistics with respect to survival probability; cf. Chap. 5.

Further developments in rocketry were now characterized by the use of probabilistic reasoning, applied statistics, and a methodical approach to developing technical, reliable solutions.

After World War II, Lusser initially moved to the USA (Operation Paperclip) to work for Werner von Braun (development of the Redstone rocket) and in 1959 became Technical Director of the Entwicklungsring (EWR) Süd (consisting of: Heinkel Flugzeugbau, Messerschmitt, Bölkow). The EWR subsequently dissolved in the course of the merger to form Messerschmitt-Bölkow-Blohm (MBB) in 1969. MBB was headed by Ludwig Bölkow, among others. Furthermore, Bölkow was involved in the development of the world's first operational and mass-produced jet-powered aircraft, the Messerschmitt Me 262 (anno 1943–1945), at the Messerschmitt company. In 1964, Bölkow was significantly involved in initiating the foundation of the Reliability and Quality Control Committee within the Association of German Engineers (VDI); cf. VDI Guideline 4001 (1998). The committee compiled and published the standard work "VDI Handbuch Technische Zuverlässigkeit". This made the specialist knowledge for every (technical) discipline—independent of military technology—accessible to a broad, civilian spectrum of users.

Peter Bitter was also active at the Entwicklungsring (EWR) Süd (reliability department). Here at the EWR, too, the idea matured of making the technical reliability expertise available to the general public. In 1967, the information program Reliability was started on behalf of the Federal Ministry of Defence. In a period of just under 2 years, about 60 "Reliability Instruction Letters" (mathematics, methods of reliability analyses, data analysis) were developed and sent to clients and industry (Bitter 1971). Based on the contents of the reliability teaching notes, Bitter et al. published the standard work "Technische Zuverlässigkeit" (publisher Messerschmitt-Bölkow-Blohm (MBB)) in 1971. The work appeared in second and third editions in 1977 and 1986 and is considered, along with the VDI Handbook mentioned above, to be one of the first basic works on technical reliability in the German-speaking world; cf. (Bitter 1986).

In addition to this brief history of the development of the discipline of technical reliability, the development of reliability methods is outlined using the example of the FTA and the FMEA in the USA.

One of the most important methods of technical reliability analysis is the Fault Tree Analysis (FTA). FTA is based on Boolean algebra and probabilistic analysis. The aim is to determine the probability of failure of a technical system. FTA was initially developed for reliability analyses in military technology.

1.2 History of the Discipline Technical Reliability

H. A. Watson, working at Bell Laboratories, developed the FTA in the 1960s to analyze the reliability and safety of the LGM-30 Minuteman Launch Control System; cf. (Watson 1961), (Rogers 1971), (Martensen and Butler 1987), (Leveson 2003). The LGM-30 Minuteman is a Boeing intercontinental ballistic missile equipped with nuclear warheads. Subsequently, the FTA was also applied to the development of civil aircraft by Boeing. With the civilian use of nuclear energy, FTA was also used in the development of nuclear power plants for safety analyses, starting in the 1970s (Ericson 1999). Subsequently, the use of FTA became established in the automotive industry by manufacturers and suppliers alongside Failure Mode and Effects Analysis (FMEA). FMEA is an inductive method for the systematic discovery of potential failures and risks in the development of products and planning of production processes. Although the FMEA is not a probabilistic method, it is one of the most widespread methods for risk analysis and risk prevention and, due to its inductive character, forms the counterpart to the deductively oriented FTA. The FMEA was also initially used in the context of the development of military products and was published in 1949 as Military Specification for Planned Maintenance MIL-P 1629 (MIL-P 1949); cf. however also (MIL 1998). In the 1960s, FMEA was applied by NASA as part of the Apollo program in the context of extensive safety engineering analyses.

Today, both Fault Tree Analysis (FTA) and Failure Mode and Effects Analysis (FMEA) are standards in most technical, civil disciplines with regard to reliability and risk analysis. Especially in automotive engineering, numerous publications can be found on the mentioned methods, e.g. at the German Association of the Automotive Industry VDA; cf. e.g. (VDA 1976), (VDA 2006) and (VDA 2019a).

Technical Reliability and Product Emergence Process

Technical reliability is a characteristic of a product or a process. The analysis of technical reliability takes place in various phases of the product emergence process or the product life cycle. Furthermore, technical reliability is elementary for risk analysis and risk prevention in the development and production of new product generations and in the assessment of product operation in the use phase. This chapter outlines the product emergence process and product life cycle with elementary phases and activities with a focus on technical reliability (Sect. 2.1). Furthermore, possibilities of influence and confidence in decision making with regard to the technical reliability of a product are shown in the context of risk prevention in development, production and field use (Sect. 2.2). The explanations are based on the example of automotive engineering, since the development and manufacture of an automobile involves a large number of technical disciplines. The relationships shown are therefore representative and easily transferable to other technically complex products.

2.1 Product Emergence Process and Product Life Cycle

2.1.1 Basic Features of the Product Emergence Process

A product emergence process or product life cycle is divided into different phases (e.g.: series development), subphases (e.g.: preparation for series production) and activities (e.g.: component testing), see DIN ISO 15226 (DIN 2017). The terms product emergence process and product life cycle differ as follows: The product emergence process comprises the phases between concept development and sales (cf. Fig. 2.1). In common usage, a part of the use phase is often included here, since field support is also a part of the field phase (after sales). If the field phase with the subphase use is mapped comprehensively up to the end of the product use, then the term product life cycle should be chosen instead of product

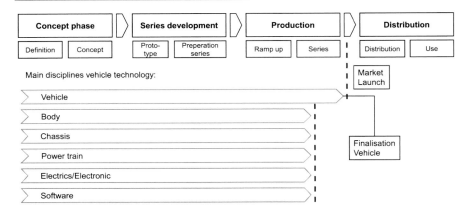

Fig. 2.1 Product emergence process using the example of automotive engineering

emergence process, since the use phase can cover a longer period than the field support.

Furthermore, the product life cycle also includes the recycling phase, which follows the use phase; cf. also (Bender and Gerricke 2021). The product emergence process is therefore part of the product life cycle. In summary, it is stated that the product life cycle, in contrast to the product emergence process, additionally includes the complete use phase as well as the recycling phase. In the development of a complex, technical product, the phases often take place in parallel in superordinate specialist disciplines. In the example of automotive engineering, this can be the six specialist disciplines of *vehicle, body, chassis, powertrain, electrics/electronics* and *software* (Fig. 2.1; cf. also Braess and Seiffert. 2013:1133). The following systems or component scopes, mentioned in excerpts, belong to the above-mentioned disciplines:

1. *Vehicle*: Subject areas that relate to the entire vehicle (e.g. fuel consumption, noise emissions; cf. further explanations) and cannot be assigned exclusively to one of the other disciplines mentioned.
2. *Body*: body frame, doors, covers, interior (assembly peripheries such as seats, interior door panels, instrument panel), exterior (add-on parts such as mirrors, trim panels).
3. *Chassis*: steering, axles, trailing arms, transverse control arm, braking system.
4. *Power train*: engine, gearbox, clutch, attachments (e.g. coolant pump, starter, intake system, exhaust manifold).
5. *Electrics/electronics*: actuators, control units, bus systems, networking.
6. *Software*: programs, algorithms, control codes, characteristic diagrams.

If a target value is defined at vehicle level (e.g. fuel consumption), several disciplines may be involved in its implementation. The following topics can be assigned to the vehicle discipline:

a. Fuel consumption: Each of the above-mentioned disciplines—body, electrics/electronics, chassis, powertrain, software—can influence fuel consumption: The drive unit through its design (e.g. power), the electrics or software, e.g. through the choice of ignition characteristics, the chassis, e.g. through tyre design, and the body, e.g. through its aerodynamic properties. In addition, there are interdependencies between the issues of fuel consumption and vehicle weight.
b. Vehicle weight: The definition of the vehicle function spectrum (equipment), the selection of materials as well as the geometric design of the components have a significant influence on the total weight of the vehicle.
c. Vehicle noise: The background noise of a vehicle is influenced, for example, by the powertrain and exhaust system (specialist discipline Powertrain), rolling noise of the wheels (specialist discipline Chassis) or also by noise emissions of individual actuators (specialist discipline Electrics/Electronics).

In Sects. 2.1.2 and 2.1.3 the phases and subphases of the product development process are outlined and supplemented with examples from automotive engineering (see also (Braess and Seiffert 2013:1133)). The focus is on activities of the subphases with a strong relation to technical reliability. Furthermore, the essential *milestones* are named with reference to the phases of the product development process, up to which the technical content of the activities is generally developed.

2.1.2 Concept and Series Development Phases

The *concept phase*, associated milestones and important activities related to technical reliability are shown in Fig. 2.2. The concept phase essentially consists of

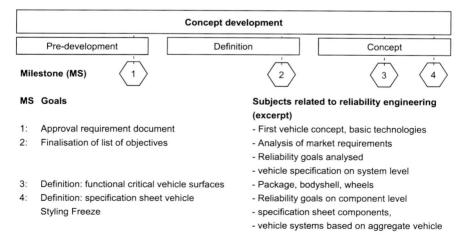

Fig. 2.2 Concept phase of the product emergence process with essential milestones

the sub-phases pre-development (development of new technologies and functional principles, respectively), definition and confirmation of the product design (concept confirmation); cf. also (Braess and Seiffert 2013). The aim of the concept phase is essentially to define the product structure, the styling and the development of the performance specifications for systems, assemblies and components on system, assembly and component level. Furthermore, a decision is made as to whether systems, assemblies, components are to be developed respectively produced in-house (OEM; Original Equipment Manufacturer) or developed respectively produced in outsourcing (supplier).

The *series development phase* comprises the two subphases of design and prototypes and series preparation, cf. Figs. 2.3 and 2.4. The objective of series development is the development of all systems, assemblies, components as well as the preparation of series production (procurement and construction of systems, machines, tools).

In addition to component development, the *design and prototype subphase* also includes simulation as well as the design and implementation of system, assembly and component testing. Simulation and testing have the goal of testing functions and reliability. The procedures include, among others, laboratory testing, Accelerated Testing, Highly Accelerated Life Test (HALT) or the vehicle test (cf. Chap. 14). Prototypes are developed with progressive design status 1 … n, which reflect the respective degree of maturity of the product. Within the *subphase series preparation* the production facilities for vehicle manufacture are implemented; approximately 95% of components are produced using series tools. If the manufacture of small or large series is planned, the activity Confirmation of the capabilities of product and process provides an essential contribution with regard to the preparation of a controlled and quality-capable manufacturing process, in particular for the realization of safety and function-critical product features (cf.

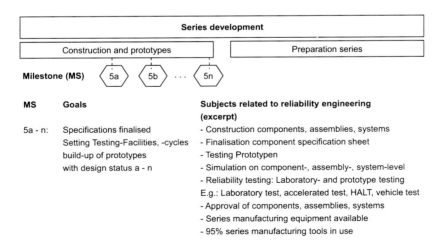

Fig. 2.3 Series development phase with milestones 5a to 5n of the product emergence process

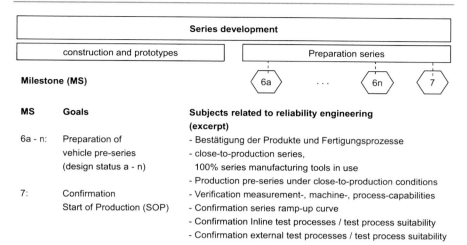

Fig. 2.4 Series development phase with milestones 6a to 6n and 7 of the product development process

Sect. 3.2). This includes, among others, the activities of measurement system and process capability, machine capability as well as process capability investigation (cf. Chaps. 15 and 16). Once series preparation has been completed, it is possible to manufacture a vehicle under near-series conditions and to use series tools.

2.1.3 Production and Distribution Phases

The *production phase* after point of time "start of production (SOP)" comprises the two subphases series ramp-up and series production, cf. Fig. 2.5. The *series ramp-up* (in short *"ramp-up"*) refers to the number of vehicles to be produced in relation to the first weeks of production. The number of units is steadily increased until the so-called operation line is reached. The operation line is the upper limit of the potential capacity utilization of a production. When the operation line is reached, *series production* starts *(short: "series")*. With regard to ensure product reliability, the activities of inline testing during series production (system, subassembly, component level), final product testing (entire product) and external testing processes (e.g. laboratory or supplier) are of particular interest here in addition to proven process capability; cf. in particular also *series preparation,* milestone 7 (Fig. 2.4). Safety–critical features are often subjected to a full inspection (also referred to as 100% inspection), provided that non-destructive testing is possible. Function-critical features can be checked by means of full inspection or sampling inspection, also in connection with Statistical Process Control (SPC; cf. Chap. 16). Safety and function-critical product characteristics can be inspected directly or indirectly (cf. Sects. 3.2 and 16.1). The production phase ends with the EOP (End of Production): The production of the vehicle ends, spare parts continue to be

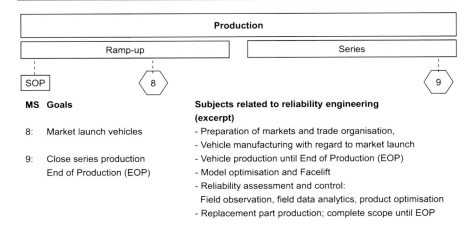

Fig. 2.5 Production phase of the product emergence process

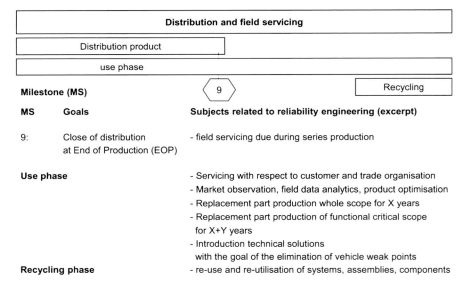

Fig. 2.6 Sales and field support phase as part of the product development process and the product life cycle respectively

manufactured for the time being. For this reason, some of the production engineering equipment continues to be operated for the supply of spare parts to the field.

The *sales and field support phase* comprises the sub-phases of the actual *product distribution* and the *field support* related to product usage, cf. Fig. 2.6. The *sales* sub-phase starts with the sale of the products usually directly after the ramp-up and ends with the discontinuation of the production of the current series (*End of Production (EOP))*. The *use phase* is understood as the complete period of product use by the customer until the end of the life cycle. The spare parts supply can be ensured for the entire scope of components for X years (automotive engineering: e.g. 15 years), then to a reduced extent – e.g. function- and safety–critical components – for X + Y years (e.g. automotive engineering: X + Y can cover 30 to 40 years). *Field support* begins with the sale and use of the first products and also includes the period after the EOP.

With regard to product reliability the use within the field phase is an important sub-phase: On the one hand, risk analyses are based on damage data (cf. e.g. Chaps. 11 and 12), which serve as a basis for product optimisations and for damage prevention. On the other hand, the field data analysis is the starting point for the development of reliable components and products of the successor generation: In addition to the elimination of product weaknesses by the introduction of technical solutions within the product fleet, the data and information on reliability topics (example: cause-effect chains with regard to product damage) can be analyzed. Parallel to the product optimization of the current series, documentation of the weak points is also essential for the optimized design of the product successor generation.

Furthermore, on the basis of the field experience with regard to the current series, an analysis can be made with regard to components as possible carry-over-parts (COP). The design statuses of carry-over-parts are carried over unchanged from the current series to the successor generation: On the one hand, proven reliable components can be identified as COPs for which the functional scope is not to be extended, and on the other hand, development efforts can be saved in this way. In addition, the environmental conditions, usage profiles and load spectra of the products in field use are of particular importance to the manufacturer (cf. Chap. 12), which can be taken into account accordingly in the successor generation with regard to the design of components and the vehicle.

2.2 Technical Reliability in the Product Emergence Process

2.2.1 Use of Technical Reliability Methods

The approaches and methods of planning, analyzing and assuring the technical reliability of products of products can be assigned to the phases and subphases of the product development process. Fig. 2.7 shows an overview of some approaches and methods of technical reliability in the form of a matrix: The temporal classification of the method/procedure in the product development process can be taken from the horizontal, the main objective of the application from the vertical. It is easy to understand that the chronological application of the approaches and methods within the matrix follows the course starting at the top left and ending at the bottom right. The reason for this is that neither planning aspects in the field deployment phase (development and production have already been completed),

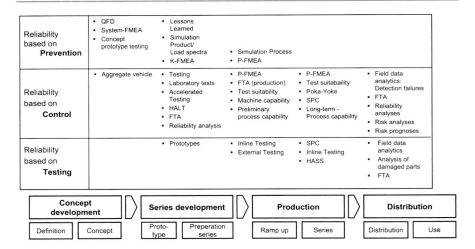

Fig. 2.7 Selection and excerpt of methods and procedures of technical reliability and assignment of these to the phases of the product emergence process

nor to assure (controlling) measures in the early sub-phases of the concept phase (there is no physical product realization yet) can take place. The methods and approaches of technical reliability mentioned represent an excerpt: In the following chapters, the contents as well as the temporal allocation are explained in detail.

The goal of any development of a technically complex product is to ensure a high level of technical reliability. In this respect, it is ideal to align the corresponding activities as preventively as possible in order to avoid rework and correction in the later phases of product emergence. Rework in the development phase can be, for example, a complex re-design and thus corrections of technical drawings or component specifications. Rework in production is understood to mean the elimination of manufacturing defects or the correction of component conditions that have arisen due to an inadequate specification.

The challenge for preventive reliability planning and the corresponding assurance is the fundamental contradiction between *decision-making certainty* and *influence* (cf. Fig. 2.8) depending on the point in time in the product emergence process. Decision certainty relates, for example, to decisions on product design, the load spectrum to be taken into account and the conditions of use in the service phase. Here, the decision-making certainty at the beginning of a product emergence process in the concept phase is rather low, since little detailed knowledge about application scenarios in the field is known. Decision certainty increases in the phases of product realization and reaches the highest level in the phase of field use. Problems that arise with the technical reliability of a product can be analyzed in detail and the corresponding decision can be made in a well-founded manner.

The curve for influencing the technical reliability of a product is diametrically opposed. Within the concept phase, product characteristics at system and assembly level can easily be adapted on the basis of new findings, since product realization

2.2 Technical Reliability in the Product Emergence Process

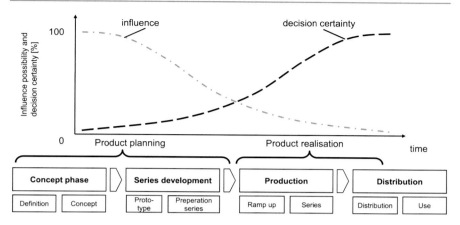

Fig. 2.8 Courses of decision certainty versus influence as a function of the product emergence process

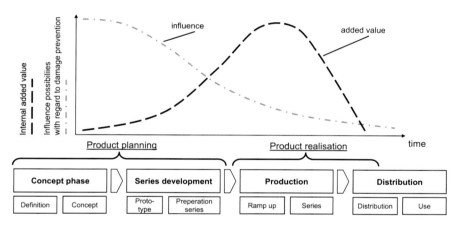

Fig. 2.9 Course of the possibility of influence versus value creation course depending on the product emergence process

has not yet begun. On the other hand, optimization measures in the phase of field use can only be realized with a very high effort, since this would require a recall with regard to the product fleet in use, explicitly for safety–critical weak points; for example, for a technical upgrade. The certainty of decision-making is high, since a specific problem is remedied in known products, but the influence is low, since the product concept cannot be changed or design features can only be changed with considerable effort.

This relationship is substantiated by a comparison of value added and decision certainty; cf. Fig. 2.9. *Value added* is generally understood as the difference between production value and intermediate inputs. With regard to a manufacturing process, value added means the transformation of an existing good (e.g. forged raw part) into a good with a higher monetary value (e.g. transformation of a raw part into a finished part through machining). The *degree of value added* is the quotient consisting of the numerator (difference between production costs and material procured externally) and the denominator production costs.

The value added with regard to a product is low at the beginning of the concept phase and reaches its peak in the current series up to the EOP (End of Production). After EOP, the value added consists of only two parts: Design changes (spare parts) and spare parts supply field. The comparable, fundamental progression of decision certainty and value creation from the concept to the production phase is obvious: the higher the detailed knowledge of the product realization, the higher the decision certainty. When value creation is complete – the entire product fleet is in the utilization phase – decision certainty reaches its highest level with regard to options for optimizing technical reliability in the event of damage.

2.2.2 Risk Elimination and Risk Prevention

Due to the contradiction outlined between the processes of decision certainty and influence in relation to the product emergence process (Fig. 2.8) and the connection between influence and value creation (Fig. 2.9), the efforts for risk elimination or risk prevention increase significantly in many projects of OEMs and/or suppliers in the realization phase (series preparation and production phase). The illustration in Fig. 2.10 shows the expenditure for risk elimination or risk prevention along the product emergence process as well as the curve for decision certainty (see also curve in Fig. 2.8). The concept phase plays a decisive role here: The basic project knowledge is available in this phase, which in many cases leads

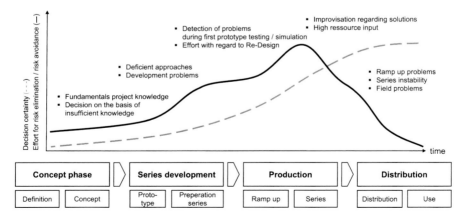

Fig. 2.10 Efforts for risk elimination and avoidance as well as decision certainty in the product emergence process with insufficient project knowledge in the concept phase

2.2 Technical Reliability in the Product Emergence Process

to decisions based on little knowledge. If these decisions are wrong, the effort for failure elimination (risk elimination) increases significantly within the phase series preparation, explicitly in the prototype phase as well as in the production of the pre-series: Through realization and testing, the product weaknesses become apparent and the effort for a re-design also increases significantly. The load peak occurs when the assembly line is reached and leads to high resource input and improvised product and process solutions. The consequences are start of production (SOP) problems, series maturity problems and field problems in the early use phase.

In the so-called *frontloading approach,* resources are increasingly used for reliability planning and for the application of technical reliability methods in the concept and series development, since a large part of the product costs is determined here with regard to production. In contrast to the outlined efforts for risk elimination with low resource input in the concept phase (Fig. 2.10), Fig. 2.11 shows the curve with regard to the efforts for risk avoidance when focusing on concept development and careful reliability planning in the concept phase. The expenses for risk avoidance and elimination are on a high level in the concept development. Furthermore, the development work is systematically accompanied in the series development by methods of the technical reliability, such as Fault Tree Analysis (FTA), Failure Mode and Effects Analysis (FMEA), Simulation, Accelerated Testing, Highly Accelerated Life Test (HALT).

This reduces the efforts for a potential re-design. Early process control and quality capability in the start of series production are made possible. The increased use of resources with well-founded reliability planning in the concept phase and the use of technical reliability methods during development can lead to a low-fault series start of production. In addition, the level of decision reliability is explicitly raised in the early phases of product development (cf. Figs. 2.10 und 2.11).

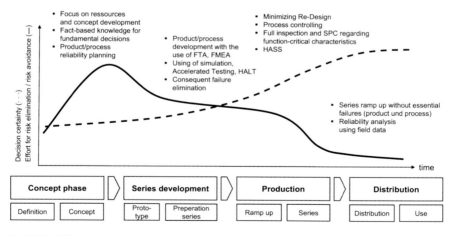

Fig. 2.11 Efforts for risk elimination and avoidance as well as decision certainty in the product emergence process with careful reliability planning in the concept phase

Elementary Terms 3

This chapter outlines the basic terms and concepts of technical reliability and the associated disciplines as a basis for understanding the following chapters. The explanation of the elementary terms is based on definitions in the relevant standard and normative works. The presentation of the terms is selectively supplemented by a technical perspective to facilitate understanding in the engineering application context. First, the terms of statistics and technical specification are explained, Sect. 3.1. The analysis of technical reliability is based on the examination of characteristics, ergo product and process characteristics as well as their characteristics are the focus of Sect. 3.2. This is followed by a detailed discussion and explanation of elementary terms of technical reliability and maintenance, Sect. 3.3. Technical analyses are usually carried out with the aid of measurement and testing processes; the basic terms are presented in Sect. 3.4. Since reliability analyses often serve to substantiate a risk quantification, the terms of risk analysis form the focus of Sect. 3.5.

3.1 Statistics and Specification

The starting point of statistical analyses is often the random experiment. A *random experiment* is an experiment under defined conditions, the outcome of which is a random result. The actual result is part of a defined set of possible, previously known results. The quantity with which the result of the random experiment is represented is the *random variable X*.

Accordingly, the random experiment is conducted under the following three conditions:

1) The random experiment and the framework conditions for carrying it out are defined.
1) The set of all potential outcomes is known before the experiment is performed.
2) The potential outcome is unpredictable both before and during the experimental procedure.

A random variable X is assigned to the random experiment, where the value x represents a realization of random variable X. The random variable X is called continuous if the value x can take on any real number in an interval as a value. Example: When measuring the diameter of a component (cf. case study engine piston rod; explained in detail in Sect. 4.1), the diameter would be the random variable X, to which the value x = 64.001 mm is assigned after a measurement.

The measured value is called the *original value* in statistics.

The *probability* that the random variable can take a value in an interval I is shown by the expression according to Eq. 3.1.

$$P(X \in I) \qquad (3.1)$$

Population and sample
The term *population* (often referred to as N) refers to the totality of all units of analysis, that can occur in a study. It is thus the entire set of all elements with characteristics of interest. A population N can be either finite or infinite. A *sample* (commonly referred to as n) is a subset of the population N, usually with a particular property. When analyzing large amounts of data, several samples n_i are often evaluated in order to be able to make a statement about the population N.

If, for example, a population with N = 10,000 engine piston rods is produced, a statement about the total production can be made with the aid of the analysis of 10 samples $n_1, n_2,...,n_{10}$ with 25 parts each (sample size), which are measured during manufacturing; cf. also Chap. 16. The population would therefore be the total production volume, the samples would be the 10 series of measurements (with 25 measurements each) and the measured values are the so-called *original values*. The characteristic of interest (e.g. diameter of the large connecting rod eye) is described by the random variable X.

Distribution function and density function
The *distribution function F(x)* indicates the probability with which the values of a random variable X are assumed, cf. Eq. 3.2.

$$F(x) = P(X \leq x) \qquad (3.2)$$

Since it is an accumulation, the distribution function is monotonically increasing with the limits given in Eqs. 3.3 and 3.4.

$$F(-\infty) = 0 \qquad (3.3)$$

$$F(\infty) = 1 \qquad (3.4)$$

For a continuous random variable, the distribution function can be obtained by integration over the *probability density function f(x)* (or just briefly named density

3.1 Statistics and Specification

function or probability density) can be determined. If integration is performed in a certain interval, e.g. to determine the probability, then Eqs. 3.5 and 3.6 are obtained.

$$P(\alpha < X \leq b) = \int_a^b f(x)dx \qquad (3.5)$$

$$P(X \leq x) = \int f(x)dx \qquad (3.6)$$

Expected value and variance
The *expected value E(X)* is the mean value of a random variable. For continuous random variables, the expected value is generally calculated according to Eq. 3.7.

$$E(X) = \int_{-\infty}^{+\infty} xf(x)dx \qquad (3.7)$$

The *variance Var(X)* indicates the range that the dispersion of the random variable X can cover with respect to the expected value E(X), Eq. 3.8. The variance is also called the second central moment.

$$Var(X) = \int_{-\infty}^{+\infty} (x - E(X))^2 f(x)dx \qquad (3.8)$$

Dispersion from a technical point of view
From a technical point of view, the term *dispersion (scatter)* refers to deviations that can be observed in several measurements of the same characteristic. The scatter can be caused, for example, by the diversity of the measured objects (deviations of different components from the specification due to the manufacturing process) or the measuring system itself. Frequently used key indicators for the dispersion (here: estimator!) are the range R, the standard deviation s or the variance s^2.

Target value/Set point, specification and tolerance
The required *target value/set point* for a product or process characteristic can be a fixed value and/or lie within a range. This requirement is called a *specification*. Both specifications—set point and range—are often found in the technical product drawing or the process specification. The range in which the setpoint should lie is often referred to as *tolerance* with its tolerance limits or simply *specification* with its specification limits. Feature or component tolerances are specified, for example, by the designer and are found on the manufacturing-oriented drawing (note: not on the function-oriented drawing). Process specifications and process specification limits are usually defined by the manufacturing planner and are found in the definitions and workflows describing a manufacturing process. The following abbreviations are often used:

1. Upper/lower tolerance limit: UTL/LTL,
2. Tolerance: T,
3. Upper/Lower Specification Limit: USL/LSL.

Target value/Set point, actual value and deviation
A *deviation* exists if the *actual value* (e.g. measured by a measurement system) differs from the required *target value* (e.g. fixed in drawing). A deviation is the difference between the target value and the actual value.

Example:

1. The diameter of a hole after measurement $x_{actual} = 64.030$ mm (actual value).
2. The specification target (nominal value) is $x_{Target} = 64.000$ mm.
3. The deviation is therefore a = 0.030 mm.

Failure
A deviation with respect to a measured value is called an *failure* if the deviation exceeds or falls below a permissible, specified value or range (cf. specification and specification limits). The measurement uncertainty should be taken into account (caused by the measuring system itself or the measuring process): If the measured value is within the specification, but could also be outside this range, taking into account the measurement uncertainty, an failure cannot necessarily be assumed. The same applies in the reverse case. The term measurement uncertainty is discussed in detail in Sect. 3.4.

3.2 Product and Process Characteristics and Their Specifity

The basis of the analysis of the properties of a product or a manufacturing process are *characteristics*. *Product characteristics* can be, for example, geometrical characteristics to describe a component (for example, length, diameter), or also function-relevant characteristics (for example, noise, switching force). Characteristics for the analysis of a manufacturing process are usually called *process characteristics* or *process parameters*. In the case of a welding process, for example, these can be current intensity and welding duration. The assessment of a product can thus be carried out *directly* via the product-related characteristics (e.g.: strength of a spot weld) or *indirectly* via the process parameters (here e.g. current intensity and welding time) which played a role in the manufacture of the product characteristic (here: joining of two sheets). The precondition is, of course, that there is a high correlation between the process parameters and the product characteristic (cf. Chap. 10).

Figure 3.1 shows an overview of different types of characteristics based on DIN 55350 – Part 12 (DIN 1989). In addition, examples are given for each type of characteristic in relation to the engine piston rod case study (cf. Sect. 4.1).

3.2 Product and Process Characteristics and Their Specifity

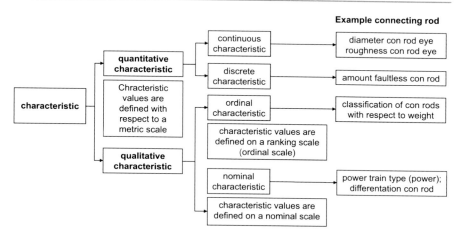

Fig. 3.1 Differentiation of types of characteristics and examples

First, a differentation is made between quantitative and qualitative characteristics (DIN 1989). *Quantitative characteristics* are based on a metric scale.

A characteristic is called *metric* (or cardinally scaled) if its expression can be expressed as a multiple of a unit. The diameter of the piston rod eye is a metric characteristic because, for example, the measured value 60 mm is a multiple of the unit millimetre [mm]. The interval length relative to the scale is the same. Ergo, the characteristics of the feature – for example, related to different components, but same feature – are different from each other and have a unique distance. By using the scale with equal interval lengths, the characteristic values also have a clearly defined distance. Furthermore, metric characteristics are subdivided as follows:

1. Ratio-scaled characteristics with a natural zero: the ratio of two expressions is easy to interpret (e.g. temperature).
2. Interval scaled features which do not have a natural zero point (e.g. space coordinates with reference point).

The quantitative characteristics are assigned continuous and discrete characteristic types. A characteristic is described as *continuous* if the values of the characteristics can be any number in an interval (for example, diameter of the piston rod eye, connecting rod weight). On the other hand, a characteristic is called *discrete* if the value can only be an integer (for example, number of defective connecting rods, number of voters for a particular party, number of paint scratches). Thus, discrete characteristics have a countably large number of values in a sample or a population.

In the case of *qualitative characteristics,* a basic distinction is made between ordinal and nominal characteristics. A characteristic is described as *ordinal* if the states of expression have only one order relationship to each other. The ordinal relationship can, for example, be "greater", "lesser" or "more", "less" or "more beautiful", "less beautiful". This means that the characteristic values are different from each other and also have a unique order. However, the distance between two characteristic values is not unambiguously defined and thus cannot be unambiguously interpreted. Measured values of a continuous characteristic can also be sorted and assigned a *rank:* In this case, the unique ordering is preserved, but the distance between the characteristic values is equalized. Another example is the formation of different connecting rod weight classes: Conrods from series production are assigned to classes depending on their measured weight. Thus, the classes are different and also have an unambiguous arrangement; at the same time, the distance between the connecting rod weights is unclear or the distance between the weight classes is equalized.

A characteristic is described as *nominal* if the characteristic value does not allow a unique arrangement. The characteristic states can only be distinguished by their relationships to each other. The states of expression are therefore different from each other, but neither the arrangement to each other is clear nor is the distance between two states of expression defined.

Example: Engine Piston Rod Manufacturing Process
In the ongoing production of engine components, series of measurements are taken which comprise various characteristics. In the case of an engine connecting rod, these can be the characteristics of diameter (large connecting rod eye and small connecting rod eye), as well as the center distance of both connecting rod eyes to each other. Furthermore, the number of connecting rods that do not meet the given specifications is counted. Finally, the connecting rods are weighed in order to use connecting rods with the same weight as far as possible when assembling the engine (the different weights can be caused, for example, by deviations in the component geometry and by inhomogeneity in the material). The connecting rod is assigned to a weight class on the basis of the measured weight. The use of connecting rods with the same weight as far as possible has an influence on the smooth running of the engine in operation.

Thus, four quantitative characteristics would be documented and monitored: Three continuous characteristics (two different diameters [mm], one length [mm]) and one discrete characteristic (production defective onnecting rods). Furthermore, one qualitative, ordinal characteristic would be recorded (connecting rod weight class).

Technical drawing, component specification and characteristic
Features that describe a technical product (component, assembly, system) are documented in series development within the framework of a technical drawing. A *technical drawing* is thus the central document for mapping the complete product specification. The technical drawing is the starting point for communication between development, manufacturing, field support (maintenance) and between OEM (Original Equipment Manufacturer) and supplier. For further reading, (Labisch 2014), (Hoischen 2016) and (Bender and Gericke 2021) is recommended.

Another central document is the *component specification*: It includes all requirements for the component in terms of design, functional scope, reliability and sustainability. The same applies to assemblies and systems.

The technical drawing and the component specification form the basis for component realisation and are often the central documents in the event of damage with regard to guarantee, warranty and recourse claims.

Function-critical characteristic and safety-critical characteristic
From the totality of all characteristics that describe a technical product, two subsets are of particular importance: function-critical characteristics and safety-critical characteristics. When analyzing the technical reliability of a product or a process, function-critical and safety-critical characteristics often represent the starting point of the analysis.

The subset of function-critical characteristics cannot be sharply delimited from all other characteristics of a unit by definition. In the context of the present explanations, *function-critical characteristics* are understood to be all characteristics that are of particular importance for the central functions and for the reliability of a component, an assembly or a system. Function critical features are often considered within the tests to be performed during component manufacture. The inspection procedures may be: random sampling, Statistical Process Control (SPC) or full inspection (Chap. 16). The measurement and test results are documented and archived.

Example: In the case of an engine pistion rod, the centre distance between the large and small connecting rod eye is functionally critical, as a deviation from the specification would result in a change in the compression ratio, for example. Furthermore, the connecting rod geometry is decisive for reliability in operation: connecting rod thickness, diameter, parallelism of the eye axes are further function-critical characteristics. For further details, please refer to Sect. 4.1 and (Basshuysen and Schäfer 2006 and 2015).

On the other hand, the differentiation of safety–critical features from the other features of a product or process can be carried out more clearly: *Safety–critical features* include all features that, in the event of a deviation from their specification, lead to a fault that endangers users respectively people. Therefore, safety–critical characteristics in series production are monitored, for example, with a (non-destructive) full inspection (Chap. 16). Measurement and test results from the full inspection are documented and archived in the same way as for the function-critical characteristics.

Example: The wheel bolting of an automobile is to be classified as safety-relevant, since a defective bolting can lead to a loosening of the wheel and thus to a vehicle accident. Therefore, the bolting is a safety–critical feature. The process parameters torque and angle of rotation of the bolted joint are monitored within the assembly with a full inspection.

If non-destructive, direct testing is not possible, a combination of destructive, direct testing based on random samples and indirect full testing based on process parameters is also frequently performed. Example: The strength of a spot weld can be tested directly destructively and indirectly via welding process parameters.

Inspection characteristic
A characteristic is referred to as an *inspection characteristic* if it is the subject of an inspection in development (e.g. prototype construction), production or maintenance. In addition to safety–critical or function-critical characteristics, these can also be characteristics that are decisive in a production process with regard to the further processing of the component.

Full inspection and sampling
In Chap. 16 Statistical Process Control (SPC), the topics of *full inspection* and *sampling* are dealt with in more detail; cf. in particular the section "Notes on inspection strategy: full inspection versus sampling inspection procedures".

3.3 Technical Reliability and Maintenance

3.3.1 Technical Reliability According to Bitter and DIN

The definition of the term *technical reliability* is available in various standard works with different characteristics and depth, so that an interdisciplinary clearness is not given. In this chapter, two different definitions are first cited and discussed. Subsequently, a definition is developed that serves as a starting point for all subsequent considerations in this book.

The classical definition of the term *reliability* is given in the standard Technical Reliability by Messerschmidt-Bölkow-Blohm GmbH, cf. Bitter (1986, p. 2; original is German language):

> "*Reliability is the probability that a unit will not fail during a defined period of time under given functional and environmental conditions.*"

This definition has also been adopted by a number of other authors, including Bertsche and Lechner (2004, p. 20) and Birolini (1997, p. 2).

Comment 1:
With this definition, the concept of reliability in relation to a technical product is directly related to the concept of probability, or more precisely: the probability of failure. Indirectly, this definition thus also refers to the probability of survival (complement of the probability of failure). It follows that the reliability of a technical product cannot be measured directly, but is characterized by a probability. On the one hand, the probability of failure can be determined empirically on the basis of damage data (for example, testing or in the field) with the help of a probability model. On the other hand, the probability of failure can also be estimated by means of a failure model and probability calculus.

Comment 2:

In practice, the period of time during which the product should be functional is sometimes also expressed or approximated using non-time-related variables. For example, the requirement for a motor vehicle may be that functionality must be ensured for 15 years. Strictly speaking, this represents a mixture of actual operating time (vehicle in operation) and downtime (vehicle not in operation); however, the downtime also has an influence on a faultless operating time; for example, due to corrosion damage. Likewise, a vehicle may also be required to function over a distance of 400,000 km. The requirement for a trouble-free service life in the field is then indirectly represented by the distance variable. The distance often correlates with the operating time, but there is no complete correlation (cf. Chap. 10). Other variables that are not directly time-related can be, for example, operating cycles, switching cycles, travel distances. Each of these variables correlates at different levels (different correlation coefficients) with the operating time or field runtime.

Interim conclusion on the discussion of the term reliability according to (Bitter 1986, p. 2):

1. The probability of default is estimated either on the basis of empirical data and a probability function or by means of a probability model.
2. The variable time is called the lifetime variable.
3. The requirement for reliability refers to the proper functioning of the product.

The definition of *reliability* based on DIN standard 40041 (1990, p. 2 [translated from original German language]) reads:

"*The condition of a unit with respect to its ability to meet reliability requirements during or after specified periods of time under specified conditions of use.*

Note 1: short form of the definition: part of the quality in terms of the behaviour of the unit during or after specified periods of time under specified conditions of use.

Note 2: instead of a time period, e.g..in hours, the specified number of operating cycles or similar can also be used.

Note 3 deals with the english term "reliability" and the relation to the german term "Zuverlässigkeit". Therefore, Note 3 is cited within the original, German language:

Note 3: In dieser Norm ist der Begriff Zuverlässigkeit wie oben definiert umfassend zu verstehen. Der Begriff reliability ist dagegen teils in der Bedeutung "Funktionsfähigkeit", teils in der Bedeutung "Überlebenswahrscheinlichkeit" definiert und daher als Übersetzung für "Zuverlässigkeit" mißverständlich"

Comment 1:
The definition according to DIN 40041 (1990, p. 2) is more abstract compared to the definition according to Bitter (1986, p. 2). As before, reliability refers to the variable time (here: specified period of time), whereby under DIN 40041 Note 2, non-time-related variables which can correlate with time were already taken into account.

Comment 2:
Another clear difference to the definition according to Bitter (1986, p. 2) is the general reference to a reliability requirement. The property of reliability is thus referred to here not only to the potential failure, but generally to a requirement to be defined beforehand. The possibilities for this are not further specified within the framework of the DIN 40041 standard; the term "reliability requirement" is only defined in general terms.

Comment 3:
The quantitative variable of the probability of failure for mapping the technical reliability of a system is not part of the definition according to DIN 40041 (1990). This leaves it up to the user to decide which variable or measurand is to be used to represent reliability. DIN 40041 Note 3 only briefly discusses the translation into English of the term reliability and the associated meaning of the probability of survival.

Interim conclusion on the discussion of the term reliability according to (DIN 40041 1990, p. 2):

1. The reliability of a technical product according to DIN 40041 (1990) refers to an unspecified reliability requirement.
2. The definition does not refer to a concrete, quantitative characteristic with which reliability is mapped.
3. The lifetime variable is the variable time or similar; respectively a period of time.

Other standards on terms from the related discipline of quality, such as DIN 55350-11 (2008), DIN 55350-12 (1989), DIN 55350-17 (1988) do not contain any further, independent definitions of the term reliability, but refer directly or indirectly to DIN standard 40041 (1990). Furthermore, no further definition can be found in the DIN 1319 series of standards (Parts 1–4; 1995, 1996, 1999, 2005) on the subject area of fundamentals of measurement technology.

3.3.2 Discussion of the Definition of Reliability According to Bitter and DIN

The advantage of the definition of reliability according to Bitter (1986) is that the focus is on the clear (reliability) requirement of the fault-free function of a system. Every technical product (component, assembly, system) fulfils one or more defined function(s): if there is a product failure, the function(s) is (are) not fulfilled and reliability is no longer given. The same applies to processes. The inclusion of the probability of failure for the quantitative assessment of reliability may also be seen as an advantage. It is a concrete quantity which can be mapped mathematically (by empiricism or estimation) on a database. The exclusive reference to the variable time may be classified as a disadvantage, since the probability of failure can also be meaningful on the basis of lifetime variables that have no direct time

3.3 Technical Reliability and Maintenance

reference, e.g. operating cycles. Particularly as there can also be a mixture of different time references (e.g. operating time, field time) and thus the lifetime variable time – as commented above – is not unambiguous in its expression.

The definition of reliability according to DIN 40041 (1990) is abstract: The reliability requirement to which the reliability of a product should refer is not specified, so that the interpretation is up to the user. This means that the reliability of technical systems can be assessed in different ways, which means that there is no comparability. If the reliability requirement is incompletely formulated, false estimates of fault-free operation may result. The same applies to the lack of a quantitative variable or value to represent technical reliability. Due to a (forced) individual design by the user, technical systems can only be compared with each other to a limited extent. An unfavourable choice of the (quantitative or qualitative) quantity can lead to considerable misunderstandings when assessing the reliability of technical systems. It may be seen as an advantage that DIN 40041 (1990) refers in its comments on the definition to a possible use of non-time-related lifetime variables.

3.3.3 Definition of Technical Reliability

On the basis of the definitions according to Messerschmidt-Bölkow-Blohm GmbH (Bitter 1986) and DIN 40041 (1990) as well as the above comments and discussion, an adapted *definition of Technical Reliability,* which combines the advantages of both definitions, explicitly includes function- and safety–critical characteristics and serves as the basis for the explanations in this book.

Definition of Technical Reliability
"Reliability is the probability that a unit will not fail with respect to its function- and safety-critical characteristics during a defined time interval or interval related to a time-correlated lifetime variable under given functional and environmental conditions."

Comment:

1) The definition refers to the lifetime variable time or a lifetime variable correlating with it.
2) In focus of the reliability requirement is the goal of failure-free operation.
3) Reliability is mapped quantitatively by means of the probability of failure.
4) The quantitative mapping of reliability is based on function- and safety–critical characteristics.
5) The quantitative, feature-based mapping of reliability allows comparability between different systems.

3.3.4 Technical Reliability and Failure Behaviour

Within the discipline of technical reliability, a number of statistical terms are used (cf. e.g. Sect. 3.1), which have been adapted to the engineering context of product or process reliability. In the following section, some of the most important terms of technical reliability are presented, which have their origin in applied statistics.

The term of reliability in relation to a technical product is closely related to the term of probability (cf. definition of "technical reliability" in Sects. 3.3.1 and 3.3.3). The reliability of a technical product cannot be measured directly, but is characterized by a probability. Of central importance are the terms of survival probability and failure probability. The *survival probability R(t)* is the probability that a product retains its functional capability after a certain time unit. The *failure probability F(t)* is the complement of the survival probability and results from the integral of the probability density function f(t). The survival probability R(t) is often called the reliability function. The failure probability F(t) or survival probability R(t) is represented with a *distribution model* e.g. a Weibull distribution model or a normal distribution model; Chap. 7. Also, instead of the general term distribution model, the term *lifetime distribution* is often found in the literature on technical reliability; cf. for example (Bertsche and Lechner 2004), (Linß 2016). Chap. 7 shows a number of distribution models, many of which are used as lifetime distributions in reliability analysis. The variable t of these distributions is usually referred to as the *lifetime variable t*.

The term *failure behaviour* of a technical product is understood to mean the time-related change in the probability of failure F(t), which can be mapped by the use of a distribution model.

Another important parameter for the reliability of a product is the *failure rate* $\lambda(t)$. The failure rate is determined by the quotient of probability density f(t) and survival probability R(t), Eq. 3.9. Ultimately, the failure rate represents the quotient of failed units and the still surviving units as a function of the lifetime variable t. The failure rate is not the same as the ratio of damaged units (cf. Sect. 3.3.6).

$$\lambda(t) = \frac{f(t)}{R(t)} = \frac{f(t)}{1 - F(t)} \qquad (3.9)$$

With the help of the failure rate, the essential phases of the failure behaviour of a system can be mapped: *Early failure behaviour, random failure behaviour (constant failure rate)* as well as *failure behaviour due to runtime*. Three different failure models are available for the three phases mentioned, which are linked in a schematic diagram in Fig. 3.2. First, the failure rate shows a degressive course, which can be caused, for example, by production-related damage causalities: The units fail after a short period of use. The failure rate is approximately constant when isolated field failures are observed. In the case of mechanical components, this is only approximately constant in the phase of random failures; in the case of electronic components, it can actually be almost constant. The essential factor here is the level of the failure rate. The failure rate exhibits a progressive behavior in

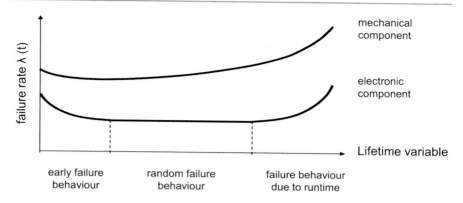

Fig. 3.2 Schematic diagram of the failure rate of a technical product related to the use phase

the phase of runtime-related failures. This is caused, for example, by wear failures, corrosion damage or embrittlement. A more detailed explanation with comprehensive examples of reliability analysis is shown in Chap. 12.

3.3.5 Data, Damage Data, Product Fleet and Censoring

Failure behavior is often mapped based on damage data. *Damage data* include information on time of damage, damage symptom (e.g. oil leakage at the crankcase/crankshaft seal complex) and cause of damage (e.g. worn radial shaft seal). Damage data can be collected, for example, as part of the testing of a group of components or as part of the use phase (field use) of a product (cf. product development process; Sect. 2). Components that are tested during series development are called *test products*, device under test (DUT) or *prototypes*. The totality of all products of the same or similar design status which are in field use during the utilization phase are often referred to as a *fleet* or *product fleet*. Operating data and load spectra of non-failed units are considered in the reliability analysis: If there is a assumption that non-failed units may be potential candidates with respect to the damage pattern being analyzed, the non-failed units are *candidates* (aspirants). In case a (damage) event occurs within the test or the product fleet which cannot be assigned to the expected failure behaviour due to its characteristics, this event is called an *outlier*. The same applies to a measured value which does not fit the distribution model of the already known series of measured values; cf. in detail Chap. 13.

If no damage data or operating data exist for some of the test vehicles (prototypes) in testing or for some of the product fleet in field use at a certain point in time, the data situation is *censored*. The starting point for an analysis at time t_A is a database (e.g. failed units and non-failed units), which refers to a specific observation period before time t_A. The following reasons can cause a censoring of the data situation:

a) At the time of the analysis, the expected events (e.g. damage) could not yet be observed for all components under consideration. The failure behaviour can take longer than the present observation period (which is limited to the time of the analysis). Example: Within a vehicle fleet in the field, a damage focus oil leakage in the power train crystallizes at a minimum mileage of $t = 30{,}000$ km. The known damage cases were detected in an observation period of two years. Since some vehicles have not yet exceeded this mileage after two years, it is obvious that further cases of damage (candidates) are to be expected.
b) The event has already occurred before the observation period on which the analysis is based. Example: A vehicle fleet has been in use in the field for several years; however, the data collection within the market observation has only been carried out for one year.
c) Products or components are removed from the observation period to which the analysis refers. Example: Decommissioning, accident or sale of a vehicle from a vehicle fleet. In the case of sale, this is accompanied by a change of market in the field.

Therefore, a differentiation is made between the following types of *censoring:*

1) *Right-censored data*: Event A will probably occur at a point in time in the future which is after the point in time of the concrete analysis.
2) *Left censored data*: The event may have occurred before the observation period on which the present analysis is based.
3) *Interval censored data*: The known events lie within the observation period on which the present analysis is based. It is possible that further events exist both before the observation period and in the future after the observation period.

In the reliability analysis of test vehicles (prototypes) in development as well as of product fleets in the use phase, there is often a censoring of the data situation. Sect. 12.3 shows procedures for the consideration of non-failed units in the context of the reliability analysis.

3.3.6 Key Indicators

The terms failure rate (Eq. 3.9) and ratio of damaged units (Eq. 3.10) are often confused in industrial applications. The *ratio of damaged units* is defined as the quotient consisting of the number of failed units n_F and the population N (e.g.: produced units), see Eq. 3.10. If there are only a few failures in relation to a large number of produced units, the failure rate and the ratio of damaged units are of course approximately equal. However, in the case of many numerical ratios, the two quantities are completely different, so care must be taken to use the terms and the quotients behind them precisely.

$$s = \frac{n_F}{N} \qquad (3.10)$$

3.3 Technical Reliability and Maintenance

Further key indicators for mapping the failure behaviour

To describe the failure behaviour of a system – explicitly in the case of multiple failures of a repairable system – the key figure MTBF is used: *Mean Time Between Failures (MTBF)*. The characteristic value MTBF is defined as the expected value of the operating time between two successive failures (failure i, start b, end e), which each lead to a standstill of the system in the interval $[t_{i,b}; t_{i,e}]$; Eqs. 3.11 or 3.12 (cf. DIN IEC 60050-191 (DIN 1994) or BS 4778–3.2 (BS 1991).

For systems that cannot be repaired, the characteristic value *Mean Time To Failure (MTTF)* is used. This refers to the expected value (mean value) of the distribution of service lives: The mean service life. In contrast, the characteristic value Mean Time To Recover respectively *Mean Time To Repair (MTTR)* denotes the average repair time that must be spent to correct a failure.

With the aid of the two key indicators MTBF and MTTF, the key indicator *availability A* of a system is formed, Eq. 3.13.

$$MTBF = \frac{\sum_{i=1}^{n} t_{i,b} - t_{i,e}}{n} \tag{3.11}$$

$$MTBF = \int_0^\infty R(t)dt = \int_0^\infty t \cdot f(x)dt \tag{3.12}$$

$$A = \frac{MTBF}{MTBF + MTTR} \cdot 100\% \tag{3.13}$$

The average time between two overhauls of technical products can be used as a further characteristic value: *Mean Time Between Overhaul (MTBO)*. The overhaul of a technical system does not necessarily require the presence of a failure.

3.3.7 Maintenance and Quality

Closely linked to the issues of describing the failure behaviour of a technical system and thus the technical reliability is the specialist discipline of *maintenance*. The maintenance of a product, component, assembly or system ensures its proper functioning and thus has a direct influence on the reliability of the system. The basic activities of maintenance are servicing, inspection, repair and improvement of a technical system. In principle, a distinction is made between three different *maintenance strategies*: Damage-dependent maintenance, time-based maintenance and condition-based maintenance. In the case of *damage-based maintenance*, the system failure is taken as the reason for performing the maintenance. In contrast, in *predictive maintenance*, maintenance work is performed at defined intervals, usually with reference to a time interval. In *condition-based maintenance*, operating parameters and operating conditions are monitored (continuously by means of production data acquisition or by means of regular inspections) and maintenance work is scheduled and performed depending on the system condition. Condition monitoring systems are often used for condition-based maintenance.

Relationship between the terms reliability and quality
Closely linked to the term of reliability is the term of quality. The term *quality* is defined as "the degree to which a set of inherent characteristics of an object fulfils requirements", see DIN EN ISO 9000:2015 (DIN 2015). Characteristics can relate, for example, to the function of a product or also to design, surface finish or other—quantitative and qualitative—properties. Reliability, on the other hand, focuses on the proper functioning of a product (Sect. 3.3.3).

If the term quality is understood as a time-related consideration, the reliability of a product can be interpreted as a time-period-related consideration of quality. Reliability is then the ability of a component, an assembly or a system to provide the fulfilment of the requirement placed on it over a defined period of time. Reliability can thus be interpreted as an integral of quality over time, cf. also comments in Bitter (1986, p. 3) and (Kamiske and Brauer 1995).

3.4 Measuring and Testing

Analyses in the context of the technical reliability of a product or process are based on the analysis of product or process characteristics, cf. Sect. 3.2. Quantitative analyses require a database, which is often composed of observed values or measured values. In the context of the technical reliability of a product, this can be measurement data, such as operating data, damage data, component geometry data or process parameters.

Measurement data are collected with the aid of measurement systems, measurement processes and/or test processes. Therefore, *measurement and test processes* as well as *measurement and test process capability* (Chap. 15) have a decisive influence on the acquisition of measurement data, which form the starting basis of a data analysis for mapping the technical reliability of a product or process.

Within the present chapter, the basic terms of the subject area of measuring and testing are explained; cf. Tab. 3.1. The definitions are quoted on the basis of DIN 1319 Parts 1–4 (DIN 1995, 2005, 1996, 1999). Furthermore, the terms measurement uncertainty and test process capability are explained and discussed separately in detail.

Measurement uncertainty: fundamentals
The term *measurement uncertainty* is of particular importance in the context of technical measurement and testing processes. On the one hand, influences caused by the measurement technology used lead to a potential scatter of the measurement result, as they become visible, for example, during a repeated measurement. This measurement uncertainty is determined regarding a measurement system, whereby as many standardized processes as possible (example: measurements on a certified standard) are carried out, or as many boundary conditions as possible are kept constant (example: avoidance of temperature influence by measurement

3.4 Measuring and Testing

Tab. 3.1 Terms of measurement and testing based on DIN 1319; (DIN 1995), (DIN 1996), (DIN 1999), (DIN 2005) using the example of engine piston rod

No	Term	Definition	Example piston rod
1	Measured variable	Physical quantity to which the measurement applies	Diameter
2	Measurement object	Carrier of the measurand	Piston rod
3	True value	Value measurand as the target of the evaluation of measurements of the measurand	$d = 64$ mm
4	Correct value	Known value for comparison purposes whose deviation from the true value is considered negligible for the comparison purpose	$d_R = 64.001$ mm
5	Measurement	Performing planned activities to quantitatively compare the measurand to a unit	Operation of a workpiece-bound measuring system
6	Testing	Determine to what extent a test object fulfils a requirement Note: Generic term for objective (by gauging or measuring) and subjective (by sense perception) assessment	Comparison of measured value and specification (technical drawing)
7	Measured value	Value that belongs to the measurand and is uniquely assigned to the output of a measuring system	$x_1 = 64.010$ mm $x_2 = 64.012$ mm …
8	Measurement result	Estimated value obtained from measurements for the true value of the measurand	$x = 64.011$ mm
9	Measurement deviation	Deviation from "measurement result" from "true value"	$a = 0.011$ mm
10	Expected value	Value to which the arithmetic mean approaches as the number of measured values increases	$x = 64.012$ mm
11	Random measurement deviation	Deviation of the uncorrected measurement result from the expected value	$s = \pm 0.002$ mm
12	Systematic measurement deviation	Deviation of the expected value from the true value	$Bi = 0.001$ mm
13	Outlier	Measured value which deviates significantly from the other measured values of a series of measured values or the observer's model assumption	$x_5 = 66.001$ mm
14	Test/measuring device	Measuring instrument intended solely for the measurement of a measurand. The measuring instrument is a test instrument if the measurement result is used to verify a claim	workpiece measuring device

(continued)

Tab. 3.1 (continued)

No	Term	Definition	Example piston rod
15	Measuring system	The totality of all measuring instruments and additional equipment used to obtain the measurement result	workpiece-bound measuring device; computer, output unit
16	Resolution	Specification for the quantitative recording of the characteristic of a measuring system, in order to distinguish unambiguously between measured values that are close to each other	RE = 0.001 mm
17	Measurement standard	Measuring instrument, measuring device or reference material having the purpose of representing, preserving or reproducing a unit or value(s) of a quantity for the purpose of communicating such value(s) to other measuring systems by comparison	Ring gauge with diameter d = 64 mm
18	Calibration	Determining the deviation of the display of a measuring system from the true or correct value of the measurand	–
19	Adjustment	Minimization of systematic measurement deviations with respect to the exact value Note: Calibration and adjustment are performed by comparison with a standard representing the value of the measurand assumed to be correct	–
20	Gauging	Setting the correct function of a measuring device, determining the scale of a displayed measuring device (adjustment) and determining the deviation of the displayed value from the true/correct value (calibration)	–

under air-conditioned conditions at 20 degrees Celsius; avoidance of operator influence by automation or definition of an operator). Later, a distinction is made between the measurement *uncertainty of the measurement system* (measurement uncertainty due to the measurement technology used) and the measurement *uncertainty of the measurement process* (measurement uncertainty due to the application of the measurement system at the installation site); cf. in detail Chap. 15.

First of all, the term of measurement uncertainty is explained in general terms in the following, cf. also DIN 1319 Part 3 (DIN 1996).

Measured value y, measurement result Y and measurement uncertainty U
The result of a measurement is a measured value. However, since every measurement process is subject to a certain degree of scatter with regard to its result, a so-called *measurement uncertainty U is* assigned to the *measured value y*.

Definition of measurement uncertainty U: The measurement uncertainty U is the range within which the true measured value can lie with a certain probability.

3.5 Risk and Risk Analysis

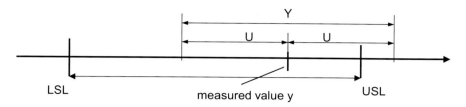

Fig. 3.3 Measured value y, measurement result Y and measurement uncertainty U in relation to the specification limits LSL and USL

Thus, the measurement uncertainty U is a measure of the possible measurement deviation of the estimated value of the measurand for which the measured value y was determined. If the determined measured value y is close to a defined specification limit, the measured value y could also lie outside the specified range. If the characteristic is inspected with regard to its specification, the component would not be evaluated without doubt as an "OK" part or an "NOK" part ("not OK"). (cf. Fig. 3.3).

The complete specification of a *measurement result Y* is done according to Eq. 3.14. In addition to the determined measurement value y, the expanded measurement uncertainty U is specified. Furthermore, the *enhancement factor k* for the calculation of the measurement uncertainty U should be stated (here: k = 2; corresponds to probability of P = 0.9545 considering a Normal distribution model). It is obligatory to indicate whether the stated measurement uncertainty refers to the measurement system or the measurement process; cf. in detail Chap. 15.

$$Y = y \pm U(k = 2) \qquad (3.14)$$

Test process suitability/capability
The analysis of the *test process suitability/capability* is an analysis of the properties of a measurement system as well as of the test process with regard to the capability to fulfil the planned test task under the specified test conditions. The result is an assessment of the capability for the planned or existing measurement or test process. For detailed explanations and information, see Chap. 15; cf. also Bracke (2016).

3.5 Risk and Risk Analysis

The reliability of a product, system, assembly or component is directly related to the subject area of risk or risk analysis. The analysis of the technical reliability of a product can be a component of a more comprehensive risk analysis. The risk analysis often also includes entrepreneurial, organizational and social aspects. This chapter explains the basic terms relating to the subject area of risk and risk analysis and is mainly based on ISO 31000:2018 (ISO 2018) and DIN ISO 31000 (DIN 2018).

The term *Risk* is defined as "Effect of uncertainty on objects" (DIN 2018). Whereby the following notes are given in accordance with (ISO 2018):

1) Effects of a risk can be positive or negative.
2) Uncertainty is estimated with probabilities.
3) Goals of an organization generally include financial goals, safety of people, property and environment.
4) A risk comprises directly occurring damaging events, the unfavourable impact of external circumstances as well as continuously progressing undesirable developments.

Brühwiler (2003) makes the following additional comments on the concept of risk:

a) Risk exists when people or organizations actively pursue goals and/or rely on internal or external circumstances that can change unexpectedly. The term risk stands for the possibility that goals will (not) be achieved.
b) The first component of risk is uncertainty. The aim of risk management is to limit or assess uncertainty and to take it into account in decision-making.
c) The second component of risk considers the impact on objectives or expectations. By applying risk management methods, these effects are measured or estimated.
d) Risk is the umbrella term for opportunity and danger or opportunity and threat.

It can be seen that the concept of risk is commented on here in general terms, and thus encompasses organisations and units (products, processes). A transfer to technical products and processes is easily possible. In the following sections, the technical perspective is taken and the focus is on risk with regard to technical products and processes.

Risk analyses are primarily serve to identify and eliminate weaknesses in systems (products or processes) and to compare alternative systems. By applying risk analysis, probabilities or causes of failure are determined that lead to the failure (failure sequence) of a system. The risk analysis allows the evaluation of a possible damage or a damage sequence. *Risk analysis* concentrates on determining the probability of occurrence of the cause of damage (e.g. product failure) or the probability of occurrence of the consequence of damage (e.g. fire event). The consequences of a system failure are considered from the system level under consideration (example: assembly level) through all levels to the product user and the environment (so-called failure sequence chain); similar to the procedure in a Failure Mode and Effects Analysis (FMEA, Sect. 11). For example, the breakage of a vehicle engine piston rod leads to engine damage. The engine damage usually results in a vehicle break down. In certain driving situations, the stalled vehicle can endanger people.

The *risk matrix* represents a way to plot several risks of an organization or system in a two-dimensional graph according to the criteria of probability (ordinate) and their impact (abscissa), see also ISO 31000:2018 (ISO 2018). With regard to

3.5 Risk and Risk Analysis

technical systems, the variables must be carefully defined: For example, the probability may refer to a system failure, a component failure, or to the occurrence of the cause of the fault. On the other hand, the effect of a system failure can be represented with the following variables, for example: Danger to persons, repair costs, warranty costs, costs per case of damage, societal damage.

Risk correlation is the interdependence of risks. The correlation can be positive (risks reinforce each other) or negative (risks neutralize each other) and can refer to both probability and impact; (ISO 2018).

The term *risk management* encompasses processes and behaviors (Coordinated activities) designed to control and monitor an organization with respect to risks, ISO 31000:2018 (ISO 2018). It therefore focuses on the systematic identification and assessment of risks and the management of responses to identified risks at each stage of the product development process.

Essentially, risk management proceeds in four process steps: *Risk identification, risk analysis, risk handling* and *risk monitoring*, ISO 31000:2018 (ISO 2018).

Risk identification serves to identify potential risks; quantitative and qualitative methods are used for this purpose, such as Fault Tree Analysis (FTA, Chap. 11) or Failure Mode and Effects Analysis (FMEA, Chap. 11). The *risk analysis* serves to determine the essential risk components of cause and consequence of damage as well as the probability of occurrence (cf. also the term risk matrix) and can also be carried out with an FTA, for example.

Once the risk has been identified and analyzed, the *risk* is handled. Here, a differentiation can be made between six handling strategies: Risk avoidance, risk reduction, risk diversification, risk shifting, risk insurance, risk assumption and risk acceptance.

Risk avoidance and *risk reduction* can be achieved in development, for example, through design changes to the product concept or the product design (component design, component geometry, material changes). In the manufacture of technical products, on the other hand, the manufacturing technology or the testing concept can contribute to risk avoidance or risk reduction. In individual cases, the product concept change must be weighed up against the system/assembly/component design with regard to economic efficiency and expenditure.

Example of Automotive Engineering: Risk Avoidance Versus Risk Reduction

The principles of risk avoidance versus risk reduction are contrasted using the conceptual design of a convertible car top.

Two concepts are available for the development of a convertible vehicle top: The Softtop concept (fabric roof; soft top materials PVC, fabric mesh) or the Retractable Hardtop concept (rigid components; steel, plastics). The softtop allows a more optimal folding of the roof in terms of installation space, but carries risks in terms of vehicle theft, leakage (e.g. water ingress), embrittlement and moss formation. With the hardtop concept, the softtop risks mentioned (in particular embrittlement and moss formation) can be avoided (risk avoidance principle) and/ or reduced (in particular theft and leakage; risk reduction principle). However, this concept requires more installation space in the vehicle package with regard to roof folding (in the case of roof stowage), as the roof folding can be designed less flexibly compared to the soft top. A soft top is generally less expensive to develop and manufacture than a hard top concept.

Risk management can also be based on the principle of *risk diversification*. In this case, the risk is distributed among several risk carriers. In the case of technical products, this is done, for example, by taking *redundancy* into account, sometimes also with simultaneous application of diversification. For example, the power supply of a drive can be ensured by various power generators connected in parallel (so-called *hot redundancy*). If an additional diesel generator is provided, this is a technical diversification, since the redundancy is ensured via a technically different system. Since the diesel generator is only switched on in an emergency, this is a so-called *cold redundancy*.

When managing a risk by means of the principle of *risk shifting* the risk is transferred to suppliers and/or subcontractors, for example, and requires the design of appropriate contracts. For example, risky technologies, components or even manufacturing processes can be shifted to suppliers with proven, higher competence in order to avoid or reduce the risk for the OEM.

Risk insurance represents a further principle of risk handling: Insurance is taken out in respect of predefined loss events. This is a safe but possibly also cost-intensive form of risk management. The actual cause or consequence of the failure is neither avoided nor reduced.

If the identified and analyzed risk is considered acceptable by the *risk owner* (decision maker of an organization or part of it, who bears the overall responsibility for success and failure, opportunities and threats; cf. ISO 31000:2018 (ISO 2018), this is the principle of *risk acceptance*.

The final step in risk management is *risk monitoring*. In the product development phase, risk monitoring can take the form of component approvals, testing and product trials, cf. Chap. 14. In the manufacturing phase, risk monitoring can take the form of product testing or the monitoring of production process parameters, cf. Chaps. 15 and 16. In the product use phase (sales phase), risk monitoring can take the form, for example, of the analysis of warranty/complaint cases or of independent testing institutions.

Descriptive Statistics: Forms of Visulization

4

The initial analysis of observed values or measurement or test data can be performed using a proper visualization, for which basic, standardized forms are suitable in order to ensure illustration and comparability. This chapter shows the basic forms of visualization as they are used in engineering sciences: Value pattern, bar diagram, histogram, cumulative representation, boxplot, mathematical papers, boxplot and X–Y plot; Sects. 4.2, 4.3, 4.4, 4.5, 4.6, 4.7 and 4.9. Finally, there is an explanation of the representation form isochrone diagram, with which, in addition to relative/absolute frequencies, temporal references can also be taken into account; Sect. 4.8. All forms of representation are explained on the basis of the engine piston rod case study, so the introduction in Sect. 4.1 contains a short introduction to the engine piston rod design, specification and an extract of measurement data from a current series production.

4.1 Introduction and Case Study Engine Connecting Rod

The basic visualization options for the representation of data are explained using a case study of engine piston rod. The piston rod forms the connection between the crankshaft (crankpin) and the piston in an internal combustion engine (see Fig. 4.1). The large connecting rod eye encloses the crankpin; a piston pin is pushed through the small connecting rod eye, via which the movable connection with the piston is realized. The manufacturing process of the connecting rod usually includes several procedures, such as forging the raw part, machining (among other things, realizing the function-critical characteristics mentioned below), drilling, tapping, cracking or sawing (dividing the connecting rod and cap). Furthermore, the pressing the bushing into the small connecting rod eye is essential. For a more in-depth look at the technology of internal combustion engines (Basshuysen and Schäfer 2006 and 2015) is recommended. The following

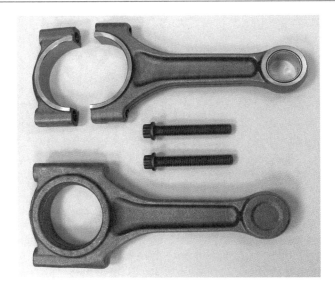

Fig. 4.1 Connecting rod of an internal combustion engine (bottom: forged blank; centre: connecting rod screws; top: Finished part with pressed-in bush)

function-critical features can be measured and tested in series production as part of process monitoring:

(a) diameter of the large connecting rod eye, d = 64 mm (+0.020),
(b) diameter of the small connecting rod eye, d = 30 mm (+0.020),
(c) centre distance (axles large/small connecting rod eye), l = 150 mm (±0.05).

A sample of n = 200 connecting rods is taken out of a series production and the function-critical features mentioned are measured. Table 4.1 shows a synthetic data set with measured values of the function-critical characteristic "diameter of large connecting rod eye".

Table 4.1 Measured value series for the characteristic large connecting rod eye diameter; synthetic measured values (original values, unit [mm]) of the sample values x_1, ..., x_n of a sample with a size of n = 200 from series production

64.007	64.006	64.008	64.001	64.011	64.005	64.004	64.006	64.003	64.008
64.002	64.004	64.007	64.009	64.006	64.005	64.012	64.002	64.009	64.009
64.002	64.008	64	64.003	64.003	64.003	64.007	64.013	64.007	64.01
64.012	64.005	64.005	64.005	64.008	64.009	64.011	64.006	64.002	64.003
64.008	64.009	64	64.002	64.001	64.007	64.005	64.002	64.012	64.003
64.007	64.003	63.999	64.009	64.007	64.005	64.002	64.006	64.003	64.002

(continued)

4.2 Value Pattern

Table 4.1 (continued)

64.003	64.007	64.003	64.006	64.008	64.003	64.009	64.007	64.003	64.009
64.007	64.002	64.011	64.01	64.008	64.008	64.004	64.006	64.006	64.007
64.009	64.006	64.003	64.008	64.005	64.009	64.002	64.006	64.007	64.003
64.005	64.006	64.01	64.007	64.002	64.009	64.005	64.004	64.014	64.003
64.007	64.009	64.006	64.01	64.011	64.008	64.003	64.007	64.003	64.001
64.008	64.006	64.008	64.009	64.003	64.006	64.003	64.008	64.001	64.003
64.007	64.008	64.007	64.002	64.008	64.007	63.999	64.001	64	64.012
64.003	64.008	64.001	64.007	63.999	64.006	64.006	64.007	64.006	64.01
64.004	64.002	64.007	64.01	64.003	64.009	64.006	64.006	64.007	64.006
64.003	64.011	64.004	64.011	64.005	64.008	64.005	64.006	64.002	64.011
64.005	64.009	63.999	64.002	64.011	64.003	64.006	64.008	64.004	64.007
64.001	64.009	64.001	64.007	64.004	64.011	64.003	64.009	64.008	64.005
64.003	64.009	64.007	64.007	64.011	64.007	64.012	64,006	64.008	64.003
64.006	64.007	64.008	64.001	64.004	64.009	64.003	64.009	64.009	64.009

4.2 Value Pattern

The value pattern is understood as the plotting of values in the order of their determination. Figure 4.2 shows a typical value visualisation: The two hundred measured values of the function-critical characteristic "diameter of large connecting

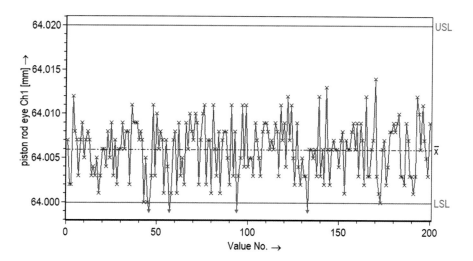

Fig. 4.2 Value pattern using the example of a series of measured values (characteristic "diameter of large connecting rod eye", case study piston rod)

rod eye" are shown; cf. Table 4.1. On the x-axis the identification of the measured component can be seen: Connecting rod number 1, 2, 3, ..., 200. The y-axis shows the measured value for the respective component. In addition, the specification limits (OSL, USL) for the characteristic "diameter of large connecting rod eye" are noted here in the diagram. The display of the value progression enables, for example, the visual recognition of jumps, trends both in the process position (mean) and in its dispersion. Obvious process changes are thus easily recognizable visually and can subsequently be analyzed in more detail. For example, jumps in the process position can indicate a shift change (e.g. operator changes machine parameters). A trend can, for example, be caused by a continuous temperature change (e.g. influenced by a non-air-conditioned machine hall). Value patterns are used, for example, in the context of statistical process control (SPC), here in the control chart; Chap. 16. Typical characteristics of value curves and possible interpretations are outlined in Sect. 16.5.

4.3 Bar Diagram

The bar diagram, also called a value beam diagram, allows a first visualisation of the frequency distribution of data. Figure 4.3 shows a bar diagram based on the engine connecting rod case study. The x-axis shows the scale to which the measured value is assigned. For the y-axis, two different forms of ablation of the values are usually chosen: The absolute frequency or the relative frequency with which a measured value appears in the measurement series (calculation of absolute/relative frequency; Sect. 6.1). The number of measured values in relation to a

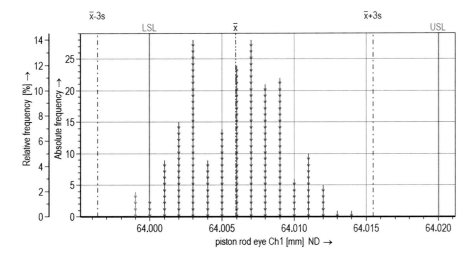

Fig. 4.3 Bar diagram using the example of a series of measured values (characteristic "diameter of large connecting rod eye", case study engine connecting rod)

specific value can be read off directly in this form of presentation, since the number is assigned directly to the particular original value. Thus, both the frequency of measured values and the distribution in the entire range of measured values can be analyzed. Furthermore, the estimators for the process parameters mean (here: arithmetic mean) and process dispersion (here: six times the standard deviation) are plotted in the diagram (estimation of mean and dispersion, Sect. 6.3). The disadvantage of this representation is that it becomes confusing with larger amounts of data (e.g. with n = 20,000 measured values), as each original value is plotted individually.

4.4 Histogram

In contrast to the bar chart, the individual original values are no longer directly recognizable within a histogram. The histogram shows the distribution of the calculated classes after a data classification. Figure 4.4 shows a typical histogram based on the connecting rod case study (Table 4.1). The x-axis shows the scale to which the classes – which contain measured values – are assigned. Furthermore, the relative frequency or absolute frequency of the respective class is shown on the y-axis. The data classes with class number and class width can be determined on the basis of different procedures for data classification. The number of classes can be seen (here: k = 10) and a constant class width w. The histogram enables a clear representation of the data frequency distribution, explicitly for large data sets, in contrast to the bar diagram. A number of data classification techniques exist, some

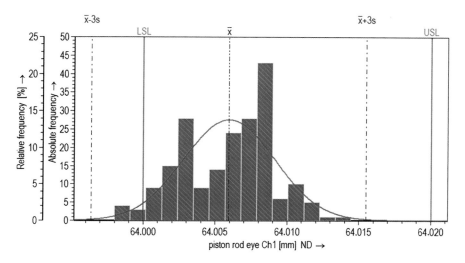

Fig. 4.4 Histogram based on the example of a series of measured values (characteristic "diameter of large connecting rod eye", case study engine connecting rod) on the basis of a classification procedure; fitted normal distribution model (cf. Chap. 7)

of which are outlined in Sect. 6.2. In addition, Fig. 4.4 shows a fitted density function (here: probability density according to Gauss) and the two estimators for the parameters mean (here: arithmetic mean) and dispersion (here: six times the standard deviation). The different data classification procedures have no influence on the calculation and visualization of the density function when the data basis is the same (of course). Depending on the classification procedure chosen, different histograms can therefore be developed with the same adjusted density function. The density function would only change if the estimators (estimation procedures) were varied or if the original values were changed.

4.5 Cumulative Diagram

The cumulative diagram form is not an independent form of visualization in the true sense, because it results from the bar diagram or histogram (incl. density function). Figure 4.5 shows a cumulative representation of the original values on the basis of the measured values of the engine connecting rod case study (Table 4.1). In contrast to the histogram, however, the relative probabilities related to the measured value ranks have been cumulated (added up) here. The summation can be done in different ways. On the one hand, classified data can be summed up. On the other hand, the estimated individual probabilities related to the ranks of the original values can be added up. The individual probabilities, with which probability a measured value occurs in a measurement series, can be approximated by a

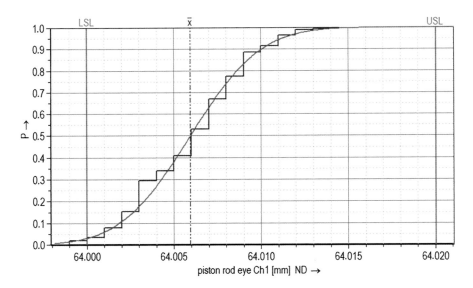

Fig. 4.5 Cumulative representation of measured value series and adapted normal distribution model (Chap. 7) using the example of a measured value series (characteristic "diameter of large connecting rod eye", case study engine connecting rod)

simple inclusion of the measurement value ranks, cf. also Chap. 12. Both ways are resulting in a staircase function. In addition, Fig. 4.5 shows a distribution function according to Gauss, which was adapted to the measured values. The adjustment of the distribution function is independent of the chosen method for the determination of the rank probabilities. The shown representation of the cumulated ranks as well as the distribution function enables the observer, for example, to estimate the probability with which a range of measured values is to be expected in relation to the entire population. Example: Measured values in the interval [64.003; 64.005] are represented with $P \sim 0.26$ according to the cumulative rank probability, and with $P \sim 0.22$ according to the adjusted normal distribution model.

4.6 Mathematical Papers: Log Representations

Mathematical papers are also used in the analysis of measured values with regard to certain properties or a certain distribution model. The *mathematical paper* contains an imprint of a defined, specific coordinate system. Known mathematical papers are for example the graph paper, logarithmic paper, polar coordinate paper or the probability paper. Mathematical papers allow, among other things, an estimation of whether observed values follow a certain statistical model deposited on the paper or a first simple analysis of the properties of a series of measured values (often due to linearisation effects, Sects. 4.6.1 and 4.6.2). Mathematical papers represent a subfield of nomograms (nomography). *Nomograms* are graphical visualizations of mathematical functions in order, for example, to be able to determine probability values without calculation for given parameters. In the recent past, nomograms were frequently used for the application of the models binomial distribution and Poisson distribution (both in Chap. 7), cf. Larson nomogram (Larson 1967) and Thorndike nomogram (Thorndike 1926). However, with the everyday use of computers, the manual use of nomograms has gradually been superseded.

Among the mathematical papers, the simple coordinate systems include, for example, the millimetre paper: The distances of the units are equidistant on the abscissa as well as on the ordinate, the distance is one millimetre. In the following, the logarithmic paper (Sect. 4.6.1) and the Weibull probability paper (Sect. 4.6.2) are discussed, since both mathematical papers are frequently used in the analysis of technical reliability.

4.6.1 Logarithmic paper

Logarithmic paper also belongs to the category of mathematical papers. Here, a distinction is made between the single logarithmic and the double logarithmic paper. In single logarithmic paper, only one axis (abscissa or ordinate) is divided. Figure 4.6 shows a single-logarithmic ablation of the measured values of the engine connecting rod case study (Table 4.1), where the ordinate is logarithmically divided and the abscissa follows the equidistant division. This representation

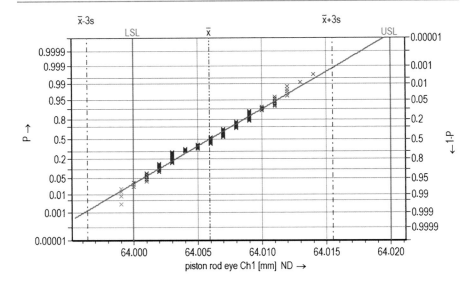

Fig. 4.6 Simple logarithmic representation using the example of a series of measured values (characteristic "diameter of large connecting rod eye", case study engine connecting rod) and adapted normal distribution model

allows a simple check whether the observed measured values follow a certain distribution model: The observed values lie almost or completely on a straight line (linearization). An exponential function according to Eq. 4.1 is visually transformed into a linear graph, since the values are plotted according to Eq. 4.2.

$$f(x) = y = a^x \qquad (4.1)$$

$$\log y = x \cdot \log a \qquad (4.2)$$

With double logarithmic plotting, both abscissa and ordinate follow a logarithmic division. This means that the curved course, when the axes are scaled equidistantly, of a power function (cf. Eq. 4.3) can also be visualized in the form of a straight line based on a double logarithmic scaling (cf. Eq. 4.4).

$$f(x) = y = ax^b \qquad (4.3)$$

$$\log y = b \cdot \log x + \log a \qquad (4.4)$$

4.6.2 Weibull Probability Paper

In reliability analysis, another special mathematical paper is often used: The representation of measured values in a Weibull probability paper. The Weibull distribution model is one of the most commonly used life distribution models. It contains an exponential approach as well as a few parameters, which can be interpreted very

4.6 Mathematical Papers: Log Representations

well with regard to technical failure behavior; cf. Chap. 12 in detail. A representation of the Weibull distribution model with linear division of both axes leads to a curve with an S-shaped course. In a Weibull probability paper, the division of the ordinate is double logarithmic and the division of the abscissa is single-logarithmic. This visualizes a Weibull distribution model (two parameters, cf. Eq. 4.5) as a straight line (cf. Fig. 4.7 left), which allows a first, simple analysis: In this representation, the parameter b corresponds to the gradient of the straight line. A Weibull distribution model with three parameters according to Eq. 4.12 only shows a curvature of the straightline (curve) to the value of the parameter t_0 (interpretation: failure-free time) in the representation using Weibull probability paper (cf. Fig. 4.7 right).

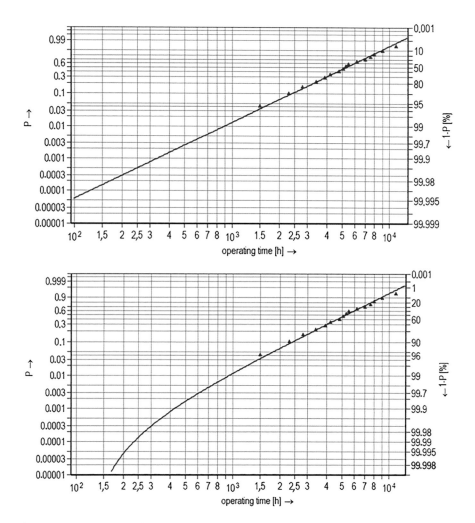

Fig. 4.7 Representation of Weibull distribution models within a Weibull probability paper; left: Fitting two-parameter model, right: Fitting three-parameter model

The starting point for this division of the abscissa or ordinate is the Weibull distribution model; Eq. 4.5. After transformation and double logarithmizing, Eq. 4.6 is obtained. The general degree equation can be seen in Eq. 4.7, followed immediately by Eq. 4.8 for the division of the abscissa, the division of the ordinate according to Eq. 4.9, the intercept according to Eq. 4.10 and for the gradient Eq. 4.11.

$$F(x) = 1 - \exp\left(-\left(\frac{x}{T}\right)^b\right) \tag{4.5}$$

$$\ln(-\ln(1 - F(x))) = b \cdot \ln x - b \cdot \ln T \tag{4.6}$$

$$f(x) = mx + c \tag{4.7}$$

$$x = \ln x \tag{4.8}$$

$$f(x) = y = \ln(-\ln(1 - F(x))) \tag{4.9}$$

$$c = -b \cdot \ln T \tag{4.10}$$

$$b = m \tag{4.11}$$

$$F(x) = 1 - \exp\left(-\left(\frac{x - t_0}{T - t_0}\right)^b\right) \tag{4.12}$$

Figure 4.7 shows the visualization of the cumulative probabilities – related to measured value strings – and fitted Weibull distribution models (Eqs. 4.5 and 4.12). The underlying data set is a case study coolant pump with 16 failure dates; Chap. 12. The parameters of the Weibull distribution models (cf. Table 4.2) were determined using maximum likelihood estimation methods; Chap. 8.

4.7 Boxplot

A boxplot comprises the representation of elementary mean and scatter parameters in relation to a series of observations or measured values. The visualization allows a first analysis and is suitable for the comparison of different samples. The visualization is based only on a number ray, thus the parallel representation and

Table 4.2 Case study coolant pump: Parameters of the fitted Weibull distribution models (two-parameter/three-parameter) in relation to Fig. 4.7; see also Chap. 12

Model	t_0	T	b
Weibull model two-parameter	Not applicable	6,191 h	2.36
Weibull model three-parameter	122.5 h	6,176 h	2.31

the comparison of series of measured values are easily possible (for example in contrast to histograms). Figure 4.8 shows an example of the representation of two boxplots with regard to the measured values of the characteristic "large connecting rod eye" of the case study engine connecting rod: The left boxplot is based on the original values; Table 4.1, Sect. 4.1 (batch 1), the right boxplot on another series of measured values related to a batch 2.

The boxplot can contain different mean and scatter parameters, the most common parameters are:

1. Box in which 50% of the data lie in the middle of the measurement series: Interquartile range,
2. Median, which divides the box into upper/lower quartile,
3. Antennas (Whisker),
4. Outliers,
5. Value axis or normalized axis scaling,
6. Entry of specification limits.

The use of the median estimator is to be preferred to the estimator arithmetic mean (in Fig. 4.8: continuous line inside the box). In contrast to the arithmetic mean, the median is not sensitive to extreme observed values (outliers). Thus, based on the position of the median in relation to the box, a conclusion can be drawn about the skewness S of the measurement data distribution (cf. Chap. 6): In the case of a pronounced skewness, the median does not lie in the middle of the box. For this

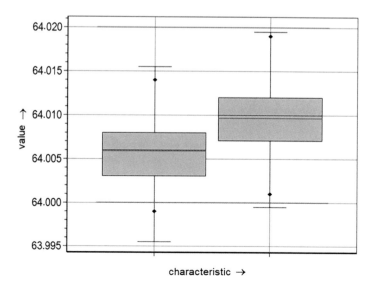

Fig. 4.8 Boxplots based on two measured value series (characteristic "diameter of large connecting rod eye", case study engine connecting rod, two batches)

reason, the arithmetic mean is used less frequently (in Fig. 4.8: dashed blue line within the box). Furthermore, the antennas (whisker) can be seen in Fig. 4.8. The definition of the antenna expression is not uniform. According to Tukey (1977), the whisker ends at a maximum of 1.5 times the interquartile range (box length), i.e. at the last observation value that does not exceed this limit. Thus, this may also be an indicator of any skewness that may be present. All further values – outside the whiskers – are plotted as extreme observation values, but without any connection to the box. The form of representation in which the ends of the whiskers represent the minimum and maximum observed values (range of the sample, cf. Sect. 3.1) is found rather more rarely; the disadvantage here is the loss of information on any extreme observations (outliers).

Figure 4.8 shows the comparison of two batches of connecting rods (here based on the representation according to Tukey (1977)), an offset of the means of both underlying series of measurements – both of the estimators for arithmetic mean or median and of the boxes – can easily be seen. Furthermore, a symmetrical distribution model (low skewness) seems to be present in each case, since box and whiskers are arranged symmetrically to the position parameter median. Outliers outside the whiskers cannot be detected.

4.8 Isochrone Diagram

With an isochrone diagram can be used to display frequencies in relation to series of measured values (example: batches, samples) with time references. In technical reliability analysis, isochrone diagrams are often used in field data analysis. In many cases, the time reference is as follows: The starting time is either the manufacturing date of the product batch (sample) or the date of placing on the market. Furthermore, the time period during which a product batch is in the use phase is used: The market observation period. The frequency shown (at different points in time within the use phase) is usually the relative frequency of incidents of damage in relation to the quantity of product produced in this batch/sample; the so-called damage rate (Sect. 3). In this case, indications of failure behaviour – explicitly early/random/runtime-related failure behaviour; cf. Sect. 3.3.2) – can be read off.

Ergo, the isochrone diagram is widely used to visualize and analyze field data for products that are already in the use phase. For this reason, Fig. 4.9 uses the example of vehicle complaint data to show the three variables production month (vehicle), ratio of damaged units and field duration (Month in Service [MIS]). The abscissa shows the production months in which the respective rejected vehicle was produced. The ordinate shows the ratio of damaged units (damaged units in relation to the number of units produced). The damage rate is determined at different field run times: Four different field run times (6–36 MIS) can be seen in the diagram, so the complaint rate can be compared from different batches at the same field run times.

Furthermore, the current complaint rate (dashed line) is plotted, whereby a comparison of different production months is not meaningful, since different field

4.8 Isochrone Diagram

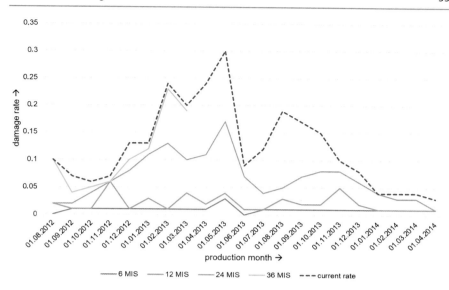

Fig. 4.9 Isochrone diagram based on field data

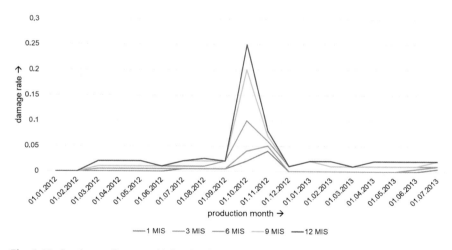

Fig. 4.10 Isochrone diagram with batch-related peak of damage rate; data basis: field data

run times are available. The connection of the simultaneous complaint rates to the different production months shows a typical stratified curve; for this reason, this type of diagram is sometimes also called a stratified line diagram in practice (instead of an isochrone diagram).

The isochrone diagram according to Fig. 4.10 reveals defect foci with regard to certain production batches. For example, the production month 10.2012 shows an increased complaint rate at different field observation times (MIS) compared

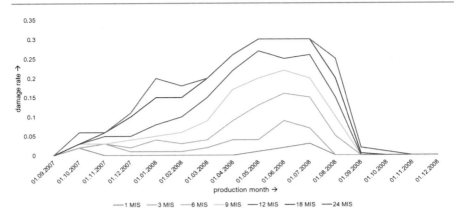

Fig. 4.11 Isochrone diagram based on field data; visualization of the effects trend as well as the impact of a product/process measure

to previous and subsequent periods. On the other hand, the isochrone diagram in Fig. 4.11 shows an initial swelling failure behavior of the complaint rates in the production batches until 07.2008, after which a measure was brought to serial use, causing the complaint rate to drop abruptly (09.2008). The decrease of the ratio of damaged units over a longer period of time (in this case: 2 months) would be typical for a measure used by a supplier. Due to a potential mixing of previous and improved component status / production process status, the decrease of the ratio of damaged units usually does not occur at the same time (supplier versus OEM), but over a longer period of time. The swelling of the complaint rates with regard to the progressing production months (example: period 11.2007–07.2008) may be interpreted as a trend, for example due to a negatively developing production process.

If the distance between two shift lines develops with disproportionate growth at a far advanced field runtime, this is a first indicator for a runtime-related failure behaviour. This effect is shown in Fig. 4.12: A disproportionate increase after 9 months in service compared to rather low complaint rates up to time 9 MIS using the example of the production months 11.2012 to 02.2013.

On the other hand, Fig. 4.13 shows characteristics of a typical early failure behaviour of product batches in the use phase. Up to the field observation time 6 MIS, high complaint rates (ratio of damaged units) are evident (e.g. in the production period 10.2012 to 01.2013); in the subsequent field observation period, the intervals between the isochrones become smaller for the same time durations.

4.8 Isochrone Diagram

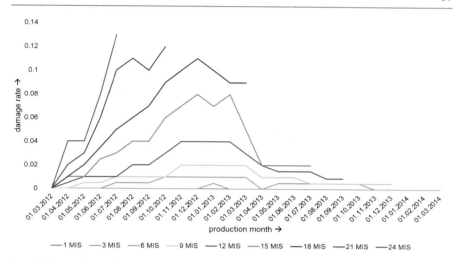

Fig. 4.12 Isochrone diagram based on field data; visualization of the effect of runtime-related failure behavior

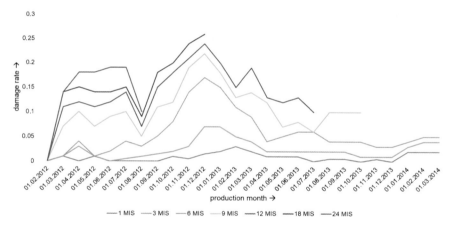

Fig. 4.13 Isochrone diagram based on field data; visualisation of the effect early failure behaviour

The isochrone diagram according to Fig. 4.14 shows the effect of changing ratio of damaged units in relation to different production batches. A typical cause may be inadequate control or quality capability of production processes (cf. Sec. 16.3.4).

Of course, the application of an isochrone diagram is not limited to field data: When evaluating a production line, for example, the three variables production month (abscissa), damage rate at final product inspection (ordinate) and machine operating time (isochrones; "shift lines") can also be plotted.

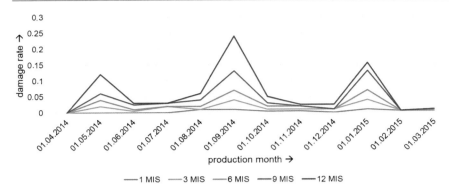

Fig. 4.14 Isochrone diagram based on field data; visualization of the effect of an unstable production process

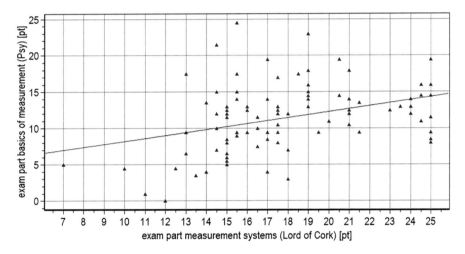

Fig. 4.15 Representation of exam results, consisting of two parts of an exam (study), in the X–Y plot as well as a linear regression model

4.9 X–Y Plot

Analysis of data is often done in bivariate form to be able to plot relationships between two variables X and Y. As examples are mentioned:

a) Variable power in relation to the variable displacement of regarding a combustion engine.
b) Variable volume of a component in relation to the ambient temperature.

4.9 X–Y Plot

The graphical visualization of a bivariate relationship can be done using an X–Y plot. Figure 4.15 shows an X–Y plot for the representation of exam results of an exam, which consists of the two exam parts: part "Basics of Measurement" (ordinate) and part "Measurement Systems" (abscissa). Each point in the diagram shows an exam result, consisting of two scores related to both exam parts. For example, a student achieved 12 points in the part "Measurement Systems (Lord of Cork)", but no point in the part "Basics of Measurement (Psy)". The position of the exam results in the X–Y plot suggests a correlation: students who achieved a high score in one part of the exam also achieved a high score in the other part. This assumption is also reflected in the linear regression model, which was determined under the assumption of a linear correlation. The detailed explanations on correlation and regression including the use of X–Y plots can be found in Chap. 10. Of course, correlations between more than two variables can also be visualized.

Probability and Random Experiments 5

The basics of probability and probability theory in the context of single- and multi-stage random experiments as well as various application examples form the core of this chapter. Within the framework of Sect. 5.1, terms, connections, axioms as well as visualizations on the topic of probability are explained. The focus of Sect. 5.2 is on pobabilistics, first presented fundamentally for the simple case of two events. Subsequently, multistage random experiments are discussed, focusing on event trees, total probability and Bayes' theorem.

5.1 Probability

In this chapter, the most important basics of probability are explained. The chapter focuses on basic terms, visualization possibilities according to Venn and Euler, links between events, the Laplace experiment and Kolmogorov's axioms.

The starting point should be a random experiment. A *random experiment* is an experiment which is carried out under previously defined conditions. The outcome of the random experiment is random (cf. also Chap. 3), although the possible results are known. That is, one of the possible outcomes will occur; but which of the possible outcomes is not known before and during the random experiment. Therefore, the outcome of the random experiment is described as random.

A simple example is the dice *experiment*: the random experiment is the throw, the possible outcomes are {1, 2, 3, 4, 5, 6}. Before the experiment, it is not known which outcome of the dice will occur.

In probability theory, the events that can occur are called *sample space* Ω. The possible events within the outcome set Ω are the *elementary events* ω_x. This relationship is shown by Eq. 5.1.

$$\Omega = \{\omega_1, \omega_2, \ldots, \omega_n\} \tag{5.1}$$

© Springer-Verlag GmbH Germany, part of Springer Nature 2024
S. Bracke, *Reliability Engineering*, https://doi.org/10.1007/978-3-662-67446-8_5

Each of the elementary events can now be assigned a probability. If event A is expected, then the probability that *event A* would be P(A). Here P(A) is a real number. If event A does not occur, then the *event complementary* to A occurs, cf. Eq. 5.2. The illustration of random experiments involving result set, events as well as basic links can be done by means of diagrams. Figure 5.1 shows event A within the event set Ω. The space not covered by event A is the event complementary to event A.

$$\bar{A} = \Omega \backslash A \qquad (5.2)$$

Example Random Experiment: The Dice Roll
Let the starting point be a six-sided die with corresponding numbers. The random experiment dice roll therefore has the result set according to Eq. 5.3.

$$\Omega = \{1, 2, 3, 4, 5, 6\} \qquad (5.3)$$

For example, possible events can now be defined as follows: The throw of the number 2 is event A (Eq. 5.4). Event B is the throw of an even number (Eq. 5.5). Event C is a dice event that is divisible by three and without remainder (Eq. 5.6). Event D is a dice result that takes a value greater than 1 (Eq. 5.7).

$$A = \{2\} \qquad (5.4)$$

$$B = \{2, 4, 6\} \qquad (5.5)$$

$$C = \{3, 6\} \qquad (5.6)$$

$$D = \{2, 3, 4, 5, 6\} \qquad (5.7)$$

The connection between event, complementary event becomes clear in the throwing experiment as follows: Let event E be the throw of an odd number. Let event B be the throw of an even number. Then B would be the complement of E (cf. Eq. 5.8) and E would be the complement of B (cf. Eq. 5.9).

$$\bar{E} = \{2, 4, 6\} = B \qquad (5.8)$$

$$\bar{B} = \{1, 3, 5\} = E \qquad (5.9)$$

The Linking of Events
The *union (OR-linkage) of* the results A and B is shown in Eq. 5.10 (cf. also Fig. 5.2), the *average (AND-linkage) of* events A and B is shown in Eq. 5.11 (cf.

Fig. 5.1 Visualization of the relationship between event A and the event complementary to A within the result set Ω

5.1 Probability

Fig. 5.2 Union of event A and B; event A has an intersection with event B within event set Ω. Ergo, event A or B or both events are valid

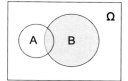

Fig. 5.3). With respect to linkage and complement, the rules according to De Morgan apply; Eqs. 5.12 and 5.13.

$$A \cup B \tag{5.10}$$

$$A \cap B \tag{5.11}$$

$$\overline{A \cup B} = \overline{A} \cap \overline{B} \tag{5.12}$$

$$\overline{A \cap B} = \overline{A} \cup \overline{B} \tag{5.13}$$

Example Random Experiment: The Dice Roll

Let event A be the throw of an odd number. Let event B be the throw of an even number (cf. Eqs. 5.14 and 5.15), so the union of both events directly yields the result set according to Eq. 5.16 and the empty set according to Eq. 5.17 when both events are averaged.

$$A = \{1, 3, 5\} \tag{5.14}$$

$$B = \{2, 4, 6\} \tag{5.15}$$

$$A \cup B = \{1, 2, 3, 4, 5, 6\} = \Omega \tag{5.16}$$

$$A \cap B = \emptyset \tag{5.17}$$

Visualization of Links According to Venn and Euler

The event linkage union or average of different events can be visualized easily. Figure 5.2 shows the union, and Fig. 5.3 shows the average of the linkage of event A and B within the result set Ω. This form of representation involves overlapping of events: If elementary events are part of event A and event B, they are entered within the intersection set. In the opposite case, the overlap is empty. This representation follows the visualization according to John Venn (mathematician, 1834–1923).

Fig. 5.3 Average of event A and B; event A has an intersection with event B within event set Ω. The average is the joint occurrence of A and B

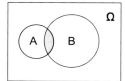

An alternative visualization is the representation according to Leonhard Euler (1707–1783) as shown in Fig. 5.4: The events A and B have no common elementary events, ergo both events are represented without intersection. The representation according to Venn would show the intersection even if the intersection is empty (i.e. no elementary events are contained).

Summary of the Visualization According to Venn and Euler
When two events occur, the diagrams according to Euler and Venn differ in the representation of the intersection. According to Venn, an intersection is always represented. If the result set contains identical elementary events, these are entered in the intersection. If there are no equal elementary events in both result sets A or B, then the overlap area remains empty, and the elementary events are drawn into the remaining space of the circles A and B. Ergo, the positions of the event sets are always the same, the statement is made by the assignment of the set elements. With Euler the positioning of the result sets is different depending on the statement. Thus by the positioning of the result sets the relations between the events are clearly expressed.

The representation according to Euler focuses on the properties of the events via their positions in relation to each other. The representation according to Venn focuses on the relations of the events to each other via the positioning of the elementary events.

Let three events A, B and C be given according to Eqs. 5.18, 5.19 und 5.20. Figure 5.5 shows the visualization of the events A (red), B (yellow), C (green) according to Euler (right) and Venn (left). In the representation according to Euler, quantities B and C show no overlap, since there are no common elementary events. On the other hand, the overlap for sets B and C can be seen in the visualization according to Venn, but without the entry of one or more elementary events.

$$A = \{1, 2, 3, 4, 5\} \tag{5.18}$$

$$B = \{1, 2, 6\} \tag{5.19}$$

$$C = \{4, 7, 8, 9\} \tag{5.20}$$

The Laplace Experiment
According to Pierre-Simon Marquis de Laplace (1749–1827), the probability of occurrence of an event A within the result set Ω is defined under the condition that it is the same for all elementary events. This is based on the Laplace experiment,

Fig. 5.4 Event A has no intersection with event B within event set Ω. According to Euler, the two events are represented separately within the event set Ω

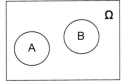

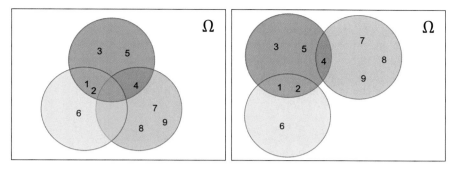

Fig. 5.5 Comparison of the representation form of events and the relations to each other according to Venn (left) and Euler (right)

in which an equally frequent occurrence of the respective elementary events can be observed with frequent repetition and can be formulated accordingly. The random experiment of throwing the dice is such a Laplace experiment, since all possible throwing results (elementary events) are equally probable. The prerequisite is that the dice is not marked.

Equation 5.21 holds for the probability of occurrence of the elementary event ω_1 given m possibilities. Moreover, if an event A consists of several elementary events (i = 1, 2, ..., m), Eq. 5.22 holds.

$$p(\omega_1) = \frac{1}{m} \qquad (5.21)$$

$$P(A) = \sum_i \rho(\omega_i) = g(A)\frac{1}{m} = \frac{g(A)}{m} \qquad (5.22)$$

Equation 5.22 may be read as follows: The quotient of the number of all elementary events g(A) favorable to event A and the number of all possible events m of the outcome set Ω is the probability that outcome A occurs. Condition is:

1) The result set Ω is finite,
2) All elementary events are equally likely in terms of their occurrence.

Example Random Experiment: Dice Roll
If A is the event "throw result even numbered of points" according to Eq. 5.23, then the probability P(A) for a die with six numbers (m = 6) is according to Eq. 5.24.

$$A = \{2, 4, 6\} \qquad (5.23)$$

$$P(A) = \frac{g(A)}{m} = \frac{3}{6} = 0.5 \qquad (5.24)$$

Example Test: Delivery of a Box with Lighting Units
The delivery of a box contains 100 light units. Of these lighting units, 75 units are defective. In the random experiment "drawing a lighting unit", the event A is defined as the drawing of a defective lighting unit from the box. Then the probability P(A) of drawing a defective light unit is defined according to Eq. 5.25; the probability of the complementary event ("drawing a working light") is shown by Eq. 5.26 and thus Eq. 5.27 holds.

$$P(A) = \frac{g(A)}{m} = \frac{75}{100} = 0.75 \qquad (5.25)$$

$$P(\overline{A}) = \frac{g(\overline{A})}{m} = \frac{25}{100} = 0.25 \qquad (5.26)$$

$$P(A) + P(\overline{A}) = 1 \qquad (5.27)$$

Axioms of Kolmogorov
The *axioms of Kolmogorov* form the basis of probability theory and were published by Kolmogorov in the work "Foundations of the Theory of Probability", cf. (Kolmogorov 1956). The axioms still form the central basis for all probability considerations in many disciplines—mathematics, physics, engineering, et cetera.

Axiom 1: The probability of an event A lies between 0 and 1, cf. Eq. 5.28.

$$0 \leq P(A) \leq 1 \qquad (5.28)$$

Axiom 2: The probability for a certain event is 1, cf. Eq. 5.29.

$$P(\Omega) = 1 \qquad (5.29)$$

Axiom 3: The addition theorem applies for mutually exclusive events $A_1, A_2 \ldots$, cf. Eq. 5.30.

$$P(A_1 \cup A_2 \cup \ldots) = P(A_1) + P(A_2) + \cdots \qquad (5.30)$$

Relation of Kolmogorov's axioms to relative frequencies in random experiments
A random experiment is performed n times, the possible event A is observed with the number n(A). Thus, the relative frequency h(A) for event A is calculated according to Eq. 5.31. The relative frequency $h_n(A)$ can therefore assume a value between 0 and 1 (cf. Eq. 5.32). If all relative frequencies of all potential events of the random experiment are added together, the result is 1. Example: The summation of all relative frequencies in a Bundestag election with respect to all elected parties results in 1 with respect to the votes cast.

5.1 Probability

If the relative frequency with respect to an event is 1 (cf. Eq. 5.33), it is called a certain event. Furthermore, the addition theorem applies to independent, mutually exclusive events (cf. Eq. 5.34). With reference to the example of the Bundestag election, this would mean that, for example, the shares of two parties may be added together, since the election results of the parties are independent of each other (and each voter may only vote for one party).

$$h(A) = \frac{n(A)}{n} \tag{5.31}$$

$$0 \leq h_n(A) \leq 1 \tag{5.32}$$

$$h_n(\Omega) = 1 \tag{5.33}$$

$$h_n(A \cup B) = h_n(A) + h_n(B) \tag{5.34}$$

Notes on the Kolmogorov Axioms in Connection with Relative Frequencies
a) If a random experiment is performed frequently, the probability of an event P(A) can be approximated by the observed relative frequency $h_n(A)$; Eq. 5.35 applies. Example: Random experiment dice; with a high number of single rolls, the relative frequency per possible outcome will approximate the outcome set $\Omega = \{1, 2, 3, 4, 5, 6\}$ at $h_n(A) = 1/6$.
b) For an impossible event cf. Axiom 2 of Kolmogorov, Eq. 5.36 holds, where Ø is the empty set.
c) The probability of the complementary event to A is according to Eq. 5.37.
d) Since the Axiom 1 probability of an event is between 0 and 1, Eqs. 5.38 and 5.39 also hold.
e) Thus, to calculate the probability of a defined event A, the summation of the probabilities of the elementary events is performed (provided that there are several elementary events associated with event A); Eq. 5.40 holds.

$$P(A) \approx h_n(A) \tag{5.35}$$

$$P(\emptyset) = 0 \tag{5.36}$$

$$P(\overline{A}) = 1 - P(A) \tag{5.37}$$

$$P_i \geq 0 \quad mit\ i = 1, 2, \ldots \tag{5.38}$$

$$\sum_{i=1}^{\infty} P_i = P_1 + P_2 + P_3 + \ldots = 1 \tag{5.39}$$

$$P(A) = \sum_i P_i \tag{5.40}$$

5.2 Probabilistics

On the basis of Kolmogorov's axioms, the addition theorem for arbitrary events is derived directly. The addition theorem is first shown for two possible events A and B.

For two independent events A and B, the *addition theorem* applies according to Eq. 5.41. For any events A and B – which are not mutually exclusive – Eq. 5.42 applies. For events which are not mutually exclusive, Eq. 5.43 applies.

$$P(A \cup B) = P(A) + P(B) \tag{5.41}$$

$$P(A \cup B) = P(A) + P(B) - P(A \cap B) \tag{5.42}$$

$$A \cap B \neq \emptyset \tag{5.43}$$

In addition to independent events or events occurring at will, there is also the situation where an event B occurs on the condition that an event A has previously occurred. Without the occurrence of event A, event B would not occur. This relationship is called conditional probability and is shown in Eq. 5.44. By rearranging Eq. 5.44, we obtain the *multiplication theorem* according to Eq. 5.45, which describes the simultaneous occurrence of events A and B.

$$P(B|A) = \frac{P(A \cap B)}{P(A)}; P(A) \neq 0 \tag{5.44}$$

$$P(A \cap B) = P(A) \cdot P(B|A) \tag{5.45}$$

In the case where events A and B are independent, Eqs. 5.46 and 5.47 hold, and Eq. 5.48 follows immediately. Thus, for independent events, the rule of calculation shown in Eq. 5.49 holds.

$$P(A|B) = P(A) \tag{5.46}$$

$$P(B|A) = P(B) \tag{5.47}$$

$$P(A \cap B) = P(A) \cdot P(B|A) = P(A) \cdot P(B) \tag{5.48}$$

$$P(A \cap B) = P(A) \cdot P(B) \tag{5.49}$$

Multi-Stage Random Experiments

In multistage random experiments, to determine probabilities, intermediate events with their associated paths must be considered as the connection between the root (input) and final event. Therefore, the following topics are discussed for mapping multistage random experiments:

1) Event trees
2) Total probability
3) Bayes theorem

5.2 Probabilistics

Event Trees

The description of successive events can be done clearly using the representation of event trees, cf. Fig. 5.6. The representation is based on the visualizations of Papula (2008). An event tree usually consists of a root, the paths, which lead via intermediate events (Fig. 5.6: A_1, A_2) to possible final events (Fig. 5.6: B_1, B_2, ..., B_7). For each path, a probability is given with which the next intermediate event or the final event is reached.

The following rules apply to the determination of probabilities within the event tree:

1) The multiplication theorem applies to the combination of probabilities that lead along a path via possible intermediate outcomes to the possible final outcome; Eqs. 5.45 and 5.49, respectively.
2) If several paths lead to the same final result, the addition theorem applies to the probabilities of the paths in question; Eq. 5.41.

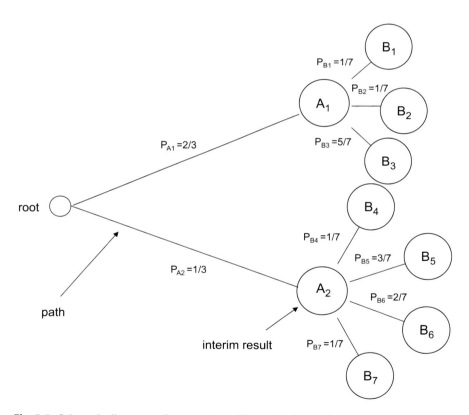

Fig. 5.6 Schematic diagramm of an event tree with root (basic event), paths, intermediate events A_i and possible end results B_j

The sum of all probabilities for parallel paths between two levels is, of course, one. In the example shown in Fig. 5.6, the sum of the probabilities for events A_1 and A_2 is one, as is the sum of B_1 to B_3 and B_4 to B_7.

Total probability

First, let the initial situation be that an event B can be reached via various paths or intermediate events (cf. Fig. 5.7). Each path presupposes the passage through an intermediate event A_j: In general, Eq. 5.45 applies to the probability P(B) via an intermediate event A. Since event B can be reached via various intermediate events A_j, all probabilities that lead to event B via intermediate event A_j must be added together. This addition of all probabilities to the possibilities leading to the final outcome is called *total probability:* The probability for the occurrence of event B is the total probability according to Eq. 5.50

$$P(B) = P(B|A_1) \cdot P(A_1) + P(B|A_2) \cdot P(A_2) + \cdots + P(B|A_n) \cdot P(A_n)$$
$$= \sum_{i=1}^{n} P(A_i) \cdot P(B|A_i) \tag{5.50}$$

Bayes' Theorem

First, Eq. 5.51 (cf. Eq. 5.44) applies to conditional probabilities, from which the multiplication theorem follows directly; Eq. 5.52. However, the following also applies: The order of two jointly occurring events does not play a role in the average, ergo Eq. 5.53 applies. By substituting Eq. 5.53 into Eq. 5.51, *Bayes' theorem* follows according to Eq. 5.54; (Thomas Bayes, 1701–1761).

$$P(A|B) = \frac{P(A \cap B)}{P(B)}; P(B) \neq 0 \tag{5.51}$$

$$P(A \cap B) = P(A|B) \cdot P(B) \tag{5.52}$$

$$P(A \cap B) = P(B|A) \cdot P(A) \tag{5.53}$$

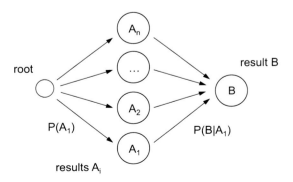

Fig. 5.7 Schematic diagram illustrating the "total probability" P(B)

$$P(A|B) = \frac{P(B|A) \cdot P(A)}{P(B)}; P(B) \neq 0 \qquad (5.54)$$

Summary
The calculation of probabilities for complex event trees (cf. Fig. 5.6) is carried out with the help of the laws of conditional probability, total probability and Bayes' theorem.

The *total probability* $P(B)$ taking into account n paths or intermediate events, each of which is the prerequisite for the occurrence of the final event B, is shown in Eq. 5.55. The probability that event B was reached via exactly intermediate event A_j is calculated according to Eq. 5.56. The symbol O indicates the path via intermediate event A_j, via which event B was reached (cf. also Fig. 5.7). *Bayes' theorem:* The conditional probability $P(A_j|B)$ is obtained by calculating the probabilities along the path $OA_j B$ and dividing this by the total probability $P(B)$. The probability $P(B)$ takes into account all possible paths in the event tree that can lead to event B.

$$P(B) = P(B|A_1) \cdot P(A_1) + P(B|A_2) \cdot P(A_2) + \cdots + P(B|A_n) \cdot P(A_n)$$
$$= \sum_{i=1}^{n} P(A_i) \cdot P(B|A_i) \qquad (5.55)$$

$$P(A_j|B) = \frac{P(OA_j B)}{P(B)} = \frac{P(A_i) \cdot P(B|A_j)}{\sum_{i=1}^{n} P(A_i) \cdot P(B|A_i)} \qquad (5.56)$$

5.3 Examples for Probabilistics

This chapter shows examples to illustrate the application of visualization options and laws of probability theory. Chap. 5.3.1 focuses on the sampling inspection within the scope of a delivery, in which functioning and defective units are detected. The urn model with the investigation of probabilities with respect to different color combinations based on different drawing strategies is discussed in Sect. 5.3.2. Sect. 5.3.3 focuses on the application of the basic principles of probabilistics and Bayes' theorem to the "Monty Hall problem" (or also: the goat problem).

5.3.1 Sampling Inspection in the Incoming Inspection

A manufacturer receives a delivery of 100 headlamps. 10 headlamps of this delivery are defective. The manufacturer subjects the delivery to a goods inward sampling inspection and pulls n = 3 headlights for inspection (inspection procedure: pull one headlight, inspect, do not put back regardless of inspection result).

Question
What is the probability of pulling three headlights that are functional?

Solution
The elementary event A_i, which is favorable for event A, is defined as: Drawing a functional headlight. Event A itself is the event of drawing three functional headlights in succession. The supply includes 100 headlamps; of these, 90 headlamps are functional, ergo, according to Laplace, for the first draw the number is 90 with respect to the cases favorable to event A_1, and the number of all possible cases is 100; from this follows for $P(A_1)$ Eq. 5.57. Thus, for the second draw the probability is given by Eq. 5.58 and for the third draw Eq. 5.59. From this, by linking the events according to Eq. 5.60, the result for the question follows.

$$P(A_1) = \frac{g_1(A_1)}{m} = \frac{90}{100} = 0.9 \qquad (5.57)$$

$$P(A_2|A_1) = \frac{g_2(A_2)}{m_2} = \frac{89}{99} = 0.89899 \qquad (5.58)$$

$$P(A_3|A_1 \cap A_2) = \frac{g_3(A_3)}{m_3} = \frac{88}{98} = 0.89796 \qquad (5.59)$$

$$P(A_1 \cap A_2 \cap A_3) = P(A_1)P(A_2|A_1)P(A_3|A_1 \cap A_2) = 0.7265 \qquad (5.60)$$

5.3.2 Urn Model: The Ball Drawing

In an urn there are 5 red, 3 green and 4 blue balls.

Question 1: Balls are drawn one after the other without putting them back. The following probabilities with regard to the drawing of ball colours in a certain order are to be visualised and determined mathematically.

a) Tree diagram for drawing two balls.
b) Probability that the first ball drawn is red.
c) Probability that the first ball is red and the second ball is green.
d) Probability that the second ball drawn is green.
e) Probability that the first two balls drawn are blue.

Question 2: 3 balls are drawn in one go. The following probabilities are to be determined with regard to the drawing of ball colours.

f) Probability that all balls are red.
g) Probability that exactly two balls are red.
h) Probability that no ball is red.
i) Probability that at least one ball is red.

5.3 Examples for Probabilistics

Solution

Figure 5.8 shows the event tree with all possible event paths for the serial drawing of two balls. From this, the probabilities can be determined directly using the addition and multiplication theorem for tasks (b) to (e), cf. Eqs. 5.61–5.64.

$$P(R) = \frac{5}{12} \tag{5.61}$$

$$P(G|R) = \frac{5}{12} \cdot \frac{3}{11} = \frac{15}{132} = \frac{5}{44} \tag{5.62}$$

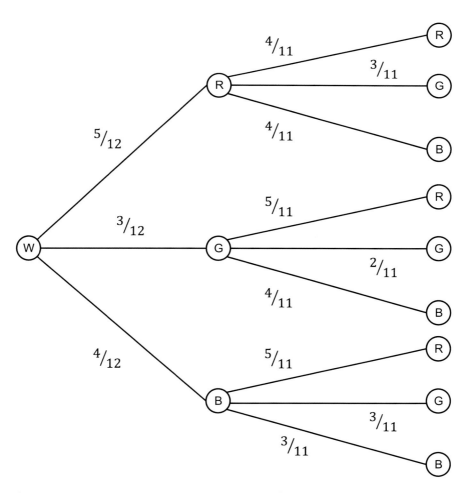

Fig. 5.8 Event tree for ball draw possibilities (urn: 5 red, 3 green, 4 blue balls)

$$P(G|R \cup G \cup B) = \frac{5}{12} \cdot \frac{3}{11} + \frac{3}{12} \cdot \frac{2}{11} + \frac{4}{12} \cdot \frac{3}{11} = \frac{15}{132} + \frac{6}{132} + \frac{12}{132} = \frac{33}{132} = \frac{1}{4} \tag{5.63}$$

$$P(B|B) = \frac{4}{12} \cdot \frac{3}{11} = \frac{12}{132} = \frac{1}{11} \tag{5.64}$$

For problem (f) to (h), the solution approaches are chosen based on the hypergeometric distribution (cf. Sect. 7.4.4), cf. Eqs. 5.65 to 5.67.

$$P(R = 3) = \frac{\binom{5}{3} \cdot \binom{7}{0}}{\binom{12}{3}} = \frac{10 \cdot 1}{220} = \frac{1}{22} \tag{5.65}$$

$$P(R = 2) = \frac{\binom{5}{2} \cdot \binom{7}{1}}{\binom{12}{3}} = \frac{10 \cdot 7}{220} = \frac{7}{22} \tag{5.66}$$

$$P(R = 0) = \frac{\binom{5}{0} \cdot \binom{7}{3}}{\binom{12}{3}} = \frac{1 \cdot 35}{220} = \frac{7}{44} \tag{5.67}$$

The second way is to determine the probability directly and then determine the solution via the complement, Eq. 5.71 and 5.72.

Problem (i) analyses the probability that at least one of the three balls drawn in one move is of red colour, i.e. $P(R>0)$. In the following, two possibilities of calculation are outlined, cf. Eqs. 5.68 to 5.72.

First option:

$$P(R > 0) = P(R = 1) + P(R = 2) + P(R = 3) \tag{5.68}$$

$$P(R = 1) = \frac{\binom{5}{1} \cdot \binom{7}{2}}{\binom{12}{3}} = \frac{5 \cdot 21}{220} = \frac{21}{44} \tag{5.69}$$

$$P(R > 0) = \frac{21}{44} + \frac{7}{22} + \frac{1}{22} = \frac{37}{44} \tag{5.70}$$

5.3 Examples for Probabilistics

Second option:

$$P(R > 0) = 1 - P(R = 0) \tag{5.71}$$

$$P(R > 0) = 1 - \frac{7}{44} = \frac{37}{44} \tag{5.72}$$

5.3.3 The Monty Hall Problem (Goat Problem)

In the context of the American game show "Let's Make a Deal" with host Monty Hall, the so-called "Monty Hall Problem" (other names are the "Goat Problem" or the "Monty Hall Dilemma") became popular. Components of the game are the host, the candidate, and three initially locked doors. Behind one of the doors is a vehicle as the main prize, and behind each of the other two doors is a goat. The principle or the procedure of the game is as follows:

1. First, one vehicle and two goats are randomly placed behind three gates.
2. All gates are locked: Vehicle and goats are not visible.
3. The candidate may select a gate, but the gate remains locked.
4. *Possibility 1:* The candidate has randomly chosen the gate with the vehicle (main prize). The moderator then opens one of the two remaining gates. A goat will appear in any case, regardless of which gate he opens. The moderator can choose freely between the remaining gates.
5. *Possibility 2:* The candidate has randomly chosen a gate with a goat. The moderator will then open the gate behind which the second goat is located and will (of course) not open the gate behind which the vehicle is located.
6. The moderator asks the candidate whether he wishes to revise his initial decision and possibly open the remaining unopened gate, or whether he wishes to stick with his first decision (first gate chosen).
7. The candidate makes the final decision (stay with the initial decision or change gate), the moderator opens the final desired gate and the candidate wins the vehicle or receives a goat.

Question
What decision should the candidate make after the moderator has opened another gate (cf. step 6)? Does the probability of winning change when the gate is changed? Or is the probability of winning independent of the candidate's decision to change the gate or to stick to his original decision?

Solution
Two solutions are presented:

a) Visualization and solution of the goat problem via event tree.
b) Estimation of the probability of winning via Bayes' theorem.

To a) Visualization and solution of the goat problem via event tree.

First, the situation "The Goat Problem" is visualized in an event tree (cf. Fig. 5.9). The event tree is shown for the case that the candidate has chosen gate 1 in the first step. The probabilities that lead from the root to the decision without or with switching are not relevant, since the candidate makes the decision consciously. Afterwards, the different action alternatives of the moderator result depending on the scenario (A = car; Z = goat). For example, the scenario A Z Z means that behind gate 1 there is a car, and behind gate 2 and 3 there is a goat. If the moderator now opens gate 2, but the candidate sticks to his decision gate 1, he would have won. The event tree thus shows the results depending on the candidate's decision (switch/no switch) and the moderator's previous action.

The visualization of the goat problem via the event tree already shows that the probability of winning is higher in the case of a goal change compared to the strategy that the candidate remains at the originally selected goal. In the end, the probability for the strategy "goal change" is P = 2/3, for the strategy "without goal change" only P = 1/3. The strategy "goal change" is recommended.

To b) Estimation of the probability of winning via Bayes' theorem.

Provided that the moderator randomly chooses a gate (no systematic influence; the moderator has, for example, no own preference regarding a gate), behind which a goat is to be found, Bayes' theorem applies according to Eqs. 5.73 and 5.74, respectively, in the case of several events taking into account the total probability.

$$P(A_j|B) = \frac{P(OA_jB)}{P(B)} = \frac{P(A_i) \cdot P(B|A_j)}{\sum_{i=1}^{n} P(A_i) \cdot P(B|A_i)} \quad (5.73)$$

$$P(B) = P(B|A_1) \cdot P(A_1) + P(B|A_2) \cdot P(A_2) + \cdots + P(B|A_n) \cdot P(A_n)$$
$$= \sum_{i=1}^{n} P(B|A_i) \cdot P(A_i) \quad (5.74)$$

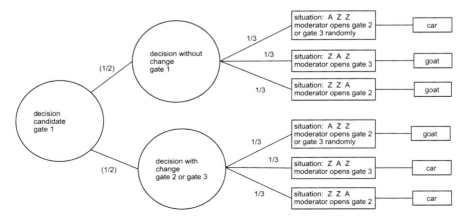

Fig. 5.9 The Monty Hall problem; Visualization of the goat problem via event tree

5.3 Examples for Probabilistics

In the present application (decision candidate: gate 1) of the Monty-Hall problem, the event („prize ") is defined as the win G_i behind gate i (i=1,2,3). From this, first the probability of a win with respect to gate 1, 2, 3 follows Eq. 5.75. Let M_j be the event "Moderator opens gate j (j=1,2,3)". Then, for example, for the probability of the moderator opening gate 3, Eqs. 5.76–5.78 follows. Thus, for the probability based on Eq. 5.73 that the win is behind gate 2 under the condition that the moderator has opened gate 3, Eq. 5.79 follows. Substituting in also yields the result P(G_2 |M_3)=2/3. Figure 5.10 shows the event tree with the three possible paths to the event M_3 outlined above. The same applies to other gate constellations that are identical in content. Ergo, the application of Bayes' theorem also yields the result (as does the analysis via event tree) that the strategy "gate change" is to be recommended.

$$P(G_1) = P(G_2) = P(G_3) = \frac{1}{3} \tag{5.75}$$

$$P(M_3|G_1) = \frac{1}{2} \tag{5.76}$$

$$P(M_3|G_2) = 1 \tag{5.77}$$

$$P(M_3|G_3) = 0 \tag{5.78}$$

$$P(G_2|M_3) = \frac{P(M_3|G_2) \cdot P(G_2)}{P(M_3|G_1) \cdot P(G_1) + P(M_3|G_2) \cdot P(G_2) + P(M_3|G_3) \cdot P(G_3)} =$$
$$= \frac{1 \cdot \frac{1}{3}}{\frac{1}{2} \cdot \frac{1}{3} + 1 \cdot \frac{1}{3} + 0 \cdot \frac{1}{3}} = \frac{2}{3} \tag{5.79}$$

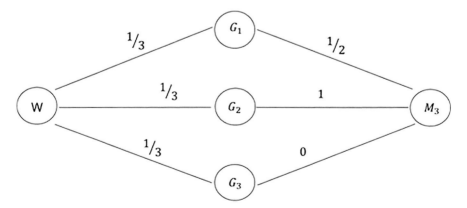

Fig. 5.10 The Monty Hall problem; event tree with three possible paths to the event gate opening M_3 in the use case that the candidate chooses gate 1 first

Data Collection and First Analysis Steps 6

The present chapter contains simple basic features of data collection as well as data compression. First, the drawing of samples (Sect. 6.1) and various methods of data classification (Sect. 6.2) are discussed. Subsequently, basic estimation methods for unknown parameters of a distribution (mean, variance, proportion) are presented in Sect. 6.3. The application of different estimators for the parameters mean and variance on the basis of available samples is presented in Sect. 6.4 and concludes with the comparison of different mean and variance estimators as well as the skewness of a measurement series. To illustrate the subject matter, the case study engine connecting rod (cf. database in Sect. 4.1) is used with a focus on the function-critical characteristic diameter of the connecting rod eye. The estimators presented here are used in basic analyses of technical reliability. Therefore, they are the subject of the present, basic-oriented chapter. Further estimators and estimation methods, also partly with reference to special distribution models, are the content of Chap. 8.

6.1 Sample

The analysis of a technical context is often based on the evaluation of a sample. Within the case study engine connecting rods (cf. Sect. 4.1), the analysis of function-critical characteristics is the focus: For example, several samples can be taken per day from a running connecting rod production and the components can be measured. The measurement is based on function-critical component characteristics such as diameter, length, width or parallelism.

The measured sample values of a characteristic are denoted as x_i, the sample size as n. The absolute frequencies n_i (i = 1, …, k) of the sample values x_i add up to the sample size n (cf. Equation 6.1). If the absolute frequencies n_i of the sample values are related to the sample size, the relative frequencies are calculated according to Eq. 6.2, where $0 < h_i \leq 1$. The sum of the relative frequencies of a sample is 1 (Eq. 6.3).

$$\sum_{i=1}^{k} n_i = n_1 + n_2 + \cdots + n_k = n \qquad (6.1)$$

$$h_i = \frac{n_i}{n} \qquad (6.2)$$

$$\sum_{i=1}^{k} h_i = h_1 + h_2 + \cdots + h_k = 1 \qquad (6.3)$$

Accordingly, for each sampling characteristic x_i there is exactly one absolute frequency n_i and exactly one relative frequency h_i (cf. Table 6.1).

The function – the frequency distribution or density function – can be represented on the basis of the relative frequencies (cf. Eq. 6.4). The sum of all frequencies that are less than or equal to x is called the distribution function or also called the cumulative frequency function (Eq. 6.5).

$$f(x) = \begin{cases} h_i & \text{for } x = x_i (i = 1, 2, \ldots, k) \\ 0 & \text{for all other x} \end{cases} \qquad (6.4)$$

$$F(x) = \sum_{x_i \leq x} f(x_i) \qquad (6.5)$$

6.2 Data Classification

In many cases of data analysis, the sample size or the number of original values to be analysed is very high, so that a clear representation in the bar diagram (cf. Sect. 4.3) does not make sense. Furthermore, the handling of the data can be cumbersome. For these reasons, large amounts of data are often classified (or grouped) and visualized accordingly in the histogram (cf. Sect. 4.4).

The frequency distribution can consequently also be carried out on the basis of a classified ("grouped") sample, whereby, strictly speaking, the data are classified. In principle, the procedure is as follows:

a) A sample with n values $x_1, x_2, \ldots, x_n$ is available (original values, observed values; e.g. measured values).
b) Selection of a method for data classification and implementation of the classification.
c) Finding absolute/relative class frequencies h_i.
d) Visualization of the class or class frequencies in a histogram.

In concrete terms, this means that each original value from the sample in the class i is assigned the class mean. This results in Eq. 6.6 for the frequency function of a classified sample and Eq. 6.7 for the distribution function (or cumulative frequency) of a classified sample.

6.2 Data Classification

Table 6.1 Sample value and absolute and relative frequency

Sample value	x_i	x_1	x_2	x_3	...	...	...	...	...	...	x_{k-1}	x_k
Absolute frequency	n_i	n_1	n_2	n_3	...	...	...	...	...	...	n_{k-1}	n_k
Relative frequency	h_i	h_1	h_2	h_3	...	...	...	...	...	...	h_{k-1}	h_k

$$f(x) = \begin{cases} h_i & \text{for } x = x_{Class,i} (i = 1, 2, \ldots, n) \\ 0 & \text{for all other } x \end{cases} \quad (6.6)$$

$$F(x) = \sum_{x_{Class,i} \leq x} f(x_{Class,i}) \quad (6.7)$$

The user can choose from a variety of methods for data classification. First of all, it should be noted that data classification is not recommended for small samples or data volumes. The rule of thumb here is: For sample sizes $n \leq 50$, the original values can be sorted according to their size and presented in the bar diagram (cf. Sect. 4.3) – ergo in unclassified form. For the classification of a larger data volume, three possibilities of data classification are shown in the following chapters.

Data Classification According to DIN 55302 (DIN 1967)
The data classification according to DIN 55,302-2 (DIN 1967 and DIN 1970) provides for a fixed number of classes depending on the sample size, cf. Table 6.2. The advantage is the possibility of a simple determination of the number of classes k without calculation. If several histograms of comparable size are created for different analyses, comparability can be ensured if the number of classes is chosen appropriately, since the number of classes is the same.

Data Classification According to Classical Procedure
With a data size of $n \geq 50$, the following approach to classifying data has proven effective in practice:

a) Determination of the smallest and the largest value x_{min} as well as x_{max} within the sample.
b) Determination of the span $R = |x_{max} - x_{min}|$.
c) Selection of the border point of the number interval (related to the class limits) with an additional decimal place compared to the documented decimal places of the measured values. This prevents a sample value to be classified from lying on the class boundary and requiring another rule for the assignment.
d) All classes are chosen to be of equal width; Eq. 6.8 shows the recommendation for class width w and Eq. 6.9 for number of classes k for $50 \leq n \leq 400$. On the other hand, Eq. 6.10 is used to determine w for $n > 400$.
e) The number of decimal places with respect to the class width w should correspond to the number of decimal places of the original values to be classified in order to avoid assignment problems.

Table 6.2 Number of classes depending on the sample size according to DIN 55302-1 (DIN 1970)

Sample size n	Number k of classes
Up to 50	No class division
Up to 100	≥ 10 classes
Up to 1000	≥ 13 classes
Up to 10,000	≥ 16 classes
Up to 100,000	≥ 20 classes

Since with this approach the last class does not always include all values of the sample, a class with the appropriate class width can be added.

$$w = \frac{R}{\sqrt{n}} \qquad (6.8)$$

$$k = \sqrt{n} \qquad (6.9)$$

$$w = \frac{R}{20} \qquad (6.10)$$

Data Classification According to CNOMO Standard
Within the framework of its standardization activities at CNOMO (Comité de NOrmalisation des MOyens de production), the French automotive industry has also developed a readily applicable procedure for classifying data (for scopes $n \geq 50$); cf. also (Dietrich and Schulze 2005).

The procedure is largely the same as the data classification according to the classical procedure, but the number of classes and the class width are determined according to Eqs. 6.11 and 6.12:

$$k = 1 + \frac{10 \log(n_{Ges})}{3} \qquad (6.11)$$

$$w = \frac{R}{k} \qquad (6.12)$$

By taking into account the logarithm in relation to the available sample size, a disproportionately small increase in the number of classes results with regard to larger sample sizes. The same relationship is also reflected in the classification according to DIN 55350-1 (DIN 1970), but here a fixed minimum is specified for the number of classes.

Data classification in the engine connecting rod case study
Figure 6.1 shows a comparison of the classification according to DIN 55302/Sheet 1 (DIN 1970) (Fig. 6.1 left) and according to the CNOMO standard (Fig. 6.1 right). This is based on the measured values of the function-critical feature "large connecting rod eye" of the engine connecting rod case study. (cf. Sect. 4.1, Table 4.1). The classified measured values x_i for the characteristic connecting rod eye are plotted on the abscissa, and both the relative and absolute frequency of class h_i are plotted on the ordinate in two scales.

6.3 Estimation Methods for Unknown Parameters Mean and Dispersion of a Distribution

In this section, the basic estimation methods – consisting of estimation function and estimator – for unknown parameters of a distribution function are shown. Specifically discussed are the estimates for the mean µ, the variance σ^2 and the proportion value p of a population.

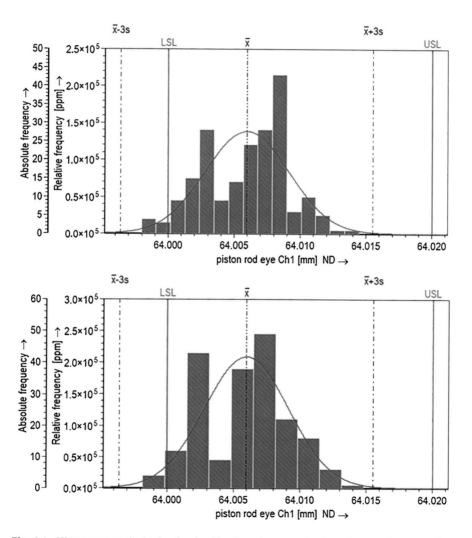

Fig. 6.1 Histograms on the basis of a classification of measured values via procedure according to DIN 55302 (left) as well as on the basis of the CNOMO standard (right) on the basis of the character "large connecting rod eye" within the case study engine connecting rod

The estimation of a parameter on the basis of an existing sample (e.g. measurement series) is carried out via point estimators and the specification of a confidence interval:

1) Estimates are determined for the unknown parameter of a distribution function. These estimates are also called point estimators.
2) Confidence intervals are determined. The confidence interval is the range in which the value of the unknown parameter will lie with a certain probability (confidence level γ) (cf. Chap. 9). The interval size is an indication of the precision of the previously determined estimated value. In linguistic usage, the term confidence interval is sometimes replaced by the word confidence range.

6.3.1 Estimators

There is a random variable Z (note: function), which is composed of n stochastically independent random variables X_1, X_2, ..., X_n (cf. Eq. 6.13). Thus, the random variable can be seen as an n-dimensional random vector.

$$Z = g(X_1, X_2, \ldots, X_n) \qquad (6.13)$$

The realization of the random vector can be represented by a concrete sample with sample values x_1, x_2, ..., x_n; e.g. measured values. The insertion of these concrete sample values provides an estimate or approximate value for the random variable Z on the basis of the n times the random experiment was carried out.

The estimator (estimation function) can generally be written down according to Eq. 6.14. The estimator is said to be accurate if the following criteria apply:

a) The estimation function Θ is *expectation-true*. This means that the expected value is equal to the parameter υ to be estimated (cf. Eq. 6.15a).
b) The estimator Θ is *consistent*: As the sample size increases, the estimator Θ converges to γ. Therefore, Eq. 6.15b holds, where ε>0.
c) The estimator Θ satisfies the criterion of *efficiency*. This means that the estimator Θ' has the smallest variance compared to all other existing estimators Θ; cf. Eq. 6.15c.

$$\Theta = g(X_1, X_2, \ldots, X_n) \qquad (6.14)$$

$$E(\Theta) = \upsilon \qquad (6.15a)$$

$$\lim_{n \to \infty} P(|\Theta - \upsilon| > \varepsilon) = 0 \qquad (6.15b)$$

$$\mathrm{Var}(\Theta') \leq \min_{\Theta} \mathrm{Var}(\Theta) \qquad (6.15c)$$

6.3.2 Estimators and Estimates for Mean M and Variance σ^2

If the random variables X_1, X_2, ..., X_n all follow the same distribution with parameters mean μ and the variance σ^2, then the estimator of the mean Eq. 6.17 applies to the expected value Eqs. 6.16 and 6.18 applies to the estimated value μ based on a sample at hand.

$$E(\overline{X}) = \mu \tag{6.16}$$

$$\overline{X} = \frac{1}{n} \cdot \sum_{i=1}^{n} X_i \tag{6.17}$$

$$\mu \approx \widehat{\mu} = \overline{x} = \frac{1}{n} \cdot \sum_{i=1}^{n} x_i \tag{6.18}$$

In addition, for the variance $Var(X) = \sigma^2$ (Eq. 6.19), the estimator S^2 is given by Eq. 6.20, and for the estimated value σ^2 based on a sample at hand, Eq. 6.21 holds.

$$E(S^2) = \sigma^2 \tag{6.19}$$

$$S^2 = \frac{1}{n-1} \sum_{i=1}^{n} (X_i - \overline{X})^2 \tag{6.20}$$

$$\sigma^2 \approx \widehat{\sigma}^2 = s^2 = \frac{1}{n-1} \sum_{i=1}^{n} (x_i - \overline{x})^2 \tag{6.21}$$

The estimator for the mean is expectation-true, consistent and efficient. The estimation function S^2 is expectation-true and consistent; but of course not efficient.

Central Limit Theorem

If the independent random variables X_i follow the same, but arbitrary distribution model, then the sum of random variables X_i approximately follows the normal distribution model. Thus, if for the expected value of X_i Eq. 6.22 and for the variance of X_i Eq. 6.23 holds, then the estimator of the mean is approximately normally distributed; Eqs. 6.24 and 6.25 hold.

$$E(X_i) = \mu \tag{6.22}$$

$$Var(X_i) = \sigma^2 \tag{6.23}$$

$$E(\overline{X}) = \mu \tag{6.24}$$

$$Var(\overline{X}) = \frac{\sigma^2}{n} \tag{6.25}$$

However, if the random variables $X_1, X_2, \ldots, X_n$ all follow the normal distribution according to Gauss, then the estimation function of the mean value also directly follows the normal distribution according to Gauss. And thus Eqs. 6.24 and 6.25 hold directly.

The estimation function for the standard deviation of the population follows directly from Eqs. 6.20 to 6.26. This estimation function is expectation-true, because when estimating the standard deviation or variance on the basis of an existing sample, the sample size minus 1 is also used in the quotient and not – comparable to the estimation of the arithmetic mean (Eqs. 6.17 and 6.18) – the sample size n (see Eqs. 6.20 and 6.21).

$$S = \sqrt{S^2} \tag{6.26}$$

6.3.3 Estimation Function and Estimation of the Proportion Value p

The estimator for the proportion value p is defined according to Eq. 6.27. Where X is the function of the occurrence of a defined event A when the Bernoulli random experiment is performed n times.

$$\widehat{P} = \frac{X}{n} \tag{6.27}$$

The random variable of the proportion follows approximately the normal distribution, if the random experiment according to Bernoulli is carried out with very frequent repetition. Thus, Eq. 6.28 may be applied for the binomially distributed random variable, expressed via the proportion value, and Eq. 6.29 for the variance of the random variable.

$$E\left(\widehat{P}\right) = p \tag{6.28}$$

$$\mathrm{Var}\left(\widehat{P}\right) = \frac{p(1-p)}{n} \tag{6.29}$$

Accordingly, in the presence of a realization in the form of an existing sample, Eq. 6.30 applies with respect to an estimated value for the proportion value p. Here, k is the number of occurrences of event A in the case of n-times execution of the random experiment according to Bernoulli.

$$p \approx \widehat{p} = h(A) = \frac{k}{n} \tag{6.30}$$

6.4 Basic Estimators and Characteristics of a Sample

In an initial analysis of a sample, various basic characteristic values are determined: First, characteristic values are collected that describe the series of measured values. Then, characteristic values are determined which, on the one hand, represent an information compression related to the observed values and, on the other hand, serve as estimators for the parameters of the underlying population from which the sample originates. Sect. 6.4.1 shows an overview of the characteristic description and basic sampling parameters using a case study of engine connecting rods. The subsequent chapters focus on the calculation of estimators for the mean (Sect. 6.4.2) and the dispersion (Sect. 6.4.3) based on a sample. Finally, Sect. 6.4.4 outlines the relationship between the mean, the dispersion and the skewness of a measurement series.

6.4.1 Case Study Engine Connecting Rod: Characteristic Description, Characteristic Values of a Sample

The basic analysis of a sample and the elementary sampling characteristics are first explained using the case study of engine connecting rod (cf. Sect. 4.1). A random sample of $n=200$ connecting rods is taken from a running series production and the function-critical characteristics are measured for each component. Table 6.3 shows technical specifications in the first section and basic characteristic values of the analyzed sample in the second section. The basis of the calculated sample characteristic values are the observed values from the case study connecting rod (cf. Sect. 4.1, Table 4.1) with regard to the function-critical characteristic diameter of the large connecting rod eye with the specification: $d=64$ mm (+0.020).

Characteristic description (cf. Table 6.3; first section)
First, the characteristic diameter to be analyzed is described by its basic properties. Furthermore, there is usually a specification: The required nominal value can be a fixed value or also lie within a range (common terms: specification and specification limits, tolerances and tolerance limits). Frequently used abbreviations for the specification limits are Upper Specification Limit (USL) or Lower Specification Limit (LSL). The difference between the two limits is usually referred to as tolerance T; cf. in detail Sect. 3.1.

Sample and sampling characteristics (cf. Table 6.3; second section)
In the engineering context, a deviation exists when the actual value (for example, a measured value) differs from the required target value (for example, drawing specification). The deviation is the difference between the target value and the actual value. In practice, the deviation is only called an failure when the actual value is outside the specification limits.

The second section of Table 6.3 shows an overview of the result of a sample analysis: basic characteristic values of the sample in relation to the specification (e.g.: proportion values in relation to the specification limits) as well as point estimators for mean value (e.g.: median) and dispersion (e.g.: standard deviation).

6.4.2 Estimator for the Mean Value Based on a Sample

The starting point for the estimation of a *mean value* – also called *centroid* – is the existence of a concrete sample, based on observed values (e.g. measured values).

Table 6.3 Basic characteristic description and associated characteristic values of a sample based on the case study engine connecting rod, characteristic connecting rod eye

Item no.	Specification	Designation	Case study engine connecting rod
1	Nominal value	S	64.000 mm
2	Upper specification limit	USL	64.020 mm
3	Lower specification limit	LSL	64.000 mm
4	Tolerance	T	0.020 mm
5	Tolerance center	T_M	64.010 mm
No	Characteristic value, designation	Characteristic value, formula symbol	Example, case study engine connecting rod
1	Sample size	n	200
2	Minimum value	x_{min}	63.999 mm
3	Maximum value	x_{max}	64.014 mm
4	Measured values within specification	n_T	196
5	Measured values above USL	$n_{>USL}$	0
	Proportional value Measured values > USL	$p_{>USL}$	0
6	Measured values below LSL	$n_{<LSL}$	4
	Proportional value Measured values < LSL	$p_{<USG}$	0.02
7	Arithmetic mean	$\bar{x}$	64.00594 mm
8	Median	$\tilde{x}$	64.00600 mm
9	Modal value	$\bar{x}_{M1}$ $\bar{x}_{M2}$	64.00700 mm 64.00300 mm
10	Standard deviation	s	$3.185 \cdot 10^{-3}$ mm
11	Variance	s^2	$1.014 \cdot 10^{-5}$ mm^2
12	Range	R	0.015 mm

The best known estimator of a mean based on a sample is the *arithmetic mean* (cf. Eq. 6.31). The measured values of the sample are added and divided by the sample size.

$$\bar{x} = \frac{x_1 + x_2 + \ldots + x_n}{n} = \frac{1}{n} \cdot \sum_{i=1}^{n} x_i \qquad (6.31)$$

The arithmetic mean has the advantage that it considers all sample values equally. However, it has the disadvantage that it can be sensitive to outliers. The estimation of the mean value based on a sample can also be done using the median. For this, the values of a sample are sorted in ascending order; the *median* corresponds to the central value and thus to the 0.5 quantile. It is therefore less sensitive to outliers. The median for an odd number of values of a measurement series is defined according to Eq. 6.32, the median for an even number of values of a measurement series results according to Eq. 6.33.

$$\tilde{x} = x_{(n+1)/2} \qquad (6.32)$$

$$\tilde{x} = \frac{x_{n/2} + x_{(n/2+1)}}{2} \qquad (6.33)$$

The simplest way to determine a mean value based on a series of measurements is the estimator *modal value* $\bar{x}_m$. The modal value is the value within a measurement series or sample that is observed most frequently. Compared to other mean values, the modal value has the advantage that it is easy to calculate. A disadvantage is that it is not always unambiguous, as the following example illustrates: if the measured values {2, 5, 7, 7, 8, 12, 16, 16, 21} are recorded in a sample, then two modal values exist: $\bar{x}_{m1} = 7$ and $\bar{x}_{m2} = 16$.

Of course, a number of other estimators for mean values exist. However, these are not discussed here within the framework of the fundamentals; for further in-depth study, e.g. with the help of (Hedderich and Sachs 2020) and (Hartung et al. 2009), is recommended.

6.4.3 Estimator for the Dispersion Based on a Sample

The term *dispersion* or also *scatter* is a characteristic value that describes the spread of values – usually around a suitable position parameter. Similar to the mean value, the different possibilities for calculating a dispersion value are differently sensitive to outliers. The starting point for all estimators presented below is a sample with size n and the observed values $x_1, x_2, \ldots, x_n$.

The simplest calculation of a dispersion is the calculation of the *range R* according to Eq. 6.34. The range R can be very sensitive to outliers due to the difference between the maximum and minimum values of a sample.

$$R = |x_{max} - x_{min}| \qquad (6.34)$$

Another common characteristic value for the calculation of a spread is the so-called standard deviation s. The *standard deviation* s is calculated from the positive root of the quotient of the sum of the squares of the deviations (related to the arithmetic mean) and the sample size (cf. Eq. 6.35). However, the far more common form for calculating the standard deviation s is shown in Eq. 6.36. It is not divided by the sample size n – analogous to the arithmetic mean – but by (n—1). Explicitly for small sample sizes (n less than or equal to 30), this estimator is considered to be more faithful to expectations compared to the estimator according to Eq. 6.34. The deviation square is understood to be the square of the difference between the measured value x_i and the arithmetic mean.

$$s = \sqrt{\frac{1}{n} \cdot \sum_{i=1}^{n}(x_i - \bar{x})^2} = \sqrt{\frac{(x_1 - \bar{x})^2 + (x_2 - \bar{x})^2 + \ldots + (x_n - \bar{x})^2}{n}} \quad (6.35)$$

$$s = \sqrt{\frac{1}{n-1} \cdot \sum_{i=1}^{n}(x_i - \bar{x})^2} = \sqrt{\frac{(x_1 - \bar{x})^2 + (x_2 - \bar{x})^2 + \ldots + (x_n - \bar{x})^2}{n-1}} \quad (6.36)$$

Furthermore, to indicate a measure of dispersion, the so-called variance s^2 (cf. Eqs. 6.37 and 6.38) is also used to indicate a measure of dispersion. The variance s^2 indicates how far the values of a sample deviate on average from the arithmetic mean. The variance is directly related to the standard deviation: the positive result of taking the square root of the variance s^2 yields the standard deviation s. When calculating the variance, it also applies that the sum of the squares of the deviations can be divided by the sample size or sample size minus one, respectively. For small sample sizes (rule of thumb: $n \leq 30$), the estimator for the variance based on Eq. 6.38 is more expectation-true with regard to expectations compared to the estimator based on Eq. 6.37.

$$s^2 = \frac{1}{n} \sum_{i=1}^{n}(x_i - \bar{x})^2 \quad (6.37)$$

$$s^2 = \frac{1}{n-1} \sum_{i=1}^{n}(x_i - \bar{x})^2 \quad (6.38)$$

6.4.4 Mean, Dispersion and Skewness of a Measured Value Series

The relationship between *mean (centroid)*, the *dispersion* and the *skewness* of a measured value distribution has already been briefly explained in the explanation of the boxplot visualisation (cf. Sect. 4.7). In the following, the effects of the use of different estimators in the determination of position and dispersion parameters are outlined.

6.4 Basic Estimators and Characteristics of a Sample

Each estimator for the mean value (Sect. 6.4.2) based on a sample has its own specific characteristics and, accordingly, the statement on the situation can vary and the interpretation in an engineering context can be very different. An example shall illustrate this: For a process engineering plant, the downtimes (loss of production) are documented within one month. After nine months, the following downtimes (sample) result, specified in the unit of operating hours. The values of the sample downtime are {2 h, 2 h, 2 h, 3 h, 8 h, 15 h, 44 h, 46 h, 152 h}. Mathematically, the mean values are as follows:

Modal value $\quad \bar{x}_m = 2$ h
Median $\quad \tilde{x} = 8$ h
Arithmetic mean value $\quad \bar{x} = 30.4444$ h

If the modal value (the value for the downtime, which is most frequently found in the sample) is used to determine the mean value, it is 2 h; this downtime also represents the minimum value of the sample. The median with value 8 h locates the focus of the measurement series rather in the lower range of the number series. In contrast to this, the outlier 152 h has the consequence that, with the aid of the arithmetic mean, the focus of the average standstill of the plant is determined at 30.4444 h in the upper range of the number series. The statement about the mean (a.k.a. average) downtime thus varies in part by a factor of 15, and possibly does not correctly reflect the technical facts, since the outlier and the accumulation of small downtimes cause the estimates to differ significantly.

The dispersion can be determined by estimating the standard deviation according to Eq. 6.36 to 6.39 and is of course not directly comparable to the dispersion estimator Range R (Eq. 6.40). If a normal distribution of the measured values without outliers were assumed, the standard deviation could be estimated using Eq. 6.41. The background would be the fact that six times the standard deviation forms the quantile 0.9973 of the assumed normal distribution model. However, it is clear that the estimated standard deviation does not match the estimate over the range assuming a normal distribution. The deviation may be an indicator of pronounced skewness of the measurement distribution.

$$s = 53,297 \text{ h} \tag{6.39}$$

$$R = 150 \text{ h} \tag{6.40}$$

$$R/6 = s = 25 \text{ h} \tag{6.41}$$

The analysis of the skewness provides an indication of the symmetrical or asymmetrical frequency distribution of measured values with regard to the position of the arithmetic mean. With the aid of the modal value, arithmetic mean and standard deviation, the slope – compare also statistical skewness of a distribution – of the value distribution can be characterized. Karl Pearson defined the *mode skewness S* according to Eq. 6.42; cf. (Hippel 2005). The skewness S can assume a value between -1 and 1. If the distribution of the measured values were symmetrical, the skewness would be zero. In the present example, as expected, a right-positive skewness results (Eq. 6.43) with a clear expression due to the outlier or the asymmetrical distribution of values.

$$S = \frac{\bar{x} - \bar{x}_m}{s} \tag{6.42}$$

$$S = 0.534 \tag{6.43}$$

In summary, the following rules will be noted:

a) Right-skewed frequency distribution: the median is larger than the mode; however, the median is smaller than the arithmetic mean.
b) Left-skewed frequency distribution: the median is smaller than the mode; however, the median is larger than the arithmetic mean.
c) Unimodal symmetrical frequency distribution: The median approximates the median or arithmetic mean.

In addition to the *skewness, kurtosis* and *excess* are used as statistical parameters to describe a frequency distribution or density function. The kurtosis describes the steepness of a distribution, the excess establishes the relation between the concretely analyzed frequency distribution and the normal distribution.

Distribution Models and Functions 7

This chapter outlines important distribution models that are used in reliability and risk analysis. First, an introduction to the general principles of a distribution model is given in Sect. 7.1, followed by an explanation of important continuous distribution models (Sect. 7.2) and test distributions (Sect. 7.3). Subsequently, Sect. 7.4 is focusing to discrete distribution models. Section 7.5 focuses on the saturation function, which can be used in the analysis of complex damage events. Section 7.6 concludes with an overview of mixed distribution models.

7.1 Fundamentals of a Distribution Model

Given an engineering problem, which is analysed on the basis of a random variable X. The first is that the random variable X can be assigned a unique value. Two things are of central importance in the application: On the one hand, a unique value can be assigned to the random variable X. On the other hand, the probability is of interest with which the random variable X can be assigned a value within a certain value range. In the case of a unique value assignment, the distribution of the random variable follows a discrete distribution model. In the case of the assignment of a value in a value range the distribution follows a continuous distribution model.

The *distribution function F(x)* is defined according to Eq. 7.1, where P is the probability that the random variable X takes on a value below x (e.g.: runtime, cycles, time).

$$F(x) = P(X \leq x) \tag{7.1}$$

Notes on the distribution function F(x):

1) F(x) is monotonically increasing.
2) If Eq. 7.2 holds, the event is called an impossible event.
3) If Eq. 7.3 holds, the event is called a certain event.
4) For discrete random variables, the following applies: The random variable X takes a value within the interval [a; b] with a probability according to Eq. 7.4.
5) For continuous random variables, the following applies: The random variable X takes a value within the range [a; b] with a probability according to Eq. 7.5.

$$\lim_{x \to -\infty} F(x) = 0 \tag{7.2}$$

$$\lim_{x \to \infty} F(x) = 1 \tag{7.3}$$

$$P(a < X \leq b) = F(b) - F(a) \tag{7.4}$$

$$P(a \leq X \leq b) = F(b) - F(a) \tag{7.5}$$

7.2 Continuous Distribution Models

If a continuous characteristic is recorded in a random experiment, then the result set consists of an infinite number of elementary events. Accordingly, the random variable X can be assigned an infinite number of values. The focus is on the probability with which the random variable assumes a value in a certain interval.

Within engineering tasks, typical, continuous characteristics can be the subject of the following tests: Component geometry tests (e.g.: characteristics diameter, length, height), functional tests (e.g.: characteristics oil pressure, leakage, system pressure, temperature) or reliability tests (e.g.: characteristics actuation cycles, mileage).

Within Sect. 7.2, besides the basics, the continuous distribution models are outlined, which are frequently used in reliability and risk analysis as well as in the investigation of production processes. Furthermore, besides the typical distribution models (Sect. 7.2.2 to 7.2.11) also extreme value distributions (Sect. 7.2.12) are outlined. Extreme value distributions are ideally suited for the investigation of extreme values within samples and thus for the analysis of extreme measurement data or events.

If a distribution model is used to represent a probability of failure or survival probability within the reliability analysis, the term lifetime distribution model is also often used instead of distribution model.

7.2.1 Basics and Fields of Application

The important basic definitions of density function and distribution function have already been explained in Chap. 3. In technical reliability and risk analytics, continuous distribution models are for example are used to represent probabilities for

7.2 Continuous Distribution Models

a damage event. A distribution model can be used to specify the probability of the event occurring as a function of a time-related variable. The probability function can be used to represent a wide variety of relationships: For example, the probability function can relate to a product (e.g., failure behavior) or a single production process step (e.g., distribution of measured values based on a sample). Furthermore, the distribution model may also refer to an entire product fleet, for example to describe a usage profile or failure behavior. Furthermore, the distribution model can also be used to map the characteristics of a production plant (Chap. 16).

Two engineering application examples will illustrate this:

Example 1: Illustration of Crack Behaviour in a Structural Component
Starting point: A component is tested on a test bench under the application of an external force. In the course of the test time, a crack develops which continues to propagate as time progresses. At a later point in time, the component breaks.

Model: The time-related crack extension could be described with a continuous distribution model: For example, the increases in crack extension could be represented with a Weibull distribution model. If the crack reaches its maximum extension (component rupture), the crack would be complete, and probability 1. The gradient of the curve could be interpreted as the propagation velocity of the crack. A high gradient indicates rapid crack growth or a rapid component fracture.

Example 2: Field Complaints Within a Product Fleet
Starting point: A manufacturer sells a batch (N = 1000) of vehicles to the market via his trade organisation. After a certain time, vehicle users complain about an oil leak. As the time of use progresses, the number of complaints increases.

The application of a continuous distribution model can be outlined as follows: First, the cumulative frequencies of the complaints can be determined. If the cumulative frequencies are represented in a corresponding diagram, a typical "staircase function" results (depending on the resolution of the time-related variables with regard to the time of the complaint). If a Weibull distribution model is now fitted to the cumulative probabilities, the model reproduces the probability of a complaint of an oil leak in the vehicle fleet during the time of use in the field. Of course, the model then only refers to the known complaints. If further vehicles are in the use phase for which no complaint has yet been made (so-called candidates), the model must be adjusted accordingly (cf. Chap. 12).

For a continuously distributed random variable the density function and distribution function are generally defined: The density function of a continuously distributed random variable is shown in Eq. 7.6, and the distribution function is illustrated in Eq. 7.7.

$$f(x) = \frac{d}{dx}F(x) \tag{7.6}$$

$$F(x) = P(X \leq x) = \int_{-\infty}^{x} f(u)du \tag{7.7}$$

Notes on a continuous density and distribution function:

1) The density function f(x) is non-negative and normalized, so Eqs. 7.8 and 7.9 hold.
2) The probability of an event occurring within an interval with bounds a and b is obtained based on Eq. 7.10.
3) The important parameters expected value E, variance Var and standard deviation SD of a continuous distribution model are shown in Eqs. 7.11, 7.12 and 7.13.

$$f(x) \geq 0 \tag{7.8}$$

$$\int_{-\infty}^{\infty} f(x)dx = 1 \tag{7.9}$$

$$F(x) = P(a \leq X \leq b) = \int_{a}^{b} f(x)dx \tag{7.10}$$

$$E(X) = \int_{-\infty}^{\infty} x \cdot f(x) \tag{7.11}$$

$$Var(X) = \int_{-\infty}^{\infty} (x - E(x))^2 \cdot f(x)dx \tag{7.12}$$

$$SD(X) = \sqrt{Var(X)} \tag{7.13}$$

Within this chapter, the continuous distribution models normal distribution, standard normal distribution, Weibull distribution, logarithmic normal distribution, folded normal distribution, Rayleigh distribution, exponential distribution, uniform distribution, gamma distribution, Erlang distribution as well as extreme value distributions are outlined. They are often used in product development (e.g.: mapping of cumulative damage mechanisms in the context of testing a prototype), product manufacturing (e.g.: analysis of measurement data – product testing or process parameters on the basis of samples) as well as in field use (e.g.: failure behavior of products within a product fleet in the utilization phase).

7.2.2 Normal Distribution

One of the most important distributions that can be used to illustrate engineering relationships or phenomena in nature is the normal distribution according to Johann Carl Friedrich Gauss (1777–1855). Equation 7.14 shows the density

7.2 Continuous Distribution Models

function, and Eq. 7.15 shows the distribution function. Since the density function is normalized (of course, since probability density), Eq. 7.16 holds.

$$f(x) = \frac{1}{\sigma\sqrt{2\pi}} \cdot e^{-\frac{(x-\mu)^2}{2\sigma^2}}, \text{ with } \sigma > 0 \qquad (7.14)$$

$$F(x) = \frac{1}{\sigma\sqrt{2\pi}} \cdot \int_{-\hbar}^{x} e^{-\frac{1}{2}\left(\frac{t-\mu}{\sigma}\right)^2} dt \qquad (7.15)$$

$$\int_{-\infty}^{\infty} f(x)dx = \frac{1}{\sigma\sqrt{2\pi}} \cdot \int_{-\infty}^{\infty} e^{-\frac{1}{2}\left(\frac{x-\mu}{\sigma}\right)^2} dx = 1 \qquad (7.16)$$

The typical, bell-shaped course of the density function of a normal distribution, which is determined by the parameters mean μ and standard deviation σ, is shown in Fig. 7.1.

Notes on the normal distribution (density function/distribution function) according to Gauss:

1) The only maximum of the probability density is at μ, and the two inflection points are symmetric about the maximum at μ ± σ.
2) The shape of the probability density is influenced by the parameters μ and σ: The parameter μ is the position parameter, the parameter σ determines the height and width of the curve: The higher the standard deviation σ, the wider the curve, the lower the maximum at μ. The smaller the standard deviation σ, the narrower the curve, the higher the maximum at μ; cf. Fig. 7.2.
3) The density function f(x) is easy to handle, with the distribution function F(x) cumulative probabilities cannot be calculated directly, because the integral is not elementary solvable. The solution is done numerically. The manual calculation of probabilities of arbitrarily normally distributed contexts is performed by means of transformation to the standard normal distribution (cf. Sect. 7.2.3).

Fig. 7.1 Schematic diagram of the density function of the normal distribution according to Gauss

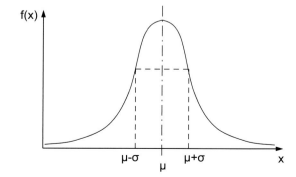

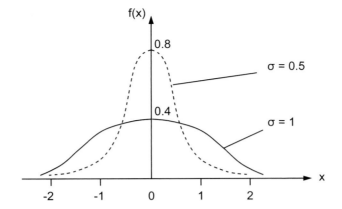

Fig. 7.2 Schematic diagram of the density function of the normal distribution with the parameters $\mu=0$ and $\sigma=0.5$ or $\sigma=1$ (cf. also special case of the standard normal distribution, Sect. 7.2.3)

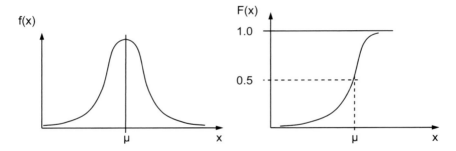

Fig. 7.3 Schematic diagram of the density function (left) and the distribution function (right) of the normal distribution model

4) The distribution function F(x) has a typical S-shaped course (cf. Fig. 7.3, right), the inflection point is determined by the position parameter μ.
5) The shape of the curve of the density function is reminiscent of a bell, ergo the term "bell curve according to Gauss" is often found in the literature.

Examples of phenomena, which follow the normal distribution model:

a) The weight of newborns (birth within the 37th and 43rd week of pregnancy).
b) Distribution of intelligence quotient to assess intellectual ability in general (general intelligence) compared to a reference group, (Wirtz et al. 2017).
c) Distribution of measured values for frequent measurements of the same characteristic on a component, e.g. a measurement standard (cf. in particular Sect. 15.3).

d) Distribution of measured values related to a product characteristic with regard to a production lot of a series production (production using a machine or production line), if the production process is capable and no systematic influencing variables are present.

7.2.3 Standard Normal Distribution and Application

The Gaussian normal distribution with the parameters µ and σ can also be specified in standardized form. One of the reasons is an easier use in calculations. The traceability of arbitrarily, randomly distributed correlations to the standard normal distribution enables, for example, the simple determination of probabilities in relation to defined events.

Furthermore, the standard normal distribution is also a component of many parametric and parameter-free statistical hypothesis tests: The test variables of statistical tests are often subject to random scatter, so that the standard normal distribution is used. Thus, the (standard) normal distribution is also one of the so-called test distributions. For didactic reasons, however, the standard normal distribution is outlined here directly after the normal distribution model (Sect. 7.2.2).

The transformation of the general normal distribution into the standard normal distribution is performed based on the following approach: The parameters of the normal distribution model are chosen according to Eqs. 7.17 and 7.18, then the random variable X is transformed using Eq. 7.19. This results in the standard normal distribution φ(u) of a continuous random variable U with density function (Eq. 7.20) as well as distribution function (Eq. 7.21). The standard normal distribution φ(u), like the normal distribution (and all probability densities), is normalized, so Eq. 7.22 holds.

$$\mu = 0 \tag{7.17}$$

$$\sigma = 1 \tag{7.18}$$

$$U = \frac{X - \mu}{\sigma} \tag{7.19}$$

$$\varphi(u) = \frac{1}{\sqrt{2\pi}} \cdot e^{-\frac{1}{2}u^2} \quad (-\infty \leq u < \infty) \tag{7.20}$$

$$\phi(u) = \frac{1}{\sqrt{2\pi}} \cdot \int_{-\infty}^{u} e^{-\frac{1}{2}t^2} dt \tag{7.21}$$

$$\int_{-\infty}^{\infty} \varphi(u) du = \frac{1}{\sqrt{2\pi}} \cdot \int_{-\infty}^{\infty} e^{-\frac{1}{2}u^2} du = 1 \tag{7.22}$$

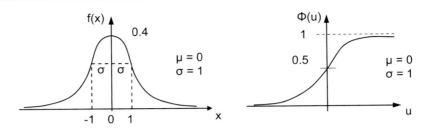

Fig. 7.4 Schemantic diagram of the graphical course of the density function (left) and the distribution function (right) of the standard normal distribution

Summary: A normal distribution with the parameterization $\mu=0$ and $\sigma=1$ is called standard normal distribution or standardized normal distribution. The graph of the density function $\varphi(u)$ is shown in Fig. 7.4 on the left, the graph of the distribution function $\phi(u)$ is shown in Fig. 7.4 on the right.

Notes on the standard normal distribution:

1) The density function $\varphi(u)$ is axisymmetric with respect to the ordinate; Eqs. 7.23 and 7.24 hold.
2) The maximum is at $u=0$; the inflection points are at $u=-1$ and $u=1$ respectively.
3) The standard normal distribution is stored in tables (cf. Appendix Table A.1). Arbitrary normally distributed correlations can be solved after transformation to the standard normal distribution model, although the integral of the distribution function of the normal distribution model (cf. Eq. 7.15) is not elementarily solvable.

$$\varphi(-u) = \varphi(u) \tag{7.23}$$

$$\phi(-u) = 1 - \phi(u) \tag{7.24}$$

The handling of the Standard Normal Distribution table (Appendix A.1) is simple, and will be briefly explained below: the table is valid for $u \geq 0$ and contains $\phi(u)$ of the standard normal distribution. Thus, given a known u, $\phi(u)$ can be read directly. Referring to an arbitrary value of a, which is in the range of any normal distribution model, this means that: the value a must first be transformed to u. For negative values of u, Eq. 7.23 and 7.24 apply.

Reading Example 1:
For $u=0.91$, the probability can be read directly, Eq. 7.25 (Appendix A.1).

$$\phi(0,91) = 0.8186 \tag{7.25}$$

The area with respect to the density function of the standard normal distribution in the interval $(-\infty, 0.91]$ is thus 0.8186; the principle of the relation is shown in Fig. 7.5.

7.2 Continuous Distribution Models

Reading Example 2:
For u = −0.35, after transforming according to Eq. 7.24, the probability is obtained by reading off the corresponding table value (Appendix A.1), cf. Eq. 7.26.

$$\phi(-0,35) = 1 - \phi(0.35) = 1 - 0.6368 = 0.3632 \qquad (7.26)$$

The area with respect to the density function of the standard normal distribution in the interval $(-\infty, -0.35]$ is thus 0.3632.

Calculation of probabilities via standard normal distribution

When calculating the probability, the following four cases are distinguished in the calculation of probabilities and area contents of arbitrarily normally distributed phenomena.

Case 1: The probability of an event described by the random variable X is sought. The probability is sought at the value x, or between the limits $-\infty$ and the value x; i.e. $P(X \leq x)$; cf. Fig. 7.5 left. A transformation of F(x) is performed, cf. Eq. 7.27. First, the random variable X is transformed into the random variable U, Eq. 7.28. If the value x is present, the transformation is accordingly performed using Eq. 7.29. Thus, the probability that the random variable X takes the value $\leq x$ is obtained from Eq. 7.30. The context of the transformation of the normally distributed random variable X into the standard normal distribution U is shown in Fig. 7.5.

$$F(x) \rightarrow \phi(u) \qquad (7.27)$$

$$U = \frac{X - \mu}{\sigma} \qquad (7.28)$$

$$u = \frac{x - \mu}{\sigma} \qquad (7.29)$$

$$P(X \leq x) = \phi \underbrace{\left(\frac{x - \mu}{\sigma}\right)}_{u} \qquad (7.30)$$

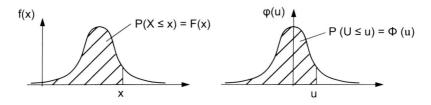

Fig. 7.5 Transformation of the normally distributed random variable X (left) into the random variable U (right) and the value x into the value u

Case 2: The probability of an event described by the random variable X is sought. The probability is sought above the value x, or between the limits x and ∞; hence P(X ≥ x), cf. Fig. 7.6.

The complement to the sought probability P(X ≤ x) is formed, so Eq. 7.31 holds, since the area is normalized. Subsequently, the random variable X or the concrete value x is transformed and the probability P(X ≥ x) can be determined.

$$P(X \geq x) = 1 - P(X \leq x) = 1 - \phi \underbrace{\left(\frac{x-\mu}{\sigma}\right)}_{u} = 1 - \phi(u) \tag{7.31}$$

Case 3: The probability of an event described by the random variable X is sought. The probability is sought that the random variable assumes a value in the interval [a; b]; thus P(a ≤ X ≤ b), cf. Fig. 7.7 (left). The random variable X is transformed to U; if interval boundaries are present, they are also transformed (cf. Fig. 7.7, right).

Thus, with transformation of the two interval boundaries a and b for the sought probability, P(a ≤ X ≤ b) is obtained according to Eq. 7.32. The areas to ϕ(b') and ϕ (a') can be determined using tabular work on the standard normal distribution, P(a ≤ X ≤ b) is calculated using the difference of the two areas.

$$P(a \leq X \leq b) = \phi \underbrace{\left(\frac{b-\mu}{\sigma}\right)}_{b'} - \phi \underbrace{\left(\frac{a-\mu}{\sigma}\right)}_{a'} = \phi(b') - \phi(a') \tag{7.32}$$

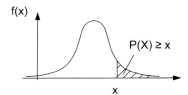

Fig. 7.6 Normally distributed random variable X; the probability above the value x (shaded area) is sought

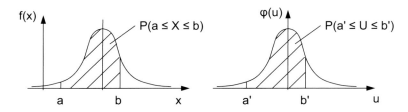

Fig. 7.7 Transformation of the normally distributed random variable X (left) into the random variable U and of the interval boundaries a and b into the boundaries a' and b' (right)

7.2 Continuous Distribution Models

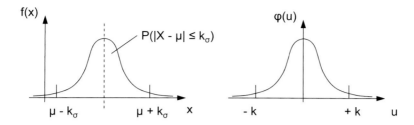

Fig. 7.8 Transformation of the normally distributed random variable X (left) into the random variable U (right) of a normally distributed random variable X as well as outlining of the area in the bounds of k times the standard deviation σ

Case 4: The probability of an event described by the random variable X is sought. The probability is sought in a range of k times the standard deviation $\pm\sigma$, cf. Fig. 7.8. This case is only a special case of the explanations for case 3. However, it is frequently used in the specification of measurement uncertainties, the test process suitability analysis (Chap. 15) as well as the statistical process control (SPC); Chap. 16), so that it is listed separately here.

The probability we are looking for is given by Eq. 7.32: By the limits of k times the standard deviation σ specially chosen here, transformation to the standard normally distributed variable U by converting yields Eq. 7.33.

$$P(|X - \mu| \leq k_\sigma) = P(\mu - k_\sigma \leq X \leq \mu + k_\sigma) = P(-k \leq U \leq k)$$
$$= \phi(k) - \phi(-k) = \phi(k) - [1 - \phi(k)] = 2\phi(k) - 1 \quad (7.33)$$

As already mentioned, variables – which describe engineering situations/events – are often normally distributed. In this respect, the multiple of the standard deviation often forms a reference value, for example, in order to be able to assess production or metrological events. For this reason, the scatter range of $\pm k\sigma$ is often specified in the general evaluation of measurement results. The probability of obtaining a measured value in the range $\pm 1\sigma$ is P=0.683 (cf. Eq. 7.34 and Appendix Table A.1).

In the evaluation of production engineering processes, the range of interest is often the range in which with a probability of approximately P=0.99 the measured values are expected with regard to a function-critical characteristic in the units produced (cf. Chap. 16). If the measured values in relation to the measured components follow a normal distribution model, six times the standard deviation is often selected as the reference value, since with a probability of P=0.9974 the measured value can be expected in this range. Furthermore, in the assessment of measurement procedures – especially in the determination of measurement uncertainty – the reference quantity is also frequently six times or four times the standard deviation (Chap. 15). However, the dispersion here refers to a repeat measurement, the test object is a setting gauge or a reference component.

Important ranges for the k-fold standard deviation and the associated probabilities are shown in Eqs. 7.34, 7.35 and 7.36; determined on the basis of Eq. 7.33 and the standard normal distribution table (Appendix A.1).

For k = 1:

$$\phi(1) = 0,8413 \rightarrow P(\mu - 1\sigma \leq X \leq \mu + 1\sigma) = 0.6826 \qquad (7.34)$$

For k = 2:

$$\phi(2) = 0,9772 \rightarrow P(\mu - 2\sigma \leq X \leq \mu + 2\sigma) = 0.9544 \qquad (7.35)$$

For k = 3:

$$\phi(3) = 0,9987 \rightarrow P(\mu - 3\sigma \leq X \leq \mu + 3\sigma) = 0.9974 \qquad (7.36)$$

7.2.4 Folded Normal Distribution

The folded normal distribution arises when the Gaussian normal distribution model is convolved at any value $x_F < \mu$. The convolution point is denoted as x_F, and the parameter μ is the mean value of the normal distribution model. In principle, the frequencies below the folding point x_F are assigned to the frequencies to the right of the folding point x_F, of course the assignment is done based on the same distance from the folding axis (read: magnitude). The density function f(x) for $x \geq 0$ as well as the distribution function of the convolved normal distribution is shown in Eqs. 7.37 and 7.38, respectively. A special case is the convolution of a normal distribution model, where the convolution point x_F corresponds to the mean value μ of the normal distribution model to be convolved, then Eqs. 7.39 and 7.40 directly result for density function and distribution function (with a as deviation from the origin).

$$f(x) = \frac{1}{\sigma\sqrt{2\pi}}\left[e^{-\frac{(x-\mu)^2}{2\sigma^2}} + e^{-\frac{(x+\mu)^2}{2\sigma^2}}\right] \qquad (7.37)$$

$$F(x) = \frac{1}{\sigma\sqrt{2\pi}}\int_o^x \left[e^{-\frac{(t-\mu)^2}{2\sigma^2}} + e^{-\frac{(t+\mu)^2}{2\sigma^2}}\right] dt \qquad (7.38)$$

$$f(x) = \frac{2}{\sigma\sqrt{2\pi}} exp\left\{-\frac{1}{2}\left(\frac{|x-a|}{\sigma}\right)^2\right\}; 0 \leq |x-a| < \infty \qquad (7.39)$$

$$F(x) = \int_0^{|x-a|} f(t)dt \qquad (7.40)$$

Figure 7.9 shows the principle of a convolution of the normal distribution model (A) at the convolution point $x_F < \mu$ and the resulting model (B). Figure 7.10 shows the convolution of a normal distribution model (A), which has the ordinate as the mirror axis as the mean with $\mu = 0$, and the resulting distribution model (B). Folded normal distribution of the form shown in Fig. 7.10 are often observed in

Fig. 7.9 Principle of convolution of the normal distribution model (**a**) at the convolution point $x_F < \mu$ and resulting model (**b**)

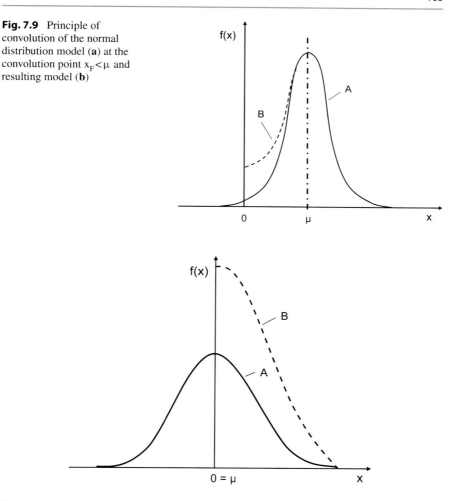

Fig. 7.10 Principle of the convolution of the normal distribution model (**a**) at the convolution point $x_F = \mu = 0$ as well as resulting model (**b**)

product realization on the basis of a characteristic (explicitly in the case of shape and position tolerances), in which the lower specification limit corresponds to the value zero (natural tolerance limit). If, for example, the requirement flatness or parallelism is defined, the deviation from the ideal value zero is checked after product realization. When measuring a sufficiently large sample, it can often be observed that the majority of the measured components are flat (deviation = 0) and the distribution of the deviation decreases to higher values. A negative deviation is not measured because the specification always refers to the absolut deviation. In this case, the distribution of the measured values can be mapped with a folded normal distribution.

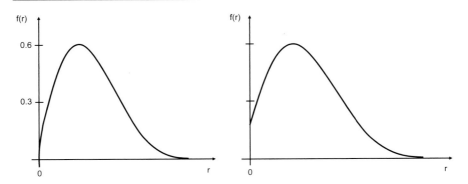

Fig. 7.11 Rayleigh distribution based on the polar coordinate r (left). Convolution of the Rayleigh distribution, e.g. with one-sided specification (right)

7.2.5 Rayleigh Distribution

Function-critical features can be composed of n ≥ 2 components. In the presence of two components, each of which follows a normal distribution model, the resulting distribution for the function-critical feature is mapped via a two-dimensional normal distribution model, Eq. 7.41. If polar coordinates are used in the measurement recording, Eq. 7.43 follows for the probability density with Eq. 7.42 (Pythagoras) as well as integration in the interval [0; 2π]; the visualization is shown in Fig. 7.11, left. The distribution model is named after John William Strutt, 3rd Baron Rayleigh (1842–1919); cf. (Rayleigh 1919). The Rayleigh distribution model is used – just like the two-parameter Weibull distribution model – for example in the description of wind speeds (starting point: two-dimensional, Cartesian coordinate system).

Notes on the Rayleigh Distribution and Possible Fields of Application:
In production process analyses and technical reliability, for example, the Rayleigh distribution model can often be observed in the realization of location positions in the plane (x/y position). In metal or plastics processing, for example, hole or spot weld positions (location positions) for housings or drive train components are part of the standard specifications. For the case of one-sided specification, the picture of a folded Rayleigh distribution can be obtained as shown in Fig. 7.11 on the right, the density function is shown in Eq. 7.44 and the distribution function is shown in Eq. 7.45, where the parameter a represents the displacement (Dietrich and Schulze 2005).

$$f(x,y) = \frac{1}{2\pi\sigma^2} \cdot e^{\frac{x^2+y^2}{2\sigma^2}} \quad (7.41)$$

$$r^2 = x^2 + y^2 \quad (7.42)$$

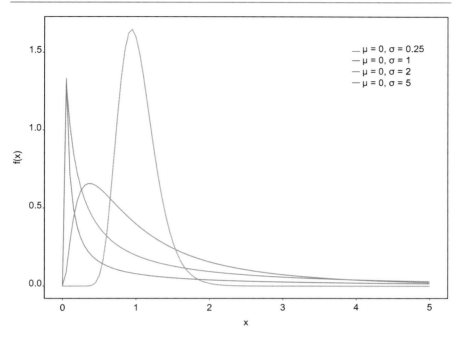

Fig. 7.12 Schematic diagram of the logarithmic normal distribution with position parameter $\mu = 0$ and different standard deviations σ

$$f(r) = \frac{r}{\sigma^2} \cdot e^{-\frac{r^2}{2\sigma^2}} \tag{7.43}$$

$$g(x_B) = \frac{x_B}{2\pi\sigma^2} e^{-\frac{1}{2\sigma^2}(a^2 + x_B^2)} \cdot \int_0^{2\pi} e^{\frac{ax_B}{\sigma^2}\cos\alpha} dx \tag{7.44}$$

$$G(x_B) = \frac{1}{2\pi\sigma^2} \cdot \int_0^{x_B} x e^{-\frac{1}{2\sigma^2}(a^2 + x^2)} \cdot \left(\int_0^{2\pi} e^{\frac{ax}{\sigma^2}\cos\alpha} d\alpha \right) dx \tag{7.45}$$

7.2.6 Logarithmic Normal Distribution

The logarithmic normal distribution is based on the normal distribution according to Gauss. A random variable X – where mean and standard deviation are elements of the real number set – is log normally distributed if the transformation according to Eq. 7.46 yields a variable U which follows the Gaussian normal distribution. The density function f(x) and distribution function are shown by Eqs. 7.47 and 7.48. It is uniquely described by the parameters μ and σ. Typical curves of the density function as a function of various standard deviations are shown in the schematic diagram in Fig. 7.12.

$$U = \ln X \tag{7.46}$$

$$f(x) = \frac{1}{x \cdot \sigma \sqrt{2\pi}} \cdot e^{-\frac{(\ln x - \mu)^2}{2\sigma^2}}, \text{ with } \sigma > 0 \qquad (7.47)$$

$$F(x) = \int_0^x \frac{1}{t \cdot \sigma \sqrt{2\pi}} \cdot e^{-\frac{(\ln t - \mu)^2}{2\sigma^2}} \, \mathrm{d}t \qquad (7.48)$$

Notes on the Log Normal Distribution and Possible Fields of Application:
Due to the typical right-skewed course, this distribution model can be used well for the representation of one-sided specified characteristics (lower specification limit corresponds to the natural limit zero). It should be noted that a large part of the characteristics defined during the development and design of a component have an upper specification limit—this includes in particular the shape and position tolerances (cf. also Sect. 16.3.3; Fig. 16.5). The lower specification limit is often the natural limit. This limit can be the zero value (e.g. feature parallelism) or a defined value (e.g. minimum diameter of a spot weld). In both cases, it is often the nominal value to which, for example, a production engineering process is designed.

Examples of approximately log normally distributed correlations:

1) The function-critical characteristics flatness, parallelism, roundness and concentricity usually have the natural limit value zero. During the production of components, it often happens that components manufactured in series have measured values close to zero – but not exactly zero – with regard to the feature types flatness, parallelism, roundness or concentricity due to production-related influences. If these deviations are negligibly small, it is of course also possible to fit a folded normal distribution.
2) In technical systems (e.g. containers, pipes) which contain or transport fluids (gases, liquids), the function-critical characteristic leakage rate (multiplication of pressure and volume throughput per unit of time; [Pa·m^3/s]) often plays a decisive role. For example, the internal pressure of a motor vehicle or bicycle wheel is not constant because the leakage rate is greater than zero. Therefore, the internal pressure decreases with time. If the leakage rate is measured on several wheels, the measured values often follow a logarithmic normal distribution model. Strictly speaking, the pressure loss is measured.
3) The early failure behavior of technical components in the utilization phase is often characterized by the fact that the component fails in the overall system after a short period of operation. The underlying failure modes pass through a short operating time (short exposure to a load spectrum) before the system fails. However, none or few component(s) fail at time t=0.

The logarithmic distribution model can be used to represent the technical characteristics mentioned very well: The function-critical characteristics flatness, parallelism, roundness or concentricity often assume values close to the nominal value zero, depending on the type of production engineering realization. The same

7.2 Continuous Distribution Models

applies to the function-critical characteristics leakage rate. Failure modes that show up in the use phase of the product after a short operating time with the symptom product failure are also distributed right-skewed and can be mapped with the logarithmic distribution model.

7.2.7 Weibull Distribution

The Weibull distribution according to Ernst Hjalmar Waloddi Weibull (1887–1979) is a form of distribution which is characterized by extreme flexibility and at the same time simple mathematical handling; cf. (Weibull 1951). In principle, it is an exponential distribution. The Weibull distribution is often used in two-parameter (density function and distribution function show Eqs. 7.49 and 7.50) or three-parameter form, the density function shows Eq. 7.51, the distribution function shows Eq. 7.52. Note that the three-parameter form is also often given in the form according to Eqs. 7.53 or 7.54, cf. e.g. (Lai 2014). Fitting the different three-parameter forms results in different position parameters and affects the interpretation of the parameter.

$$f(x) = \frac{b}{T} \cdot \left(\frac{x}{T}\right)^{b-1} \cdot e^{-\left(\frac{x}{T}\right)^b} \tag{7.49}$$

$$F(x) = 1 - e^{-\left(\frac{x}{T}\right)^b} \tag{7.50}$$

$$f(x) = \frac{b}{T - t_o} \cdot \left(\frac{x - t_o}{T - t_o}\right)^{b-1} \cdot e^{-\left(\frac{x - t_o}{T - t_o}\right)^b} \tag{7.51}$$

$$F(x) = 1 - e^{-\left(\frac{x - t_o}{T - t_o}\right)^b} \tag{7.52}$$

$$f(x) = \frac{b}{T} \cdot \left(\frac{x - t_o}{T}\right)^{b-1} \cdot e^{-\left(\frac{x - t_o}{T}\right)^b} \tag{7.53}$$

$$F(x) = 1 - e^{-\left(\frac{t - t_o}{T}\right)^b} \tag{7.54}$$

Figure 7.13 (left) shows a schematic sketch of various density functions, Fig. 7.13 (right) shows the distribution function of the two-parameter Weibull distribution as a function of various shape parameters b as well as the characteristic life span T = 1

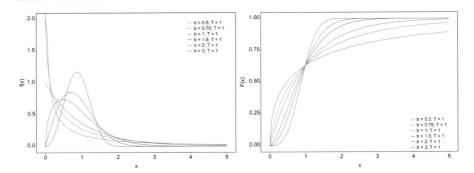

Fig. 7.13 Schematic diagram of the two-parameter density function of the Weibull distribution model (left) and distribution function (right) as a function of various shape parameters b

Notes on the Weibull Distribution and Possible Fields of Application:

a) The two-parameter form contains the parameters b (shape parameter) and T (location parameter); the three-parameter form additionally contains the parameter t_0 (location parameter; threshold parameter).
b) The *shape parameter b* describes the shape of the curve (example: gradient of the distribution function); for the *location parameter T*, the distribution function yields the value $P = 0.633$; independent of the shape parameter b.
c) The *threshold parameter t_0* indicates a shifted increase of the distribution function along the abscissa. Depending on the fitted three-parameter distribution model (Eqs. 7.51 or 7.53), this leads to different results and interpretations regarding the parameters.
d) The Weibull distribution is normalized (like any probability density): the area under the density function is 1.
e) The Weibull distribution is mathematically easy to handle, since the primitive function can be formed in a simple way and, accordingly, probabilities related to certain interval boundaries are easy to determine.
f) With appropriate parameterization, other common distribution models can be approximated; for example: Gaussian normal distribution, Logarithmic normal distribution and Exponential distribution. Thus, with a shape parameter arround $b = 3.5$, the curve of a Gaussian normal distribution model can be approximated.
g) In reliability analysis, the Weibull distribution serves as the standard distribution with which the failure behavior of parts, components or systems can be mapped and interpreted, see for example (VDA 2019a) or (Birolini 1997). In general, the interpretation can refer to the typical failure characteristics early failure behaviour, random failure behaviour or a runtime-related failure behaviour.
h) In the reliability analysis, the shape parameter b can be used to interpret whether the failure behavior is early failure ($b < 1$), random failure ($b \approx 1$) or

runtime failure (b>1). The location parameter T is called *characteristic lifetime* T and is also a typical parameter to describe the component failure behavior. The parameter t_0 (threshold) can be used in the context of reliability analysis as earliest possible failure time, incubation time or failure-free time (for a detailed explanation of the parameter see Chap. 12).

i) The visualization of the distribution function can be done in a double-logarithmic representation in a meaningful way: The distribution model appears as a straight line (cf. Sect. 4.6.2). This simplifies the interpretation even more, since the shape parameter b now represents the gradient of the straight line. In the analysis of the failure behaviour – e.g. of a component or different production batches – a different gradient (or corresponding shape) of the distribution model can be interpreted with regard to different causes of failure.

Chapter 12 deals in detail with the application of the Weibull distribution model in technical reliability analysis.

7.2.8 Exponential Distribution

The exponential distribution is very suitable as a life model for components, assemblies or systems. The density function is shown in Eq. 7.55 and the distribution function is shown in Eq. 7.56. It should be emphasized that only the parameter λ needs to be determined to complete the model, thus allowing a simple interpretation of the model. The parameter λ represents the number of expected events per unit interval. Figure 7.14 shows the schematic diagram of different density functions depending on different values of the parameter λ.

$$f(x) = \lambda \cdot e^{-\lambda x} \quad (7.55)$$

$$F(x) = 1 - e^{-\lambda t} \quad (7.56)$$

Both density function and distribution function are valid for $x \geq 0$. For $x<0$, $f(x)=0$ and $F(x)=0$, respectively.

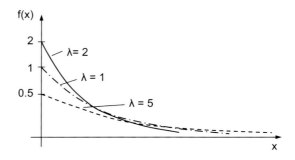

Fig. 7.14 Schematic diagram of the exponential distribution for different parameters λ

The exponential distribution is applicable if the criteria of a Poisson process are fulfilled. Equation 7.55 shows the simplest form of the exponential distribution; there are numerous similarly defined functions whose starting point is the exponential distribution, such as the Weibull distribution.

Notes on the Exponential Distribution and Possible Fields of Application:

a) The probability for the length x until a subsequently occurring event is $\exp(-\lambda x)$.
b) The area under the density function is normalized, so it is 1.
c) The expected value of the exponential distribution is $1/\lambda$.
d) The median of the exponential distribution is $(\ln 2)/\lambda$.
e) The variance of the exponential distribution is $1/\lambda^2$.
f) If the exponential distribution is used as a lifetime distribution to describe the reliability of components, assemblies, systems, then Eq. 7.57 applies.
g) For the failure rate h(x), substituting Eq. 3.9 (cf. Sect. 3.3.2) yields Eq. 7.58. The usual formula symbol λ for the failure rate has been replaced here by h to avoid confusion with the parameter λ of the exponential distribution.
h) Relationship to the Weibull distribution: If the shape parameter of the Weibull distribution assumes b=1, the Weibull distribution results in the exponential distribution (1/T becomes λ). Thus, the failure rate is constant and is referred to as the random failure range in the context of reliability analysis.

$$P(X < x) = 1 - F(x) = e^{(-\lambda x)} \qquad (7.57)$$

$$h(x) = \lambda \qquad (7.58)$$

7.2.9 Uniform Distribution

If the probability density is constant over an interval [a, b], then a uniform distribution exists. The density function is shown in Eq. 7.59, defined for all x in the interval [a, b], for all other x f(x)=0. The distribution function is shown in Eq. 7.60, also valid for all x in the interval [a, b]. Outside the interval [a, b], Eqs. 7.61 and 7.62 apply to the distribution function. A schematic diagram for different parameters a and b, respectively, is shown in Fig. 7.15. The probability that a random variable takes a value in an interval [c, d], which is part of the interval [a, b], is shown in Eq. 7.63.

$$f(x) = \frac{1}{b-a} \quad \text{for } a \leq x \leq b \qquad (7.59)$$

$$F(x) = \frac{x-a}{b-a} \quad \text{for } a \leq x \leq b \qquad (7.60)$$

$$F(x) = 0 \quad \text{for } x < a \qquad (7.61)$$

7.2 Continuous Distribution Models

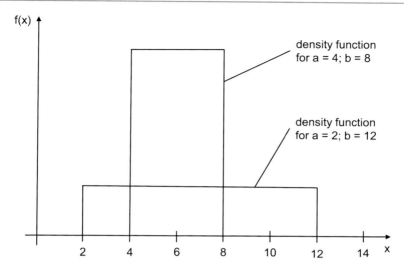

Fig. 7.15 Schematic diagram of the probability density of a uniform distribution for different parameters a, b

$$F(x) = 1 \quad \text{for } x > b \tag{7.62}$$

$$P(c \leq X \leq d) = F(d) - F(c) = \frac{d-c}{b-a} \tag{7.63}$$

Notes on Uniform Distribution and Possible Fields of Application:

a) In the context of industrial applications, the name rectangular distribution is often used for the present distribution model instead of the name uniform distribution.
b) The area below that of the density function is normalized – as with any probability density – so it is 1.
c) The expected value as well as the median of the uniform distribution result from Eq. 7.64, the variance shows Eq. 7.65 and from this Eq. 7.66 shows the standard deviation.
d) The uniform distribution is often used in metrology to conservatively estimate measurement uncertainties whose distribution model is not known. The interval limits can, for example, be estimated from empirically available knowledge or on the basis of a few experiments, so that the uniform distribution model for the uncertainty component under investigation can be derived directly.

$$E(x) = \frac{a+b}{2} \tag{7.64}$$

$$Var(X) = \frac{1}{12}(b-a)^2 \qquad (7.65)$$

$$\sigma = \frac{b-a}{2\sqrt{3}} \qquad (7.66)$$

7.2.10 Gamma Distribution

The gamma distribution describes the so-called Poisson process named after Siméon Denis Poisson (1787–1840). In essence, it is a renewal process: an event occurs at each successive stage n in time. The increments over the k stages follow a Poisson distribution model. Equation 7.67 shows the density function of the Gamma distribution. The parameter λ represents the intensity of the process: Exactly λ increments are expected per unit time (the parameter λ is also the expected value of the Poisson distribution; cf. Sect. 7.4.3). If $\lambda = 1$, the standard Gamma distribution is present. The parameter k represents the count of events (the "kth" event). Thus, the increment at the occurrence of the kth event is always one. The time periods between the occurrence of the events follow the exponential distribution model. Due to the connection between the time and the counting of the events and the influence on the shape of the distribution, the parameter k is also called the shape parameter. The Poisson process is discrete, referring to the continuous variable time. The gamma distribution includes the gamma function $\Gamma(n)$ according to Euler; cf. Eq. 7.68. The gamma function $\Gamma(n)$ may be taken as a generalization of the factorial, since Eq. 7.69 holds. Further properties of the gamma function are shown by Eqs. 7.70–7.72. Equation 7.70 gives, for example, $\Gamma(1) = \Gamma(2) = 1$ as well as $\Gamma(3) = 2$. Figure 7.16 shows the gamma function. In technical reliability, increases in operating data or also increases in failure behavior can be represented with the aid of the Poisson process. The Erlang distribution model is often used for this purpose, which in turn is a special case of the gamma distribution; cf. Sect. 7.2.11.

$$f(x) = \frac{\lambda^k}{\Gamma^k} x^{k-1} e^{-\lambda x} \quad \text{for } x > 0, \lambda > 0 \text{ and } k > 0 \qquad (7.67)$$

$$\Gamma(n) := \int_0^\infty t^{n-1} e^{-t} dt \quad \text{for } n > 0 \qquad (7.68)$$

$$\Gamma(n+1) = n\Gamma(n) = n! \quad \text{for } n = 1, 2, 3, \ldots \qquad (7.69)$$

$$\Gamma(n) = (n-1)! \qquad (7.70)$$

$$\Gamma(0) \text{ is not defined} \qquad (7.71)$$

$$\Gamma(\infty) = \infty \qquad (7.72)$$

7.2 Continuous Distribution Models

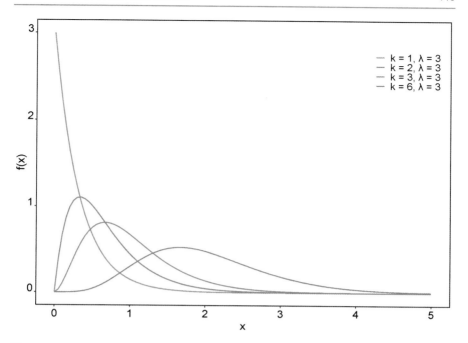

Fig. 7.16 Gamma function for process intensity $\lambda = 3$ and different shape parameters k

7.2.11 Erlang Distribution

The Erlang distribution is a probability distribution which is on the one hand a special case of the gamma distribution (Sect. 7.2.10) and on the other hand a general form of the exponential distribution (cf. Sect. 7.2.8). The Erlang distribution was developed by Agner Krarup Erlang (1878–1929) in the course of his work on queuing problems in telecommunications. The underlying problem was related to estimating the number of employees or telephone lines required in a telephone exchange as a function of incoming calls, waiting times and call durations. Equation 7.73 shows the density function, Eq. 7.74 the distribution function of the Erlang model. Expected value and variance are shown in Eqs. 7.75 and 7.76.

The starting point of the Erlang distribution is the passage of a time interval or localization distance x (runtime variable). After passing through the distance x, the nth event occurs under the condition that the number λ of events per unit is expected. Here, the probability up to the nth event is smaller than the mean of the Erlang distribution. The background is that events occur more often at shorter distances from each other than events have longer distances from each other. Figure 7.17 shows the density function as a function of the nth step. If the distances are sorted by size, the distribution of the frequencies corresponds to an exponential distribution.

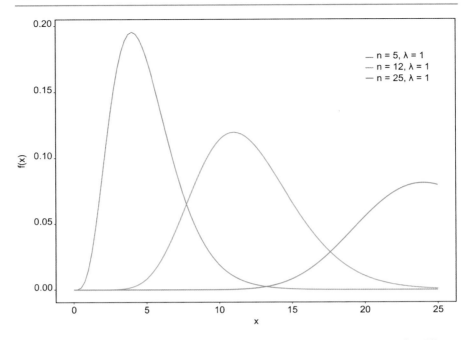

Fig. 7.17 Schematic diagram of the probability density of the Erlang distribution for different events n

In technical reliability, a staged failure behaviour can be mapped by means of an Erlang distribution. The product failure (event) occurs in the nth stage. The failure rate is shown in Eq. 7.77. It should be noted that the failure rate is subject to continuous growth, but converges to a limit at infinite runtime, hence Eq. 7.78. In contrast, the failure rates based on the distribution models already outlined do not converge to a limit: e.g. the failure rate based on a Weibull distribution model converges to infinity.

$$f(x) = \frac{\lambda^n x^{n-1}}{(n-1)!} e^{-\lambda x} \tag{7.73}$$

$$F(x) = 1 - e^{-\lambda x} \sum_{i=0}^{n-1} \frac{(\lambda x)^i}{i!} \tag{7.74}$$

$$E(x) = \frac{n}{\lambda} \tag{7.75}$$

$$Var(X) = \frac{n}{\lambda^2} \tag{7.76}$$

7.2 Continuous Distribution Models

$$\lambda(x) = \frac{\lambda^n x^{n-1}}{(n-1)!} \left(\sum_{i=0}^{n-1} \frac{(\lambda x)^i}{i!} \right)^{-1} \tag{7.77}$$

$$\lim_{x \to \infty} \lambda(x) = \lambda \tag{7.78}$$

7.2.12 Extreme Value Distribution Models

The extreme value theory deals with the potential minima and maxima of a sample. The three extreme value distributions according to Gumbel, Frèchet and Weibull as well as the underlying general extreme value distribution are based on the Fisher-Tippett-Gnedenko theorem; cf. (Fisher and Tippett 1928), (Gumbel 1958) and also (Coles 2001:45–49).

The Fisher-Tippett-Gnedenko Theorem
The independent random variables X_1, X_2, ..., X_n are present, which follow the same distribution model. Within the present sample, the random variables are realized by concrete real numbers. Let M_n be the maximum of the sample. Furthermore, let F(x) be the distribution function of M_n; cf. Eq. 7.79, and let G be a non-degenerate function (note: non-degenerate means that G is not a function that can take only one value).

Now, if it holds that constants a_n and b_n exist, so that the convergence according to Eq. 7.80 is fulfilled, then G can only consist of the three distribution families according to Gumbel distribution (type I), Frèchet distribution (type II), Weibull distribution (type III) can originate. The condition can also be formulated by substituting Eqs. 7.80 into 7.79; directly follows Eq. 7.81 and transforming gives Eq. 7.82.

$$F(x) = P(M_n \leq x) \tag{7.79}$$

$$F(b_n + a_n x) \to G(x) \tag{7.80}$$

$$F(b_n + a_n x) = P(M_n \leq b_n + a_n x) \to G(x) \tag{7.81}$$

$$F(b_n + a_n x) = P\left(\frac{M_n - b_n}{a_n} \leq x \right) \to G(x) \tag{7.82}$$

Generalized Extreme Value Distribution
The general extreme value distribution—also Fisher-Tippett distribution—is shown in Eq. 7.83. With parameter ξ (shape parameter or extreme value index) is used to describe the shape of the distribution. The parameter μ is a location parameter, the parameter σ is a scale parameter (with σ>0).

$$G(x) = \exp\left[-\left(1 + \xi\left(\frac{x-\mu}{\sigma}\right)\right)^{-\frac{1}{\xi}}\right] \qquad (7.83)$$

If shape parameter ξ is chosen according to Eq. 7.84, the *Gumbel distribution* (Eq. 7.87) is obtained from Eq. 7.83. If shape parameter ξ is chosen according to Eq. 7.85, the *Fréchet distribution* (Eq. 7.88) is obtained with $G_\alpha(1 + 1/\alpha \cdot x)$. If Eq. 7.86 is chosen for shape parameter ξ, the *Weibull distribution* (Eq. 7.89) with $G_\alpha(-1 + 1/\alpha \cdot x)$ is obtained.

$$\xi = 0 \qquad (7.84)$$

$$\xi = \frac{1}{\alpha} > 0 \qquad (7.85)$$

$$\xi = -\frac{1}{\alpha} < 0 \qquad (7.86)$$

Gumbel Distribution (Type I)
The Gumbel distribution is shown by Eq. 7.87, with scale parameter a (a>0) and location parameter b. The Gumbel distribution has exponential edges; ergo, one field of application is, for example, hydrology with descriptions of the extremes of flood and drought; cf. (Singh 1998).

$$G(x) = \exp\left\{-\exp\left[-\left(\frac{x-b}{a}\right)\right]\right\}, \text{for} -\infty < x < \infty \qquad (7.87)$$

Fréchet Distribution (Type II)
The Fréchet distribution is shown by Eq. 7.88, with scale parameter a (where a>0), location parameter b, and shape parameter α. The Fréchet distribution has heavy edges (decrease in density slower than in the exponential case) with polynomial expressions. Fields of application of the Fréchet distribution are, for example, the mapping of survival times after diseases or phenomena in hydrology; cf. (Ramos et al. 2017) and (Ramos et al. 2020).

$$G(x) = \begin{cases} 0, \text{for } x \leq b \\ \exp\left[-\left(\frac{x-b}{a}\right)^{-\alpha}\right], \text{for } x > b \end{cases} \qquad (7.88)$$

Weibull Distribution (Type III)
The Weibull distribution is shown in Eq. 7.89, with scale parameter a (a>0), location parameter b and shape parameter α (where α>0). Substituting t = b − x for the lifetime variable x gives the shape of the Weibull distribution used in reliability analysis (cf. Sect. 7.2.7). It is a distribution with slight margins and a finite upper bound. The Weibull distribution can also be used in hydrology to represent smaller values or minima (e.g. the phenomenon of low tide); cf. (Singh 1998).

7.2 Continuous Distribution Models

$$G(x) = \begin{cases} \exp\left[-\left(-\left(\frac{x-b}{a}\right)^\alpha\right)\right], \text{for } x < b \\ 1, \text{for } x \geq b \end{cases} \quad (7.89)$$

Figure 7.18 shows the extreme value distribution functions, Fig. 7.19 shows the probability densities according to Gumbel, Fréchet as well as Weibull with shape parameter $\alpha = 1$.

7.2.13 Distribution Model and Technical Reliability Analysis

A number of distribution functions were presented in the context of Sects. 7.2.2 to 7.2.11. Within the framework of reliability analysis and risk analysis these functions are used to describe the behavior of products or to represent the characteristic with respect to the manufacturing process. In principle, failure behavior can be mapped with the help of many distribution functions. However, some distribution models have proven to be particularly suitable, hence these distribution models in technical reliability are also called life time distributions.

The following distribution models are particularly suitable with respect to the mapping of the failure behavior of technical components and products:

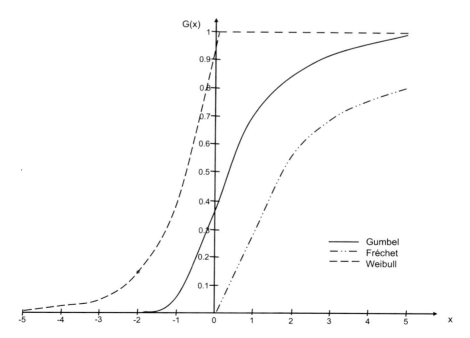

Fig. 7.18 Extreme value distribution functions according to Gumbel, Fréchet, Weibull (form $\alpha = 1$)

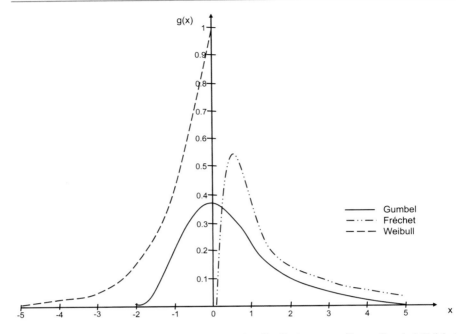

Fig. 7.19 Probability densities of the extreme value distributions according to Gumbel, Fréchet as well as Weibull (form $\alpha = 1$)

1) Weibull distribution
2) Exponential distribution
3) Normal distribution
4) Logarithmic normal distribution
5) Erlang distribution
6) Extreme value distributions in general

In the manufacture of components, the following distribution models are particularly suitable for mapping measured value distributions:

1) Normal distribution
2) Logarithmic normal distribution
3) Folded normal distribution
4) Rayleigh distribution

In the context of measurement uncertainty analysis and measurement/test process suitability analysis, e.g. the following distribution models are of particular interest:

1) Normal distribution
2) Equal distribution

The following distribution models are suitable for the analysis of samples with regard to the analysis of development and production processes as well as for field data analysis:

1) Binomial distribution
2) Poisson distribution
3) Hypergeometric distribution

The basic terms density function and distribution function have already been explained in detail in Chap. 3 and Sect. 7.1. Fundamental to engineering contexts and problems is the density function f(x). The density function can be used, for example, to represent the distribution of failure times based on the damage data of a product. If the failure behaviour of a product is in focus, the density function f(x) would be applied to estimate the failure probability at a certain value related to a runtime-related variable X. The distribution function F(x), on the other hand, describes the cumulative probabilities based on the density function f(x); Eq. 7.90. The distribution function F(x) is therefore also referred to in technical reliability as the failure probability, the complement (Eq. 7.91) as survival probability R(x). The quotient of the failure probability at a certain value related to the runtime-related variable X and the survival probability at the same value is the failure rate λ(x); Eq. 7.92. Further reliability parameters are discussed in more detail in Chap. 12 (in particular Sect. 12.1.4).

$$f(x) = \frac{d}{dx}F(x) \tag{7.90}$$

$$R(x) = 1 - F(x) \tag{7.91}$$

$$\lambda(x) = \frac{f(x)}{R(x)} \tag{7.92}$$

7.3 Test Distributions

This chapter explains important distribution models that are used in the context of calculating confidence intervals and performing statistical significance tests. Such distribution models are subsumed under the term test distributions.

7.3.1 t-Distribution According to Gosset (Respectively Student)

One of the best known distribution models for studying small sample sizes is the t-distribution. It was published by an author named "Student" (Student 1908). But this is a pseudonym behind which the statistician William Sealy Gosset (1876–1937) is hidden. Gosset did research in the laboratory of Karl Pearson

(1857–1936) at the Dublin brewery Arthur Guinness & Son and especially investigated small sample sizes. Since Guinness did not permit publication under his own name, Gosset published under the pseudonym Student. In this respect, the t-distribution is known in the literature under the term "Student's t-distribution" or "t-distribution according to Student", but not under the name of Gosset.

The t-distribution describes the distribution of a sample mean when the variance of the population is unknown and estimated on the basis of the available sample. The distribution of a sample mean is therefore not normally distributed, although for larger sample sizes (rule of thumb: $n \geq 30$) an approximation via a normal distribution model is possible.

Base of operations: Independent variables X (assumption: standard normally distributed) and Y (assumption: χ^2-distributed with f degrees of freedom). The random variable T is formed from the random variables X and Y, Eq. 7.93.

$$T = \frac{X}{\sqrt{\frac{Y}{f}}} \tag{7.93}$$

The density function of the random variable T represents the t-distribution according to Gosset (resp. Student), Eq. 7.94. The only parameter of the t-distribution is the degree of freedom f, for which $f > 0$ holds. The quotient, which contains the gamma function is often written down separately in a normalization constant A_n; cf. Eq. 7.95. This gives the simple notation of the t-distribution to Eq. 7.96 and the distribution function to Eq. 7.97. Important parameters of the t-distribution are expected value (Eq. 7.98) and variance for $f > 2$ (Eq. 7.99). The gamma function $\Gamma(n)$ is presented in Sect. 7.2.10. Figure 7.20 shows the schematic diagram of a t-distribution (parameter $f = 2$) and, in comparison, the standard normal distribution (parameter $\mu = 0; \sigma = 1$).

$$f(t) = \frac{\Gamma\left(\frac{f+1}{2}\right)}{\sqrt{f \cdot \pi} \cdot \Gamma\left(\frac{f}{2}\right)} \left(1 + \frac{t^2}{f}\right)^{-\frac{(f+1)}{2}} \tag{7.94}$$

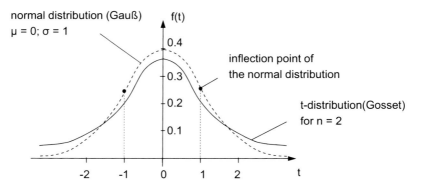

Fig. 7.20 Schematic diagram comparing t-distribution and standard normal distribution

$$A_n = \frac{\Gamma\left(\frac{f+1}{2}\right)}{\sqrt{f \cdot \pi} \cdot \Gamma\left(\frac{f}{2}\right)} \tag{7.95}$$

$$f(t) = A_n \frac{1}{\left(1 + \frac{t^2}{f}\right)^{\frac{(f+1)}{2}}} \tag{7.96}$$

$$F(t) = A_n \cdot \int_{-\infty}^{t} \frac{du}{\left(1 + \frac{u^2}{f}\right)^{\frac{(f+1)}{2}}} \tag{7.97}$$

$$E(T) = 0 \tag{7.98}$$

$$Var(T) = \frac{f}{f-2} \tag{7.99}$$

Notes on the t-distribution and Possible Fields of Application:

a) The distribution is completely determined by the parameter f (degree of freedom) under the condition f>0. The degree of freedom f of the distribution corresponds to the sample size n reduced by 1: f=n−1.
b) The t-distribution is normalized, the area under the curve is 1.
c) The t-distribution is axisymmetric to the ordinate, the maximum is at t=0.
d) Compared to the standard normal distribution, the t-distribution is narrower in the center and wider in the outgoing regions.
e) The t-distribution has heavy edges (the density falls more slowly compared to the exponential distribution).
f) For large sample sizes, the t-distribution can be approximated by the standard normal distribution (rule of thumb: n ≥ 30).
g) The t-distribution is often used to determine the confidence interval for the mean when sample sizes are small and when comparing means based on multiple samples using significance testing.
h) For ease of use, the t-distribution is presented in a table; see Annex Table A.3.

7.3.2 χ^2-Distribution

The χ^2 distribution was developed by Friedrich Robert Helmert (1843–1917; geodesist and mathematician) and Karl Pearson (1857–1936; mathematician).

There are several independent random variables X_1, X_2, X_3 …, X_n under the condition that they follow the standard normal distribution model. From the random variables X_1, X_2, X_3 …, X_n the sum of squares is formed; Eq. 7.100. The

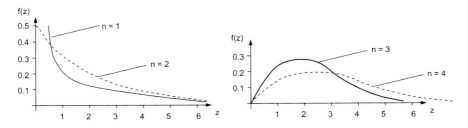

Fig. 7.21 The density function of the χ^2-distribution for $n \leq 2$ (left) as well as $n > 2$ (right)

random variable Z follows the χ^2 distribution; the density function is shown in Eq. 7.101. The quotient, which includes the gamma function, is often also written down separately in a normalization constant A_n; cf. Eq. 7.102. This gives the simple notation of the χ^2-distribution to Eq. 7.103 and the distribution function to Eq. 7.104. Important characteristics of the χ^2-distribution are expected value (Eq. 7.105) and variance (Eq. 7.106). Figure 7.21 (left) shows the χ^2-distribution for $n \leq 2$, Fig. 7.21 (right) shows the χ^2-distribution for $n > 2$.

$$Z = \chi^2 = X_1^2 + X_2^2 + \ldots + X_n^2 \tag{7.100}$$

$$f(z) = \begin{cases} \frac{1}{2^{\frac{n}{2}} \cdot \Gamma(\frac{n}{2})} \cdot z^{\left(\frac{n-2}{2}\right)} \cdot e^{-\frac{z}{2}} & \text{for } z > 0 \\ 0 & \text{for } z \leq 0 \end{cases} \tag{7.101}$$

$$A_n = \frac{1}{2^{\frac{n}{2}} \cdot \Gamma(\frac{n}{2})} \tag{7.102}$$

$$f(z) = \begin{cases} A_n \cdot z^{\left(\frac{n-2}{2}\right)} \cdot e^{-\frac{z}{2}} & \text{for; } z > 0 \\ 0 & \text{for; } z \leq 0 \end{cases} \tag{7.103}$$

$$F(z) = A_n \cdot \int_o^z u^{\left(\frac{n-2}{2}\right)} \cdot e^{-\frac{u}{2}} du \ (z > 0) \tag{7.104}$$

$$E(X) = n \tag{7.105}$$

$$Var(X) = 2n \tag{7.106}$$

Notes on the χ^2-distribution and possible fields of application:

a) The distribution is completely determined by the parameter n (here: sample size). The sample size corresponds to the degree of freedom f of the distribution.

b) The density function of the χ^2-distribution is normalized, the area under the curve is 1.
c) For $n \leq 2$ the function is strictly monotonically decreasing (cf. Fig. 7.21 left).
d) For $n > 2$ the function has a maximum at $z = n - 2$ (cf. Fig. 7.21 right).
e) For large sample sizes, the density function can be approximated by a Gaussian normal distribution (rule of thumb: $n > 100$), with $\mu = n$ and $\sigma^2 = 2n$ for the parameters μ and σ.
f) The χ^2-distribution is used, for example, to determine the confidence interval for the variance or standard deviation when a sample is available and when comparing variances based on several samples using a significance test.
g) The χ^2-distribution is also used, for example, to determine the confidence interval for the position parameter T (characteristic lifetime) of the Weibull distribution model (cf. Chap. 9).
h) Fig. 7.21 shows the schematic diagram of a χ^2-distribution for $n \leq 2$ (parameters $n = 1$ and $n = 2$) and for $n > 2$ (parameters $n = 3$ and $n = 4$).
i) The χ^2-distribution is presented in a table for ease of use; see Annex Table A.5.

7.3.3 Fisher's F-Distribution

Another important test distribution, which is used for example in the context of significance tests, is the Fisher distribution or also called the F-distribution for short. The prerequisite is an F-distributed random variable. An F-distributed random variable is obtained when there is a quotient of two random variables, each of which follows the χ^2-distribution; Eq. 7.107. Here m and n respectively denote the respective degrees of freedom of the χ^2-distributions; ergo $n = n_{Ges} - 1$ or $m_{Ges} = m - 1$. Equation 7.108 shows the density function of the F-distribution for $x > 0$, which contains the gamma function (Eq. 7.109) and depends only on the parameters (degrees of freedom) m and n of the two random variables. Important parameters of the distribution are the expected value (Eq. 7.110) and variance (Eq. 7.111). Figure 7.22 shows different F-distributions depending on the degrees of freedom m, n of the underlying random variables, which each follow the χ^2-distribution. For ease of use, the F-distribution is presented in a table; see Appendix Table A.4.

$$F_{m,n} = \frac{\frac{\chi_m^2}{m}}{\frac{\chi_n^2}{n}} \quad (7.107)$$

$$f(x) = m^{\frac{m}{2}} \cdot n^{\frac{n}{2}} \cdot \frac{\Gamma\left(\frac{m}{2} + \frac{n}{2}\right)}{\Gamma\left(\frac{m}{2}\right)\Gamma\left(\frac{n}{2}\right)} \cdot \frac{x^{\left(\frac{m}{2} - 1\right)}}{(mx + n)^{\left(\frac{m+n}{2}\right)}} \quad (7.108)$$

$$\Gamma(z) = \int_0^\infty t^{z-1} e^{-t} dt \quad (7.109)$$

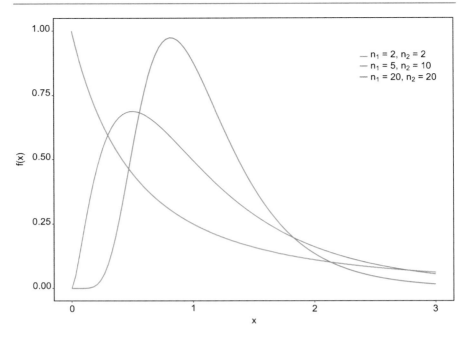

Fig. 7.22 Schematic diagram of the density function of the F-distribution for different degrees of freedom

$$E(X) = \frac{n}{n-2} \quad (7.110)$$

$$Var(X) = \frac{2n^2(m+n-2)}{m(n-2)^2(n-4)} \quad (7.111)$$

7.4 Discrete Distribution Models

Let a random experiment be given as a starting point, in the center of which a discrete characteristic is recorded. The result set of the experiment then comprises finitely many elementary events, so that the random variable X can take on finitely many values. Discrete probability distributions give the relationship between the number of events for a discrete characteristic (e.g.: number of damages or number of defective units) and the corresponding probability. Within Sect. 7.4 first the basics (Sect. 7.4.1) and then the discrete distribution models binomial distribution (Sect. 7.4.2), Poisson distribution (Sect. 7.4.3) and hypergeometric distribution (Sect. 7.4.4) are outlined.

7.4 Discrete Distribution Models

7.4.1 Basics

In the context of technical reliability analysis, discrete distribution models are used to evaluate are used, for example, to assess defective proportions of product fleets in the field or of units produced during manufacturing on the basis of random samples. Two application cases shall illustrate this:

Example 1: Complaints within a product fleet in field use.
A product fleet is in field use worldwide. Within a reference market, an incidence of damage in the product fleet with a certain damage causality is determined. This is a discrete characteristic ("damage exists"/"damage does not exist"), the reference market represents the sample. With a discrete distribution model, the probabilities for proportion values (field complaint) could be mapped on the basis of the reference market (sample).

Example 2: Defective components within series production.
Series production of engine connecting rods for an internal combustion engine also includes a final inspection of the function-critical features: The final inspection is carried out by means of a random sample inspection. If a measured value related to a characteristic does not meet the requirements (specification limits), a distinction can be made between components within specification and components outside specification. A discrete distribution model could be used to describe the probabilities for proportional values (connecting rods outside specification) on the basis of the final inspection (sample).

The density function f(x) represents the probabilities for certain numbers of events (e.g.: probability of finding exactly 20 defective connecting rods in a sample of size n=1000). The distribution function F(x), on the other hand, describes the cumulative probabilities based on the density function f(x) (e.g. probability of finding up to 20 defective connecting rods in a sample with a size of n=1000).

For a discretely distributed random variable the density function is defined according to Eq. 7.112 and the distribution function according to Eq. 7.113.

$$f(x) = \begin{cases} p_i & \text{for } x = x_i \\ 0 & \text{for all other } x \end{cases} \qquad (7.112)$$

$$F(x) = P(X \leq x) = \sum_{x_i \leq X} f(x_i) \qquad (7.113)$$

Notes on a Discrete Distribution Model:

1) For the density functon holds: $f(x_i) \geq 0$
2) Like all probability densities, the density function f(x) is normalized, so Eq. 7.114 holds.
3) For discrete random variables, the following applies: The random variable X takes a value within the range [a; b] with the probability according to Eq. 7.115.

4) The important characteristics of a continuous distribution function are mean (expected value E(X); Eq. 7.116), variance Var(X) (Eq. 7.117) and standard deviation SD (Eq. 7.118).

$$\sum_{i=1} f(x_i) = 1 \tag{7.114}$$

$$P(a < X \leq b) = F(b) - F(a) \tag{7.115}$$

$$E(X) = \sum_i x_i \cdot f(x_i) \tag{7.116}$$

$$Var(X) = \sum_i (x_i - E(X))^2 \cdot f(x_i) \tag{7.117}$$

$$SD(X) = \sqrt{Var(X)} \tag{7.118}$$

Within this chapter, the discrete distribution models binomial distribution, Poisson distribution and hypergeometric distribution commonly used in technical reliability and risk analysis are outlined. They are often used in product development (e.g.: evaluation of prototype tests), product manufacturing (e.g.: sampling in manufacturing) as well as in analyses in the use phase (e.g.: analysis of failure modes in products in the field).

7.4.2 Binomial Distribution

The binomial distribution is one of the most important and common discrete distribution models and was practically visualized by the Galton board (named after

Fig. 7.23 Schematic diagram of a Galton board for visualisation the binomial distribution model

7.4 Discrete Distribution Models

Francis Galton; 1822–1911) (see Fig. 7.23). The Galton board consists of obstacles arranged at regular intervals. It is perpendicular to the plane, and at the top end there is an opening through which a ball can be thrown in. When the ball is thrown in, the ball can be directed to the right or left when it hits an obstacle. Once all levels of the Galton board have been passed through, the ball is caught in a container after the last level. The experiment is repeated as often as desired; the distribution of the balls in the catching containers can be represented by a binomial distribution.

The density function and distribution function of the binomial distribution are shown in Eqs. 7.119 and 7.120, with the parameters sample size n, proportion of characteristic carriers p and, for this, the complement $q = 1 - p$.

$$f(x) = P(X = x) = \binom{n}{x} p^x \cdot q^{n-x} \tag{7.119}$$

$$F(x) = P(X \leq x) = \sum_{k \leq x} \binom{n}{k} p^k \cdot q^{n-k} \tag{7.120}$$

Notes on the Binomial Distribution Model and possible fields of application:

a) In applying the binomial distribution model, it is assumed that the population N is infinite.
b) The distribution model is used to analyze a sample with regard to mutually exclusive events. For example, measurement series based on alternative decisions can be mapped with the following differentiation criteria: "Characteristic is present" or "Characteristic is not present". In this way, the number of characteristic carriers is counted. In reliability analysis or in production process analytics, the distinction can be, for example: "component defective" or "component defect-free", "complaint" or "no complaint" as well as "cause of damage XYZ" or "damage-free".
c) The density function f(x) can be used to determine the probability that exactly x of n units exhibit a certain characteristic ("one defect per unit"); the term unit can, for example, stand for component, assembly, system or product.
d) The principle of "drawing with laying back" applies, which is sketched on the basis of the urn model: A ball is drawn n times from an urn with red and white balls. In each process, the colour is documented and the number of feature carriers (example: number of drawn balls of red colour) resulting from n-fold repeated draws is counted. However, the drawn ball is put back into the urn after each draw. The consequence of this procedure is that there is no change in the population due to sampling with regard to the proportions of red and white balls. With a small population (example $N = 75$), caution is therefore advisable, as sampling does have an influence on the probability distribution (cf. comments on the hypergeometric distribution model; Sect. 7.4.4). On the other hand, the influence is negligible in the case of a large population (example: population $N = 100,000$; sample $n = 400$).

e) Strictly speaking, the principle of "drawing with laying back" (cf. urn model) is not given in many engineering contexts. If, during the final inspection within a series production of engine connecting rods, it were found that there was a defective connecting rod, the inspector would not return the component to the population to be inspected. However, depending on the existing numerical ratio of the population, sample size and number of characteristic carriers, the binomial distribution can often be used as a model with sufficient accuracy.

f) The hypergeometric distribution model takes into account the population, sample size and number of characteristic carriers as well as the principle of "drawing without laying back"; cf. Sect. 7.4.4. In contrast to the binomial distribution model, the hypergeometric distribution model thus takes all information into account. However, an approximation of the hypergeometric distribution model (with $p = M/N$) by the binomial distribution model is possible if Eqs. 7.121–7.122b are fulfilled.

$$0.1 < \frac{M}{N} < 0.9 \qquad (7.121)$$

$$n < 0.05N \qquad (7.122a)$$

$$n > 10 \qquad (7.122b)$$

Note: Binomial Coefficient
The binomial distribution includes the binomial coefficient. The binomial coefficient can be used to determine the number of ways in which k units can be selected from a set of n units. The binomial coefficient is defined according to Eq. 7.123 (read: "n over k"). If n is a positive number and the number of units $n \geq k$ then Eq. 7.124 holds.

$$\binom{n}{k} = \frac{n \cdot (n-1) \cdot \ldots \cdot (n-(k-1))}{k!} = \prod_{i=1}^{k} \frac{n+1-i}{i} \qquad (7.123)$$

$$\binom{n}{k} = \frac{n!}{k!(n-k)!} \qquad (7.124)$$

For certain parameter k of a binomial coefficient holds:

1) For $k = 0$ holds: $\binom{n}{0} = 1$

2) For $k = 1$ holds: $\binom{n}{1} = n$

3) For $k = n$ holds: $\binom{n}{n} = 1$

7.4 Discrete Distribution Models

Application example Binomial distribution: Evaluation of a sample size within the scope of a paint inspection with defect pattern "inclusion of foreign particle "

Within the scope of an "End of Line" inspection the painting of particle attachments at a supplier is visually inspected. In the process, the component paint is checked for foreign particle inclusions and a distinction is made only between faultless and non-faultless parts. With a daily production of 1000 painted parts, a random sample of 65 parts is inspected daily. It is known from past production years that with $n_{Ges} = 200{,}000$ parts using a 100% inspection, a total of $n_F = 4900$ defective parts were counted. It is required that at least 3 defective parts are detected with a probability of at least $P = 0.9$ in a random sampling of $n = 65$ parts.

Procedure:

a) To visualize the binomial distribution model, the probabilities of detecting 0, 1, 2, ..., 9 faulty part(s) in a sample are determined.
b) Determination of the probability of detecting at least three defective components in the sample.
c) assessment of the sample size.

Solution:
For a) The proportion of expected error p is determined with Eq. 7.125, the complement q follows directly with Eq. 7.126. The probability density function of the binomial distribution (Eq. 7.127) results in Eq. 7.128 by insertion p and q, e.g. for $x = 0$. Thus the probabilities for $x = 1$, ..., 9 can be determined (Eq. 7.129). Figure 7.24 shows the probability density.

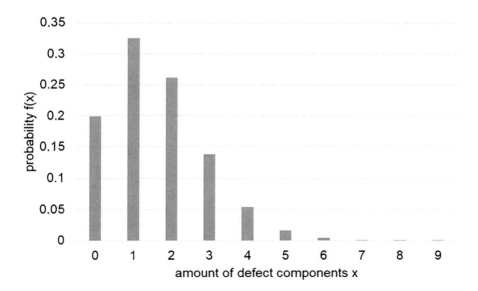

Fig. 7.24 Binomial distribution, density function f(x) with $p = 0.0245$ and $n = 65$

$$\widehat{p} = \frac{n_F}{n_{Ges}} = \frac{4,900}{200,000} = 0.0245 \qquad (7.125)$$

$$q = 1 - p = 0.9755 \qquad (7.126)$$

$$f(x) = P(X = x) = \binom{n}{x} p^x \cdot q^{n-x} \qquad (7.127)$$

$$f(x = 0) = P(X = 0) = \binom{65}{0} 0.0245^0 \cdot 0.9755^{65-0} = 0.1994 \qquad (7.128)$$

$$f(x = 1) = 0.3256; f(x = 2) = 0.2616; f(x = 3) = 0.1380; f(x = 4)$$
$$= 0.0537; f(x = 5) = 0.0165; f(x = 6) = 0.0041; f(x = 7)$$
$$= 0.0009; f(x = 8) = 0.0002; f(x = 9) = 0.00003$$
$$(7.129)$$

For (b), the probability of finding at least three defective parts in the sample n = 65 is given by the complement to P $(X \geq 3)$; cf. Eq. 7.130.

$$P(X \geq 3) = 1 - P(X < 3) = 1 - [f(0) + f(1) + f(2)] = 0.2134 \qquad (7.130)$$

For (c), the requirement that at least 3 parts are detected with a probability of at least P = 0.9 for a sampling of n = 65 parts is not satisfied, because Eq. 7.131 holds.

$$P(X > 3) = 0.2134 \leq 0.9 \qquad (7.131)$$

7.4.3 Poisson Distribution

The Poisson distribution was first described by Siméon Denis Poisson (French mathematician and physicist; 1781–1840) for the first time. Density function and distribution function is defined according to Eqs. 7.132 and 7.133.

$$f(x) = P(X = x) = \frac{\mu^x}{x!} \cdot e^{-\mu} \quad \text{for} \quad (x = 0, 1, 2, \ldots) \qquad (7.132)$$

$$F(x) = P(X \leq x) = e^{-\mu} \cdot \sum_{k \leq x} \frac{\mu^k}{k!} \qquad (7.133)$$

where μ is the mean value of the function, it is completely determined by the same. The distribution function F(x) is valid for $\mu > 0$, for $\mu \leq 0$ F(x) = 0. Equation 7.134 and Eq. 7135 are valid for variance and standard deviation.

$$\sigma^2 = \mu \qquad (7.134)$$

$$\sigma = \sqrt{\mu} \qquad (7.135)$$

7.4 Discrete Distribution Models

Notes on the Poisson Distribution Model and possible fields of application:

a) When applying the Poisson distribution model, it is assumed that the population N is infinite.
b) The distribution model is used to analyze a sample with regard to mutually exclusive events. The examination of the observed values of a sample is carried out with regard to a distinguishing criterion: "characteristic expression is present" or "characteristic expression is not present". Thus, the number of characteristic carriers is counted in relation to a unit. In the reliability analysis, the distinction can be, for example: "component defective" or "component defect-free"; "complaint" or "no complaint". (cf. also notes on binomial distribution).
c) Furthermore, the Poisson distribution is suitable for answering questions in which several characteristic values exist for each characteristic holder in relation to a unit. This means that a characteristic value can be observed more frequently in total than there are units in the investigated sample. Example 1: Number of defects per prototype in a test of a group of prototypes. Example 2: Number of paint defects per vehicle in a series production. Example 3: Number of complaints per vehicle in a fleet of vehicles in field use.
d) The density function f(x) can be used to determine the probability that exactly x characteristic values of n units are observed. The term unit can, for example, stand for batch, component, assembly or system or a group of components.
e) The principle of "draw with laying back" applies (cf. urn model; cf. remarks on binomial distribution).
f) Strictly speaking, the principle of "drawing with laying back" (cf. urn model) is not given in many engineering contexts (cf. remarks on the binomial distribution).
g) Approximation of the binomial distribution by the Poisson distribution: Using $\mu = np$ both models show similar distribution shapes, if the boundary conditions according to Eqs. 7.136 and 7.137 are fulfilled: These are simple rules of thumb; in many engineering tasks, the probabilities mapped based on Poisson or binomial distribution differ only in the fourth or fifth decimal place, respectively.

$$np \leq 10 \tag{7.136}$$

$$n \geq 1500p \tag{7.137}$$

Application example Poisson distribution: Evaluation of a process for the assembly of windscreens (windshields) in the automotive industry.

In vehicle technology, windscreens (windshields) are automatically glued into the body frame. In the course of production (480 vehicles/day), the gluing process is largely error-free. However, minimal adhesive residues are detected at approximately 2400 locations per day (mostly in the course of several complaints per vehicle). There is a requirement for a direct run rate of 5% per day (vehicles without faults) in relation to perfect bonding.

Procedure:

a) Calculation of the probability that a vehicle has a perfect bond.
b) Verification of the direct run rate.
c) Determination of the number of faults detected on a vehicle with the highest probability and visualization of the probability density in the interval [0; 10].

Solution:
(a) The distribution of defects is represented by a Poisson distribution; Eq. 7.138. The mean value μ (defects per vehicle) is determined by Eq. 7.139. This gives the probability that a vehicle is fault-free according to Eq. 7.140.

$$f(x) = \frac{\mu^x}{x!} \cdot e^{-\mu} \tag{7.138}$$

$$\mu = \frac{2400}{480} = 5 \tag{7.139}$$

$$f(x=0) = \frac{\mu^x}{x!} \cdot e^{-\mu} = 0.0067 \tag{7.140}$$

Re b) The requirement for the direct run rate (p=0.05) is not met, since Eq. 7.141 follows directly from a).

$$f(x=0) = 0.0067 \leq 0.05 \tag{7.141}$$

For c) Eq. 7.138 directly yields the probabilities for x=1, ..., 10; cf. Fig. 7.25. The highest probability is obtained for x=4 and x=5 (complaints per vehicle), cf. Eq. 7.142.

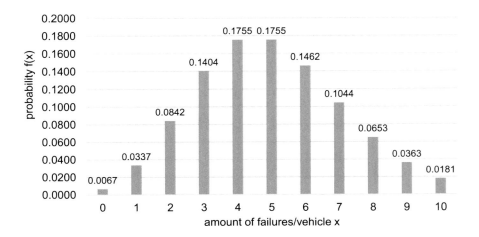

Fig. 7.25 Poisson distribution, density function f(x) with $\mu = 5$

$$f(4) = f(5) = 0.1755 \tag{7.142}$$

7.4.4 Hypergeometric Distribution

The density function of the hypergeometric distribution is defined according to Eq. 7.143, and the distribution function is defined according to Eq. 7.144. The hypergeometric distribution model takes the following information into account when mapping probabilities based on an existing sample: population N, number of characteristic carriers M in the population, sample size n, number of characteristic carriers x contained in the sample. The characteristic values mean and dispersion are defined according to Eqs. 7.145 and 7.146.

$$f(x) = P(X = x) = \frac{\binom{M}{x} \cdot \binom{N-M}{n-x}}{\binom{N}{n}} \tag{7.143}$$

$$F(x) = \sum_{k \leq x} \frac{\binom{M}{k} \cdot \binom{N-M}{n-k}}{\binom{N}{n}} \tag{7.144}$$

$$\mu = n \frac{M}{N} \tag{7.145}$$

$$\sigma^2 = \frac{nM(N-M)(N-n)}{N^2(N-1)} \tag{7.146}$$

Notes on the Hypergeometric Distribution Model and possible field of application:

a) When using the hypergeometric distribution model, it is assumed that the population N and also the number of characteristic carriers M in the population are known. Thus, discretely distributed characteristics can be mapped more accurately compared to the binomial distribution or Poisson distribution.
b) The principle of "drawing without laying back" applies, which is sketched on the basis of the urn model: A ball is drawn n times from an urn with red and white balls. For each procedure, the colour is documented and the number of feature carriers (example: number of drawn balls of red colour) resulting from n-fold repetition of the draw is counted. However, the drawn ball is not put back into the urn after each draw. The consequence of this procedure is that there is a change in the population due to sampling with respect

to the proportions of red and white balls. This influence is negligible when the population is large and the sample is small (therefore the binomial distribution model is used as an approximation with assumption $N \to \infty$; taking into account the rules according to Eqs. 7.121–7.122b).

Application Example Hypergeometric Distribution: In a small batch production, seals are produced for a sealing complex of a compressor. The small series comprises $N=50$ parts. After a full inspection, $M=10$ defective parts are detected. If a defect is detected three times, a systematic defect is assumed. For the design of future inspection scopes for small series, a sample size of $n=5$ parts is analyzed.

Question:
The probability is to be determined with which at least one faulty ($x \geq 1$) or exactly $x=3$ faulty parts are detected in the sample.

Solution:
The population is small, so that the principle of "drawing without laying back" must be taken into account. Using the hypergeometric distribution model (Eq. 7.147), the probability of detecting at least one defective seal can be determined. Substituting the population $N=50$, the number of defective units $M=10$ and a sample size of $n=5$ directly yields the probability of not detecting any defective seal for $x=0$; Eq. 7.148. Thus, Eq. 7.149 follows for the probability of detecting at least one defective seal. Directly detecting defective parts for a sampling $x=3$ can be calculated by directly substituting them into the probability density (Eq. 7.150).

$$f(x) = \frac{\binom{M}{x} \cdot \binom{N-M}{n-x}}{\binom{N}{n}} \tag{7.147}$$

$$f(x=0) = \frac{658,008}{2,118,760} = 0.3106 \tag{7.148}$$

$$P(X > 0) = 1 - f(0) = 0.6894 \tag{7.149}$$

$$f(x=3) = 0.0442 \tag{7.150}$$

Detection of at least one defective part with a sample size of $n=5$ would only be guaranteed with a certainty of $P=0.6894$. Exactly three defective parts are detected with a probability of $P=0.0442$. The corresponding hypergeometric distribution model is shown in Fig. 7.26.

7.5 Growth Process and Saturation Function 137

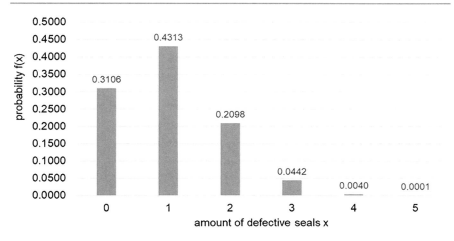

Fig. 7.26 Hypergeometric density function with parameters N = 50, M = 10, n = 5

7.5 Growth Process and Saturation Function

The focus of this chapter is on the growth process and its representation in a logistic model. Growth processes are well known, for example, from biology; the associated models can certainly also be applied to the analysis of damage processes in the context of technical reliability analysis.

7.5.1 Logistic Function

The logistic function belongs to the so-called sigmoid functions (or S-functions). The general sigmoid function (Eq. 7.151) is characterised by an S-shaped curve.

$$f(x) = \frac{e^x}{1 + e^x} \qquad (7.151)$$

The *logistic function* with run-time variable x is shown in Eq. 7.152 and includes the parameters intercept a (where: 0<a<S; cf. also Eq. 7.154), saturation limit S (where S>0) and proportionality factor k (where k>0). The parameters influence the stretching or shifting of the curve, which can be seen as a schematic diagram in Fig. 7.27. The parameters can be interpreted very well with regard to growth and saturation processes (see also remarks in this chapter), so that they are very well suited to represent complex topics in biology, economics and engineering. Strictly speaking, Eq. 7.152 shows a variation of the original saturation function (Eq. 7.153), which is published by Verhulst, cf. (Verhulst 1838) and (Verhulst 1845): The saturation limit is described by the parameter γ (Eq. 7.155). The parameter β – which is related to the initial population (i.e. comparable to the intercept a within Eq. 7.152) – describes the location. The relationship between

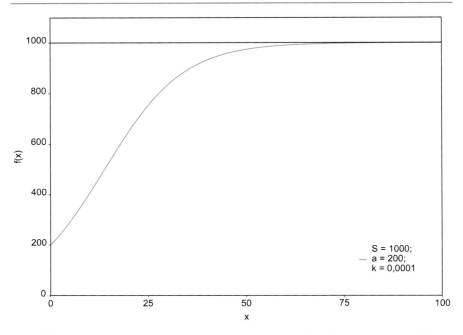

Fig. 7.27 Saturation function with the parameters saturation $S = 1000$, intercept $a = 200$ and proportionality factor $k = 0.0001$ based on Eq. 7.152

the two notations is shown in Eq. 7.156. The parameter α represents growth and is the analogue of the parameter k involving the saturation limit; Eq. 7.157. Verhulst's equation with notation according to Eq. 7.153 is included here because the estimation of the parameters via the approaches of Tintner (1958) and Rhodes (1940) refer to this notation; cf. Chap. 8.

$$f(x) = \frac{a \cdot S}{a + (S - a) \cdot e^{-Skx}} \quad (7.152)$$

$$f(x) = \frac{\gamma}{1 + \beta \cdot e^{-\alpha x}} \quad (7.153)$$

$$f(x = 0) > 0 \quad (7.154)$$

$$S = \gamma \quad (7.155)$$

$$a = \frac{S}{\beta + 1} \quad (7.156)$$

$$k = \frac{\alpha}{S} \quad (7.157)$$

7.5 Growth Process and Saturation Function

Notes on the Logistical Function and Potential Fields of Application:

a) The logistic function describes a growth which runs against a certain limit (S = growth limit); reference is Eq. 7.152.
b) The function value is greater than 0 at the beginning of the growth (e.g. time-related variable x = 0); cf. Eq. 7.154.
c) Example Biology (1): The saturation function can be used to represent the growth behaviour of bacteria. The bacterial culture medium is limited (growth limit S), the parameter a represents the initial stock of bacteria. The proportionality factor k may be interpreted as the growth rate, which influences the slope of the curve.
d) Example biology (2): The saturation function can be used to represent the growth behaviour of water lilies in a pond. The spatial extent of the water lilies is limited to the pond area (growth limit S). The parameter a represents the initial population of water lilies. The proportionality factor k represents the reproduction rate of the water lilies.
e) Example sociology: With the saturation function, the spread of a rumor in a finite group of people ("word-of-mouth") can be mapped. The spread of the rumor is restricted to the limited number of people, e.g. on a ship. The parameter a represents the number of people who know or formulate the rumor. The proportionality factor k represents the speed at which the rumor spreads.
f) Example Technical Reliability (1): The saturation function can be used to represent a corrosion process. The extent of the corrosion is limited by an area (e.g.: corrosion of a vehicle door outer skin; corrosion of an anode in an electrolysis process). The parameter a represents the area which is affected at the beginning of the corrosion process. The proportionality factor k represents the progression of the corrosion.
g) Example Technical Reliability (2): The saturation function can be used to map the failure behavior of products in a product fleet during the utilization phase. A component has an inherent weakness and fails after a certain period of use—caused by a certain load spectrum (example: oil leakage in a crankcase seal complex). The maximum number of products affected is either the product fleet as a whole or a subset of it; so that a saturation limit exists. The parameter a represents the number of complaints at the beginning. The proportionality factor k represents the increase in the number of complaints in the fleet (cf. in particular Sect. 7.5.2).

7.5.2 Application Example Verhulst Model: Corrosion Process

In this section, the application of the Verhulst saturation model is outlined using the example of a corrosion process. A detailed example for the application of the Verhulst saturation model can be found in Chap. 12 for the illustration of a sensor failure behavior: In addition to the parameter estimation by means of regression analysis, a comparison to the Weibull distribution model fit is also carried out.

Baseline:
The corrosion of a painted steel sheet is to be represented with the aid of the logistic function. The corrosion process extends over several months, so that the unit month is defined for the variable time t. The size of the steel sheet is $S = 1000$ cm^2. At the beginning of the observation of the corrosion process $t = 0$ the area $a = 200$ cm^2 is already corroded. Furthermore, a second measuring point is known: At $t = 25$ months, the steel sheet is already corroded on an area of $f(t = 25) = 752.82$ cm.^{2}

Solution:
The corrosion behaviour can be fully described by the Verhulst model (Eq. 7.158) if the proportionality factor k can be determined. The parameter a (intercept) is known. Thus, transforming Eq. 7.158 and inserting the values of the second measuring point directly yields the result for the proportionality factor k; Eq. 7.159. The saturation function follows directly, cf. Eq. 7.160. The visualization of the function is shown in Fig. 7.27.

$$f(x) = \frac{a \cdot S}{a + (S - a) \cdot e^{-Skx}} \tag{7.158}$$

$$k = \frac{1}{-St} \ln\left(\frac{1}{S-a}\left(\frac{aS}{f(t)}\right) - a\right) = 0.0001 \tag{7.159}$$

$$f(x) = \frac{200,000}{100 + 199,900 \cdot e^{-0.1x}} \tag{7.160}$$

7.6 Mixed Distribution Models

Engineering events or problems cannot always be represented and described with the aid of a simple distribution model. In more complex situations, the combination of different distribution models enables the description of the situation within one (!) model. An example of this is the mapping of the failure behaviour of components over the product life cycle, i.e. the entire service life. The failure rate related to the entire service life has already been explained in Sect. 3.3.2 (cf. Fig. 3.2). The three areas of early failure behaviour, random failure behaviour and lifetime-related failure behaviour are each represented by a distribution model. If the default behaviour were to be represented with only one equation – e.g. one Weibull distribution model for each failure phase – one would speak of three alternative default scenarios and accordingly adapt a mixed distribution model.

Of course, in the presence of a mixed population, a mixed distribution model – alternatively (cf. Sect. 7.6.1) or competitively (cf. Sect. 7.6.2) – must be not necessarily adaptpted: Individual areas of the mixed population can also be considered separately and represented with a simpler model in each case. The advantage of separation consists in the simple interpretation of the model parameters with a reduced number of the same for the respective, separated areas. A

7.6 Mixed Distribution Models

procedure for separating the areas, which can be each described with a distribution model, is shown in (Bracke 2009). The advantage of a mixed distribution model is the representation of the failure behavior in a single model.

7.6.1 Alternative Models

The alternative model is characterized by the combination of i different functions $F_i(t)$ (different function types or the same function type), cf. Eq. 7.161. The different functions are weighted differently via factors q_i, whereby the sum of all weighting factors must add up to one (cf. Eq. 7.162).

$$F(t) = q_1 F_1(t) + q_2 F_1(t) + \cdots + q_k F_k(t) \tag{7.161}$$

$$q_{Ges} = \sum_{i=1}^{K} q_i = 1 \tag{7.162}$$

If the factors can be related to subsets of a data set, the weighting factors can also be expressed via quotients. These represent the separate portions or ranges of the data series that are mapped with the respective model $F_i(t)$; Eq. 7.163.

$$F(t) = \frac{n_1}{n} F_1(t) + \frac{n_2}{n} F_2(t) + \cdots + \frac{n_k}{n} F_k(t) \tag{7.163}$$

Application Example: Alternative Mixed Distribution Model
Within a product fleet, n = 45 damage events are known. At the respective time of failure, the operating hours are documented, cf. Table 7.1. First, the cumulative frequencies F_i are determined, cf. Eq. 7.164 (here approximation for n < 50) and also Chap. 12. Subsequently, a two-parameter Weibull distribution model is fitted (Eq. 7.165), whereby the parameters were determined by means of maximum likelihood estimation procedures (b = 1.9; T = 16.801). In Fig. 7.28, a discrepancy of up to P ~ 0.15 between distribution model and cumulative probabilities is evident. Likewise, a kink in the cumulative probabilities can be seen at about 18,000 operating hours. It can be assumed that there is an alternative failure behaviour which is based on two damage causalities. Ergo, a simple distribution model is not

Table 7.1 Damage events (n = 45); failure times [h] of products within a product fleet

1500	10,699	21,732	19,531	17,892	5300	21,210	20,641
2300	11,629	22,255	19,976	18,983	5500	8000	20,797
2800	12,598	22,594	20,008	19,292	6200	9000	20,928
3400	13,736	23,484	20,411	19,355	7000	28,691	16,839
3900	15,910	27,099	20,465	19,527	7600	29,080	17,065
4200	15,999	28,551	4800	5090			

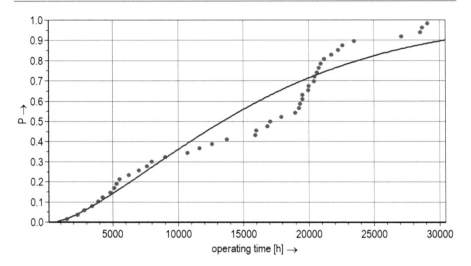

Fig. 7.28 Alternative failure behaviour based on two damage causalities: Discrepancies between cumulative probabilities and fitted two-parameter Weibull distribution model

suitable for the representation of the failure behaviour, therefore, a mixed distribution model is adapted according to Eq. 7.166.

$$F(t) = \frac{i - 0.3}{n + 0.4} \qquad (7.164)$$

$$F(x) = 1 - e^{-\left(\frac{x}{T}\right)^b} \qquad (7.165)$$

$$F(t) = \frac{n_1}{n}\left(1 - e^{-\left(\frac{t}{T_1}\right)^{b_1}}\right) + \frac{n_2}{n}\left(1 - e^{-\left(\frac{t}{T_2}\right)^{b_2}}\right) \qquad (7.166)$$

The mixed distribution model consists of two Weibull distribution models, which are weighted via quotients n_x/n. The fitting of the mixed distribution model can be performed using, for example, the maximum likelihood estimation procedure; simple manual estimation is not readily available. The parameters of the Weibull distribution models are determined to be $b_1 = 2.5$; $b_2 = 4.51$; $T_1 = 5,733$ h, $T_2 = 21,514$ h. The quotients are determined as $n_1/n = 0.32$ and $n_2/n = 0.68$. The visualization of the mixed distribution model (Eq. 7.166) with the mentioned parameters is shown in Fig. 7.29. The mixed distribution model is easy to interpret: The first Weibull distribution model represents the first part of the failure behaviour (in fact, approx. 34% of the damage cases are to be assigned to damage causality 1; the quotient is $n_1/n = 0.32$). Correspondingly, the second distribution model represents the second failure behaviour caused by a second damage causality (in fact, approx. 66% of the damage cases are to be assigned to damage causality 2; the quotient is $n_1/n = 0.68$). Accordingly, the weights of the distribution

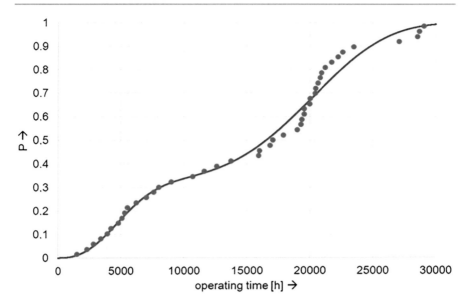

Fig. 7.29 Alternative mixed distribution model based on two Weibull distribution models to represent a failure behaviour caused by two damage causalities

models are carried out using the quotients. The deviation between cumulative probabilities and mixed distribution model is small.

Comments on the Alternative Mixed Distribution Model
If the damage causalities are known through a technical damage part analysis, the weighting quotients can be determined directly accordingly. The parameters of the Weibull distribution models are then to be determined via a parameter estimation procedure. If only the failure times are known, but the damage causalities are not known, the quotients must also be calculated via a parameter estimation procedure. In this case, the buckling points detected via the weighting are a potential indicator for the presence of two different damage causalities and thus for the staged failure behavior. Of course, the data set can also be divided (Bracke 2009), and subsequently a separate adjustment of the failure models can be carried out.

7.6.2 Competing Models

The competing model is used when a component can fail due to different damage causalities. The damage causalities are not sequential in time, as in the alternative model. Rather, the damage causalities are "in competition" with each other. The component fails because of one of the potential damage causalities. Thus, the competing model is characterized in terms of failure behavior by the addition of distribution models, with the intersection of probabilities being subtracted. This is based on the addition theorem of probability theory (cf. Sect. 5.2). Eqs. 7.167 and 7.168 apply accordingly to the probability of failure and the probability of survival respectively.

$$F(t) = 1 - \prod_{i=1}^{n}(1 - F_i(t)) \tag{7.167}$$

$$R(t) = \prod_{i=1}^{n} R_i(t) \tag{7.168}$$

Outline Example:
Let a number of damage events with their respective failure times be known. First, the cumulative probabilities F_i are determined on the basis of the failure times ordered in ascending order (ranks i). Competitive failure behavior is assumed based on a damage part analysis: the component fails due to damage causality A or B. A simple distribution model would not be able to represent the failure behavior. Therefore, based on Eq. 7167, a mixed distribution model with two distribution models is assumed, and Eq. 7.169 follows. A Weibull distribution model is chosen as the distribution model (of course, other types of functions can be used), so Eq. 7.170 follows directly. Equation 7.170 has four parameters, a manual adjustment of the model to the damage data is not useful, but can easily be done, for example, using the maximum likelihood estimator (Chap. 8).

$$F(t) = 1 - (1 - F_1(t)) \cdot (1 - F_2(t)) = F_1(t) + F_2(t) - F_1(t) \cdot F_2(t) \tag{7.169}$$

$$F(t) = \left(1 - e^{-\left(\frac{t}{T_1}\right)^{b_1}}\right) + \left(1 - e^{-\left(\frac{t}{T_2}\right)^{b_2}}\right) - \left(1 - e^{-\left(\frac{t}{T_1}\right)^{b_1}}\right) \cdot \left(1 - e^{-\left(\frac{t}{T_2}\right)^{b_2}}\right) \tag{7.170}$$

Distribution Models: Parameter Estimation

8

The following explanations focus on the estimation of parameters with regard to the fitting of a distribution model to a series of measured values. The robust and manually simple procedure of regression analysis, which is suitable for many distribution models, is shown in Sect. 8.1. A frequently used iterative procedure is the maximum likelihood estimator (MLE) according to Fisher (1912) shown in Sect. 8.2. The MLE is characterized by its universal applicability.

In addition, further estimation methods with reference to distribution models, which are used in the technical reliability analysis, are outlined in Sect. 8.3. Estimation procedures with indirect reference to the Weibull distribution model are shown in Sects. 8.3.1 and 8.3.2: Dubey's (1967) estimation of a failure-free time (threshold parameter of a Weibull distribution model) can serve, among other things, as a starting point for fitting a three-parameter Weibull distribution model; Sect. 8.3.1. Section 8.3.2 outlines Gumbel's (1958) procedure for estimating shape and location parameters of a Weibull distribution model. Section 8.3.3 shows estimation procedures for the determination of the parameters of the growth and saturation model according to Verhulst (1838 and 1845), cf. also (Chap. 7), which can also be traced back to regression analysis.

The above-mentioned regression analysis and MLE methods are standard estimators that are frequently used in reliability analysis. Of course, there are many other estimation methods, such as the moment method (Czado and Schmidt 2011), the Levenberg–Marquardt algorithm (Moré 1978) or the trust region method (Conn et al. 2000).

In technical reliability analysis, the Weibull distribution is a frequently used model for mapping the failure behavior of technical components, systems or products. Due to this, the mentioned methods Regression Analysis, MLE, Dubey, Gumbel – both in combination and in comparison – are applied for the adjustment of parameters using the example of a Weibull distribution model; cf. in detail Chap. 12.

The estimation of mean and dispersion on the basis of a concrete series of measured values is of course also a parameter estimation. These fundamental estimation procedures have already been explained in Chap. 6 within the framework of the basic data analysis and are therefore not the subject of the present chapter.

8.1 Regression Analysis

The procedure of rank regression has already been sketched in outline in Sect. 4.6 (Mathematical Papers: Log Representation), cf. section Weibull Probability Paper. The regression analysis can, of course, be performed for all linear relationships. If an exponential relationship is present, the regression analysis can be performed after linearization (by logarithmization) for simple parameter estimation. In this chapter, regression analysis is used to determine model parameters using the Weibull distribution model as an example.

The starting point for dividing the abscissa and ordinate in the Weibull probability paper (a form of mathematical paper) is the Weibull distribution model (Eq. 8.1). By transforming and double logarithmizing, the Weibull distribution model is shown as a straight line, cf. Fig. 4.7 in Chap. 4.

The transformation and the double logarithmizing of the Weibull distribution model (Eq. 8.1) leads to Eq. 8.2. The general equation of the straight line is shown by Eq. 8.3. If Eq. 8.4 applies to variable x and Eq. 8.5 applies to the function value f(x), then the intercept can be represented according to Eq. 8.6. Equation 8.7 applies to the gradient of the straight line equation: This means that the shape parameter b of the Weibull distribution model can be determined directly.

$$F(t) = 1 - \exp\left(-\left(\frac{t}{T}\right)^b\right) \tag{8.1}$$

$$\ln(-\ln(1 - F(t))) = b \cdot \ln t - b \cdot \ln T \tag{8.2}$$

$$f(x) = mx + c \tag{8.3}$$

$$x = \ln t \tag{8.4}$$

$$f(x) = y = \ln(-\ln(1 - F(t))) \tag{8.5}$$

$$c = -b \cdot \ln T \tag{8.6}$$

$$b = m \tag{8.7}$$

The subject of technical reliability analysis are often product failures which are observed at certain runtime points t_i. The probability of the failure $F(t_j)$ related to all known failures can be easily estimated via an approximation (Eq. 8.8; cf. also Chap. 12.1). With the help of Eqs. 8.4 and 8.5 the value pairs (t_i; $y = F(t_j)$)

can be transformed. Since Eq. 8.7 holds, the gradient of the straight line equation and thus the shape parameter b can be determined using a regression approach (Eq. 8.9), where the mean values refer to the transformed original values x_i and y_i, respectively. The characteristic lifetime T of the two-parameter Weibull distribution model is then determined by transforming the Weibull distribution function; Eq. 8.10.

$$F(x_i) = \frac{i - 0.3}{n + 0.4} \tag{8.8}$$

$$\widehat{b} = \frac{\sum_{i=1}^{n} x_i y_i - n\overline{xy}}{\sum_{i=1}^{n} x_i^2 - n\overline{x}^2} \tag{8.9}$$

$$\widehat{T} = e^{-\left(\frac{\overline{y} - \widehat{b}\,\overline{x}}{\widehat{b}}\right)} \tag{8.10}$$

If a failure-free time t_0 is taken into account when mapping a failure behaviour, the Weibull distribution model is used in three-parameter form, cf. Eq. 8.11. First, the parameter t_0 is estimated, e.g. using the method according to Dubey (cf. Sect. 8.3.1), then the regression analysis can be carried out as outlined above, whereby the operating times with respect to the failures (value pairs t_i; $F(t_i)$) are corrected by the failure-free time ($t_{corr} = t_i - t_0$) as well as the characteristic life ($T_{corr} = T + t_0$) and thus $F(t) = F(t_i - t_0)$ applies. A detailed example can be found in Chap. 12.

$$F(t) = 1 - \exp\left(-\left(\frac{t - t_0}{T - t_0}\right)^b\right) \tag{8.11}$$

8.2 Maximum Likelihood Estimation Method (Fisher)

The objective of the Maximum Likelihood Estimator [MLE] according to Fisher (1912) is to fit a distribution model to the measured values of a given sample. The focus is on the likelihood function L (Eq. 8.12). The function L consists of the product of the probabilities $f(x_i)$ at the point x_i. If the function L is maximized, then the likelihood expressed by it is maximized with respect to the fact that the function f represents the distribution of the measured data of the concrete sample at hand. The maximum is determined by forming the partial derivatives of the likelihood function L according to the parameters of the chosen function f and setting them equal to zero. This results in a (non-linear) system of equations, which is usually solved numerically. Often the log-likelihood function is also maximized due to its simpler handling; this procedure is also used in the example of the present chapter.

Since the parameters of the selected function f are determined in order to estimate the maximum, the prerequisite for this approach is that the available measured values of the concretely available sample are representative of the population.

Thus, by forming the maximum, the model with the highest probability would be found. The principle is shown in Fig. 8.1: The function f describes the distribution of measured values, which is visualized by a histogram. The more comprehensive the series of measured values, the more accurately the frequency distribution is represented: For each known x_i there is an f_i recognizable in Fig. 8.1 by the class probability.

$$L = \prod_{i=1}^{n} f_i(x_i) \tag{8.12}$$

Note: The assumption of a distribution function f that represents the distribution of the measured values, while at the same time assuming that the sample at hand represents the population, can also have a disadvantageous effect. If the assumption is not true, the maximum likelihood procedure may—not always—lead directly to an inconsistent result.

Application of the maximum likelihood estimation procedure using the weibull distribution as an example:
In the following, the maximum likelihood estimation procedure is outlined: The Weibull distribution model is chosen as the function f for which the maximum is to be searched. The Weibull distribution model is one of the central distribution models in reliability analysis.

The three-parameter Weibull distribution model is shown in Eq. 8.13. In addition to the runtime variable x, the model contains three parameters (parameter failure-free time (threshold) t_0; position parameter T; shape parameter b; cf. Chap. 7).

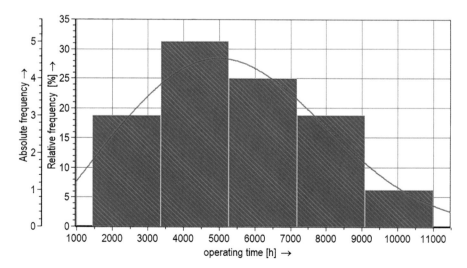

Fig. 8.1 Adjusted Weibull distribution model (probability density; failure data cf. Table 12.2, Chap. 12)

8.2 Maximum Likelihood Estimation Method (Fisher)

For different times i related to the running time variable x the likelihood function L results according to Eq. 8.14. By logarithmizing the function L a sum results from the product; Eq. 8.15. Afterwards the maximum of the function L is determined by forming the partial derivatives of the likelihood function L related to the parameters of the underlying model; the system of equations results according to Eqs. 8.16 to 8.18. The system of equations can be solved with a suitable numerical method.

$$f(x|t_0, T, b) = \frac{b}{T - t_o} \cdot \left(\frac{x - t_o}{T - t_o}\right)^{b-1} \cdot e^{-\left(\frac{x - t_o}{T - t_o}\right)^b} \qquad (8.13)$$

$$L = \prod_{i=1}^{q} f_i(x_i) = f_1(x_1|t_0, T, b) \cdot f_2(x_2|t_0, T, b) \cdot \ldots \cdot f_q(x_q|t_0, T, b) \cdot \qquad (8.14)$$

$$\ln L = \sum_{i=1}^{q} f_i(x_i) = f_1(x_1|t_0, T, b) + f_2(x_2|t_0, T, b) + \cdots + f_q(x_q|t_0, T, b) \cdot \qquad (8.15)$$

$$\frac{\delta L}{\delta T} = 0 \qquad (8.16)$$

$$\frac{\delta L}{\delta b} = 0 \qquad (8.17)$$

$$\frac{\delta L}{\delta t_0} = 0 \qquad (8.18)$$

For illustration, the assumption $t_0 = 0$ is made (no threshold; ergo, no failure-free time) so that Eq. 8.13 directly yields the two-parameter Weibull distribution model (Eq. 8.19), which is optimal at various points i. For the likelihood function L, Eq. 8.20 is obtained. For the likelihood function L, we obtain Eq. 8.20. Substituting the chosen function Eqs. 8.19 into 8.20 yields the likelihood function L according to Eq. 8.21. Logarithmizing Eq. 8.21 yields the sum according to Eq. 8.22 and simplifies the following operations: The log-likelihood function L is partially derived according to parameters b and T (Eqs. 8.23 and 8.24) and thus Eq. 8.25 is obtained for parameter T and Eq. 8.26 is obtained for parameter b, respectively, which are solved numerically.

$$f(x|T, b) = \frac{b}{T} \cdot \left(\frac{x}{T}\right)^{b-1} \cdot e^{-\left(\frac{x}{T}\right)^b} \qquad (8.19)$$

$$L = \prod_{i=1}^{q} f_i(x|T, b) \qquad (8.20)$$

$$L = \prod_{i=1}^{q} \left[exp\left\{-\left(\frac{x_i}{T}\right)^b\right\} \cdot \left(\frac{x_i}{T}\right)^{b-1} \cdot \frac{b}{T} \right] \quad (8.21)$$

$$\ln L = q \cdot \ln\left(\frac{b}{T}\right) - \sum_{i=1}^{q} \left(\frac{x_i}{T}\right)^b + (b-1) \cdot \sum_{i=1}^{q} \ln\left(\frac{x_i}{T}\right) \quad (8.22)$$

$$\frac{\delta L}{\delta T} = 0 \quad (8.23)$$

$$\frac{\delta L}{\delta b} = 0 \quad (8.24)$$

$$T = \left[\frac{1}{q} \cdot \sum_{i=1}^{q} x_i^b \right]^{1/b} \quad (8.25)$$

$$\frac{q}{b} + \sum_{i=1}^{q} \ln x_i - \frac{q \cdot \sum_{i=1}^{q} \left(x_i^b \cdot (\ln x_i) \right)}{\sum_{i=1}^{q} x_i^b} = 0 \quad (8.26)$$

8.3 Special Parameter Estimators with Reference to Specific Distribution Models

For many concrete contexts and distribution models, special parameter estimators can be applied—in addition to universal approaches such as the maximum likelihood estimator. In this chapter, the following special estimation methods are outlined:

a) Estimation of threshold (failure-free time) according to Dubey (Sect. 8.3.1),
b) Estimation of the parameters of a Weibull distribution model according to Gumbel (Sect. 8.3.2),
c) Estimation of the parameters of a saturation model according to Tintner and Rhodes (Sect. 8.3.3).

8.3.1 Dubey

The estimation of a failure-free time t_0 (threshold parameter) can be based on the approach of Dubey (1967). The method according to Dubey can thus be used in combination with the methods according to Gumbel (Sect. 8.3.2) or the regression analysis (Sect. 8.1) to estimate all parameters of a three-parameter Weibull

8.3 Special Parameter Estimators with Reference to Specific ...

distribution model. An application example of the Dubey approach is shown in Sect. 12.2.

Dubey's method is based on an approach according to Mandel (1964:250), which includes a method for estimating the displacement parameter x_0 for a general function according to Eq. 8.27 with the three parameters A, B, x_0. The parameter x_0 is estimated with the help of an existing measurement series with ascending sorted observation values. Minimum and maximum values are required, as well as a third (mean) value within the measurement series.

$$y = A(x - x_0)^B \tag{8.27}$$

Dubey carried out the application of the Mandel approach with regard to the three-parameter Weibull distribution. The three required points in time t_1, t_2, t_3 are determined on the basis of a concrete series of measurements (example: product failures at the points in time t_1 to t_n). The principle is illustrated in Fig. 8.2: The time t_1 is assigned to the first failure, the time t_3 (actually t_n for n failures) to the last failure. The increases z_{1-2} and z_{2-3} are (approximately) identical (Eq. 8.28) with respect to the failure probability F.

$$z_{1-2} + z_{2-3} = 2z \tag{8.28}$$

This results in the following procedure for estimating the threshold parameter t_0 (failure-free time; theoretical first failure time):

1. Ascending sorting of known downtimes t_1 to t_n (with ordinal number i).
2. Determination of the probability for the cumulative frequency via ordinal number i, e.g. approximately with Eq. 8.29.

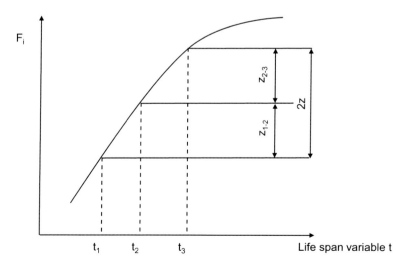

Fig. 8.2 Schematic diagram for determining the time points t_1, t_2, t_3 when estimating the parameter failure-free time (threshold) t_0

3. Determine the failure probability F(t$_2$) using Eq. 8.30, where the increments z$_{1-2}$ and z$_{2-3}$ are approximately identical; cf. Eq. 8.28.
4. Estimation of the mean time t$_2$ on the basis of F(t$_2$) with the help of the sorted downtimes t$_1$ to t$_n$ (the increases related to the probability t$_1$ to t$_2$ respectively t$_2$ to t$_3$ are approximately equal, the index 2 does not refer to the ordinal numbers!).
5. Use of the Dubey estimator according to Eqs. 8.31 or 8.32.

$$F(t_i) = \frac{i - 0.3}{n + 0.4} \tag{8.29}$$

$$F(t_2) = 1 - \exp(-(-\ln(1 - F(t_1))(-\ln(1 - F(t_3)))))^{\frac{1}{2}} \tag{8.30}$$

$$t_0 = \frac{t_1 t_3 - t_2^2}{t_1 + t_3 - 2t_2} \tag{8.31}$$

$$t_0 = t_2 - \frac{(t_3 - t_2) \cdot (t_2 - t_1)}{(t_3 - t_2) - (t_2 - t_1)} \tag{8.32}$$

8.3.2 Gumbel

Gumbel shows an approach to the estimation of shape and position parameters in his "Theory of Extreme Values, Statistical Theory of Fatigue Phenomena" (Gumbel 1958). The starting point here is the three limiting probabilities according to Fisher and Tippett (1928); cf. Chap. 7: Type III (Weibull distribution) is interesting here. If the threshold parameter is set to zero, the logarithms of the measured values of the present sample lie on a straight line (cf. Chap. 4, mathematical papers, double logarithmic form). Using an extreme value approach, the estimators for shape and location are derived for the two-parameter case of the type III distribution (Weibull distribution; Eqs. 8.33), cf. Eqs. 8.34 and 8.35 (and after logarithmizing, Eq. 8.36). The symbolism is adapted to the representation of the Weibull distribution model in Chap. 7. Standard deviation σ_N and mean y_N refer to the present series of measurements with extent N and are also functionally related. For the estimation of shape parameter b and T, standard deviation σ_N and mean value y_N can be taken from Table A.9 (Appendix). If the number of measured values converges to infinity, the mean value y_N converges to the Euler-Mascheroni constant $\gamma = 0.57721$.

$$F(x) = 1 - e^{-\left(\frac{x}{T}\right)^b} \tag{8.33}$$

$$b = \frac{\sigma_N}{2.30258 s_{log}} \tag{8.34}$$

8.3 Special Parameter Estimators with Reference to Specific ...

$$\log T = \overline{\log x} + \frac{\bar{y}_N}{2.30258 \cdot b} \quad (8.35)$$

$$T = 10^{\left[\frac{\left(\sum_{i=1}^{n} \log(x_i)\right)}{n} + \frac{\bar{y}_N}{2.30258b}\right]} \quad (8.36)$$

Note that the standard deviation $s_{\log}$ is determined from the logarithmized original values (observed values). First, Eq. 8.37 applies in general to the estimation of the standard deviation. According to Steiner's theorem Eq. 8.38 holds, thus Eq. 8.39 follows directly for the standard deviation $s_{\log}$.

$$s_{\log}^2 = \frac{1}{n-1} \sum_{i=1}^{n} \left(\log x_i - \overline{\log x}\right)^2 \quad (8.37)$$

$$s_{\log}^2 = \frac{1}{n-1} \left[\left(\sum_{i=1}^{n} (\log x_i)^2\right) - \frac{1}{n}\left(\sum_{i=1}^{n} \log(x_i)\right)^2\right] \quad (8.38)$$

$$s_{\log} = \sqrt{\frac{1}{n-1} \left[\sum_{i=1}^{n} (\log x_i)^2 - \frac{1}{n}\left(\sum_{i=1}^{n} \log(x_i)\right)^2\right]} \quad (8.39)$$

When fitting a three-parameter Weibull distribution model (Chap. 7 or Eq. 8.13), the parameter threshold t_0 would have to be estimated beforehand, e.g. using the Dubey approach (Sect. 8.3.1). Subsequently, the observed values as well as the position parameter T have to be transformed, comparable to the transformation when performing the parameter estimation by means of regression analysis (cf. example in Sect. 12.2).

8.3.3 Tintner and Rhodes

The approaches according to Tintner and Rhodes allow an estimation of the parameters of the saturation function according to Verhulst (Sect. 7.5); Eq. 8.40. The saturation function cannot be linearized directly, ergo the procedure of regression analysis cannot actually be used. Tintner's (1958) approach uses regression analysis for the inverse of the saturation function (Eq. 8.40; observed value y(x) at location x); thus here for the relationship between $1/y_x$ and $1/y_{x-1}$. Following the approach of Tintner (1958), the estimate of the growth parameter α is carried out according to Eq. 8.41, and the saturation limit γ can be estimated with Eq. 8.42; cf. also in more detail (Hartung 2009; p. 644).

$$y(x) = \frac{\gamma}{1 + \beta \cdot e^{-\alpha x}} \quad (8.40)$$

$$\hat{\alpha} = -\ln \frac{\sum_{x=2}^{n} \frac{1}{y_x \cdot y_{x-1}} - \frac{1}{n-1} \sum_{x=2}^{n} \frac{1}{y_{x-1}} \sum_{x=2}^{n} \frac{1}{y_x}}{\sum_{x=2}^{n} \left(\frac{1}{y_{x-1}}\right)^2 - \frac{1}{n-1}\left(\sum_{x=2}^{n} \frac{1}{y_{x-1}}\right)^2} \quad (8.41)$$

$$\widehat{\gamma} = \frac{(n-1)\left(1-e^{-\widehat{\alpha}}\right)}{\left(\sum_{x=2}^{n}\frac{1}{y_x} - e^{-\widehat{\alpha}}\sum_{x=2}^{n}\frac{1}{y_{x-1}}\right)} \quad (8.42)$$

The approach according to Rhodes (1940) allows an estimation of the location parameter β as a function of the growth parameter α and the estimated saturation limit γ, Eq. 8.43. The condition is that all observed values $y_1, \ldots, y_n$ lie below the estimated saturation limit γ. If some of the observed values lie above the saturation limit γ, Eq. 8.44 is used.

$$\widehat{\beta} = e^{(n+1)\frac{\widehat{\alpha}}{2} + \frac{\left[\sum_{x=1}^{n}\ln(\widehat{\gamma}/y_x - 1)\right]}{n}} \quad (8.43)$$

$$\widehat{\beta} = \frac{\sum_{x=1}^{n}\frac{\widehat{\gamma}-y_x}{y_x}e^{-\widehat{\alpha}x}}{\sum_{x=1}^{n}e^{-2\widehat{\alpha}x}} \quad (8.44)$$

Confidence Intervals 9

This chapter deals with the determination of confidence intervals for various parameters and for adjustment models. First, the basics of confidence intervals are presented in an introductory chapter (cf. Sect. 9.1). Subsequently, procedures for the determination of confidence intervals of the parameters mean and dispersion in the case of a normally distributed population are presented (Sect. 9.2) as well as the confidence interval for a proportion value in the case of a binomially distributed population (Sect. 9.3). Section 9.4 focuses on the determination of confidence intervals for parameters of the Weibull distribution model. Section 9.5 concludes with the procedure for determining the confidence interval for an adjustment function or adjusted distribution model.

9.1 Introduction Confidence Interval

The starting point is a data set, which results, for example, from measurement data within a sample with a size of n. The data set is used as a basis for the estimation of parameters. With the help of this data set, characteristic values or parameters are estimated. These estimated values are approximate values, the actual value cannot be determined, since the base of operations is only a sample from a population. Ergo, there is a very high probability that the estimated value deviates from the actual, unknown value. Especially with small sample sizes, this range of potential deviation between the estimated value and the actual, unknown value is high. This range of potential deviation can be determined and is called confidence interval. The confidence interval is a range in which the unknown parameter is contained with a certain probability – the confidence level γ. The complement to the confidence level γ is the probability of error α.

Figure 9.1 shows the relationship between the unknown parameter ν and a confidence interval with its limits c_u and c_o on a number ray. Within the confidence

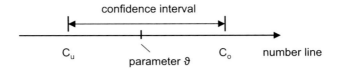

Fig. 9.1 Schematic diagramm of a confidence interval with respect to a parameter ϑ

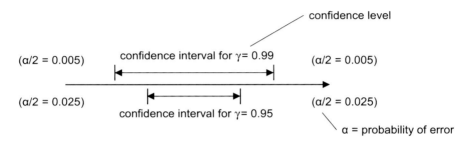

Fig. 9.2 Relationship between confidence interval, confidence level and error probability α

interval the actual value of the parameter ϑ is expected with confidence level γ, outside the confidence interval with error probability α.

Depending on the confidence level γ chosen, the confidence interval is narrower or wider; Fig. 9.2. If the confidence level γ is chosen large, the confidence interval is correspondingly wider. Ultimately, the uncertainty (probability of error) is smaller, but the potential scatter range is larger, since the relationship according to Eq. 9.1 between confidence level γ and probability of error α applies.

$$\gamma = 1 - \alpha \tag{9.1}$$

Basic notes on the confidence interval:

1. The higher the confidence level, the wider the confidence interval and the smaller the probability of error.
2. The larger the underlying, representative (!) sample size, the more precise the parameter estimate and the smaller the confidence interval.
3. The size of the confidence interval is not proportionally related to the sample size investigated. Thus, it is not true: doubling the sample size leads to a halving of the confidence interval.
4. When calculating a confidence interval, the confidence level (and thus the probability of error) is freely chosen. In the engineering context, it is common to choose a confidence level of $\gamma = 0.9$ (with probability of error $\alpha = 0.1$), or $\gamma = 0.95$ (with $\alpha = 0.05$) or $\gamma = 0.99$ (with $\alpha = 0.01$).
5. The confidence interval limits often lie approximately symmetrically on the number ray in relation to the estimated value of the parameter for large sample sizes (cf. also Fig. 9.1). However, the confidence interval may also be

asymmetric with respect to the estimated value: For example, the confidence interval for the proportion value p of a certain characteristic carrier in a sample close to zero is not symmetric, since the proportion value cannot be negative. At the same time, however, the confidence interval is correspondingly large for small sample sizes (cf. Sect. 9.3).

6. Figures 9.1 and 9.2 show a two-sided confidence interval, the probability of error is correspondingly α/2 below and above the limits c_u and c_o, respectively. The confidence interval can also be determined one-sided. In case of a one-sided confidence interval, only one limit exists: the confidence interval above or below this limit extends over an infinite length. Accordingly, the probability of error refers to the other side of the boundary. Example: For welded constructions, a minimum diameter is specified by the designer on the technical drawing for the characteristic spot weld diameter. If a sample of measured values is now examined with regard to spot weld diameters, the mean value of the spot weld diameters could be estimated and the confidence interval delimited to the left (due to the minimum requirement) on the numerical grade; cf. also example in Sect. 9.2.4.

7. There are procedures for calculating the confidence interval on the basis of a sample for a number of parameters that can be part of distribution models. Thus, for example, confidence intervals for estimators of mean values and dispersion parameters in the presence of a normally distributed population are presented in this chapter. Likewise, procedures are also presented with which the confidence intervals of parameters of a Weibull distribution can be calculated.

8. When analysing series of measurements, it is often advisable to specify a confidence interval and the probability of error, as the confidence in the estimate of the parameter is quantified. For example, the summarized analysis only based on an estimation of the arithmetic mean value of a connecting rod diameter on the basis of a produced batch (sample) is unfavorable, since it is not transparent how large the potential scatter range (confidence interval) of the mean value or the error probability can be.

9.2 Confidence Intervals for Estimators with Normally Distributed Population

The present chapter focuses on confidence intervals for the unknown parameters mean and variance of a normally distributed population; cf. Sects. 9.2.1 to 9.2.3. Furthermore, the application of the outlined procedures is explained in a case study; Sect. 9.2.4.

9.2.1 Confidence Interval for Unknown Mean with Known Variance

First, the determination of a confidence interval for the arithmetic mean value in the presence of a normally distributed population: A normally distributed random

Fig. 9.3 Schematic diagram of the two-sided confidence interval related to a standardized, normally distributed random variable

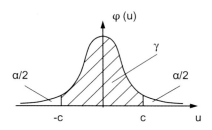

variable X is present. The variance σ^2 is known, the arithmetic mean value μ is unknown. In practice, the variance σ^2 is often not known (cf. Sect. 9.2.2), but the procedure shown here forms the methodological basis for its determination. The realization is available in the form of measurement data within a sample of size n. The confidence interval is set to two-sided. The confidence interval is determined two-sided (cf. Fig. 9.3), so that in the standardized representation the two limits c lie symmetrically about the ordinate. The probability of error refers to the ranges above and below the limits +c and −c, respectively.

The determination of a two-sided confidence interval is performed as follows:
Step 1: Choice of confidence level γ.
Examples: $\gamma = 0.9$ ($\alpha = 0.1$), $\gamma = 0.95$ ($\alpha = 0.05$), or $\gamma = 0.99$ ($\alpha = 0.01$).
Step 2: Transformation of the random variable X to the standard normally distributed random variable U (Eq. 9.2). Calculation of the limits c of the confidence interval with respect to the standard-normally distributed random variable U (Eq. 9.3) with the aid of the quantiles of the standard normal distribution (cf. Appendix Tables A.1 and A.2). The estimation function X refers to the unknown mean μ of the normally distributed population. The standard deviation σ is assumed to be known in the present procedure and refers directly to the underlying normally distributed population. The parameter n denotes the sample size of the realization.

$$U = \frac{\overline{X} - \mu}{\sigma/\sqrt{n}} \tag{9.2}$$

$$P(-c \leq U \leq c) = 2\phi(c) - 1 = \gamma \tag{9.3}$$

Step 3: Estimate arithmetic mean based on available sample (Eq. 9.4).

$$\overline{x} = \frac{x_1 + x_2 + \ldots + x_n}{n} = \frac{1}{n} \cdot \sum_{i=1}^{n} x_i \tag{9.4}$$

Step 4: Determine the confidence interval with Eq. 9.5.

$$\overline{x} - c\frac{\sigma}{\sqrt{n}} \leq \mu \leq \overline{x} + c\frac{\sigma}{\sqrt{n}} \tag{9.5}$$

9.2 Confidence Intervals for Estimators with Normally Distributed ...

Fig. 9.4 Schematic diagram for the calculation of a one-sided confidence interval with upper limit

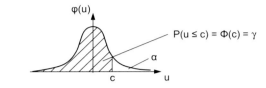

Fig. 9.5 Schematic diagram for the determination of a one-sided confidence interval with lower limit

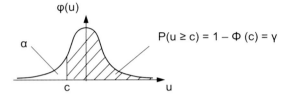

Step 5: Interpretation of the result.
The true value of the previously estimated mean μ will lie within the confidence interval at a confidence level γ based on the available knowledge (measurement data within a sample of size n). The probability of error α indicates the probability of this not being true. Since a two-sided confidence interval was specified here, the error probability with $\alpha/2$ refers to the range above and below the calculated limits, respectively. The length d of the confidence interval can be easily determined by the difference between the two limits.

Determination of a one-sided confidence interval
If the problem requires the determination of a one-sided confidence interval, this is done either with the boundary to the right or to the left with respect to the number ray; cf. Figs. 9.4 and 9.5. The calculation of the one-sided confidence interval with boundary to the right (with respect to the number ray) is done according to Eq. 9.6 as well as 9.7 and is outlined in Fig. 9.4.

$$P(u \leq c) = \phi(c) = \gamma \tag{9.6}$$

$$\mu \leq \bar{x} + c \frac{\sigma}{\sqrt{n}} \tag{9.7}$$

The determination of a one-sided confidence interval with delimitation to the left (related to the number degrees) is carried out according to Eq. 9.8 as well as 9.9 and is shown in Fig. 9.5.

$$P(u \geq c) = 1 - P(u \leq c) = 1 - \phi(c) = \gamma \tag{9.8}$$

$$\mu \geq \bar{x} + c \frac{\sigma}{\sqrt{n}} \tag{9.9}$$

Fig. 9.5b Schematic diagram for the two-sided confidence interval, related to the standardized, t-distributed random variable

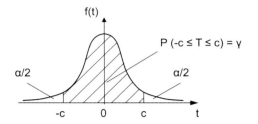

9.2.2 Confidence Interval for Unknown Mean with Unknown Variance

In Sect. 9.2.1, the calculation of a confidence interval for the mean value in the presence of a normally distributed population under the following conditions: The variance σ^2 is known, the mean μ is unknown. In application, the variance σ^2 is also likely to be frequently unknown, ergo the present chapter shows the procedure for determining a confidence interval for the unknown mean μ when the variance σ^2 is unknown. The procedures are similar, but now not the standard normal distribution is used to determine the confidence interval, but the t-distribution according to Student (respectively Gosset); cf. Student (1908), cf. Chap. 7.

Let the following be the *base of operations*:

1. The population of the random variable X under consideration is normally distributed.
2. A sample with size n and the measured values $x_1, x_2, \ldots, x_n$ is available.
3. Both mean μ and variance σ^2 are unknown.

The two-sided confidence interval is determined as follows:
Step 1: Choice of confidence level γ.
Examples: $\gamma = 0.9$ ($\alpha = 0.1$), $\gamma = 0.95$ ($\alpha = 0.05$), or $\gamma = 0.99$ ($\alpha = 0.01$).
Step 2: Transformation of the random variable X to the t-distributed random variable T (Eq. 9.10). Calculate the limits c of the confidence interval with respect to the t-distributed random variable T (Eq. 9.11; Fig. 9.5b) using the quantiles of the t-distribution (cf. Appendix Table A.3). The estimator X refers to the unknown mean μ of the normally distributed population. S refers to the estimator function of the (unknown) standard deviation σ of the underlying normally distributed population. The parameter n denotes the sample size of the realization.

$$T = \frac{\overline{X} - \mu}{S / \sqrt{n}} \tag{9.10}$$

$$P(-c \leq T \leq c) = \gamma \tag{9.11}$$

The required t-distribution according to Student (respectively Gosset) is presented in detail in Chap. 7. The t-distribution is completely determined by the degree of

9.2 Confidence Intervals for Estimators with Normally Distributed ...

freedom f and is directly related to the sample size, Eq. 9.12. The quantiles of the t-distribution can be taken from the appendix in Table A.3.

$$f = n - 1 \tag{9.12}$$

Step 3: Estimation of the arithmetic mean and the variance on the basis of the available sample (cf. Eqs. 9.13 and 9.14).

$$\bar{x} = \frac{x_1 + x_2 + \ldots + x_n}{n} = \frac{1}{n} \cdot \sum_{i=1}^{n} x_i \tag{9.13}$$

$$s^2 = \frac{1}{n-1} \cdot \sum_{i=1}^{n} (x_i - \bar{x})^2 \tag{9.14}$$

Step 4: Calculate the confidence interval based on the present sample using Eq. 9.15, the length of the interval d is obtained from the difference of the limits or Eq. 9.16.

$$\bar{x} - c\frac{s}{\sqrt{n}} \leq \mu \leq \bar{x} + c\frac{s}{\sqrt{n}} \tag{9.15}$$

$$d = \frac{2cs}{\sqrt{n}} \tag{9.16}$$

Step 5: Interpretation of the result.
The true value of the previously estimated mean μ will lie within the calculated confidence interval at a confidence level γ based on the available knowledge (sample with size n). The probability of error α indicates the probability of this not being true. Since a two-sided confidence interval was calculated here, the error probability $\alpha/2$ refers to the range above or below the calculated limits.

Notes:

1. For samples with size $n > 30$, the standard normal distribution can also be used instead of the t-distribution to determine the limits c.
2. The calculation of a one-sided confidence interval is analogous to the procedure already presented in Sect. 9.2.1.

9.2.3 Confidence Interval for Unknown Variance

The present chapter focuses on the calculation of a confidence interval for the unknown variance σ^2 in the presence of a normally distributed population. Based on the available sample, a range is determined in which the unknown variance σ^2 will lie. The mean value of the underlying population does not necessarily have to be known.

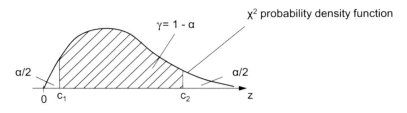

Fig. 9.6 Schematic diagram for the two-sided confidence interval, related to the standardized, χ^2-distributed random variable Z

Baseline:

1. The population of the random variable X under consideration is normally distributed.
2. A sample with size n and the measured values $x_1, x_2, \ldots, x_n$ is available.
3. Mean μ is known or unknown; variance σ^2 is unknown.

The two-sided confidence interval is determined as follows:
Step 1: Choice of confidence level γ.
Examples: $\gamma = 0.9$ ($\alpha = 0.1$), $\gamma = 0.95$ ($\alpha = 0.05$), or $\gamma = 0.99$ ($\alpha = 0.01$).
Step 2: Transformation of the random variable to the standardized random variable Z (cf. Eq. 9.17), which follows the χ^2 distribution and calculation of the bounds of the confidence interval with respect to the χ^2-distributed random variable (cf. Eq. 9.18), cf. also Fig. 9.6. The estimator S^2 is used to estimate the unknown variance σ^2 of the normally distributed population. The sample size n refers to the present sample.

$$Z = (n-1)\frac{S^2}{\sigma^2} \tag{9.17}$$

$$P(c_1 \leq Z \leq c_2) = \gamma \tag{9.18}$$

Equations 9.19 and 9.20 can also be used to determine the confidence interval bounds: F(z) (here with $F(c_1)$ and $F(c_2)$ directly related to the confidence interval limits) is the distribution function of the χ^2-distribution with degrees of freedom f (Eq. 9.21); cf. quantiles of the F-distribution in the appendix, Table A.5.

$$F(c_1) = \frac{1}{2}(1-\gamma) \tag{9.19}$$

$$F(c_2) = \frac{1}{2}(1+\gamma) \tag{9.20}$$

$$f = n-1 \tag{9.21}$$

Step 3: Estimation of the variance on the basis of the available sample (cf. Eq. 9.22).

$$s^2 = \frac{1}{n-1} \cdot \sum_{i=1}^{n} (x_i - \bar{x})^2 \qquad (9.22)$$

Step 4: Determine the confidence interval based on the available sample using Eq. 9.23. The length d of the confidence interval is obtained from the magnitude of the difference of the interval limits or by applying Eq. 9.24.

$$\frac{(n-1)s^2}{c_2} \leq \sigma^2 \leq \frac{(n-1)s^2}{c_1} \qquad (9.23)$$

$$d = \frac{(n-1)(c_2 - c_1)s^2}{c_1 c_2} \qquad (9.24)$$

Step 5: Interpretation of the result.
The true value of the previously estimated variance σ^2 will lie within the calculated confidence interval at a confidence level γ based on the available knowledge (realization in the form of the sample with size n). The probability of error α indicates the probability of this not being true. Since a two-sided confidence interval was calculated here, the probability of error α/2 applies to the areas above and below the calculated limits.

Note: The calculation of a one-sided confidence interval follows the logic already presented in Sect. 9.2.1 and is illustrated by an example in Sect. 9.2.4.

9.2.4 Application Example: Confidence Interval for Mean and Variance

The bodyshell of an automobile is joined from a large number of steel profiles and sheets. In the classic production of a steel body, the focus is on the welding process: frame parts are joined with spot welds. The production of the vehicle frame in accordance with specifications is decisive for the safety and reliability of the vehicle body. Therefore, e.g. weld spots are function-critical features, so that, among other things, the reproducibility of the weld spot geometry is decisive as a characteristic feature. For this purpose, a function-critical spot weld position is analyzed for eight vehicles.

For spot welds with respect to the vehicle frame, the requirement for the minimum diameter of the spot welds is 3.5 mm, on average the spot weld diameter should be 4 mm. In addition, a **minimum diameter** of 3.9 mm is desirable here on average, since the strength of the spot weld would be the more critical case in the deviation to a smaller diameter. In addition, a maximum variance of 0.0144 mm² is required (target standard deviation). A random sample with a extent of n = 8 vehicles is taken from a pre-series production and the spot weld diameters are measured in relation to the same, functionally critical position. The results of the diameter measurement (in [mm]) are: 3.76; 3.77; 3.78; 3.83; 3.83; 4.00; 4.06; 4.17.

Procedure:

a) The analysis of the scatter of the mean value of the spot weld diameter related to the specification is performed using a normal distribution model (even if the sample size with n<30 argues against this approach).
b) The analysis of the scatter of the mean value of the spot weld diameter with respect to the specification is performed using a t-distribution model.
c) The analysis of the scatter of the variance of the spot weld diameter related to the specification is performed using the χ^2-distribution.

Solution:
The minimum requirement for the diameter of each spot weld is fulfilled: A review of the concrete measured values available shows that no spot weld diameter is below 3.5 mm. The following analyses therefore refer to the mean and the variance of the spot weld diameter.

To a) On average, the spot weld diameter shall be 4 mm. On the basis of the available sample, the confidence interval of the mean value can be calculated. A confidence level of $\gamma=0.95$ (probability of error $\alpha=0.05$) is chosen. Since the mean and standard deviation of the population are unknown, the approach to determine the confidence interval is carried out according to Sect. 9.2.2. Mean and standard deviation are estimated according to Eqs. 9.25 and 9.26. Assuming a normal distribution model, the quantile results in Eq. 9.27, from which follows the critical limit in Eq. 9.28 (standard normal distribution; cf. Appendix Table A.2a). The two-sided confidence interval (Eq. 9.29) is determined and includes the nominal value 4 mm (Eq. 9.30). Thus, the observed values do not contradict the requirement, even if the estimated mean value (Eq. 9.25) is below the requirement (the nominal value).

$$\bar{x} = \frac{1}{n} \cdot \sum_{i=1}^{n} x_i = 3.9 \text{ mm} \qquad (9.25)$$

$$s = \sqrt{\frac{1}{n-1} \cdot \sum_{i=1}^{n} (x_i - \bar{x})^2} = 0.155 \text{ mm} \qquad (9.26)$$

$$\phi(c) = \frac{1}{2}(1+\gamma) = 0.975 \qquad (9.27)$$

$$c = u_{0.975} = 1.96 \qquad (9.28)$$

$$\bar{x} - c\frac{s}{\sqrt{n}} \leq \mu \leq \bar{x} + c\frac{s}{\sqrt{n}} \qquad (9.29)$$

$$3.79 \text{ mm} \leq \mu \leq 4.01 \text{ mm} \qquad (9.30)$$

However, a **minimum spot weld diameter** of 4 mm would be desirable. The minimum requirement leads to a one-sided limitation of the confidence interval to be determined, ergo the probability of error is also only located on one side (cf. Eq. 9.31). At a confidence level of $\gamma = 0.95$ (probability of error $\alpha = 0.05$), the confidence interval is determined (Eqs. 9.32–9.34). Although in principle the observed values do not contradict the minimum requirement, since the mean value can be above 3.9 mm (minimum requirement). However, as values between 3.81 mm and 3.9 mm would also be possible, the minimum requirement is not guaranteed.

$$\bar{x} - c \frac{s}{\sqrt{n}} \leq \mu \qquad (9.31)$$

$$\phi(c) = 0.95 \qquad (9.32)$$

$$c = u_{0.95} = 1.645 \qquad (9.33)$$

$$3.81 \text{ mm} \leq \mu \qquad (9.34)$$

For b) If the sample size is small, the application of the t-distribution is useful. The calculation path is analogous to step (a) of the solution outlined above. The difference is the use of the t-distribution in determining the quantile according to Eq. 9.35 and the resulting critical limit c; Eq. 9.36 and table of quantiles of the t-distribution, see Appendix Table A.3). The two-sided confidence interval results from Eqs. 9.37 and 9.38, respectively. Also using the t-distribution, the requirement regarding the weld spot mean is fulfilled.

$$\phi(c) = 0.975 \qquad (9.35)$$

$$c = u_{0.975} = 2.365 \quad mit \ f = n - 1 = 7 \qquad (9.36)$$

$$\bar{x} - c \frac{s}{\sqrt{n}} \leq \mu \leq \bar{x} + c \frac{s}{\sqrt{n}} \qquad (9.37)$$

$$3.77 \text{ mm} \leq \mu \leq 4.03 \text{ mm} \qquad (9.38)$$

Also, the requirement for a **minimum spot weld diameter** of 3.9 mm based on the present sample would not be contradicted in principle, since the calculation of the one-sided confidence interval (confidence level of $\gamma = 0.95$; probability of error $\alpha = 0.05$) covers the minimum requirement (Eq. 9.42). The calculation procedure is analogous to step (a), again using the t-distribution; Eqs. 9.39–9.41. However, since values between 3.80 mm and 3.9 mm (minimum requirement) would also be possible, the minimum requirement is not guaranteed with certainty based on the available measured values.

$$\bar{x} - c \frac{s}{\sqrt{n}} \leq \mu \qquad (9.39)$$

$$\Phi(c) = 0.95 \qquad (9.40)$$

$$c = u_{0.95} = 1.895 \quad mit \quad f = 7 \tag{9.41}$$

$$3.80 \text{ mm} \leq \mu \tag{9.42}$$

To c) Regarding the spot weld diameters, a max. variance of 0.0144 mm² is required. Since this is a one-sided requirement, the confidence interval is designed one-sided and calculated with a confidence level of $\gamma = 0.95$ (error probability $\alpha = 0.05$); Eqs. 9.43–9.46. The critical limit is determined on the basis of the χ^2 -distribution for the degree of freedom f = 7 (Eq. 9.46 as well as Appendix Table A. 5). The confidence interval also includes the maximum claim, so based on the available observed values the claim cannot be confirmed: The variance could also be above the maximum requirement.

$$F(c_1) = (1 - \gamma) = 0.05 \tag{9.43}$$

$$f = n - 1 = 7 \to c_1 = 2.17 \tag{9.44}$$

$$\sigma^2 \leq \frac{(n-1)s^2}{c_1} \tag{9.45}$$

$$\sigma^2 \leq 0.0775 \text{ mm}^2 \tag{9.46}$$

If the requirement were not a maximum requirement but a target requirement on the variance that had to be achieved, the confidence interval would be constructed two-sided. The calculation is outlined in Eqs. 9.47–9.51. The target value $\sigma^2 = 0.0144$ mm² lies within the calculated confidence interval (Eq. 9.52), so that such a requirement would be fulfilled or could not be contradicted on the basis of the available observed values.

$$F(c_1) = \frac{1}{2}(1 - \gamma) = 0.025 \tag{9.47}$$

$$F(c_2) = \frac{1}{2}(1 + \gamma) = 0.975 \tag{9.48}$$

$$f = 7 \to c_1 = 1.69 \tag{9.49}$$

$$f = 7 \to c_{s2} = 16.01 \tag{9.50}$$

$$\frac{(n-1)s^2}{c_2} \leq \sigma^2 \leq \frac{(n-1)s^2}{c_1} \tag{9.51}$$

$$0.0105 \text{ mm}^2 \leq \sigma^2 \leq 0.0995 \text{ mm}^2 \tag{9.52}$$

9.3 Confidence Interval for Proportion Value with Binomially Distributed Population

First, an approximation procedure for the determination of a confidence interval for a proportion value related to a sample from a binomially distributed population is explained. The procedure is based on an approximation by means of a normal distribution and is also called standard interval or Wald interval, Sect. 9.3.1. Subsequently, an application example is shown in Sect. 9.3.2. Section 9.3.3 shows the clearly more exact procedure for determining the confidence interval limits according to Clopper and Pearson (1934).

9.3.1 Approximation via Normal Distribution

If the analysis of samples focuses on the proportion of characteristic carriers, it is advisable to calculate the confidence interval in relation to the proportion value p. In the context of this chapter, the starting point is a population which follows the binomial distribution (cf. Sect. 7.4.2). Accordingly, the analysis of a sample from an infinite population is examined with regard to the expression of a characteristic ("characteristic fulfilled" versus "characteristic not fulfilled"). The confidence interval for the unknown proportion value p represents the interval in which the proportion value p will lie at a certain confidence level γ (cf. Fig. 9.7). An example of this would be the examination of cast components for possible defects (blowholes, pores, cracks, etc.), so that the decision alternatives are: cast component usable or not usable.
Baseline:

1. The population of the random variable X under consideration is binomially distributed.
2. A sample with size n and the measured values $x_1, x_2, \ldots, x_n$ is available.
3. The proportion value p of the characteristic carriers (a certain event A has occurred) in the population is unknown.
4. The parameter p is estimated based on the available sample with sample size n; number of feature carriers k, Eq. 9.53. The estimate is the probability of event A occurring.

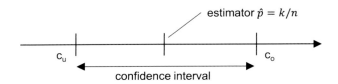

Fig. 9.7 Schematic diagram of the confidence interval of a proportion value p

Fig. 9.8 Confidence interval with the limits −c and c of the standardized random variable U ps as well as confidence level γ and probability of error α

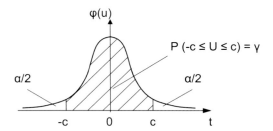

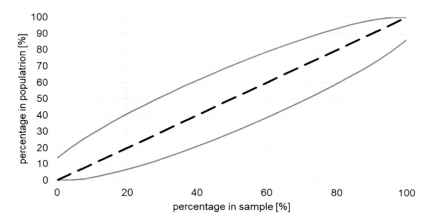

Fig. 9.9 Clopper-Pearson diagram for visualizing the confidence interval limits with respect to a proportion value p in a binomially distributed population; sample size n = 25

$$\widehat{p} = \frac{k}{n} \qquad (9.53)$$

Approximation method for determining a confidence interval for proportion value p

The approximation method by means of a normal distribution is also called standard interval or Wald interval named after Abraham Wald (1902–1950).

The underlying assumption is that the random variable X under consideration is approximately normally distributed; therefore, Eqs. 9.54 and 9.55 apply to the expected value and variance. For this, the proportion of feature carriers should satisfy the rule of thumb $k \geq 50$ and the sample size at hand should satisfy the rule of thumb $n - k \geq 50$. Similarly, the proportion of objects with present character should not approximate the sample size itself. This arises from the fact that the confidence interval in the above-mentioned ranges is asymmetrical with respect to the proportion of objects with present character and thus the scattering behavior cannot be represented by a normal distribution (cf. also Clopper-Pearson diagram, Fig. 9.9). The assessment of whether a sample is sufficiently large can also be

9.3 Confidence Interval for Proportion Value with Binomially ...

checked according to the rule of thumb in Eq. 9.56. The standardization of the random variable X is then carried out via Eq. 9.57 and leads to the standardized random variable U. The estimation function for the unknown parameter p is presented in Eq. 9.58 and leads to Eq. 9.59 after transformation and insertion in Eq. 9.57.

$$E(x) = \mu = n \cdot p \tag{9.54}$$

$$Var(x) = \sigma^2 = np(1-p) \tag{9.55}$$

$$n\widehat{p}(1-\widehat{p}) > 9 \tag{9.56}$$

$$U = \frac{X-\mu}{\sigma} = \frac{X-np}{\sqrt{np(1-p)}} \tag{9.57}$$

$$\widehat{P} = \frac{X}{n} \tag{9.58}$$

$$U = \frac{X-np}{\sqrt{np(1-p)}} = \frac{n\widehat{P}-np}{\sqrt{np(1-p)}} \tag{9.59}$$

The limits of the confidence interval at a confidence level γ can be determined for the standardized random variable U according to Eq. 9.60. Substitution of the standardized random variable U (Eq. 9.59) yields Eqs. 9.61 and 9.62, respectively, by transformation.

$$P(-c \leq U \leq c) = \gamma = 1 - \alpha \tag{9.60}$$

$$-c \leq U \leq c \Leftrightarrow -c \leq \frac{n\widehat{P}-np}{\sqrt{np(1-p)}} \leq c \tag{9.61}$$

$$\widehat{P} - \frac{c}{n}\left(\sqrt{np(1-p)}\right) \leq p \leq \widehat{P} + \frac{c}{n}\sqrt{np(1-p)} \tag{9.62}$$

Replacing the estimator $\widehat{P}$ by the estimator of the proportion value p (Eq. 9.53) on the basis of a realization via the available sample with size n to determine the limits of the confidence interval related to the proportion value p yields Eq. 9.63.

$$\widehat{p} - \frac{c}{n}\sqrt{n\widehat{p}(1-\widehat{p})} \leq p \leq \widehat{p} + \frac{c}{n}\sqrt{n\widehat{p}(1-\widehat{p})} \tag{9.63}$$

Stepwise procedure for the approximate calculation of a confidence interval related to a proportion value p within a sample from a binomially distributed population.

The starting point is a random variable X of a binomially distributed population. There is a sample with size n and the measured values $x_1, x_2, ..., x_n$, the sample size is sufficiently large in relation to the proportion of characteristic carriers (e.g. the rules of thumb $k \geq 50$ and $n - k \geq 50$ as well as Eq. 9.56 are fulfilled).

Step 1: Choice of confidence level γ.
Examples: γ=0.9 (α=0.1), γ=0.95 (α=0.05), or γ=0.99 (α=0.01).
Step 2: Transformation of the random variable X to the standardized random variable U (cf. Eq. 9.64), which follows the standard normal distribution and calculation of the limits of the confidence interval with respect to the normally distributed random variable U (Eq. 9.65), cf. also Fig. 9.8.

$$U = \frac{n\widehat{P} - np}{\sqrt{np(1-p)}} \qquad (9.64)$$

$$P(-c \leq U \leq c) = \gamma \qquad (9.65)$$

Step 3: Calculation of the estimated value for the proportion value p of the characteristic carriers on the basis of the available sample (Eq. 9.66).

$$p \approx \widehat{p} = \frac{k}{n} \qquad (9.66)$$

Step 4: Calculate the confidence interval for the unknown proportion value p (Eq. 9.67) and determine the length d of the confidence interval (Eq. 9.68). The length d can also be determined by the difference of the limits of the confidence interval.

$$\widehat{p} - \frac{c}{n}\sqrt{n\widehat{p}(1-\widehat{p})} \leq p \leq \widehat{p} + \frac{c}{n}\sqrt{n\widehat{p}(1-\widehat{p})} \qquad (9.67)$$

$$d = 2\frac{c}{n}\sqrt{n\widehat{p}(1-\widehat{p})} \qquad (9.68)$$

Step 5: Interpretation of the result.
The proportion value will lie within the calculated confidence interval at a confidence level γ based on the available knowledge, sample with size n. The probability of error α indicates the probability that this will not be the case. The probability of error α indicates the probability of this not being true. Since a two-sided confidence interval was calculated here, the proportion value could lie above or below the calculated limits with probability α/2 due to the error probability.

The determination of a one-sided confidence interval follows the logic already outlined in Sect. 9.2.1 for the unknown mean of a normally distributed population.

9.3.2 Application Example: Confidence Interval for a Proportion Value p

A housing cover is manufactured in a casting process and then the sealing surface is produced by machining. The sealing surface is part of a sealing complex in the assembly, the reliability of which must be ensured in operation. However, damage cannot be ruled out during the production of the cover due to a high cycle time. A

9.3 Confidence Interval for Proportion Value with Binomially ...

reject rate of $p_V = 1.5\%$ is assumed. A sample of 201 prototypes is tested as a basis for planning an inline inspection. The testing of the sample results in 191 defect-free parts. The proportion of defect components of the sample is analyzed under consideration of scatter effects with regard to the assumed defect share.

Procedure:
The assumed reject rate $p_V = 1.5\%$ is relatively close to the reject rate p_T (Eq. 9.69), which was observed during the testing of the prototypes, explicitly against the background of the manageable testing sample size. As a result, the confidence interval of the error rate p_T is determined for different confidence levels ($\gamma = 0.95$ and $\gamma = 0.99$).

Solution:
First, the proportion of failures p_T of the population is estimated on the basis of the testing of $n = 201$ parts as well as the detected $k = 10$ faulty parts, and the complement is determined; Eqs. 9.69 and 9.70. The assessment of whether the sample is sufficiently large to be able to perform an approximation using a normal distribution in the following is carried out according to Eq. 9.71: The condition is fulfilled.

$$\widehat{p}_T = \frac{k}{n} = 0.0498 \qquad (9.69)$$

$$q = 1 - \widehat{p}_T = 0.9502 \qquad (9.70)$$

$$n\widehat{p}_T(1 - \widehat{p}_T) = 9.5 > 9 \qquad (9.71)$$

The confidence interval is first determined for a confidence level $\gamma = 0.95$.

The confidence interval for the transformed random variable U is determined using Eq. 9.72; transforming leads via Eqs. 9.73 to 9.74 and thus subsequently to the sought quantile of the standard normal distribution. Using the tables for the standard normal distribution (Appendix Table A.2), the critical limit c is determined; Eq. 9.75. This allows the confidence interval to be determined using Eq. 9.76 for the proportion value based on the present realization; Eq. 9.77.

$$P(-c \leq U \leq c) = \gamma = 0.95 \qquad (9.72)$$

$$P(-c \leq U \leq c) = \phi(c) - \phi(-c) = \phi(c) - (1 - \phi(c)) = 2\phi(c) - 1 = 0.95 \qquad (9.73)$$

$$\phi(c) = 0.975 \qquad (9.74)$$

$$c = u_{0.975} = 1.960 \qquad (9.75)$$

$$\widehat{p} - \frac{c}{n}\sqrt{n\widehat{p}(1-\widehat{p})} \leq p \leq \widehat{p} + \frac{c}{n}\sqrt{n\widehat{p}(1-\widehat{p})} \qquad (9.76)$$

$$0.0197 \leq p \leq 0.079 \qquad (9.77)$$

The confidence interval determined does not cover the presumed proportion value $p_V = 0.015$. Accordingly, the result of the testing would initially indicate a significantly higher error proportion compared to the made assumption, but with a probability of error of $\alpha = 0.05$.

In a further step, the confidence interval for a confidence level $\gamma = 0.99$ is determined.

The approach Eq. 9.78 leads to the quantile of the standard normal distribution according to Eq. 9.79 and to the critical limit c according to Eq. 9.80. Thus, based on Eq. 9.76, the confidence interval is obtained (Eq. 9.81).

$$P(-c \leq U \leq c) = \gamma = 0.99 \tag{9.78}$$

$$\phi(c) = 0.995 \tag{9.79}$$

$$c = u_{0.975} = 2.5758 \tag{9.80}$$

$$0.0103 \leq p \leq 0.0893 \tag{9.81}$$

With the choice of the confidence level of $\gamma = 0.99$, the determined confidence interval covers the presumed proportion value $p_V = 0.015$. Under this condition, the previously made assumption cannot be refuted due to the testing (error probability $\alpha = 0.01$) – in contrast to the implementation outlined above with $\gamma = 0.95$ ($\alpha = 0.05$).

Conclusion: When analysing scatter effects by means of confidence intervals, the derived statement must always be placed in the overall context of sample size, confidence level and probability of error. The length of the confidence interval is not linear in relation to the selected confidence level (or the probability of error). This can lead to different significance statements (at different confidence levels) for the same facts.

9.3.3 Approaches Based on Clopper-Pearson and Fisher

The Clopper-Pearson approach (1934) is a method for determining the confidence interval limits for a proportion value related to a sample from a binomially distributed population. It is considered to be the most accurate method possible. However, since the binomial distribution model is a non-continuous distribution model, no defined interval can completely cover exactly a proportion of the population, yet it is often referred to as an exact procedure in the literature.

The two-sided confidence interval for an unknown proportion value p is given by Clopper-Pearson according to Eq. 9.82. In addition to the beta distribution model, the parameters hit number x (objects with marker expression), sample size n and the estimator of the proportion value $p = x/n$ are required. A visualization for representative sample sizes is shown in Fig. 9.9 or Appendix Table A.10.

$$Bt\left(\frac{\alpha}{2}; x; n-x+1\right) \leq p \leq Bt\left(1 - \frac{\alpha}{2}; x+1; n-x\right) \tag{9.82}$$

9.3 Confidence Interval for Proportion Value with Binomially ...

The beta distribution model Bt(p;q)
The density function of the beta distribution model with parameters p and q is shown in Eq. 9.83. The density function holds for the interval $0 \leq x \leq 1$ as well as p>0 and q>0. Outside this interval, f(x)=0. The inverse of the beta function within the density function is used for normalization; the beta function is shown in Eq. 9.84 and is based on the gamma function $\Gamma(n)$ (cf. Sect. 7.2.10). The distribution function of the beta distribution model for the interval $0 \leq x \leq 1$ is shown by Eq. 9.85. For $x \leq 0$, F(x)=0 and $x \geq 1$, F(x)=1. The beta binomial distribution unlike the simple binomial distribution, does not assume that the probabilities of the events occurring are equal. Rather, the probabilities of the events scatter around a certain value.

$$f(x) = \frac{1}{B(p,q)} x^{p-1}(1-x)^{q-1} \text{ ; for } 0 \leq x \leq 1 \text{ and } p,q > 0 \quad (9.83)$$

$$B(p,q) = \int_0^1 t^{p-1}(1-t)^{q-1} dt = \frac{\Gamma(p)\Gamma(q)}{\Gamma(p+q)} \quad (9.84)$$

$$F(x) = \frac{1}{B(p,q)} \int_0^x t^{p-1}(1-t)^{q-1} dt \text{ ; for } 0 \leq x \leq 1 \text{ and } p,q > 0 \quad (9.85)$$

The confidence interval can also be determined with the help of the F-distribution according to Fisher (cf. 7.3.3), Eq. 9.86.

$$\frac{x}{x + (n-x+1)F_{f_1=2(n-x+1);f_2=2x;\alpha/2}} \leq p \leq \frac{(x+1)F_{f_1=2(x+1);f_2=2(n-x);\alpha/2}}{n - x + (x+1)F_{f_1=2(x+1);f_2=2(n-x);\alpha/2}} \quad (9.86)$$

As already outlined in Sect. 9.3.1, the confidence interval for the unknown parameter p can also be determined approximately on the basis of the normal distribution model. Explicitly at the edges, however, when the proportion value p is approximately or directly 0 or approximately or directly 1, the approximation by means of normal distribution is extremely inaccurate, cf. also Fig. 9.9.

Confidence interval for margins: Estimated proportion values p=0 or p=1.
For an estimate of the proportion value p=x/n=0, Eq. 9.86 directly gives the (one-sided) confidence interval limit with confidence level $\gamma = 1 - \alpha$ to Eq. 9.87.

$$p \leq \frac{F_{f_1=2;f_2=2n;1-\alpha}}{n + F_{f_1=2;f_2=2n;1-\alpha}} \quad (9.87)$$

For an estimate of the proportion value p=x/n=1, Eq. 9.86 directly gives the (one-sided) confidence interval limit to Eq. 9.88.

$$\frac{n}{n + F_{f_1=2;f_2=2n;1-\alpha}} \leq p \quad (9.88)$$

Approximation approach for confidence interval margins according to Hanley and Lippmann-Hand (1983)
For large sample sizes with n>50, the simple determination equations according to Hanley and Lippmann-Hand (1983) can be used. For an estimate of the proportion value at 0, the confidence interval limit is given by Eq. 9.89. For an estimate of the proportion value at 1, the confidence interval limit can be estimated by using Eq. 9.90.

$$p \leq \frac{3}{n} \qquad (9.89)$$

$$1 - \frac{3}{n} \leq p \qquad (9.90)$$

Visualization of the confidence interval according to clopper-pearson
Figure 9.9 shows the visualization of the confidence interval for different proportions of characteristic carriers in the sample (abscissa) as well as in the population (ordinate) using the sample size n=25 as an example. The confidence intervals were determined according to the Clopper-Pearson approach; Eq. 9.82. The dashed line shows the proportion value, the two solid lines show the upper and lower confidence interval limits, respectively. The asymmetries of the confidence interval with respect to the feature proportion are clearly visible in the marginal areas (small or large feature proportions). In these marginal areas, the approximation method based on the normal distribution should certainly not be used; cf. explanation in Sect. 9.3.1. Furthermore, the confidence intervals for the margins for proportion values p=0 and p=1 are also clearly visible. Still, the appendix Table A.10 shows the respective confidence intervals according to Clopper-Pearson for representative sample sizes.

Example: Comparison of Confidence Interval Length for p=0; n=25

Even though the sample size of n=25 is clearly below the precondition (n>50) for an approximate solution according to Hanley and Lippmann-Hand (1983), the confidence interval lengths are determined in the following based on the two outlined approaches; Eqs. 9.91 and 9.92. The results of course show a deviation (cf. not fulfilled precondition), but are relatively close to each other and illustrate the link, which can be verified with Fig. 9.9.

Clopper-Pearson (1934) approach:

$$p \leq 0.1372 \qquad (9.91)$$

Hanley and Lippmann-Hand (1983) approach:

$$p \leq 0.12 \qquad (9.92)$$

9.4 Weibull Distribution Model: Confidence Intervals of Parameters

In the context of technical reliability analysis, the two- or three-parameter Weibull distribution model is often used; see Sect. 7.2.7. Therefore, special approaches for the determination of confidence intervals with reference to the parameters of the Weibull distribution model are presented in this chapter. Approaches to determine the confidence interval of the shape parameter (Sect. 9.4.1), the characteristic lifetime (Sect. 9.4.2) and the threshold parameter (Sect. 9.4.3) are explained.

9.4.1 Shape Parameter

The two-sided confidence interval for the shape parameter b of a Weibull distribution model can be approximated by Eq. 9.93 using the sample size n and the standard normal distribution, cf. (Graebig 2006). The limits related to the standard normal distribution are determined via the error probability α; e.g. for $\alpha = 0.05$ ($\gamma = 0.95$) the limit $u = 1.960$ can be determined from the table of quantiles of the standard normal distribution, see Appendix Table A.2.

$$\widehat{b} - u_{1-\frac{\alpha}{2}} \frac{0.78 \cdot \widehat{b}}{\sqrt{n}} \leq b \leq \widehat{b} + u_{1-\frac{\alpha}{2}} \frac{0.78 \cdot \widehat{b}}{\sqrt{n}} \qquad (9.93)$$

The approach according to (VDA 1976) is more exact, Eq. 9.94: In addition to the sample size n, the confidence level is expressed via the parameter k (cf. Table 9.1).

$$\widehat{b} \frac{1}{1+\sqrt{\frac{k}{n}}} \leq b \leq \widehat{b}\left(1 + \sqrt{\frac{k}{n}}\right) \qquad (9.94)$$

9.4.2 Characteristic Life Span

The two-sided confidence interval for the location parameter T (characteristic life span) of a Weibull distribution model can be approximated on the basis of the standard normal distribution (Graebig 2006). Equation 9.95 shows the approximation using the sample size n as well as the shape parameter b of the Weibull distribution model, which has to be estimated beforehand. The limits related to the standard normal distribution are determined via the error probability α, e.g. for

Table 9.1 Parameter k as a function of the selected confidence level γ; cf. Eq. 9.94

Parameter k	Confidence level γ
1.4	0.90
2.0	0.95
3.4	0.99

$\alpha = 0.05$ the limit $u = 1.960$ can be determined from the table of quantiles of the standard normal distribution, see Appendix Table A.2.

$$\widehat{T} - u_{1-\frac{\alpha}{2}} \frac{1.052 \cdot \widehat{T}}{\widehat{b}\sqrt{n}} \leq T \leq \widehat{T} + u_{1-\frac{\alpha}{2}} \frac{1.052 \cdot \widehat{T}}{\widehat{b}\sqrt{n}} \qquad (9.95)$$

A more precise estimate is provided by the approach according to Bitter (1986), Eq. 9.96: The approach is based on the χ^2-distribution (Sect. 7.3.2). The quantiles of the χ^2-distribution related to the confidence level can be taken from Appendix Table A.5. Furthermore, in addition to the sample size n, a prior estimate of the shape parameter b of the Weibull distribution model is required.

$$\widehat{T} \left(\frac{2n}{\chi^2_{2n,1-\frac{\alpha}{2}}} \right)^{\frac{1}{b}} \leq T \leq \widehat{T} \left(\frac{2n}{\chi^2_{2n,\frac{\alpha}{2}}} \right)^{\frac{1}{b}} \qquad (9.96)$$

9.4.3 Threshold (Failure-Free Time)

The confidence interval with respect to the threshold parameter (failure-free time) t_0 can be calculated with the approach of Mann and Fertig (1975). Here, the time of first failure is the upper natural limit of the confidence level, and the lower limit of the confidence level is the focus of Mann and Fertig's approach, derived from a Weibull goodness-of-fit test (three-parameter). The estimation is shown in Eqs. 9.97 and 9.98, respectively.

The calculation of the confidence limit λ is based on the leaps of the available measured values. Two sums are set in relation, which comprise different numbers of values depending on the sample size n and the probability of error α. For this purpose, a further parameter k is introduced, which is tabulated in Mann and Fertig (1975). The quotient of the two sums is equated as the test variable $P^*_{k,m}$ with the percentile of the distribution of Mann et al. (1971), cf. Eq. 9.97. The jumps l^*_i (cf. Eq. 9.98) result from the available measured values (times of the failures T) and the lower limit of the confidence interval λ that is sought. Moreover, the difference of the expected values E enters into another quantity, the reduced extreme value statistic Y, which depends on the shape parameter and the characteristic lifetime and is also tabulated in Mann et al. (1971). The concrete calculation of the confidence limit is done numerically by iterating this confidence limit to optimize the test variable with respect to the specified percentile value.

$$P^*_{k,m}(\lambda) = \sum_{i=k+1}^{m-1} l^*_i \bigg/ \sum_{i=1}^{m-1} l^*_i \qquad (9.97)$$

$$l^*_i = \frac{\ln(D_{i+1,n} - \lambda) - \ln(D_{i,n} - \lambda)}{E(Y_{i+1,n}) - E(Y_{i,n})} \qquad (9.98)$$

9.4 Weibull Distribution Model: Confidence Intervals of Parameters

Parameters:

P = percentile of the distribution of Mann et al. (1971).
λ = lower limit of the confidence level $\rightarrow$ searched for.
m = maximum rank (for censored data, otherwise m = n).
n = sample size.
k = parameter, depending on n and α, tabulated according to Mann and Fertig (1975).
i = rank (running index).
l = jumps (cf. Eq. (9.98)).
D = Time of failure (Damage).
E = expected value, difference also tabulated in Mann et al. (1971).
Y = reduced extreme value statistics, depending on shape parameter and characteristic life span, tabulated in Mann et al. (1971).

Note: Given is a data set consisting of the damage data of a failure focus (defined damage causality) at a time t_1 related to a product fleet in field use. In the context of a data analysis, a three-parameter Weibull distribution model is fitted to the damage data. Using the approach of Mann and Fertig (1975), the confidence interval to the estimated parameter threshold t_0 can be determined. However, a expanded scattering of the parameter t_0 depending on the increase in knowledge (further damage data with progression of the field observation time at time t_2) is to be expected. This scatter effect is clearly more dominant compared to the scatter that results from model fitting (and is represented by the Mann and Fertig (1975) confidence interval). This scatter effect, which results from the increase in knowledge, is discussed in detail in (Bracke and Puls 2020), (Bracke 2020).

9.5 Confidence Interval for a Function (Distribution Model)

For a mapping of the failure behaviour of defective components on the basis of damage data, a distribution model is adapted, e.g. Weibull distribution, normal distribution or exponential distribution. Ultimately, this is a balancing function which is fitted to the estimated, summed failure probabilities (cf. in detail Chap. 12). The possible dispersion of the fitted model – the adjustment function – can also be represented by a confidence interval. The confidence interval shows the range in which the model lies with a defined confidence level. Figure 9.10 shows an example of the failure behavior of a coolant pump (see Chap. 12; damage data in Table 12.2), mapped with an adapted Weibull distribution model as well as the limits of the confidence interval related to the model (green boundary).

The beta binomial distribution is used to determine the confidence interval (see also Sect. 9.3.3). The basic idea is that at each point in time there is a quasi random sample with respect to the runtime variable t. Figure 9.11 shows the principle

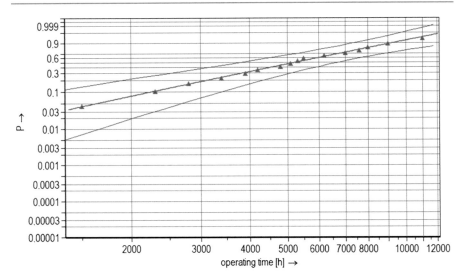

Fig. 9.10 Failure behaviour of a coolant pump (data set: Chap. 12; Table 12.2); cumulative failure probabilities (▲), fitted Weibull distribution model and confidence interval (green boundary); double logarithmic scale

on the basis of Fig. 9.10 and the underlying data set, extended by a third axis. The dispersion of the probability of failure at time t_i is described by the beta-binomial distribution according to Eq. 9.99. The sample size n, the rank number i (related to the sorted damage data; cf. Chap. 12) as well as the probability of failure F_i at rank i (resp. time i) are required. As a consequence, e.g. for a confidence level of $\gamma = 0.95$ for the determination of the upper confidence limit Eq. 9.100 and for the lower confidence limit Eq. 9.101 result. Obviously, the equations cannot be solved manually in a trivial way, since the conversion according to the searched probability of default F_i at time t_i is not possible. Accordingly, a numerical solution has to be chosen.

$$P_A = \sum_{k=i}^{n} \frac{n!}{k!(n-k)!} F^k (1-F)^{n-k} \qquad (9.99)$$

$$0.975 = \sum_{k=i}^{n} \frac{n!}{k!(n-k)!} F^k (1-F)^{n-k} \qquad (9.100)$$

$$0.025 = \sum_{k=i}^{n} \frac{n!}{k!(n-k)!} F^k (1-F)^{n-k} \qquad (9.101)$$

An alternative to the approach based on the beta binomial distribution is offered by the approach according to (Hedderich and Sachs 2020:386). On the basis of

9.5 Confidence Interval for a Function (Distribution Model)

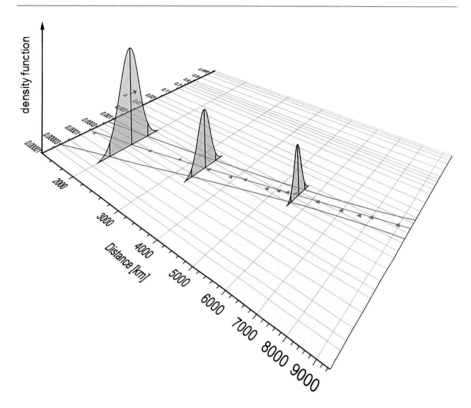

Fig. 9.11 Failure behaviour of a coolant pump (data set: Chap. 12; Table 12.2); cumulative failure probabilities (▲), adjusted Weibull distribution model, confidence interval (cf. also Fig. 9.10) and exemplary scattering models at three points in time t_i

the F-distribution (cf. Sect. 7.3.3), the limits of the confidence interval are determined, whereby sample size n, the rank number i (related to the sorted damage data; cf. Chap. 12) and the probability P related to the confidence interval (e.g. P = 0.95) are required. This approach can be solved more easily with the aid of a set of tables for the quantiles of the F-distribution (Appendix Table A.4) compared to the approach based on the beta-binomial distribution. Equation 9.102 shows the approach for determining the upper confidence interval limit (ul = upper limit) $CI_{i,\text{ul}}$ with respect to the rank number i. Eq. 9.103 accordingly shows the determination of the lower confidence interval limit (ll = lower limit) $CI_{i,\text{ll}}$ with respect to the rank number i.

$$CI_{i,\text{ul}} = 1 - \frac{1}{1 + \frac{i}{n-i+1} F_{2i,2(n-i+1),\alpha/2}} \qquad (9.102)$$

$$CI_{i,\text{ll}} = \frac{1}{\frac{n-i+1}{i} F_{2(n-i+1),2i,1-\alpha/2} + 1} \qquad (9.103)$$

Correlation and Regression 10

In technical reliability analysis, many studies focus on the dependence or correlation of various function- or safety–critical characteristics on each other: correlation analysis and the regression analysis that often goes hand in hand with it can be used to map the relationship between two or more variables. Section 10.1 shows the basics and outlines some fields of application. Subsequently, Sect. 10.2 presents bivariate correlation analysis in detail. Section 10.3 is devoted to bivariate regression analysis for the determination of a functional relationship. The application of correlation and regression analysis within various case studies as well as the interpretation of the results are shown in Sects. 10.4 and 10.5 (apparent or pseudo causality and correlation).

10.1 Basics of Correlation and Regression

When evaluating technical correlations, the analysis of the dependency or the connection (or relationship), the so-called correlation of different characteristics to each other, is of high interest. If a correlation can be shown, for example, between two variables, then there could also be a causal relationship between these variables. The examples briefly outlined in this chapter show a correlation between the variables when the corresponding series of measurements are evaluated. In all examples, a technical causal relationship also exists. Nevertheless, it may also occur that a correlation between two or more variables can be demonstrated, but that there is certainly no (technical) causal relationship; the detailed explanation with example can be found in Sect. 10.5.

The measure of the dependence of characteristics can be calculated via the correlation coefficient r and the coefficient of determination r^2. The focus of the *correlation analysis* can be, for example, product characteristics, production process parameters and environmental parameters. The dependency can be examined on the data basis of series of measured values or observed values in any combination.

It should be noted that the simple methods presented here involve the expression of a linear correlation.

In a *regression analysis* a functional relationship between two or more variables is estimated. For example, the functional relationship can be used to predict values of the dependent variable(s). Furthermore, the functional relationship can be used to interpret characteristics of the technical relationship. In industrial practice, simple linear regression analysis is often carried out first, but of course other function types can also be adapted. The following examples illustrate typical fields of application of correlation and regression analysis:

Example 1: The dependence of two product characteristics.
For the evaluation of the surface topography of a ground surface, the dependence of roughness (roughness parameter Rz or Ra) and gloss (gloss value GU) on each other is investigated.

Example 2: The dependence of product characteristics and manufacturing process parameters.
In a grinding process, the dependence between roughness (roughness parameter Rz or Ra) and feed (v) or contact force (F) of the grinding wheel on the workpiece is examined.

Example 3: The dependence of manufacturing process parameters and workpiece.
In a grinding process, the dependence between feed (v) and workpiece temperature (T) is investigated.

Example 4: The dependence of product characteristics and environmental parameters.
For a vehicle, the dependence between fuel consumption (V [l/km]) and ambient temperature [T] is investigated.

The analysis of the feature correlation can be carried out by means of parametric as well as parameter-free procedures. Most frequently, linear dependence is analyzed in the first step. Before starting the actual correlation analysis, the underlying measured values (observed values) should be examined for their distribution. If the data are normally distributed, the procedure according to Bravais-Pearson (cf. Sect. 10.2.1) can be carried out. If the data are not normally distributed, a parameter-free procedure can be chosen, which does not explicitly require a distribution model (cf. Sect. 10.2.2; Spearman approach); also (Bracke 2016). The choice of the appropriate correlation coefficient can thus be made after performing a statistical significance test with regard to a distribution model (cf. for example Kolmogorov–Smirnov goodness-of-fit test or χ^2 goodness-of-fit test; Chap. 13).

10.2 Correlation Analysis

In the following sections, methods of bivariate correlation analysis are discussed. The starting point is a random experiment in which a series of measured values (sample with size n) is documented. Two variables X and Y are observed, ergo the pairs of measured values $(x_1; y_1), (x_2; y_2),\ldots, (x_n; y_n)$ are documented. The visualization with plotted pair of measured values (x_i, y_i) is shown in Fig. 10.1.

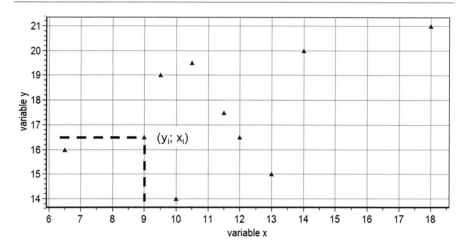

Fig. 10.1 Visualization of pairs of measured values in the X–Y plot

To determine a correlation coefficient using the Bravais-Pearson method, the arithmetic mean and variance (cf. Sect. 6.4) of the sample must be estimated on the basis of the original values of both random variables X and Y. In the Spearman procedure, these characteristic values are to be determined on the basis of the ranks of the sorted measured values after rank assignment. The empirical covariance is calculated by including the arithmetic mean values according to Eq. 10.1. The empirical *correlation coefficient r* results from the *empirical covariance* and the individual variances $s_x \neq 0$ and $s_y \neq 0$ according to Eq. 10.2. The coefficient of determination r^2 is shown in Eq. 10.3. The *coefficient of determination* indicates the influence of the random variable X on the dispersion of the random variable Y based on linear regression.

$$s_{xy} = \frac{1}{n-1}\left(\sum_{i=1}^{n} x_i y_i - n\overline{xy}\right) \tag{10.1}$$

$$r = \frac{s_{xy}}{s_x \cdot s_y} \tag{10.2}$$

$$r^2 = \frac{s_{xy}^2}{s_x^2 \cdot s_y^2} \tag{10.3}$$

Notes on the interpretation of the correlation coefficient:

(a) The correlation coefficient r can take values between -1 and 1.
(b) r = 1; there is a complete, positive (linear) correlation. When determining a regression grade (linear regression analysis, normally distributed variables X

and Y assumed, Bravais-Pearson coefficient), all points would lie on a straight line whose gradient is positive.
(c) $r=-1$; there is a complete, negative (linear) correlation. When determining a regression grade (linear regression analysis, normally distributed variables X and Y assumed, Bravais-Pearson coefficient), all points would lie on a straight line whose gradient is negative.
(d) Calculating a linear regression model as a supplementary analysis to a rank-based correlation analysis (e.g. Spearman) is not meaningful, since the correlation coefficient would refer to the ranks, but the regression model would refer to the measured values. Usually a linear regression model is calculated if the correlation analysis was done by parametric methods.
(e) $r=0$; there is no (linear) correlation. The determination of a linear regression model (linear regression analysis) is not meaningful with regard to a functional correlation.
(f) When assessing correlations in engineering, Table 10.1 may serve as an orientation (rule of thumb) for the strength of the correlation of variables X and Y. Within other disciplines, such as sociology, the values or intervals of the correlation coefficient are interpreted differently with regard to their meaning. And certainly there is no mathematically justifiable derivation in relation to the interval limits. However, if there is a technically causal relationship between the variables X and Y, Table 10.1 may serve as a first orientation and classification with regard to the coefficient r or the strength of the relationship.
(g) If Table 10.1 is used for orientation, the calculation method (parametric and parameter-free) should also be included in the technical interpretation. Naturally, parameter-free correlation coefficients turn out to be lower than parametrically determined coefficients, since rank-based calculation generally involves a loss of information (cf. Sect. 14.2.2). In this respect, the calculated strength of the correlation will often be lower.

Table 10.1 Orientation for the evaluation of an engineering correlation mean correlation coefficient between two variables X and Y

Interval	Orientation for interpretation
$0 < r \leq 0.25$	Very weak correlation
$0.25 < r \leq 0.5$	Weak correlation
$0.5 < r \leq 0.75$	Medium strong correlation
$0.75 < r \leq 0.9$	Strong correlation
$0.9 < r < 1$	Very strong correlation

Note: When assessing engineering relationships, the intervals serve as a guide (rule of thumb) for the strength of the relationship of variables X and Y. The discipline and the selected method (original value-based or rank-based) must be taken into account: Compare detailed remarks in Sect. 10.2

10.2.1 Correlation According to Bravais-Pearson

The classical correlation analysis using the correlation coefficient r_{BP} according to Bravais-Pearson assumes bivariate normally distributed measurement series; cf. Eq. 10.4 or, in a similar form, Eq. 10.5. The calculation of the Bravais-Pearson correlation coefficient r_{BP} is based on the observed values (here: the two measurement series of a sample related to the random variables X and Y). The parameter n refers to the size of the series of measurements (the sample) with its pairs of measurements (x_i; y_i). For the estimators $\bar{x}$ and $\bar{y}$ are the arithmetic means of the observed values x_i and y_i.

$$r_{BP} = \frac{\sum_{i=1}^{n}(x_i - \bar{x})(y_i - \bar{y})}{\sqrt{\sum_{i=1}^{n}(x_i - \bar{x})^2 \sum_{i=1}^{n}(y_i - \bar{y})^2}} \tag{10.4}$$

$$r_{BP} = \frac{\sum_{i=1}^{n} x_i y_i - n\bar{x}\bar{y}}{\sqrt{\left(\sum_{i=1}^{n} x_i^2 - n\bar{x}^2\right)\left(\sum_{i=1}^{n} y_i^2 - n\bar{y}^2\right)}} \tag{10.5}$$

10.2.2 Correlation According to Spearman

If the observed values to be analyzed do not follow a normal distribution model, the dependence of the characteristics can be calculated via the correlation coefficient using a parameter-free, rank-based procedure. It should be noted with rank-based methods that the information on the distances between the observed values is lost, which means that the rank-based correlation coefficient is often lower than the Bravais-Pearson correlation coefficient.

An alternative to the Bravais-Pearson correlation is the rank correlation according to Spearman (1904), whose rank correlation coefficient r_{SP} (cf. Eq. 10.6 or 10.7) is a parameter-free measure of correlation (cf. Büning and Trenkler (1994), Hedderich and Sachs (2020) also Bracke (2016)). The computational procedure is identical to Bravais-Pearson, but it does not examine the observed values (x_i;y_i) of the sample with size n, but the assigned ranks R(x_i) and R(y_i), respectively. In the case that ties occur—i.e. equal ranks must be assigned based on the measured values—average ranks are used. If there are no ties (cf. also Chap. 13) between the observation values, i.e. if there are no identical observations with identical values within the respective observations x_i or y_i, Eq. 10.7 can also be used. The result is the same when using Eq. 10.6 or 10.7. In Eq. 10.7, the rank differences d_i are formed from the ranks R(x_i) and R(y_i), respectively.

The Spearman correlation coefficient does not presuppose a specific probability distribution for the observed values analyzed, since the distances of the ranks R_i of the measured values from the rank means are taken into account and not the measured values themselves. $\overline{R(x)}$ are taken into account and not the measured values themselves.

$$r_{SP} = \frac{\sum_{i=1}^{n} \left(R(x_i) - \overline{R(x)}\right)\left(R(y_i) - \overline{R(y)}\right)}{\sqrt{\sum_{i=1}^{n} \left(R(x_i) - \overline{R(x)}\right)^2 \cdot \sum_{i=1}^{n} \left(R(y_i) - \overline{R(y)}\right)^2}} \quad (10.6)$$

$$r_{SP} = 1 - \frac{6 \sum_{i=1}^{n} d_i^2}{n\left(n^2 - 1\right)} \quad (10.7)$$

Another way of determining the strength of a relationship between two random variables is provided by the correlation coefficient based on the approach of Kendall and Stuart (1967); cf. Kendall's τ in Kendall and Stuart (1967), Kendall and Gibbons (1990:260).

10.3 Regression Analysis

In a regression analysis a functional relationship between two or more variables is estimated. This chapter focuses on bivariate regression analysis.

The simplest form of regression analysis is the determination of a functional linear relationship; this case can still be solved well manually for small sample sizes. When investigating larger sample sizes or choosing non-linear models (e.g. polynomial function), the manual effort can be high and the use of appropriate software/hardware is recommended.

Figure 10.2 shows the functional relationship between two variables using two examples: First, a linear approach to describe the relationship between the variables weight [kg] and height [cm] of students (Fig. 10.2, left) is shown. On the other hand, a non-linear approach to describe the relationship between the variables temperature [°C] and water volume [m³], Fig. 10.2, right) is shown.

In the following the method of least squares according to Gauss is presented in order to calculate a regression model which represents a functional relationship. There are two central considerations:

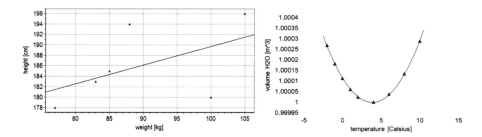

Fig. 10.2 Regression analyses on the variables weight and height of students (*left*) and the variables temperature and water volume (*right*)

1. The random difference r between measurement point y_i and best-fit curve $f(x_i)$ according to Eq. 10.8.
2. The sum S of all squared differences r^2 related to all measuring points $(x_i; y_i)$ respectively of the function value of the regression model $f(x_i)$, Eq. 10.9.

$$r = y_i - f(x_i) \tag{10.8}$$

$$S = \sum_{i=1}^{n} r_i^2 = \sum_{i=1}^{n} (y_i - f(x_i))^2 \tag{10.9}$$

Regression analysis: General procedure
The following describes a general procedure for determining an equation that outlines the functional relationship between two (or more) random variables.

Baseline:
Given is a measurement series of a random experiment with two random variables X and Y with n pairs of measurements $(x_i; y_i)$ with (i = 1,2, ..., n).

Step 0: Visualization of the measured value pairs in an X–Y plot (Sect. 4.9).
First, the available pairs of measurements $(x_i; y_i)$ are visualized in an X–Y plot with the aim of being able to make an assumption regarding the potential functional relationship, Fig. 10.3. Of course, a functional relationship can also be assumed on the basis of existing knowledge, without visualization.

Step 1: Selection of a suitable function type for the regression analysis
Based on the visualization of the pairs of measured values and/or empirically available knowledge about the investigated engineering context, a suitable

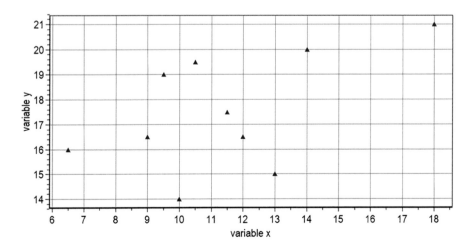

Fig. 10.3 Schematic visualization of pairs of measured values $(x_i; y_i)$ in an x–y plot

Table 10.2 Examples of functions for fitting within a regression analysis

Function	y = f(x)	Parameter
Linear function	$y = ax + b$	a, b
Quadratic function	$y = ax^2 + bx + c$	a, b, c
Exponential function	$y = a \cdot c^{bx}$	a, b, c
Exponential function to base e	$y = ae^{bx}$	a, b
Logarithm function	$y = a \cdot ln(bx)$	a, b
Polynomial function of degree n	$y = a_n x^n + a_{n-1} x^{n-1} + \cdots + a_0$	$a_0, a_1, \ldots, a_n$
Power function	$y = a \cdot x^b$	a, b
Saturation function	$y = \frac{a \cdot S}{a + (S-a) \cdot e^{-Sks}}$	a, S, k

function type for the determination of the functional relationship is selected. Table 10.2 shows examples of function types that can be used to represent simple, bivariate relationships.

Step 2: Forming the sum of the distance squares
Using the selected function type, the sum of the squared differences (distance squares) is formed (cf. Eq. 10.10). The sum depends on parameters of the previously selected function type: Ergo, the sum is a function of the parameters a, b, c, et cetera.

$$S(a; b; \ldots) = \sum_{i=1}^{n} r_i^2 = \sum_{i=1}^{n} (y_i - f(x_i))^2 \qquad (10.10)$$

Step 3: Determination of the parameters via partial derivatives
The function of the sum of squared differences is partially derived according to the parameters. For example, three parameters a, b, c result in Eqs. (10.11–10.13). With the aid of this system of equations, the same parameters can be determined. It follows inevitably that the number of observed measuring points must be greater than the number of parameters of the previously selected function type.

$$\frac{\partial S}{\partial a} = 0 \qquad (10.11)$$

$$\frac{\partial S}{\partial b} = 0 \qquad (10.12)$$

$$\frac{\partial S}{\partial c} = 0 \qquad (10.13)$$

Regression analysis: fitting a linear function
The starting point for fitting a linear function in the bivariate case is as follows:

10.3 Regression Analysis

1. There is a random experiment with variables X, Y.
2. Variable X is independent, variable Y is dependent on X (definitely also subject to scatter effects).
3. Observations are available in the form of a series of measured values with pairs of measured values $(x_i; y_i)$, where $i = 1, 2, \ldots, n$.
4. The measured value series with sample size n comprises more pairs of measured values compared to the number of parameters of the previously selected function type. Ergo, a linear function must have at least three pairs of measured values.

A linear approach is chosen as the function (cf. Table 10.2; see Eq. 10.14). According to Gauss (Eq. 10.10), Eq. 10.15 is obtained, which is to assume a minimum. For this purpose, the partial derivatives are formed according to parameters a and b (Eqs. 10.16 and 10.17). Inserting, deriving and rearranging results in Eq. 10.18 for parameter a and Eq. 10.19 for parameter b. Likewise, inserting Eq. 10.19 into Eq. 10.14 yields Eq. 10.20 and thus the manageable Eq. 10.21. Figure 10.4 shows the schematic visualization of the linear function (regression model) through the pairs of measured values. A linear regression model always runs through the center of gravity (Eq. 10.22) of the point cloud in the X–Y plot; Fig. 10.4.

$$y = ax + b \qquad (10.14)$$

$$S(a; b) = \sum_{i=1}^{n} (y_i - ax_i - b)^2 \qquad (10.15)$$

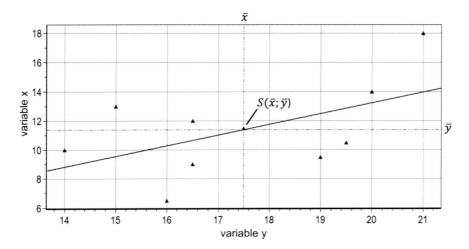

Fig. 10.4 Visualization of a regression line by pairs of measured values of the random variables X and Y with centroid S (arithmetic mean values of the measured values related to random variables X and Y)

$$\frac{\partial S}{\partial a} = 0 \qquad (10.16)$$

$$\frac{\partial S}{\partial b} = 0 \qquad (10.17)$$

$$a = \frac{n \cdot \sum_{i=1}^{n} x_i y_i - \left(\sum_{i=1}^{n} x_i\right)\left(\sum_{i=1}^{n} y_i\right)}{n \cdot \sum_{i=1}^{n} x_i^2 - \left(\sum_{i=1}^{n} x_i\right)^2} = \frac{\sum_{i=1}^{n} x_i y_i - n\overline{xy}}{\sum_{i=1}^{n} x_i^2 - n\overline{x}^2} \qquad (10.18)$$

$$b = \frac{\left(\sum_{i=1}^{n} x_i^2\right)\left(\sum_{i=1}^{n} y_i\right) - \left(\sum_{i=1}^{n} x_i\right)\left(\sum_{i=1}^{n} x_i y_i\right)}{n \cdot \sum_{i=1}^{n} x_i^2 - \left(\sum_{i=1}^{n} x_i\right)^2} = \overline{y} - a\overline{x} \qquad (10.19)$$

$$y = ax + \overline{y} - a\overline{x} \qquad (10.20)$$

$$y - \overline{y} = a(x - \overline{x}) \qquad (10.21)$$

$$S = (\overline{x}; \overline{y}) \qquad (10.22)$$

10.4 Case Studies on Correlation and Regression Analysis

This chapter presents the following case studies for conducting correlation and regression analysis, the emphasis is on the study of bivariate correlations:

1. Analysis of fuel consumption in automobiles: Test bench versus real drive operation.
2. Investigation of the dependence of mercury volume on ambient temperature.
3. Analysis of the turning machining process: Surface topography versus turning process parameters.

Example 1: Analysis of Fuel Consumption in Automobiles; Test Bench Versus Real Drive Operation

The car manufacturer H determines an average fuel consumption on a test bench for various vehicles of class XYZ on the basis of a standardised driving cycle. The result includes the consumption of fuel [litres] per 100 km. As part of an objective assessment by an independent institute I, the fuel consumption (litres/100 km) in real drive operation is measured for various vehicles of category XYZ produced by manufacturer H. Manufacturer H and institute I stating different results (see Table 10.3). The fuel consumption in real drive operation seems to be higher than under standardised test conditions.

10.4 Case Studies on Correlation ...

Table 10.3 Fuel consumption of vehicles of class XYZ (vehicle type = index i) on the basis of manufacturer specifications (test bench) and real drive operation (institute I)

Index i	1	2	3	4	5	6	7
Fuel consumption, measurement manufacturer H [l/100 km]	5	5.9	7.1	5.4	7	6.9	4.9
Fuel consumption, measurement institute I [l/100 km]	6.4	6.9	8.4	6.9	7.5	8.1	5.9

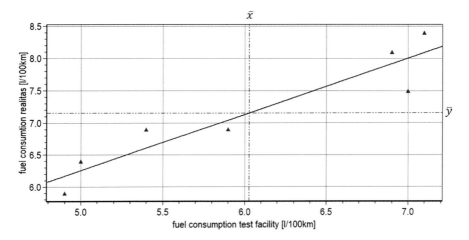

Fig. 10.5 Regression analysis vehicle fuel consumption; functional relationship between the variables fuel consumption test bench (manufacturer) and fuel consumption real operation (institute)

Questioning:
With the help of a correlation and regression analysis, the dependency as well as the functional relationship between both series of measurements will be determined.

Solution:
First, the strength of the correlation is analysed by means of the Bravais-Pearson correlation coefficient (independent series of measurements; assumption: normally distributed measured values). The mean values are calculated (Eqs. 10.23 and 10.24), then the correlation coefficient r_{BP} (Eq. 10.25) is determined. The correlation coefficient r_{BP} shows a very strong correlation (Table 10.1). A functional linear correlation is assumed (Table 10.2), the parameters a and b can be calculated directly according to Eqs. 10.26 and 10.27, so for the linear approach (Eq. 10.28), the functional correlation (Eq. 10.29) and visualization follows according to Fig. 10.5. Based on the functional relationship, the real consumption

of a comparable vehicle can now be estimated if the manufacturer's specification is known. If, for example, the manufacturer's specification for fuel consumption is 6.5 l/100 km, the real consumption should be approx. 7.57 l/100 km.

$$\bar{x} = \frac{1}{n}\sum_{i=1}^{n} x_i = 6.0286 \tag{10.23}$$

$$\bar{y} = \frac{1}{n}\sum_{i=1}^{n} y_i = 7.157 \tag{10.24}$$

$$r_{BP} = \frac{\sum_{i=1}^{n} x_i y_i - n\bar{x}\bar{y}}{\sqrt{\left(\sum_{i=1}^{n} x_i^2 - n\bar{x}^2\right)\left(\sum_{i=1}^{n} y_i^2 - n\bar{y}^2\right)}} = 0.9378 \tag{10.25}$$

$$a = \frac{n \cdot \sum_{i=1}^{n} x_i y_i - \left(\sum_{i=1}^{n} x_i\right) \cdot \left(\sum_{i=1}^{n} y_i\right)}{n \cdot \sum_{i=1}^{n} x_i^2 - \left(\sum_{i=1}^{n} x_i\right)^2} = \frac{\sum_{i=1}^{n} x_i y_i - n\bar{x}\bar{y}}{\sum_{i=1}^{n} x_i^2 - n\bar{x}^2} = 0.8721 \tag{10.26}$$

$$b = \frac{\left(\sum_{i=1}^{n} x_i^2\right)\left(\sum_{i=1}^{n} y_i\right) - \left(\sum_{i=1}^{n} x_i\right)\left(\sum_{i=1}^{n} x_i y_i\right)}{n \cdot \sum_{i=1}^{n} x_i^2 - \left(\sum_{i=1}^{n} x_i\right)^2} = \bar{y} - a\bar{x} = 1.8998 \tag{10.27}$$

$$f(x) = ax + b \tag{10.28}$$

$$f(x) = 0.8721x + 1.8998 \tag{10.29}$$

Example 2: Investigation of the dependence of the mercury volume on the ambient temperature

The liquid mercury has been used for centuries in medicine and technology. The physicist Daniel Gabriel Fahrenheit recognised the linear relationship between temperature and the change in mercury volume in barometers (cf. Table 10.4) and thus replaced alcohol with mercury in measuring systems. Mercury was used for decades in fever thermometers, for example, before it was banned by law because of its harmful effect on health if the thermometer broke. With the aid of a correlation and regression analysis, the dependence and functional relationship between mercury volume and temperature will be investigated.

Table 10.4 Pairs of measured values for the variables temperature and volume of mercury

Index	i	1	2	3	4	5	6
Temp. [°C]	x_i	−10	−5	0	5	10	15
Vol. [m³]	y_i	0.9982	0.9991	1	1.0009	1.0018	1.0027

10.4 Case Studies on Correlation …

Solution:
The strength of the correlation is calculated by means of the Bravais-Pearson correlation coefficient (independent measurement series). The arithmetic mean values result in Eqs. 10.30 and 10.31. The correlation coefficient r_{BP} can then be determined (Eq. 10.32). The correlation coefficient shows a direct correlation with $r_{BP}=1$, a complete linear correlation is present. There are no scatter effects, all pairs of measured values lie on a straight line. For a functional correlation, the linear correlation is determined according to Eq. 10.33. The parameters a and b can be calculated directly according to Eqs. 10.34 and 10.35, so that Eq. 10.36 follows for the functional correlation and thus as visualization Fig. 10.6. Using the functional relationship, the volume of mercury can now be calculated if the ambient temperature is known. The complete correlation with $r_{BP}=1$ is reasonable with regard to the function of a thermometer, since the same temperature should be displayed for each repeated measurement under the same ambient condition.

$$\bar{x} = \frac{1}{n}\sum_{i=1}^{n} x_i = 2.5 \tag{10.30}$$

$$\bar{y} = \frac{1}{n}\sum_{i=1}^{n} y_i = 1 \tag{10.31}$$

$$r_{BP} = \frac{\sum_{i=1}^{n} x_i y_i - n\bar{x}\bar{y}}{\sqrt{\left(\sum_{i=1}^{n} x_i^2 - n\bar{x}^2\right)\left(\sum_{i=1}^{n} y_i^2 - n\bar{y}^2\right)}} = 1 \tag{10.32}$$

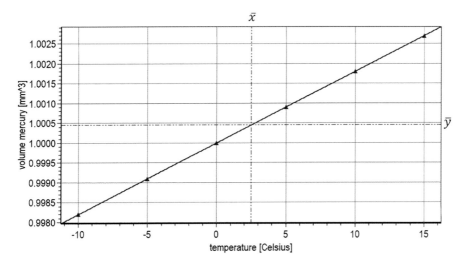

Fig. 10.6 Regression analysis thermometer; functional relationship with respect to the measurement series for the variables mercury volume and temperature

$$f(x) = ax + b \tag{10.33}$$

$$a = \frac{n \cdot \sum_{i=1}^{n} x_i y_i - \left(\sum_{i=1}^{n} x_i\right) \cdot \left(\sum_{i=1}^{n} y_i\right)}{n \cdot \sum_{i=1}^{n} x_i^2 - \left(\sum_{i=1}^{n} x_i\right)^2} = \frac{\sum_{i=1}^{n} x_i y_i - n\overline{xy}}{\sum_{i=1}^{n} x_i^2 - n\overline{x}^2} = 0.0002 \tag{10.34}$$

$$b = \frac{\left(\sum_{i=1}^{n} x_i^2\right)\left(\sum_{i=1}^{n} y_i\right) - \left(\sum_{i=1}^{n} x_i\right)\left(\sum_{i=1}^{n} x_i y_i\right)}{n \cdot \sum_{i=1}^{n} x_i^2 - \left(\sum_{i=1}^{n} x_i\right)^2} = \overline{y} - a\overline{x} = 1 \tag{10.35}$$

$$f(x) = 0.0002x + 1 \tag{10.36}$$

Example 3: Turning Machining Process; Assessment of Surface Topography as a Function of Turning Process Parameters

Turning is a machining process for the production of rotationally symmetrical parts. The workpiece (the cylindrical blank) is clamped on a lathe and rotates around the axis of rotation at a speed U. A turning tool is brought up to the workpiece and the chip is lifted off on contact. For example, different outside diameters can be produced. The process parameters (variables) speed and feed rate are the main factors influencing the surface topography in the machining process. The surface topography is characterized with characteristic values, such as Rz or Ra. The characteristic value Rz is determined by means of arithmetic averaging of the maximum difference height/depth related to five individual measuring distances. It can be assumed that the surface has a higher roughness if the machining is carried out at a high speed U, cf. pairs of measured values in Table 10.5. With the aid of a correlation and regression analysis, the dependence and the functional relationship between the process parameters turning process speed U [1/min] and roughness Rz [μm] are to be investigated.

Solution:
The strength of the correlation is calculated by means of the Bravais-Pearson correlation coefficient (independent measurement series). The arithmetic mean values are obtained from Eqs. 10.37 and 10.38. The correlation coefficient r_{BP} can then be determined (Eq. 10.39). The correlation coefficient shows a strong correlation.

Table 10.5 Turning process; pairs of measured values for speed of cutting process and roughness Rz

Index		1	2	3	4	5	6
Speed [1/min]	x_i	1.500	1.600	1.700	1.800	1.900	2.000
Rz [μm]	y_i	2.7	3.05	3.25	3.3	3.55	3.4

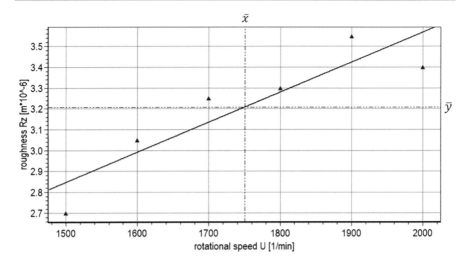

Fig. 10.7 Machining process turning; functional relationship between the process parameters (variables) speed and roughness

For a functional correlation, a linear function is assumed according to Eq. 10.14 (for small speed ranges). The parameters a and b can be calculated directly according to Eqs. 10.40 and 10.41, so that Eq. 10.42 follows for the functional correlation; cf. also Fig. 10.7. Using the functional correlation, the roughness parameter Rz of the surface can now be estimated if the speed of the machining process is known. The result of the regression analysis confirms the assumption: A higher rotational speed has an influence on the surface topography achieved, the roughness increases.

$$\bar{x} = \frac{1}{n}\sum_{i=1}^{n} x_i = 1.750 \tag{10.37}$$

$$\bar{y} = \frac{1}{n}\sum_{i=1}^{n} y_i = 3.2083 \tag{10.38}$$

$$r_{BP} = \frac{\sum_{i=1}^{n} x_i y_i - n\overline{xy}}{\sqrt{\left(\sum_{i=1}^{n} x_i^2 - n\bar{x}^2\right)\left(\sum_{i=1}^{n} y_i^2 - n\bar{y}^2\right)}} = 0.9027 \tag{10.39}$$

$$a = \frac{n \cdot \sum_{i=1}^{n} x_i y_i - \left(\sum_{i=1}^{n} x_i\right) \cdot \left(\sum_{i=1}^{n} y_i\right)}{n \cdot \sum_{i=1}^{n} x_i^2 - \left(\sum_{i=1}^{n} x_i\right)^2} = \frac{\sum_{i=1}^{n} x_i y_i - n\overline{xy}}{\sum_{i=1}^{n} x_i^2 - n\bar{x}^2} = 0.0014 \tag{10.40}$$

$$b = \frac{\left(\sum_{i=1}^{n} x_i^2\right)\left(\sum_{i=1}^{n} y_i\right) - \left(\sum_{i=1}^{n} x_i\right)\left(\sum_{i=1}^{n} x_i y_i\right)}{n \cdot \sum_{i=1}^{n} x_i^2 - \left(\sum_{i=1}^{n} x_i\right)^2} = \bar{y} - a\bar{x} = 0.6833$$
(10.41)

$$f(x) = 0.0014 \cdot x + 0.6833 \tag{10.42}$$

10.5 Spurious Causality and Correlation

When investigating the correlation (relationship) of two or more variables on the basis of measurement series, it can happen that a correlation between two or more variables can be demonstrated, but there is certainly no (technical) causal relationship. In German language, the term „Scheinkorrelation" is usually used, although the term „Scheinkausalität" would be the correct one. In English, however, the term spurious causality or pseudo-causality is used. A proven correlation can be an indicator of a (technical) relationship, but it is not a proof. Often, correlations between two (or more) variables are explained by another unknown variable (confounder), so that spurious causalities can be the result.

Table 10.6 Data collection on stork pairs and the birth rate of the human population in Europe; taken from (Matthews 2001)

Country	Area (km^2)	Storks (couples)	People (10^6)	Birth rate (10^3 / year)
Albania	28,750	100	3.2	83
Belgium	30,520	1	9.9	87
Bulgaria	111,000	5,000	9.0	117
Denmark	43,100	9	5.1	59
Germany	357,000	3,300	78	901
France	544,000	140	56	774
Greece	132,000	2,500	10	106
Netherlands	41,900	4	15	188
Italy	301,280	5	57	551
Austria	83,860	300	7.6	87
Poland	312,680	30,000	38	610
Portugal	92,390	1,500	10	120
Romania	237,500	5,000	23	23
Spain	504,750	8,000	39	439
Switzerland	41,290	150	6.7	82
Turkey	779,450	25,000	56	1,576
Hungary	93,000	5,000	11	124

10.5 Spurious Causality and Correlation

Example 1: Investigation of the Birth Rate of Humans in Relation to the Number of Stork Pairs

A case study by Matthews (2001) examines whether a relationship exists between storks and birth rates in humans. The underlying question, which is often asked in children's books, is: „Does the stork bring the children?". The starting point is a survey of data on the number of stork pairs, humans, and the birth rate with respect to humans in Europe (cf. Table 10.6; taken from (Matthews 2001)). Based on the variables birth rate (variable X) and number of stork pairs (variable Y) a correlation and regression analysis is performed. First, the mean values are calculated, Eqs. 10.43 and 10.44. Based on Eq. 10.45, the correlation coefficient $r_{BP}=0.6089$ is calculated (Bravais-Pearson, assumption of normally distributed measurements). There is a pronounced correlation (Table 10.1) between the variables birth rate (variable X) and number of stork pairs (variable Y). Based on the assumption of a linear correlation (Eq. 10.48), a regression analysis (Eqs. 10.46 and 10.47) yields the parameters a and b, so that the functional relationship between the variables birth rate and number of stork pairs can be determined (Eq. 10.49). The relationship is shown in Fig. 10.8.

The conclusion that with proven correlation also the causal relationship „The stork brings the children" exists is probably wrong:-). Rather, the third variable area/land [km^2] (confounding variable) leads to the fact that this correlation between variable stork pairs and variable birth rate is provable, but no causality exists (spurious causality). If the correlation analyses are carried out for the variables area and number of people (correlation coefficient $r=0.8122$) and for the variables stork pairs and area (correlation coefficient $r=0.5793$) respectively, the influence of the third variable area (per country) becomes clear. The causal

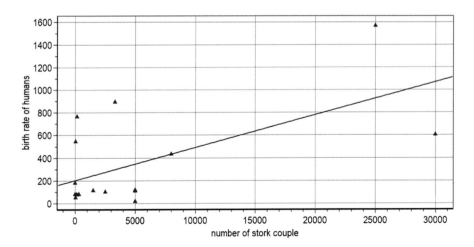

Fig. 10.8 Apparent causality; functional relationship between variable stork pairs and variable birth rate in humans

relationships „The larger the land, the more people" and „The larger the land, the more stork pairs" are plausible.

$$\bar{x} = \frac{1}{n}\sum_{i=1}^{n} x_i = 348.65 \; births \cdot 10^3/year \tag{10.43}$$

$$\bar{y} = \frac{1}{n}\sum_{i=1}^{n} y_i = 5059.35 \; stork \; pairs \tag{10.44}$$

$$r_{BP} = \frac{\sum_{i=1}^{n} x_i y_i - n\overline{xy}}{\sqrt{\left(\sum_{i=1}^{n} x_i^2 - n\bar{x}^2\right)\left(\sum_{i=1}^{n} y_i^2 - n\bar{y}^2\right)}} = 0.6089 \tag{10.45}$$

$$a = \frac{n \cdot \sum_{i=1}^{n} x_i y_i - \left(\sum_{i=1}^{n} x_i\right) \cdot \left(\sum_{i=1}^{n} y_i\right)}{n \cdot \sum_{i=1}^{n} x_i^2 - \left(\sum_{i=1}^{n} x_i\right)^2} = \frac{\sum_{i=1}^{n} x_i y_i - n\overline{xy}}{\sum_{i=1}^{n} x_i^2 - n\bar{x}^2} = 0.02894 \tag{10.46}$$

$$b = \frac{\left(\sum_{i=1}^{n} x_i^2\right)\left(\sum_{i=1}^{n} y_i\right) - \left(\sum_{i=1}^{n} x_i\right)\left(\sum_{i=1}^{n} x_i y_i\right)}{n \cdot \sum_{i=1}^{n} x_i^2 - \left(\sum_{i=1}^{n} x_i\right)^2} = \bar{y} - a\bar{x} = 202.25 \tag{10.47}$$

$$f(x) = ax + b \tag{10.48}$$

$$f(x) = 0.02894x + 202.25 \tag{10.49}$$

Example 2: Investigation of the Relationship Between Global Warming and the Number of Pirates

In another case study, the correlation between the increasing global average temperature (variable X) and the decreasing number of pirates on earth (variable Y) is examined. The underlying figures are taken from Henderson (2006), cf. Table 10.7. A correlation and regression analysis is carried out on the basis of the variables global average temperature (variable X) and the number of pirates (variable Y). First, the mean values are calculated, Eqs. 10.50 and 10.51. Based

Table 10.7 Data collection and estimation of the number of pirates n_p and the global average temperature T in the years 1820 to 2000; data basis (Henderson 2006)

Index	i	1	2	3	4	5	6	7
Number of pirates n_p	x_i	35,000	45,000	20,000	15,000	5,000	400	17
Global average temperature T [Celsius]	y_i	14.2	14.3	14.6	14.9	15.2	15.6	15.9
Year		1820	1860	1880	1920	1940	1980	2000

10.5 Spurious Causality and Correlation

on Eq. 10.52, the correlation coefficient $r_{BP} = -0.9261$ is calculated (Bravais-Pearson, assumption of normally distributed measured values). A very strong (Table 10.1), negative correlation exists between the variables global average temperature (variable X) and pirate count (variable Y). Based on the assumption of a linear correlation (Eq. 10.55), a regression analysis (Eqs. 10.53 and 10.54) yields the parameters a and b, so that the functional relationship between variables global average temperature and the amount of pirates can be determined (Eq. 10.56) and Fig. 10.9. The conclusion that with proven correlation also the causal relationship „The pirates influence the climate, because a falling pirate number causes a rising global average temperature (global warming)" (cf. also Fig. 10.9) may be doubted (= is false). Rather, a number of other confounding variables (climate variability, ship routing) mean that there is a correlation here, but of course there is no causality between the variables number of pirates and global average temperature.

$$\bar{x} = \frac{1}{n}\sum_{i=1}^{n} x_i = 17,202.43 \approx 17.203 \ pirates \tag{10.50}$$

$$\bar{y} = \frac{1}{n}\sum_{i=1}^{n} y_i = 14.96 \ degree \ Celsius \tag{10.51}$$

$$r_{BP} = \frac{\sum_{i=1}^{n} x_i y_i - n\bar{x}\bar{y}}{\sqrt{\left(\sum_{i=1}^{n} x_i^2 - n\bar{x}^2\right)\left(\sum_{i=1}^{n} y_i^2 - n\bar{y}^2\right)}} = -0.9261 \tag{10.52}$$

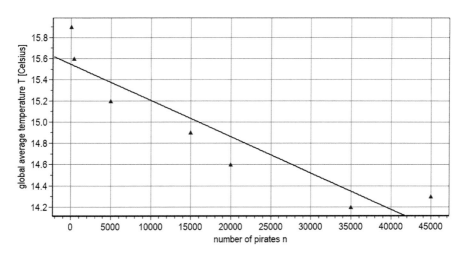

Fig. 10.9 Regression analysis; functional relationship with respect to the measurement series pirate number n_P and global average temperature T

$$a = \frac{n \cdot \sum_{i=1}^{n} x_i y_i - \left(\sum_{i=1}^{n} x_i\right) \cdot \left(\sum_{i=1}^{n} y_i\right)}{n \cdot \sum_{i=1}^{n} x_i^2 - \left(\sum_{i=1}^{n} x_i\right)^2} = \frac{\sum_{i=1}^{n} x_i y_i - n\overline{xy}}{\sum_{i=1}^{n} x_i^2 - n\overline{x}^2} = -3.42 \cdot 10^{-5} \tag{10.53}$$

$$b = \frac{\left(\sum_{i=1}^{n} x_i^2\right)\left(\sum_{i=1}^{n} y_i\right) - \left(\sum_{i=1}^{n} x_i\right)\left(\sum_{i=1}^{n} x_i y_i\right)}{n \cdot \sum_{i=1}^{n} x_i^2 - \left(\sum_{i=1}^{n} x_i\right)^2} = \overline{y} - a\overline{x} = 15.5459 \tag{10.54}$$

$$f(x) = ab + b \tag{10.55}$$

$$f(x) = -3.42 \cdot 10^{-5} x + 15.5459 \tag{10.56}$$

System Analysis: Function, Fault Tree and Failure Mode and Effects

11

In the context of this chapter, the focus is on the analysis of the function and failure of technical systems. One of the most important methods of technical reliability analysis is Fault Tree Analysis (FTA). The FTA is based on the algebra according to Boole and probability theory. The aim is to determine the probability with regard to a specific failure of a technical system. First, the fundamentals of fault tree analysis (FTA), Boolean algebra and importance parameters are presented; see Sect. 11.2. Then, in Sect. 11.2, the application examples of automotive braking systems, bridge circuits, and double bridge configurations for function and fault tree analysis are discussed. Finally, the Failure Mode and Effects Analysis (FMEA) is presented, Sect. 11.3. The FMEA is a method for the systematic discovery of possible risks in the development of products and planning of production processes. Strictly speaking, the FMEA focuses on the methodical procedure for detecting possible errors, whereby no special knowledge of technical statistics or data analysis is required. Nevertheless, the FMEA is mentioned here: The procedure of the FMEA is characterized by its inductive character and thus represents the counterpart to the Fault Tree Analysis (FTA) with its deductive character.

11.1 Basics of Function and Fault Tree Analysis

The basics of function and fault tree analysis are structured as follows: First, the basics of the fault tree (development and visualisation) are outlined in Sect. 11.1.1. Then the basics of algebra according to George Boole are explained in Sect. 11.1.2. The mapping of a system function and the counterpart of the failure function are the focus of Sect. 11.1.3. The analysis of the importance of a component in a system with the aid of importance parameters is shown in Sect. 11.1.4.

11.1.1 Fault tree: Development and Visualisation

This chapter outlines the basics of fault tree analysis (FTA). To ensure easy access to the topic, the basics are explained using the example of a disc of a brake system from automotive engineering.

Figure 11.1 (left) shows a schematic diagram of a disc brake and a photo of a worn brake disc (Fig. 11.1 right). Essentially, a disc brake consists of the components brake caliper (here: fixed caliper), brake disc and brake pads. Inside the brake caliper there is a hydraulically actuated piston system. The brake disc shows the typical damage pattern of wear: scoring due to foreign bodies as well as a reduction of the brake disc thickness, e.g. recognisable by the burr formation at the disc edge (Fig. 11.1 right).

The operation of a brake can be simplified as follows: By operating the brake pedal, pressure is built up in the brake line system and the pistons inside the brake caliper are actuated. The pistons press the brake lining against the brake disc, which in driving operation brakes the rotating disc due to the resulting friction and thus slows down the vehicle; see also Sect. 11.2.1.

The fault tree analysis
Let there be a technical system consisting of parts and assemblies (components). When analyzing damage scenarios and causes of damage, a distinction is made between three different types of failure: primary failure, secondary failure, commanded failure.

Primary failure is the failure of a component due to an inherent weakness. Example: In a brake system, a brake disc rupture caused by an incorrect geometric design would be referred to as a primary failure. A *secondary failure* occurs when a component fails due to operating conditions or environmental conditions. Example: A secondary failure of a brake system can be caused by foreign bodies between the brake lining and the brake plate. In this case, foreign bodies lead to increased scoring and thus to increased wear of the brake disc or brake pads. The

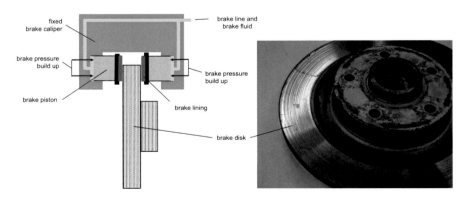

Fig. 11.1 Schematic diagram of a disc brake system (left) of a motor vehicle and brake disk photo (right)

11.1 Basics of Function and Fault Tree Analysis

commanded failure on the other hand, describes a failure of the technical system caused by a faulty or missing control. The system fails despite functioning components. Example: Incorrect actuation of the brake system can lead to insufficient brake pressure being built up. The brake system is incorrectly controlled and does not develop the desired effect.

Figure 11.2 shows an excerpt of a fault tree based on the example of the brake system. Standard inputs (example: negated inputs x_1, x_2), comments describing the fault branches (example: "Wear lining"), links (example: "Wear lining" or [link] "Oily lining" leads to "Brake lining failure") and the undesired event "Reduction in vehicle braking power" (with output negated y) can be seen. The fault tree can be read bidirectionally; both from the fault symptom (negated output y) to the potential fault causes (negated inputs x_i), and from the fault cause to the fault symptom. With the aid of this visualisation, cause-effect chains and their links that lead to an undesired event can be clearly displayed.

Procedure for the development of a fault tree
The development of a fault tree is based on the steps outlined below, exemplified by the technical system dual-circuit braking system; cf. also Fig. 11.2:

1. Determination of the "undesirable event", ergo the fault symptom, damage or risk scenario; example: reduction in vehicle braking performance,
2. Failure of subsystem(s); example: failure of braking circuit 1,
3. Failure of assembly(s); example: failure of the front left brake (VL) on the front axle, driver's side,
4. Component(s) failure; example: brake pad failure,
5. Consideration of failure modes (primary, secondary, commanded); example: brake lining wear (primary failure mode).

The structure of the fault tree is deductive: The development of the fault tree branches is carried out starting from the effect on the technical system (technically complex product or complex process) to the possible fault causes in the component or sub-process. Fault trees can be applied to both technical products and processes (manufacturing processes, workflows). The fault tree is thus constructed in the opposite direction to Failure Mode and Effects Analysis (FMEA; cf. Sect. 11.3), which is characterized by inductive implementation.

The following standard symbols are used in the development of fault trees: *Standard input* (Fig. 11.3, left); *transmission input/output* (Fig. 11.3, center); *comment, links* (Fig. 11.3, right).

Event Link
The symbols shown in Fig. 11.4 are used for the links between events. The "AND" link (*conjunction*) is shown in Fig. 11.4 (left), x_1 and x_2 must occur for y to occur. Figure 11.4 (middle) shows the conjunction "OR" (*disjunction*): x_1 or x_2 must occur for y to occur. Furthermore, the conjunction "NOT" (*negation*) is shown in

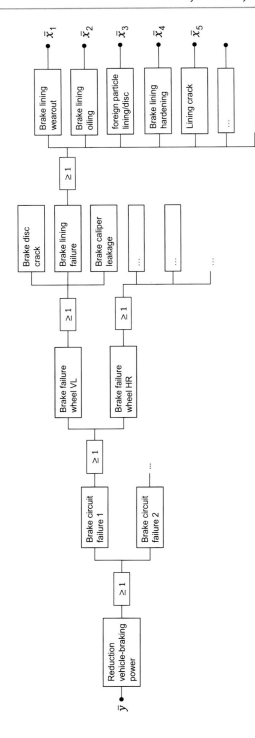

Fig. 11.2 Extract of a fault tree for the technical system dual-circuit brake system of a motor vehicle

11.1 Basics of Function and Fault Tree Analysis

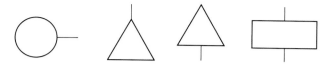

Fig. 11.3 FTA standard icons Standard input (left); transmission input/output (center); comment, links (right)

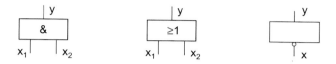

Fig. 11.4 "AND"/conjunction (left), "OR"/disjunction (middle), "NOT"/negation (right)

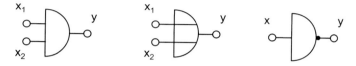

Fig. 11.5 Alternative representations of the conjunctions conjunction (left), disjunction (middle), negation (right)

Fig. 11.6 Visualization of the principle of negation using Euler circles

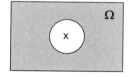

Fig. 11.4 (right): No signal is present at input x, so y occurs. Alternatively, in various literature references, the forms of representation shown in Fig. 11.5 can be found for the three essential types of logic operation.

Note that Fig. 11.4 and 11.5 show the linking of events in the "positive logic", which is the same for an event tree (Event Tree Analysis) would be chosen. If the symbols are used in a fault tree analysis, inputs and outputs would be negated accordingly, as in Fig. 11.2.

11.1.2 Boolean Algebra

The basic links between events—negation, disjunction, conjunction—are presented in the following by means of regularities, derivations from the same as well as Euler circles. Note that the occurrence of the event in Figs. 11.6 to 11.8

Fig. 11.7 Visualization of the principle of disjunction using Euler circles

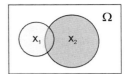

Fig. 11.8 Visualization of the principle of conjunction using Euler circles

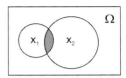

is visualized in gray. The occurrence of the event is assigned the value one to the corresponding variable and the value zero to the complement.

Negation
The negation is shown in Eq. 11.1 and related Fig. 11.6. With Eq. 11.1, Eqs. 11.2 and 11.3 also hold.

$$y = \bar{x} \tag{11.1}$$

$$y = 1 \tag{11.2}$$

$$x = 0 \tag{11.3}$$

Disjunction
Equation 11.4 shows disjunction (cf. also Fig. 11.7), and Eq. 11.5 also holds if Eq. 11.6 or Eq. 11.7 or Eq. 11.8 are satisfied.

$$y = x_1 \vee x_2 = x_1 + x_2 \tag{11.4}$$

$$y = 1 \tag{11.5}$$

$$x_1 = 1 \tag{11.6}$$

$$x_2 = 1 \tag{11.7}$$

$$x_1 = x_2 = 1 \tag{11.8}$$

Conjunction
Equation 11.9 shows the conjunction (cf. also Fig. 11.8). Equation 11.10 holds directly if Eq. 11.11 is satisfied. Similarly, Eq. 11.12 holds directly if Eq. 11.13 and Eq. 11.14 or Eq. 11.15 and Eq. 11.16 are satisfied.

$$y = x_1 \wedge x_2 = x_1 \cdot x_2 \tag{11.9}$$

11.1 Basics of Function and Fault Tree Analysis

$$y = 1 \tag{11.10}$$

$$x_1 = x_2 = 1 \tag{11.11}$$

$$y = 0 \tag{11.12}$$

$$x_1 = 1 \tag{11.13}$$

$$x_2 = 0 \tag{11.14}$$

$$x_1 = 0 \tag{11.15}$$

$$x_2 = 1 \tag{11.16}$$

Algebra according to Boole Theorems and axioms
In the following section theorems and axioms of Boolean algebra are outlined. First of all, the basic laws of algebra apply: commutative law, associative law, distributive law.

a) *Commutative law* (law of interchange) according to Eq. 11.17 as well as 11.18.

$$x_1 \wedge x_2 = x_2 \wedge x_1 \tag{11.17}$$

b) *Associative law* (connection law) according to Eq. 11.19 as well as 11.20.

$$x_1 \vee x_2 = x_2 \vee x_1 \tag{11.18}$$

$$x_1 \vee (x_2 \vee x_3) = (x_1 \vee x_2) \vee x_3 \tag{11.19}$$

c) *Distributive law* (law of distribution) according to Eq. 11.21 as well as 11.22.

$$x_1 \wedge (x_2 \wedge x_3) = (x_1 \wedge x_2) \wedge x_3 \tag{11.20}$$

$$x_1 \vee (x_2 \wedge x_3) = (x_1 \vee x_2) \wedge (x_1 \vee x_3) \tag{11.21}$$

$$x_1 \wedge (x_2 \vee x_3) = (x_1 \wedge x_2) \vee (x_1 \wedge x_3) \tag{11.22}$$

Algebra according to Boole: Postulates
The following postulates apply concerning the existence of a zero-element and a one-element (cf. Eqs. 11.23 and 11.24). Furthermore, Eqs. 11.25 and 11.26 hold for the existence of a complement. The *idempotent law* is shown by Eqs. 11.27 and 11.28, Eqs. 11.29 and 11.30 show the law of absorption.

$$x \vee 0 = x \tag{11.23}$$

$$x \wedge 1 = x \tag{11.24}$$

$$x \wedge \bar{x} = 0 \tag{11.25}$$

$$x \vee \bar{x} = 1 \tag{11.26}$$

$$x \vee x = x \tag{11.27}$$

$$x \wedge x = x \tag{11.28}$$

$$x_1 \vee (x_1 \wedge x_2) = x_1 \tag{11.29}$$

$$x_1 \wedge (x_1 \vee x_2) = x_1 \tag{11.30}$$

One of the most important laws relates to the negation of an expression and the associated inversion of the operator; cf. *De Morgan's law* according to Eqs. 11.31 and 11.32. De Morgan is used for the inversion of system function (function tree) to negated system function (fault tree). The double complement of an element results in the original element; Eq. 11.33. If the OR-operation connects an element x to the one, then the one holds, Eq. 11.34. If the element x is connected to a zero via an AND-operation, then this results in a zero, Eq. 11.35.

$$\overline{x_1 \vee x_2} = \overline{x_1} \wedge \overline{x_2} \tag{11.31}$$

$$\overline{x_1 \wedge x_2} = \overline{x_1} \vee \overline{x_2} \tag{11.32}$$

$$\bar{\bar{x}} = x \tag{11.33}$$

$$x \vee 1 = 1 \tag{11.34}$$

$$x \wedge 0 = 0 \tag{11.35}$$

11.1.3 Illustration of System Function and Failure Analysis

The event tree can be used to clearly describe and display cause effect chains of events. This applies to events of all kinds, both positive events (Event Tree Analysis (ETA) (e.g. planned operating states) as well as damaging events (Fault Tree Analysis (FTA) (e.g.: malfunction). The block diagram, on the other hand, is oriented towards the function of an assembly or a system; it shows the action and interconnections of various components in a system.

Figure 11.9 shows the comparison of a component with two inputs and one output as part of a fault tree (Fig. left) and a block diagram (Fig. right). The representation of the fault tree focuses on the relationships between cause of damage, damage and symptom of damage. Ergo, the focus is on the failure probability F_S,

Fig. 11.9 Element of a fault tree (left) and a block diagram (right)

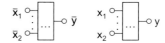

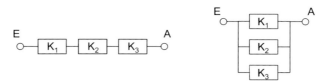

Fig. 11.10 Series structure (left) and parallel structure (right) of three components K_x as well as input E and output A

with negated input(s) and output(s). The representation of the block diagram (Fig. right) focuses on the component function ergo the survival probability R_S is the focus. The relationship between the survival probability R_S (x) (R=Reliability) and the failure probability F_S (x) (F=Fault) of a system S, i.e. the relationship between fault and function of the system, is shown in Eq. 11.36. The failure probability F_S (x) is the complement of the survival probability R_S (x).

$$R_S(x) = 1 - F_S(x) \qquad (11.36)$$

Application of Boole's algebra in reliability analysis of an engineering system.

First, the prerequisites for the application of Boole's algebra to technical systems are outlined.

General Requirements:

a) The technical system cannot be repaired. Failure of the system means that the life span of the product or process has ended.
b) There are two states with regard to the component (element of the system): Functional (state 1) or non-functional (state 0).
c) The components of the product (elements of the system)—and thus the function or malfunction—are independent of each other.

Furthermore, a distinction is made between the two different basic structures: series structure (cf. block diagram in Fig. 11.10, left) and parallel arrangement (cf. block diagram in Fig. 11.10, right). Of course, mixed forms of series and parallel structure are also possible (cf. block diagram in Fig. 11.11).

On the basis of the principles of probability theory shown in Chap. 5 and the fundamentals of Boolean algebra (Sect. 11.1.2), the laws outlined below can be used to calculate the probability of survival (reliability) or the probability of failure of a technical system. Every technical system can be represented in a

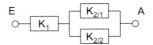

Fig. 11.11 Parallel structure of two components $K_{2/1}$ and $K_{2/2}$ in combination with K_1 in a series structure as well as input E and output A

combination of parallel and series structures (cf. Fig. 11.10). Cross-connections within a technical system present a difficulty here and may have to be separated.

Assumptions and framework conditions for mapping the survival probability or reliability of a technical system without redundancies:

a) The survival probability of the technical system is always lower than the component with the lowest reliability installed in the system.
b) Each additionally installed component reduces the system reliability.
c) If a component is affected by a failure, a system stop must also be assumed.
d) The components of a system do not necessarily have to interact directly from a technical point of view.

The reliability of a technical system when its components are arranged in series is represented by the survival probability R_s according to Eq. 11.37. Formulated with Boolean algebra, the result is Eq. 11.38. The reliability of a technical system with a parallel arrangement of ist components is represented according to Eq. 11.39. The representation using Boolean algebra is shown by Eq. 11.40.

$$R_s(t) = \prod_{i=1}^{n} (R_i(t)) \tag{11.37}$$

$$y = x_1 \wedge x_2 \wedge \ldots \wedge x_n \tag{11.38}$$

$$R_s(t) = 1 - \prod_{i=1}^{n} (1 - R_i(t)) \tag{11.39}$$

$$y = x_1 \vee x_2 \vee \ldots \vee x_n \tag{11.40}$$

The failure probability of a technical system when ist components are arranged in series is represented by the failure probability F_S is shown by Eq. 11.41, according to Boole Eq. 11.42. The failure probability of a technical system with a parallel arrangement of components shows Eq. 11.43; according to Boole Eq. 11.44.

$$F_s(t) = 1 - \prod_{i=1}^{n} (1 - F_i(t)) \tag{11.41}$$

11.1 Basics of Function and Fault Tree Analysis

$$\bar{y} = \bar{x}_1 \vee \bar{x}_2 \vee \ldots \vee \bar{x}_n \tag{11.42}$$

$$F_s(t) = \prod_{i=1}^{n} (F_i(t)) \tag{11.43}$$

$$\bar{y} = \bar{x}_1 \wedge \bar{x}_2 \wedge \ldots \wedge \bar{x}_n \tag{11.44}$$

Example: Engine redundancy
The simple application example in Fig. 11.12 (left) shows the fault tree with regard to the damage symptom aircraft crash due to the simultaneous failure of two engines, which are in a redundant arrangement. The block diagram in Fig. 11.12 (right) shows the redundant structure (parallel structure) of the jet engines T_1 and T_2 with input E and output A.

If now the failure probability per engine at time t_1 would be $F_{T1} = F_{T2} = 0.2$, then according to Eq. 11.45 for the survival probabilities R_{T1} and R_{T2} the result according to Eq. 11.46 and 11.47 is obtained.

$$R(t = t_1) = 1 - F(t = t_1) \tag{11.45}$$

$$R_{T1}(t = t_1) = 0.8 \tag{11.46}$$

$$R_{T2}(t = t_1) = 0.8 \tag{11.47}$$

The survival probability of the system aircraft R_S is obtained on the basis of the addition theorem for independent events (here: survival probability for $n = 2$ engines) on the basis of Eq. 11.48. The failure probability F_S of the aircraft is also obtained directly, Eq. 11.49.

$$R_s = 1 - \prod_{i=1}^{n}(1 - R_i(t)) = 1 - (1 - R_{T1}(t = t_1))(1 - R_{T2}(t = t_1))$$
$$= R_{T1}(t = t_1) + R_{T2}(t = t_1) - R_{T1}(t = t_1) \cdot R_{T2}(t = t_1) = 0.96 \tag{11.48}$$

$$F_s = 1 - R_s = 0.04 \tag{11.49}$$

Of course, the same result is also obtained if the respective failure probability of the engines is used as a starting point. On the basis of the multiplication theorem

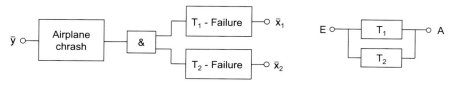

Fig. 11.12 Fault tree (left) and block diagram (right) for aircraft engine redundancy

(Eq. 11.50) for events occurring in parallel (AND operation), the failure probability F_S (Eq. 11.51) and the survival probability R_S of the aircraft system with n = 2 engines are obtained (Eq. 11.52).

$$F_s = \prod_{i=1}^{n} (F_i(t)) \qquad (11.50)$$

$$F_s = F_{T1}(t = t_1) \cdot F_{T2}(t = t_1) = 0.2 \cdot 0.2 = 0.04 \qquad (11.51)$$

$$R_s = 1 - F_s = 0.96 \qquad (11.52)$$

Evaluation of a technical system
The evaluation of a system can be done in three ways:

a) Minimal cut sets,
b) Minimum success paths,
c) Separation.

Application examples for carrying out a fault tree analysis (FTA), the use of Boole's algebra within the reliability analysis of technical systems and the evaluation of a system are presented in Sect. 11.2. The procedures via minimum cut set, minimum success path as well as via separation are shown on the basis of case studies.

11.1.4 Importance Parameters

A technical system consists of a large number of components, which can be arranged in series or in parallel. Thus, the reliability of the system is influenced on the one hand by component i itself, and on the other hand by the arrangement of component i in the overall system. The analysis of the importance of component i can be carried out with the help of importance parameters (word origin Latin: importare; word meaning: to introduce). In the following, three basic importance parameters—based on (VDA 2019a:137)—are outlined:

1. Structural importance
2. Marginal importance
3. Fractional importance

Structural importance refers to the importance of a component due to its positioning in the overall system. Thus, the structure of the system is decisive for the importance of the component. Example: A component is assigned a different importance depending on its positioning within a series structure compared to a parallel structure. Since the system structure is central, structural importance is a time-independent quantity. The structural importance can be obtained from the marginal importance by setting the failure probability $F_i(x) = 0.5$. The general formulation of structural importance is shown in Eq. 11.53, where $(\cdot, \overline{x})$ is the critical vector and i is the number of components in the overall system.

$$I_{\overline{\phi}}(i) = \frac{1}{2^{n-1}} \sum_{(\cdot, \underline{x})} \left(\overline{\phi}\right)(1_i, \overline{x}) - \overline{\phi}(0_i, \overline{x}) \qquad (11.53)$$

While structural importance refers to the positioning of the component in the system, *marginal importance* focuses on the probability of failure of the component focuses on the failure probability of the component, which in turn has an influence on the system. Ergo, marginal importance is determined by the partial derivative of the probability of failure with respect to the component i under consideration; cf. Equation 11.54.

$$I_m(i) = \frac{\partial F(q)}{\partial q_i} \qquad (11.54)$$

The *fractional importance* is formed from the product of marginal importance (Eq. 11.53) and the probability of failure of the component i under consideration; Eq. 11.55. Thus, the fractional importance expresses the importance of component i in relation to the relative change in the probability of failure.

$$I_f(i) = I_m(i) \cdot q_i. \qquad (11.55)$$

11.2 Application Examples: Function and Fault Tree Analysis

This chapter is dedicated to the application of function and fault tree analysis to technical systems. In Sect. 11.2.1, the braking system of a motor vehicle is analysed on the basis of the main components and their arrangement. Sect. 11.2.2 focuses on the bridge circuit, which is often used in safety-related control systems. Furthermore, a double bridge configuration is analysed with regard to function and failure behaviour; cf. Sect. 11.2.3.

11.2.1 Braking System of a Motor Vehicle

A vehicle brake system is designed according to the principle of a dual-circuit brake system (cf. Fig. 11.13 and also 11.1). In the following text, the functional principle—simplified here—is briefly explained with regard to a braking process. The essential components are named with abbreviations in square brackets.

Fig. 11.13 Schematic diagram of a dual-circuit braking system on a vehicle, design crosswise

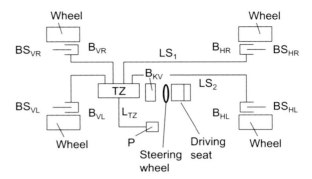

Simplified functional principle of a braking process:

1. The driver pushes the brake pedal [P].
2. Actuating the brake pedal [P] actuates the tandem brake cylinder [TZ] via the line set [L_{TZ}].
3. The brake booster [B_{KV}] supports the build-up of force on the pistons of the tandem brake cylinder [TZ].
4. Via the two pistons of the tandem brake cylinder [TZ], brake fluid is moved within two line sets [LS_1] and [LS_2].
5. The moving brake fluid actuates pistons inside the calipers on each wheel (calipers B_{VL}, B_{VR}, B_{HL}, B_{HR}), which in turn actuate brake pads.
6. The brake pads are pressed against the brake discs of each wheel (brake discs [BS_{VL}], [BS_{VR}], [BS_{HL}], [BS_{HR}]) so that each individual wheel is braked by the resulting friction and thus the vehicle comes to a standstill (see also Fig. 11.1).

Annotation:
The principle of the dual-circuit braking system ensures that if one brake circuit fails, the vehicle can still be brought to a standstill with the second brake circuit. The braking distance is, of course, extended. The brake circuits are usually laid out "crosswise", i.e. the first brake circuit brakes the front left wheel (abbreviation VL) and the rear right wheel (abbreviation HR), the second brake circuit brakes the front right wheel (abbreviation VR) and the rear left wheel (abbreviation HL). The design of the brake circuits can also refer to the axles: The first brake circuit includes the wheels of the front axle, the second brake circuit the wheels of the rear axle.

The block diagram of the dual-circuit braking system based on the above-mentioned essential components is shown in Fig. 11.14. This is a cross-circuit design (1st brake circuit VL/HR; 2nd brake circuit VR/HL). The block diagram shows the following essential features:

a) The arrangement of the components pedal [P], line set [L_{TZ}], tandem brake cylinder [TZ] and brake booster [BKV] is serial.
b) The two brake circuits [LS_1] and [LS_2] are arranged in parallel. Both brake circuits together are arranged within the series (including the series structure of point (a)) between input E and output A.

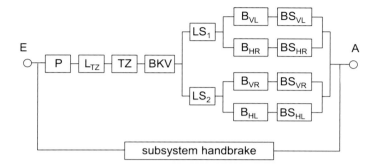

Fig. 11.14 Block diagram of a motor vehicle brake system in simplified form with essential main components in two brake circuits and a separate handbrake circuit

c) The two brake calipers $[B_{VL}]$ and $[B_{HR}]$ as well as the brake discs $[BS_{VL}]$ and $[BS_{HR}]$ are also in parallel arrangement.

d) The same applies to the brake callipers $[B_{VR}]$, $[B_{HL}]$ and the brake discs $[BS_{VR}]$ and $[BS_{HL}]$.

Due to the parallel arrangement of the main components, the braking process can still be carried out if one of the brake circuits or one of the brake calipers has failed. In contrast to Fig. 11.13, the block diagram in Fig. 11.14 shows the handbrake subsystem as a further system (which consists of various components, which are not listed here). Most vehicles have a separately operated handbrake, which can also be used to initiate a braking process in an emergency—independent of the dual-circuit braking system. The braking performance of the overall system is reduced if only one brake circuit can develop its effect while the second brake circuit has failed. The same applies to the hand brake, which does not have the same efficiency as the dual-circuit brake system and only serves as an emergency brake.

The reliability of the braking system BA (without the handbrake subsystem) is mapped according to Eq. 11.56 on the basis of Eq. 11.37 and Eq. 11.39. A dual-circuit braking system and a handbrake system are arranged in parallel within the vehicle (cf. Figure 11.14), ergo Eq. 11.57 results for both systems from Eq. 11.56 with Eq. 11.39 final. For the mapping of the survival probability with respect to each individual component, for example, the Weibull distribution model can be used (cf. Chap. 7). If, for example, a Weibull distribution model is known for each component (data basis: test/trial, simulation or field verification), the survival probability of the braking system at a selected point in time can be determined with the aid of Eqs. 11.56 and 11.57.

$$R_{BA} = R_P \cdot R_{L_{TZ}} \cdot R_{TZ} \cdot R_{BKV} \qquad (11.56)$$
$$\cdot \left[1 - \left(1 - R_{LS_1} \cdot \left(1 - \left(1 - \left(R_{B_{VL}} \cdot R_{BS_{VL}}\right)\right)\left(1 - \left(R_{B_{HR}} \cdot R_{BS_{HR}}\right)\right)\right)\right)\right.$$
$$\left. \cdot \left(1 - R_{LS_2} \cdot \left(1 - \left(1 - \left(R_{BS_{VR}} \cdot R_{B_{VR}}\right)\right)\left(1 - \left(R_{B_{HL}} \cdot R_{BS_{HL}}\right)\right)\right)\right)\right]$$

Fig. 11.15 Block diagram of a bridge circuit with two-way connection

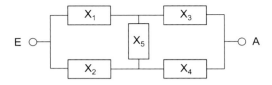

Fig. 11.16 Failure sections (paths) of a bridge circuit

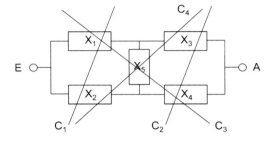

$$R_{BA+HB} = 1 - \prod_{i=1}^{n}(1 - R_i(t)) = 1 - [1 - R_{HB}][1 - R_{BA}] \qquad (11.57)$$

11.2.2 Bridge Circuit

A common circuit design for ensuring reliable operation of a technically complex product is the bridge circuit (cf. Fig. 11.15). The configuration requires that the control of element 5 is possible via component X_1 as well as X_2. This means that the element X_5 is of particular importance.

The evaluation of a system can be done in three ways:

a) Minimal cut sets
b) Minimum success paths
c) Separation

In the focus of the assessment of a system by means of *minimum cut sets* is the question of the paths along which the system could fail. In the evaluation of the bridge configuration, there are four possible paths C_1 to C_4, which could lead to failure (cf. Fig. 11.16).

Based on Boole's algebra (Sect. 11.1.2), the negated system function is Eq. 11.58. The fault tree for the bridge circuit is shown in Fig. 11.17, using the symbols from Sect. 11.1.1.

$$\bar{y} = (\bar{x}_1 \wedge \bar{x}_2) \vee (\bar{x}_3 \wedge \bar{x}_4) \vee (\bar{x}_1 \wedge \bar{x}_4 \wedge \bar{x}_5) \vee (\bar{x}_2 \wedge \bar{x}_3 \wedge \bar{x}_5) \qquad (11.58)$$

In the evaluation by means of *minimum success paths* the paths that could ensure the functionality of the system are examined. In the evaluation of the bridge

Fig. 11.17 Fault tree of a bridge circuit

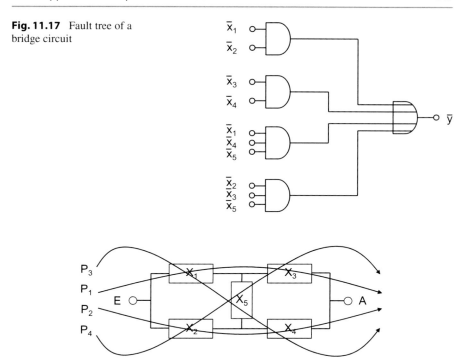

Fig. 11.18 Success paths P_1 to P_4 of a bridge circuit

configuration, there are four possible paths that could ensure functionality (cf. Fig. 11.18).

On the basis of Boole's algebra (Sect. 11.1.2), Eq. 11.59 is obtained as the system function; it includes all possible success paths of the bridge circuit.

$$y = (x_1 \wedge x_3) \vee (x_2 \wedge x_4) \vee (x_1 \wedge x_4 \wedge x_5) \vee (x_2 \wedge x_3 \wedge x_5) \quad (11.59)$$

The *function tree for the bridge circuit* is shown in Fig. 11.19, using the symbols from Sect. 11.1.1. The function tree is based on links similar to those of the fault tree, but the inputs and the output are different and not negated.

The evaluation of a system using the principle of *separation* is first to divide the system into different, easy-to-handle subsystems. Subsequently, the system functions of the subsystems are linked again on the basis of the previously performed separation. The subsystems are first analyzed with respect to the possible success paths that can ensure the functionality of the system.

In the analysis of the bridge circuit, the separation is based on component X_5. The separation means here: Either the system operates under permanent function of X_5 or a permanent failure of X_5 is assumed. Both variants of the separation are to be considered with an OR operation in the evaluation of the overall system. On

Fig. 11.19 Function tree of a bridge circuit

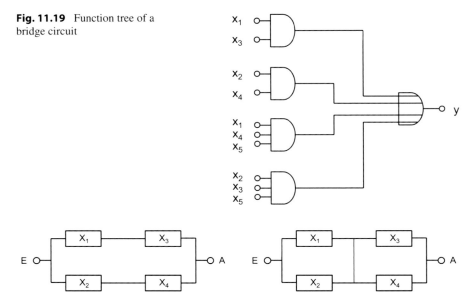

Fig. 11.20 Separation of the bridge configuration; on the left, omission of component X_5 based on the assumption of continuous failure, on the right, replacement of component X_5 with a continuous link based on the assumption of continuous functionality

the one hand, the analysis of the bridge configuration is performed without component X_5 (cf. Fig. 11.20, left), on the other hand with a continuous connection (cf. Fig. 11.20, right).

First, the full functionality of the component X_5 is assumed. From this follows for subsystem T1 according to Fig. 11.20 (right): On the basis of Boole's algebra, the functional capability of the system is mapped according to Eq. 11.60, which can be transferred to Eq. 11.61 by applying the distribution law and commutative law. Then, the assumption is made that component X_5 is subject to continuous failure, yielding subsystem T2 as shown in Fig. 11.20, left. Based on this assumption, Eq. 11.62 is obtained; note that component X_5 is negated here.

$$\begin{aligned} y_{T1} &= x_5 \wedge [(x_1 \wedge x_3) \vee (x_1 \wedge x_4) \vee (x_2 \wedge x_3) \vee (x_2 \wedge x_4)] \\ &= x_5 \wedge [(x_1 \wedge (x_3 \vee x_4)) \vee (x_2 \wedge (x_3 \vee x_4))] \\ &= x_5 \wedge [((x_3 \vee x_4) \wedge x_1) \vee ((x_3 \vee x_4) \wedge x_2)] \end{aligned} \quad (11.60)$$

$$y_{T1} = x_5 \wedge [(x_3 \vee x_4) \wedge (x_1 \vee x_2)] \quad (11.61)$$

$$y_{T2} = \bar{x}_5 \wedge [(x_1 \wedge x_3) \vee (x_2 \wedge x_4)] \quad (11.62)$$

From Eqs. 11.61 and 11.62, Eqs. 11.63 (assumes full functionality of component X_5) and 11.64 (assumes failure of component X_5) are obtained for the survival probabilities of the separated subsystems T1 and T2.

$$R_{T1} = R_5 \cdot [(1 - (1 - R_3)(1 - R_4)) \cdot (1 - (1 - R_1)(1 - R_2))] \quad (11.63)$$

$$R_{T2} = (1 - R_5) \cdot [(1 - (1 - R_1 \cdot R_3)(1 - R_2 \cdot R_4))] \quad (11.64)$$

Both subsystems are now merged in order to be able to map the system function of the bridge circuit. The system functions of the subsystems T1 and T2 are combined with an OR operation, since the component X_5 can assume the two states functional or failed. Thus, Eq. 11.65 is obtained for the system function of the bridge circuit and Eq. 11.66 for the survival probability R_{total} of the bridge circuit.

$$y = y_{T1} \lor y_{T2} \quad (11.65)$$

$$R_{total} = R_{T1} + R_{T2} \quad (11.66)$$

11.2.3 Double Bridge Configuration

A double bridge configuration is used to realise a system function (cf. Fig. 11.21). The double bridge configuration is operated within the framework of a system. Based on empirical data, the failure probability of the components can be estimated after two years of operation (cf. Table 11.1; see notation of component designation analogous to Chap. 11.2.2).

The analysis of the double bridge configuration focuses on the determination of the failure probability as well as the survival probability after two years of operation using the "minimum success path" approach.

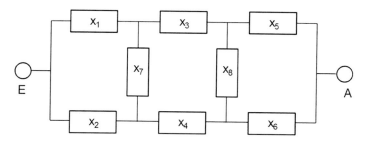

Fig. 11.21 Block diagram of a double bridge configuration

Table 11.1 Probability of failure P_i for components $X_1 - X_8$ (empirical data; 2 years of operation)

$P_1 = 0.1$	$P_2 = 0.12$	$P_3 = 0.3$	$P_4 = 0.1$
$P_5 = 0.4$	$P_6 = 0.2$	$P_7 = 0.8$	$P_8 = 0.04$

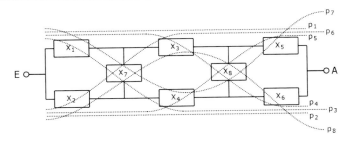

Fig. 11.22 Potential success paths P_1 to P_8 for a double bridge configuration in a block diagram

a) Determination of the system function based on Boole's logic.
b) Comparison of the individual success paths: Determination of the path with the greatest robustness.
c) Determination of the negated system function on the basis of Boole's algebra with the aid of De Morgan's law.
d) Development of the fault tree of the bridge configuration.
e) Determine the probability of failure and survival of the bridge configuration.

Analysis of the double bridge configuration (solution)
To a) The potential success paths P_1 to P_8 of the double bridge configuration are shown by Eqs. 11.67 to 11.74 as well as Fig. 11.22. By applying Boole's laws of logic (cf. Chap. 11.1.2), the system function according to Eq. 11.75 is obtained.

$$P_1 = \{x_1, x_3, x_5\} \tag{11.67}$$

$$P_2 = \{x_2, x_4, x_6\} \tag{11.68}$$

$$P_3 = \{x_1, x_7, x_4, x_6\} \tag{11.69}$$

$$P_4 = \{x_1, x_3, x_8, x_6\} \tag{11.70}$$

$$P_5 = \{x_2, x_4, x_8, x_5\} \tag{11.71}$$

$$P_6 = \{x_2, x_7, x_3, x_5\} \tag{11.72}$$

$$P_7 = \{x_1, x_7, x_4, x_8, x_5\} \tag{11.73}$$

$$P_8 = \{x_2, x_7, x_3, x_8, x_6\} \tag{11.74}$$

$$\begin{aligned} y &= (x_1 \wedge x_3 \wedge x_5) \vee (x_2 \wedge x_4 \wedge x_6) \vee (x_1 \wedge x_7 \wedge x_4 \wedge x_6) \vee (x_1 \wedge x_3 \wedge x_8 \wedge x_6) \\ &\vee (x_2 \wedge x_4 \wedge x_8 \wedge x_5) \vee (x_2 \wedge x_7 \wedge x_3 \wedge x_5) \vee (x_1 \wedge x_7 \wedge x_4 \wedge x_8 \wedge x_5) \vee (x_2 \\ &\wedge x_7 \wedge x_3 \wedge x_8 \wedge x_6) \end{aligned} \tag{11.75}$$

11.2 Application Examples: Function …

To b) The application of the multiplication law with respect to the success paths of the double bridge configuration as well as the insertion of the numbers according to Table 11.1 results in Eqs. 11.76–11.83. A comparison of the survival probabilities shows that the path P_2 has the highest robustness.

$$R_{p_1} = R_1 \cdot R_3 \cdot R_5 = 0.9 \cdot 0.7 \cdot 0.6 = 0.378 \tag{11.76}$$

$$R_{p_2} = R_2 \cdot R_4 \cdot R_6 = 0.88 \cdot 0.9 \cdot 0.8 = 0.634 \tag{11.77}$$

$$R_{p_3} = R_1 \cdot R_7 \cdot R_4 \cdot R_6 = 0.9 \cdot 0.2 \cdot 0.9 \cdot 0.8 = 0.13 \tag{11.78}$$

$$R_{p_4} = R_1 \cdot R_3 \cdot R_8 \cdot R_6 = 0.9 \cdot 0.7 \cdot 0.96 \cdot 0.8 = 0.484 \tag{11.79}$$

$$R_{p_5} = R_2 \cdot R_4 \cdot R_8 \cdot R_5 = 0.88 \cdot 0.9 \cdot 0.96 \cdot 0.6 = 0.456 \tag{11.80}$$

$$R_{p_6} = R_2 \cdot R_7 \cdot R_3 \cdot R_5 = 0.88 \cdot 0.2 \cdot 0.7 \cdot 0.6 = 0.074 \tag{11.81}$$

$$R_{p_7} = R_1 \cdot R_7 \cdot R_4 \cdot R_8 \cdot R_5 = 0.9 \cdot 0.2 \cdot 0.9 \cdot 0.96 \cdot 0.6 = 0.093 \tag{11.82}$$

$$R_{p_8} = R_2 \cdot R_7 \cdot R_3 \cdot R_8 \cdot R_6 = 0.88 \cdot 0.2 \cdot 0.7 \cdot 0.96 \cdot 0.8 = 0.095 \tag{11.83}$$

Re c) The application of De Morgan's law (Eqs. 11.31 and 11.32) with respect to the system function of the double-bridge configuration (Eq. 11.75) directly yields the negated system function to Eq. 11.84.

$$\begin{aligned}\overline{y} &= (\overline{x}_1 \vee \overline{x}_3 \vee \overline{x}_5) \wedge (\overline{x}_2 \vee \overline{x}_4 \vee \overline{x}_6) \wedge (\overline{x}_1 \vee \overline{x}_7 \vee \overline{x}_4 \vee \overline{x}_6) \wedge (\overline{x}_1 \vee \overline{x}_3 \vee \overline{x}_8 \vee \overline{x}_6) \\ &\wedge (\overline{x}_2 \vee \overline{x}_4 \vee \overline{x}_8 \vee \overline{x}_5) \wedge (\overline{x}_2 \vee \overline{x}_7 \vee \overline{x}_3 \vee \overline{x}_5) \wedge (\overline{x}_1 \vee \overline{x}_7 \vee \overline{x}_4 \vee \overline{x}_8 \vee \overline{x}_5) \wedge (\overline{x}_2 \vee \overline{x}_7 \vee \overline{x}_3 \vee \overline{x}_8 \vee \overline{x}_6)\end{aligned} \tag{11.84}$$

Re d) The fault tree in Fig. 11.23 shows the interdependency of cause of damage, damage and damage symptom of the bridge configuration.

Re e) To determine the failure and survival probability of the bridge configuration, the *separation* approach is selected, cf. Sect. 11.2.2.

First, the assumption is made that components X_7 and X_8 always function properly, resulting in the separation I according to the block diagram in Fig. 11.24 (left) and the associated event tree in Fig. 11.24 (right). Thus for the system function of subsystem T1 follows Eq. 11.85 and the survival probability to Eq. 11.86 and with Eq. 11.87 by substituting Eq. 11.88. Substituting the known values (Tab. 11.1) yields Eq. 11.89 and the result Eq. 11.90.

$$y = [(x_1 \vee x_2) \wedge (x_3 \vee x_4) \wedge (x_5 \vee x_6)] \wedge x_7 \wedge x_8 \tag{11.85}$$

$$R_I = [(1 - (1 - R_1)(1 - R_2))(1 - (1 - R_3)(1 - R_4))(1 - (1 - R_5)(1 - R_6))] \cdot R_7 \cdot R_8 \tag{11.86}$$

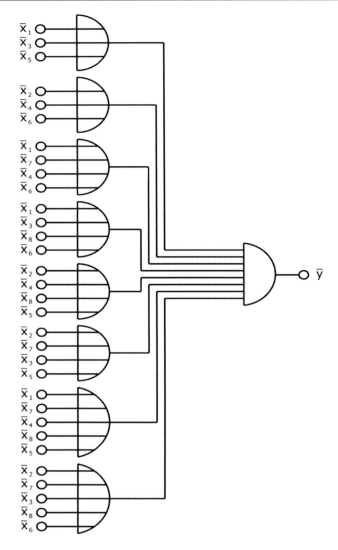

Fig. 11.23 Fault tree for double bridge configuration

$$R_i = 1 - P_i \tag{11.87}$$

$$R_I = [(1 - P_1 P_2)(1 - P_3 P_4)(1 - P_5 P_6)](1 - P_7)(1 - P_8) \tag{11.88}$$

$$R_I = [(1 - 0.012)(1 - 0.03)(1 - 0.08)](1 - 0.8)(1 - 0.04) \tag{11.89}$$

$$R_I = 0.1693 \tag{11.90}$$

11.2 Application Examples: Function ...

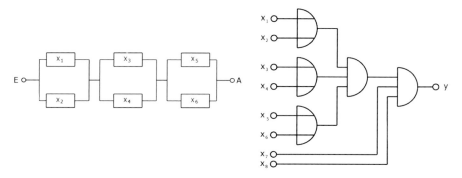

Fig. 11.24 Double bridge configuration based on Fig. 11.21 in the block diagram (left) and event tree (right), but assuming X_7 and X_8 working properly

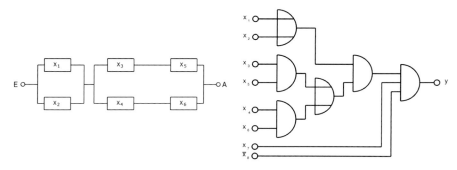

Fig. 11.25 Double bridge configuration (starting point Fig. 11.21) in the block diagram (left) and event tree (right); however, assuming that X_7 is fully functional and X_8 is permanently down

Separation II is performed with the assumption that component X_7 is fully functional and component X_8 permanently fails. The Separation II approach leads to block diagram (Fig. 11.25, left) and event tree (Fig. 11.25, right).

Thus, based on Separation II, Eq. 11.91 is obtained for the system function II, Eq. 11.92 is obtained for the survival probability R_{II}, and Eq. 11.93 and substituting the known values (Table 11.1) yield Eqs. 11.95 and 11.96.

$$y = [(x_1 \vee x_2) \wedge ((x_3 \wedge x_5) \vee (x_4 \wedge x_6))] \wedge x_7 \wedge \bar{x}_8 \qquad (11.91)$$

$$R_{II} = [(1 - (1 - R_1)(1 - R_2))(1 - (1 - R_3 R_5)(1 - R_4 R_6))] \cdot R_7 \cdot (1 - R_8) \qquad (11.92)$$

$$R_i = 1 - P_i \qquad (11.93)$$

$$R_{II} = [(1 - P_1 P_2)(1 - (1 - (1 - P_3)(1 - P_5))(1 - (1 - P_4)(1 - P_6)))] \\ \cdot (1 - P_7) \cdot P_8 \qquad (11.94)$$

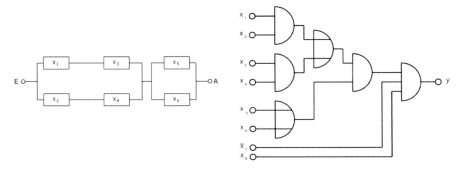

Fig. 11.26 Double bridge configuration (starting point Fig. 11.21) in the block diagram (left) and event tree (right); however, assuming that X_7 is permanently down and X_8 is perfectly functional

$$R_{II} = [(1 - 0.012)(1 - (1 - 0.7 \cdot 0.6)(1 - 0.9 \cdot 0.8))] \cdot 0.2 \cdot 0.04 \quad (11.95)$$

$$R_{II} = 0.0066 \quad (11.96)$$

Separation III is performed with the assumption that component X_8 is fully functional and component X_7 fails permanently. The approach of separation III leads to block diagram according to Fig. 11.26 (left) and event tree (Fig. 11.26, right).

Thus, based on Separation III, Eq. 11.97 is obtained for the system function III, Eq. 11.98 is obtained for the survival probability R_{III}, and Eq. 11.99 and substituting the known values (Table 11.1) yield Eq. 11.100 to 11.102.

$$y = [((x_1 \wedge x_3) \vee (x_2 \wedge x_4)) \wedge (x_5 \vee x_6)] \wedge \bar{x}_7 \wedge x_8 \quad (11.97)$$

$$R_{III} = [(1 - (1 - R_1 R_3)(1 - R_2 R_4))(1 - (1 - R_5)(1 - R_6))](1 - R_7) \cdot R_8 \quad (11.98)$$

$$R_i = 1 - P_i \quad (11.99)$$

$$R_{III} = [(1 - (1 - (1 - P_1)(1 - P_3))(1 - (1 - P_2)(1 - P_4)))(1 - P_5 P_6)] \cdot P_7 \cdot (1 - P_8) \quad (11.100)$$

$$R_{III} = [(1 - (1 - 0.9 \cdot 0.7)(1 - 0.88 \cdot 0.9))(1 - 0.08)] \cdot 0.8 \cdot 0.96 \quad (11.101)$$

$$R_{III} = 0.652 \quad (11.102)$$

Separation IV is performed with the assumption that component X_8 and component X_7 permanently fail. The approach of separation IV leads to block diagram (Fig. 11.27, left) and event tree (Fig. 11.27, right).

Thus, based on Separation IV, Eq. 11.103 is obtained for the system function IV, Eq. 11.104 is obtained for the survival probability R_{IV}, and with Eq. 11.105

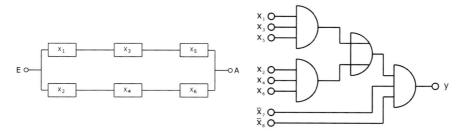

Fig. 11.27 Double bridge configuration (starting point Fig. 11.21) in the block diagram (left) and event tree (right), but assuming X_7 and X_8 permanently failed

and by substituting the known values (Table 11.1), Eqs. 11.106 to 11.108 are obtained.

$$y = [(x_1 \wedge x_3 \wedge x_5) \vee (x_2 \wedge x_4 \wedge x_6)] \wedge \bar{x}_7 \wedge \bar{x}_8 \quad (11.103)$$

$$R_{IV} = [1 - (1 - R_1 R_3 R_5)(1 - R_2 R_4 R_6)] \cdot (1 - R_7)(1 - R_8) \quad (11.104)$$

$$R_i = 1 - P_i \quad (11.105)$$

$$R_{IV} = [1 - (1 - (1 - P_1)(1 - P_3)(1 - P_5))(1 - (1 - P_2)(1 - P_4)(1 - P_6))] \cdot P_7 \cdot P_8$$
$$(11.106)$$

$$R_{IV} = [1 - (1 - 0.9 \cdot 0.7 \cdot 0.6)(1 - 0.88 \cdot 0.9 \cdot 0.8)] \cdot 0.8 \cdot 0.04 \quad (11.107)$$

$$R_{IV} = 0.0247 \quad (11.108)$$

Thus, with the aid of the survival probabilities R_I-R_{IV}, which result from the separation approaches I–IV, the survival probability R_{total} of the overall system double-bridge configuration can be determined by means of the addition law according to Eq. 11.109. Directly, the failure probability F_{ges} of the double-bridge configuration is obtained by forming the complement, Eq. 11.110.

$$R_{total} = R_I + R_{II} + R_{III} + R_{IV} = 0.1693 + 0.0066 + 0.652 + 0.0247$$
$$= 0.8526 \quad (11.109)$$

$$F_{total} = 1 - R_{total} = 0.1474 \quad (11.110)$$

11.3 Failure Mode and Effects Analysis (FMEA)

The Failure Mode and Effects Analysis (FMEA) is a method for the systematic detection of possible risks in the development of products and planning of production processes. The focus of the application of the method is the detection and evaluation of failures, failure root causes and failure symptoms. On the basis of the detected potential errors, avoidance measures are defined. When performing an FMEA, no methods of technical statistics are used; only statistical key indicators are used to evaluate the potential errors. Nevertheless, the FMEA is outlined here, since it is a counterpart to the Fault Tree Analysis (FTA) in terms of the procedure. The FMEA was initially developed within the context of military engineering in 1949 and later applied to the civilian sector within the U.S. Apollo program. Subsequently, the method became established in the automotive industry in the 1970s and later in many technical disciplines. The FMEA belongs to the state of the art in reliability engineering in the development and manufacture of technically complex products.

The counterpart becomes clearest in the procedure of the analysis of the technical product or the manufacturing process on the basis of an FMEA: The potential root cause of the failure in the component or in the sub-process and the effect on the overall product (assembly [ZSB]) or on the entire manufacturing process is analyzed. In contrast, the FTA analyzes the defect symptom at the overall product level or at the manufacturing process level with regard to possible causes of damage. When performing an FTA, the damage is present at the product/manufacturing process level (or is defined as an undesired event) and the analyse of the root cause structure is in the focus. When the FMEA is performed, potential defects are analyzed at the component level/sub-process level with regard to their effect on the overall product or manufacturing process. This makes the preventive character of the FMEA clear. Consequently, FMEA is used in the early phases of product development (cf. Chap. 2) or production process planning for function-critical components and process steps to analyze potential defects.

There are three basic types of FMEA:

1. *System FMEA (S-FMEA)*: As part of the concept phase of product development, the interaction of the systems or modules of a product is analyzed. Example: A rear axle type (trailing arm versus transverse arm principle) is examined with regard to possible defects and their effect on the driving behavior of the overall vehicle.
2. *Design FMEA (K-FMEA)*: In the early phase of series development (design phase), potential functional faults (fault cause-effect chains) of components or assemblies are analysed with regard to design and layout. Example: The geometry of an engine connecting rod eye is examined with regard to its proper function in the crank mechanism of an engine.
3. *Process FMEA (P-FMEA)*: As part of series preparation (production planning), potential faults (fault cause-effect chains) are analysed in individual process steps of a production process: Example: A welding process for fastening door hinges is examined with regard to possible effects on the surrounding sheet steel areas (e.g.: weld spatter).

In the literature there are further names for different types of FMEA, such as Design FMEA, Product FMEA, Hardware FMEA (focus: hardware of electronic products), Software FMEA (focus: program codes). In the end, the naming of each method was an attempt to define the system boundary of the component scope of the investigation. However, the methodology itself is largely identical. In the context of this chapter, the classic designations S-FMEA, K-FMEA and P-FMEA are used.

The *procedure for performing an FMEA* is the same regardless of whether it is a System, Design or Process FMEA:

1. System analysis - Definition of the system elements,
2. Definition of functions and functional structures,
3. Failure analysis (failure root cause, failure symptoms, fault cause-effect chains),
4. Risk analysis,
5. Definition of measures and implementation of optimizations.

FMEA Step 1: System Analysis - Define System Elements and System Structure
The system analysis is used to map the structure of the scope of the examination. In a technical system, this includes all system elements and subsystems. In relation to the system FMEA or design FMEA, this means the mapping of the components, assemblies and modules as well as the relationships to each other. The basis of the system analysis is thus formed, for example, by design/explosion drawings, component specifications. In relation to the process FMEA, the structure of the manufacturing process steps and work sequences is mapped within the framework of the system analysis. The basis of the system analysis of a P-FMEA is formed, for example, by production flow charts, assembly instructions, but also component load specifications, design and exploded drawings.

FMEA Step 2: Determination of functions and functional structures
Based on the system structure, functions and function structures are defined for the system elements. This step is also an important prerequisite for the subsequent failure analysis, since every negation of a function represents a potential failure. In a K-FMEA, the functions of the components, assemblies and modules are defined. In some cases, functions are stored in the component specifications, but in many cases they are defined indirectly by the component design. In a P-FMEA, the functions of the sub-process steps are defined; information can be taken from the production flow charts.

FMEA Step 3: Failure Analysis (failure root cause, failure symptoms, fault cause-effect chains)
The core of failure analysis is the definition of potential failure causes, failures and failure symptoms related to the system elements of the system to be analyzed. In relation to a system element, this means that potential failures of a system element (cf. FMEA step 1) are the negated functions of the same (cf. FMEA step 2).

It should be noted that faults are analysed here at the same level of system elements within the system structure and that the system depth does not vary. The cause of a fault can be defined as a fault with increasing system depth or in the subordinate system structure, with further potential causes of faults in the subordinate level. If the failure symptom of a failure is analyzed, the effect of the failure down to the overall product or manufacturing process must be investigated, since finally the damage during product use is of interest. Therefore, the following applies to the performance of the failure analysis:

1. The potential causes of failures are the conceivable failures (or malfunctions) of the subordinate system elements of the system structure.
2. The possible failure symptoms are the failures or malfunctions of higher-level system elements.
3. The failure sequence is analyzed as a failure sequence chain through the higher-level system elements to the overall system (S-FMEA/ K-FMEA: overall product; P-FMEA: manufacturing process).

FMEA Step 4: Risk analysis

The aim of the risk analysis is the objective evaluation of the determined failure scenarios (see step 3). The evaluation is carried out via the key figure B (importance of the failure sequence), key figure A (probability of occurrence of the failure cause), and key figure E (probability of detection of the failure cause). It should be noted that for each of the key figures, scores are assigned on the basis of empirically determined intervals. These empirically determined intervals may well be different for different disciplines; Table 11.2 shows an example from automotive engineering; cf. (VDA 2006). In detail, the key figures are defined as follows:

$B =$ *Measure of the importance of the failure symptom; score [1-10].*

The more serious the consequence of the failure, the higher the score to be awarded. For example, a score of 10 would be assigned for a potential safety risk or possible environmentally relevant damage scenario. A score of 8 or 9 would be assigned, for example, for function-critical failures that lead to product failure.

$A =$ *Measure of the probability of occurrence of a failure cause; score [1-10].*

The higher the probability of occurrence of the failure cause, the higher the score to be awarded.

$E =$ *Measure of the probability of detection of the potentially occurring failure cause; score [1-10].*

The higher the probability of detection, the lower the score to be awarded.

Note: The term probability is not used here in the classical sense (cf. Sect. 5.1; Kolmogorov), but rather as a general assessment with respect to the occurrence of an event.

On the basis of the key figures B, A, E, the *risk priority number (RPN)* is formed by means of simple multiplication, cf. Eq. 11.111. Thus the risk priority number can assume a number in the interval [1; 1000].

$$RPN = B \cdot A \cdot E \qquad (11.111)$$

11.3 Failure Mode and Effects Analysis (FMEA)

Table 11.2 K-FMEA; scoring numbers for the key figure B (importance of the failure sequence) depending on the potential risk in accordance with (VDA 2006)

Points	Risk, potential damage scenario or damage symptom
9, 10	Very high; safety risk, non-compliance with legal requirements, stranded vehicle (break down)
7, 8	High; vehicle functionality severely restricted, immediate workshop visit absolutely necessary, function restriction of important subsystems
4, 5, 6	Moderate; limited functionality of the vehicle, immediate workshop visit not absolutely necessary, limited functionality of important operating and comfort systems
2, 3	Minor; functional impairment of the vehicle, elimination at the next scheduled workshop visit, functional restriction of operating and comfort systems
1	Very low; Very low functional impairment, only detectable by qualified personnel

As an example from the automotive industry, Table 11.2 shows the criteria for assigning the score for key figure B as a function of typical damage scenarios or damage symptoms at the overall vehicle level when a K-FMEA is carried out; based on (VDA 2006). Further examples for the allocation of scores in a K-FMEA for the key figures A and E as well as for the key figures B, E, A within a P-FMEA can be found in the Annex, Table A.11.

Note: Terms Risk Priority Number RPN and Task Priority TP
Theoretically, according to the classical approach, the same RPNs could arise with different combinations of point allocations (B; A; E) according to the procedure outlined above. In order to avoid mathematically possible, equal risk scores with different key figures B, E, A, in the explanations of the AIAG VDA FMEA manual (VDA 2019b), the task priority (TP) was used instead of the risk priority number (RPN). The task priority (TP) is also based on the risk assessment by using the key figures B, E, A. But the task priority additionally includes a logical aid, which determines the task priority (TP) with respect to the individual point ratings of the key figures B, E, A. Implicitly, the task priority TP via the key figures B, E, A is already included in the classic assessment, but the applicability has been slightly simplified by the standard (VDA 2019b).

FMEA Step 5: Definition of measures and implementation of optimizations
If a risk is identified within the scope of an FMEA on the basis of a high risk priority number (e.g.: $RPN \geq 125$) or high evaluations of the individual key figures (e.g.: B or A or $E \geq 8$), decisions can be made for prevention with regard to the failure and the failure symptom(s):

1. Measures to avoid the failure cause; ergo, measures to reduce the probability of occurrence probability A.
2. Measures to detect the cause of the fault; ergo, measures to increase the probability of detection probability E.

Product and process optimizations can be used which can be applied according to the following prioritization:

1. Change of concept with regard to the exclusion of the potential failure cause; or to minimize the importance of the failure symptom.
2. Increasing product reliability by changing the concept with the goal of minimizing the probability of occurrence of the failure cause.
3. Implementation of measures to increase the probability of detection of the failure cause.

Example vehicle development: risk assessment within a K-FMEA
In the course of prototype construction for series development, a vehicle manufacturer discovers that some prototypes (n = 3) ignite during the refuelling process. The cause of the damage is determined to be a static charge, which was caused by a faulty assembly (ZSB) of an insulating device. However, the supplier of the assembly damping device delivered faultless products in 50 vehicle projects already completed. A K-FMEA is carried out in the series development in which the risk is to be evaluated.

Risk assessment
For the key figure B (measure of the importance of the failure symptom), the score 10 is assigned, since there is a safety risk (damage symptom vehicle fire), cf. Table 11.2. The probability of occurrence A of the failure cause is assessed with a low score (A = 2) (Appendix Table A.11), since the supplier has already supported 50 vehicle projects without any complaints. The probability of detection E of the cause of the defect is also assessed with a low score (E = 2) (Appendix Table A.11), since in prototype construction every product realisation is tested and therefore the probability of detection is high. Thus, the risk priority number is RPN = 40. Since the RPN score can be in the interval [1; 1000], the risk seems low at first. However, the high evaluation of the individual figure B leads to the implementation of a constructive optimization measure in order to avoid the failure cause, especially since it is a safety risk. To avoid static charging of the vehicle during the refuelling process, a voltage dissipation thread is provided in the vehicle design so that the vehicle is earthed. In addition, the proper discharge of the vehicle is checked during prototype testing.

Analysis of the Failure Behaviour of Components, Assemblies, Systems

12

This chapter outlines procedures for performing a reliability analysis based on damage data. The damage data refers to components, assemblies or systems and can be used in the context of a prototype test (development phase), during the end-of-line test (production) or on the basis of a product fleet in field use. Sect. 12.1 presents the general approach. Sect. 12.2 shows the exemplary application of data analysis on the basis of a given damage data set. The outlined procedure is based on approximation methods for parameter estimation and their confidence intervals, respectively, in order to ensure simple comprehensibility. In addition, hints are given as to which further – for the most part iterative – estimation methods can be used for data analysis. The focus of Sect. 12.3 is on different methods for candidate prediction in prototype testing and in the field (usage phase). Candidate prediction by correction methods can be understood as an prognosis – explicitly in the case of damage cases in the field – since finally the population (product fleet) is inferred from a sample (known damage cases). The presentation of the RAPP procedure in Sect. 12.4 for the risk analysis of serial damages in the field forms the finish. The procedure includes candidate prognosis, fleet proportions at risk as well as the consideration of a potential saturation behaviour with respect to the failure behaviour.

12.1 Reliability Analysis Based on Damage Data

This chapter outlines a procedure for the evaluation of damage data and the analysis of the failure behaviour of technically complex systems. This is an exemplary procedure using a selection of the models and procedures explained in Chap. 4–10.

12.1.1 Base of Operations and Overview

Damage data is available of product failures of a product fleet are available. The product fleet can, for example, be tested within the series development or already be in the use phase in the field. The documentation includes damage data with reference to one or more runtime-related variable(s). Thus, the times of the damage occurrences are recorded.

Reliability Analysis: Procedure
The analysis of the available database or damage data can be carried out according to the following steps 1–7 (detailing the steps; cf. Sect. 12.1.2 to 12.2.8):

1. Review of claims data; selection of claims to be analysed and operating-related variables X.
2. Mapping of the failure behaviour with the help of a distribution model F(x).
3. Determination of further key indicators to characterize the failure behavior.
4. Estimation of confidence intervals for parameters and ratios.
5. Determine the functions on survival probability R(x) and failure rate $\lambda(x)$.
6. Interpretation of the parameters and key indicators with regard to the failure behaviour.
7. Prediction or correction of the failure behaviour, if there are further candidates related to a potential failure (cf. in particular Sect. 12.3).

12.1.2 Data Review

The data review (Step 1) of the reliability analysis includes the analysing of the damage data and the selection of the damaged products to be analyzed as well as the operating related variable X.

In the event of a product failure, various damage data is recorded, relating to the damage itself, the damage symptom and the damage cause. Furthermore, time-related information regarding the damage is available (e.g. damage point of time, operating data). If several damages within a product fleet can be assigned to the same cause of damage, the failure behavior of the component can be mapped and analyzed with a statistical model. If the accumulation of damage is related to the damage symptom, statistical modeling can also be performed, whereby effects such as mixed distributions related to the damage data can occur, since different causes of damage can lead to the same damage symptom(s). If the damage symptom (e.g. noise complaint "howling" in a gearbox) is represented by means of a model, this is more of a "black box" approach (e.g. several causes of damage can lead to the same gearbox noise "howling").

The basis for the mapping of a failure behaviour is formed by the operating related variable X: A time of failure x_i is assigned to each damage. Examples of a runtime-related variable X:

a) Operating time of the product until damage occurs.
b) Number of actuation or switching cycles completed before damage occurs.
c) Distance travelled by a vehicle until damage occurs.

12.1.3 Mapping of the Failure Behaviour

Step 2 of the reliability analysis comprises the mapping of the failure behaviour with the aid of a distribution model F(x).

First, the failure probabilities are determined by mapping the sum probabilities (median ranks). A good approximation of the sum probabilities results from the median according to Eq. 12.1, which can be used for a damage data scope $n \leq 50$ damage cases. If $n > 50$ damage cases are available, Eq. 12.2 can be used. The result is a cumulative probability related to the damage case with rank i.

$$F(x_i) = \frac{i - 0.3}{n + 0.4} \tag{12.1}$$

$$F(x_i) = \frac{i}{n + 1} \tag{12.2}$$

Selection of a distribution model for mapping the failure behaviour
Various distribution models were outlined in Chap. 7, which can be used to map the failure behaviour of components, assemblies or systems. The following distribution models are frequently used:

a) Weibull distribution model (two/three parameter form)
b) Normal distribution model
c) Logarithmic normal distribution model
d) Exponential distribution
e) Erlang distribution

Once a distribution model has been selected, the parameters of the model can be determined using various estimation methods (cf. in detail Chap. 8). Known methods for estimating unknown parameters are:

a) Regression analysis of ranks (e.g. for Weibull distribution model)
b) Maximum likelihood estimator (MLE); for any distribution model
c) Estimation according to Gumbel explicit related to the Weibull distribution model
d) Dubey estimation, explicitly for threshold parameter of the Weibull distribution model
e) Method of moments (for any distribution model)

In reliability analysis, the use of a *Weibull distribution model* is the standard in reliability analysis in order to represent different types of failures: Early, random, long-term failure behavior (cf. Chap. 7). The maximum likelihood estimator (MLE) is one of the standard methods to estimate the parameters, but the equations should be solved numerically. Compared to the MLE, the Gumbel approach or the regression analysis are easy—and also manually—to perform; cf. Chap. 8.

In the following, the approach according to Gumbel (1958) is outlined. A detailed application example can be found in Sect. 12.2: Here, the parameter estimates according to Gumbel, maximum likelihood estimator as well as regression analysis (also in combination with Dubey (1967)) are compared.

Application of the approach according to Gumbel (1958)
The starting point is a damage case with failure data for several product failures. A two-parameter Weibull distribution model is chosen to represent the failure behaviour, Eq. 12.3 (also Sect. 7.2.7). Gumbel's estimation of the shape parameter b and the position parameter T (characteristic life) are shown in Eqs. 12.4 and 12.5, respectively. Let N be the extent of known damage events. The parameters σ_N and y_N are available as a set of tables (cf. Appendix Table A.9). When the scope of known damage cases takes a value close to infinity, $\sigma_N = 1.28255$ is set.

$$F(x) = 1 - e^{-\left(\frac{x}{T}\right)^b} \tag{12.3}$$

$$b = \frac{\sigma_N}{2.30258 s_{log}} \tag{12.4}$$

$$T = 10^{\left[\frac{\left(\sum_{i=1}^{N} \log(x_i)\right)}{n} + \frac{\bar{y}_N}{2.30258 b}\right]} \tag{12.5}$$

The standard deviation s_{log} is determined from the logarithmized original values. The general estimator for the variance is shown in Eq. 12.6; according to Steiner's law, Eq. 12.7 applies. Thus, for simplicity, Eq. 12.8 can be used for the standard deviation s_{log}. Of course, Eq. 12.6 can also be used directly using the logarithmized original values.

$$s^2 = \frac{1}{n-1} \sum_{i=1}^{n} (x_i - \bar{x})^2 \tag{12.6}$$

$$s^2 = \frac{1}{n-1} \left[\left(\sum_{i=1}^{n} x_i^2\right) - \frac{1}{n}\left(\sum_{i=1}^{n} x_i\right)^2 \right] \tag{12.7}$$

$$s_{log} = \sqrt{\frac{1}{n-1} \left[\sum_{i=1}^{n} (\log x_i)^2 - \frac{1}{n}\left(\sum_{i=1}^{n} \log(x_i)\right)^2 \right]} \tag{12.8}$$

12.1 Reliability Analysis Based on Damage Data

Note: Interpretation of the parameters of the Weibull distribution model from a technical point of view:

In the reliability analysis it can be interpreted via the shape parameter b can be used to interpret whether the failure is early failure (b<1), random failure (b ≈ 1) or runtime failure (b>1). The position parameter T is called characteristic lifetime and is a typical parameter to describe the component failure behavior. In a three-parameter Weibull distribution model, the threshold parameter t_0 can be interpreted as the earliest possible failure time (incubation time or failure-free time). For further details, Sect. 12.1.7 and 7.2.7 are recommended.

12.1.4 Key Indicators for the Characterisation of the Failure Behaviour

Step 3 of the reliability analysis focuses on the basic characterization of the failure behavior based on key indicators and parameters, which are determined with respect to the analyzed damage data. These include in particular:

a) Arithmetic mean; Eq. 12.9.
b) Median (odd data range, Eq. 12.10; even data range, Eq. 12.11).
c) Dispersion parameter: Standard deviation (Eq. 12.12) and range (Eq. 12.13).
d) B_Z-Parameter (cf. Equation 12.14 ff.).

Ad (a), (b) and (c):

$$\bar{x} = \frac{1}{n} \sum_{i=1}^{n} x_i \qquad (12.9)$$

$$\bar{x} = x_{(n+1)/2} \qquad (12.10)$$

$$\tilde{x} = \frac{x_{n/2} + x_{(n/2+1)}}{2} \qquad (12.11)$$

$$s = \sqrt{\frac{1}{n-1} \sum_{i=1}^{n} (x_i - \bar{x})^2} \qquad (12.12)$$

$$R = x_{max} - x_{min} \qquad (12.13)$$

Notes: Interpretation of mean and dispersion with respect to the underlying damage data.

1. If the arithmetic mean deviates significantly from the median in relation to the specific damage data available, this is an indication of possible outliers with

regard to the damage times available (cf. explanation in Chap. 6 and significance test for outliers in Sect. 13.3.4).
2. Assuming that the dispersion of the damage data was approximately normally distributed, the multiple of the standard deviation ±3 s would correspond to the range in which, with a probability of P = 0.9973, the known damage cases lie (cf. in detail Chap. 13; if no normally distributed damage data are available). If the range R deviates significantly from this value, this would be an indicator with regard to possible outliers (significance test; Sect. 13.3.4). The range R represents the bandwidth in which damage cases are known.

Ad d) The B_Z-parameters
For many technical products, the so-called B-parameter is specified as a service life characteristic value. The B_Z *parameter* represents the point in relation to the lifetime-related variable (e.g. time with regard to field time) at which Z percent of the scope under consideration has failed. The B_Z parameter refers to the distribution model used to represent the failure behaviour. For example, in the case of inexpensive mass-produced goods (e.g. lamps, resistors, self-tapping screws) from the non-safety–critical area, the B_3 parameter is used to determine the probability of early component failure. In addition to the B_3 parameter, B_{10} or B_{90} parameters are also commonly used. The number of the respective B parameter indicates the probability; thus, B_3 is the indication of the probability P = 0.03. This results from the probability function, which is adjusted to the cumulative failure probabilities of the failures. Accordingly, if the B_3 parameter = 4000 h of operation, the probability P = 0.03 that a component will fail by this time of 4000 h of operation.

Using the example of the two-parameter Weibull distribution the B-parameters can be easily calculated with the approach according to Eq. 12.14 and Eq. 12.15. Thus, for example, the results for the B_{10} parameter are shown in Eq. 12.16. Here, parameter Z is the lifetime-related quantity, b and T are the parameters of the Weibull distribution model, and B_Z is the B-parameter for the failure probability Z.

$$Z = 1 - e^{-\left(\frac{B_Z}{T}\right)^b} \Leftrightarrow 1 - Z = e^{-\left(\frac{B_Z}{T}\right)^b} \Leftrightarrow -ln(1-Z) = \left(\frac{B_Z}{T}\right)^b \quad (12.14)$$

$$T[-ln(1-Z)]^{\frac{1}{b}} = B_Z \quad (12.15)$$

$$B_{10} = T[-ln(1-0.1)]^{\frac{1}{b}} \quad (12.16)$$

Note: In a Weibull distribution model, the $B_{63.3}$ parameter corresponds to the characteristic lifetime T (position parameter).

12.1.5 Confidence Intervals for Parameters, Key Indicators and Model

Step 4 of the reliability analysis comprises the determination of the confidence intervals for parameters, indicators as well as model of a reliability analysis; cf. also Chap. 9.

12.1 Reliability Analysis Based on Damage Data

The confidence intervals with respect to the calculated parameters give an indication of the certainty of the interpretation of the distribution model from a technical point of view. The confidence intervals for the parameters of the Weibull distribution model b (Eq. 12.17) and T (Eq. 12.18) give an indication of the certainty of the interpretations made about the failure behaviour: Early failure, random failure or runtime-related failure behaviour; cf. in detail step 5 (Sect. 12.1.6) as well as Chap. 7. In the case that the damage data are approximately normally distributed, the confidence intervals for the arithmetic mean (with unknown mean and variance related to the population; Eq. 12.19) and the variance (Eq. 12.20) can be determined with respect to the available damage data. Hereby, the scatter effects with respect to the estimated centroid and the scatter of the damage data, respectively, are figured out. Furthermore, the confidence interval can be estimated with respect to the distribution model itself; Eqs. 12.21 and 12.22.

$$\widehat{b}\frac{1}{1+\sqrt{\frac{k}{n}}} \leq b \leq \widehat{b}\left(1+\sqrt{\frac{k}{n}}\right) \tag{12.17}$$

$$\widehat{T}\left(\frac{2n}{\chi^2_{2n,1-\alpha/2}}\right)^{\frac{1}{b}} \leq T \leq \widehat{T}\left(\frac{2n}{\chi^2_{2n,\alpha/2}}\right)^{\frac{1}{b}} \tag{12.18}$$

$$\bar{x} - c\frac{s}{\sqrt{n}} \leq \mu \leq \bar{x} + c\frac{s}{\sqrt{n}} \tag{12.19}$$

$$\frac{(n-1)s^2}{c_2} \leq \sigma^2 \leq \frac{(n-1)s^2}{c_1} \tag{12.20}$$

$$KI_{i,\text{ul}} = 0.975 = \sum_{k=i}^{n} \frac{n!}{k!(n-k)!} F^k (1-F)^{n-k} \tag{12.21}$$

$$KI_{i,\text{ll}} = 0.025 = \sum_{k=i}^{n} \frac{n!}{k!(n-k)!} F^k (1-F)^{n-k} \tag{12.22}$$

12.1.6 Survival Probability and Failure Rate

Step 5 of the reliability analysis focuses on determining the survival probability $R(x)$ and failure rate $\lambda(x)$ functions.

The failure behavior of the component was modeled using a Weibull distribution model (see Step 2 and Eq. 12.3). The probability density function (cf. Eq. 12.23) and survival probability of the system R(x) are easily determined; Eq. 12.24. The quotient of the density function and the survival probability to the same point is defined as the failure rate $\lambda(x)$ (Eq. 12.25). For example, using a two-parameter Weibull distribution model as the lifetime model (Eq. 12.3) directly yields Eq. 12.26 for the failure rate λ(x).

$$f(x) = \frac{d}{dx} F(x) \qquad (12.23)$$

$$R(x) = 1 - F(x) \qquad (12.24)$$

$$\lambda(x) = \frac{f(x)}{R(x)} = \frac{f(x)}{1 - F(x)} \qquad (12.25)$$

$$\lambda(x) = \frac{b}{T} \cdot \left(\frac{x}{T}\right)^{b-1} \qquad (12.26)$$

12.1.7 Interpretation of Key Indicators and Model Parameters

Step 6 of the reliability analysis includes the interpretation of the key indicators, parameters and confidence intervals related to failure behaviour and product reliability.

From an engineering point of view, the interpretation of the parameters of the fitted distribution models is of particular importance. Explicitly for the Weibull distribution model, on the one hand, the parameter values are indicators for the characteristic of the failure behaviour of the component, on the other hand, typical parameter values are known for numerous components and an associated failure behaviour.

General information on failure behaviour: Interpretation of different phases of the failure behaviour
The failure behaviour of a product (component, assembly, system) can generally be divided into three elementary phases: Early, random and run-time related failure behaviour (Fig. 12.1). The early failure behaviour is based on failures that

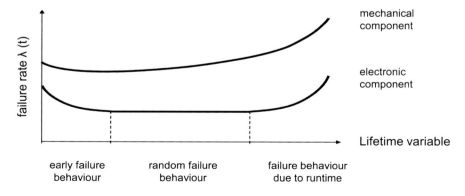

Fig. 12.1 Schematic visualisation for mapping the failure behaviour with failure rate λ(t) of a product for different phases related to the use phase

occur after a short operating time of a product. These include failures caused by design weaknesses (e.g. insufficient sealing of a sealing complex) or production-related faults (e.g. machining residues in a coolant pump). These causes of failure lead to an early failure of the product in operation. Early failures of this type are sometimes also called "teething troubles" in an industrial context. Random failures occur sporadically after the early failure phase; no significant increases in the failure rate are observed in relation to production batches or specific field operation times. On the other hand, the term runtime-related failure behaviour is understood to mean an increasing failure rate in relation to the population of products in field use. If the failure rate of a product in field use is mapped over the entire expected life span, the result is often the curve shown schematically in Fig. 12.1, based on three different models. The three phases of early failure, random failure and failure due to runtime are clearly visible.

Weibull distribution model: interpretation of the shape parameter b
Via the shape parameter b can be interpreted whether it is an early failure behaviour (b<1), random failure behaviour (b $\approx$ 1) or a runtime-related failure behaviour (b>1). The schematic illustration of these three phases is shown in Fig. 12.1. The failure rate $\lambda(t)$ decreases in the early failure behavior phase, remains at an approximately constant level in the random failure behavior phase, and increases in the runtime-induced failure behavior. Ergo, Fig. 12.1 shows the combination of three default behaviors; ergo: three underlying distribution models. This is true both when fitting a two- or three-parameter Weibull distribution model within the respective phase.

The length of the confidence interval for the parameter b depends on the size of the underlying data set: The larger the data set, the smaller the confidence interval. If, for example, the shape parameter is estimated to be b=1.2 and the confidence interval is determined to be [0.8; 1.6], the failure behavior cannot be reliably interpreted as a failure behavior caused by a runtime-related failure behaviour. If in a further analysis – e.g. at a later point of time – further damage data are used as a basis and the confidence interval is calculated as [1.1; 1.3], the failure behaviour can be interpreted as failure behaviour related to the running time. Furthermore, it should be noted that different methods for estimating the shape parameter (cf. step 2 as well as Chap. 8) can also lead to deviating results. Both sample size and confidence interval have to be checked carefully, explicitly for the range of values close to b $\approx$ 1 since slight deviations here may promote an erroneous interpretation. A value of b $\approx$ 1 means that there is no relationship between failures and lifetime variable. Figure 12.2 and Table 12.1 show overviews of the various failure behaviour and shape parameters with technical examples. Figure 12.3 shows an example of the two-parameter Weibull distribution model for mapping the failure behaviour depending on different shape parameters b: Fig. 12.3 (left) shows the distribution functions F(t), Fig. 12.3 (right) the failure rates $\lambda(t)$.

Weibull distribution model: Interpretation of the characteristic life T
The location parameter T is called characteristic life and is a typical characteristic value for describing the component failure behaviour. For the location parameter

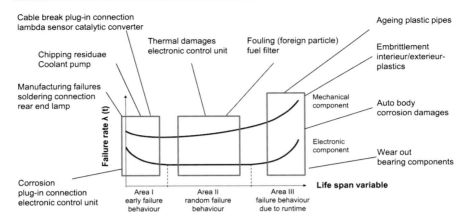

Fig. 12.2 Typical examples of product failures related to the different phases of the failure behaviour of a product in the use phase

Table 12.1 Failure behaviour, shape parameter b of the Weibull distribution model, examples

Failure behaviour	Early failure behavior	Random failures	Failure behaviour due to runtime
Shape parameter	b < 1	b ≈ 1	b > 1
Character	conditionally runtime-related	Independently of runtime	runtime-related
Examples	Manufacturing defects during joining, e.g. welding, gluing, soldering, assembly	Sporadic failures of electronic components	Oil leakage (sweating): b ≈ 1.2–1.5 Rolling bearing: b ≈ 1.5 Ball bearing: b ≈ 2 Corrosion, erosion: b ≈ 3–4 Rubber belt: b ≈ 2.5 Stress corrosion, brittle fracture: b > 4

T, the probability of a component failure is $P = 0.633$. It thus corresponds to the $B_{63,3}$ parameter (cf. Eq. 12.16). Using the confidence interval for the characteristic life T a statement can be made about the potential range in which the actual location parameter lies. High value ranges of the position parameter T in relation to the expected total lifetime of the component are more likely to be expected in case of a runtime related failure behaviour. Low value ranges of the position parameter T are more likely for an early failure behaviour; cf. also the visualisation of representative examples in Fig. 12.2.

Weibull distribution model: interpretation of the threshold parameter t_0
Prerequisite for the determination of the parameter t_0 is the fitting of a three-parameter Weibull distribution model. The failure behavior begins after the

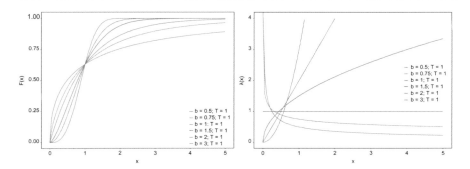

Fig. 12.3 Two-parameter Weibull distribution model for mapping the failure behaviour depending on different shape parameters b; distribution functions F(t) [left], failure rates λ(t) [right]

threshold value (threshold) in contrast to a two-parameter Weibull distribution model: The parameter t_0 characterizes the threshold of the distribution function related to the abscissa. The parameter t_0 is interpretated as earliest possible failure point of time, incubation time or failure-free time in the context of the reliability analysis. In the case of damage causalities which presuppose a certain operating time with regard to the product failure (e.g. mechanical wear), the parameter can thus be interpreted as a theoretically possible first failure time. If the determination of parameter t_0 results in a value close to or equal to zero, this is an indicator that the product can theoretically also fail directly without operating time or after a minimum operating time. The confidence interval for the parameter t_0 indicates the range from which the actual value of t_0 can originate (cf. Chap. 9).

Key characteristics mean, median, standard deviation and range
With arithmetic mean and median the centroid of the known product failures can be determined. If there are large deviations from the arithmetic mean to the median, this is an indication of the presence of outliers. The confidence interval to the arithmetic mean (prerequisite: approximately normally distributed damage data) shows the range in which the actual failure centre will lie at a selected confidence level. On the other hand, standard deviation and range are important key indicators for assessing the spread of failures. The confidence interval of the standard deviation gives the range to which the standard deviation itself is subject. It should be noted that the governing equations for the confidence intervals of centroid and dispersion in this chapter apply only to normally distributed populations. The range is less meaningful here, as it only indicates the maximum possible distance between the first and last known failure.

All the above-mentioned parameters are mostly independent of the distribution model (here, for example, the Weibull distribution model), which was adapted. Ergo, differences between the arithmetic mean or median and the B_{50} parameter are understandable. If all three key figures/parameters are close together, this indicates that there is probably no outlier and that the distribution model reflects the focus of the damage data well.

12.1.8 Candidate Prognosis

Step 7 of the reliability analysis includes the mapping or prognosis of the failure behaviour within the product fleet in the event that further candidates (non-failed units) are present. Accordingly, strictly speaking, the forecast (prognosis) refers to the candidates.

If the failure behaviour is mapped with a distribution model on the basis of an existing sample, it naturally only refers to the concrete damage data available for the known damage. Explicitly in the field data analysis the situation exists that damage data for a failure case is known, but only a percentage of the product fleet is affected at the time of the damage data collection in the field. The units that have not failed are referred to as candidates. With the help of assumptions, the candidates are taken into account within a correction procedure when determining the cumulative failure probabilities, so that a distribution model adapted to this reflects the failure behaviour of the product fleet, also called candidate prognosis. The term extrapolation or projection—explicitly in the case of damage cases in the field—is often used as a synonym for the term candidate prognosis through correction procedures, since finally the expected failures in the population (the product fleet) are inferred from a sample (concretely existing damage cases). Figure 12.4 shows this situation: On the basis of $n = 20$ failure times, the cumulative failure probabilities were calculated and a Weibull distribution model was fitted (cf. Step

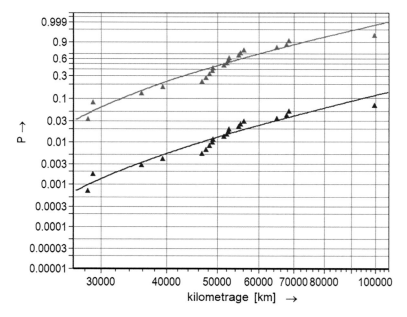

Fig. 12.4 Fitting of Weibull distribution models on the basis of a concrete sample (red colouring) and a candidate prognosis (blue colouring) according to (Johnson 1964)

2; Sect. 12.1.3). The model (distribution function in red colouring) shows the failure behaviour on the basis of the present sample. Using correction methods according to (Johnson 1964) (cf. Sect. 12.3), the default probabilities were corrected and a Weibull distribution model was again fitted (distribution function in blue colouring), which shows the failure behaviour in relation to the product fleet (N = 1,000 units).

The following principles and correction procedures for candidate prediction are commonly used:

1. In principle: Transfer of the Sudden Death Testing to the field data analysis
2. Johnson ranking
3. Procedure according to Nelson
4. Kaplan and Meier method
5. Eckel method

A detailed description of these procedures as well as a method for forecasting endangered fleet shares on the basis of the RAPP procedure can be found in Sect. 12.4.

12.2 Case Study: Analysis of Failure Behaviour of a Coolant Pump

Within the scope of the present chapter, a reliability analysis on the basis of field data related to a fleet of vehicles in the use phase is performed. Failures of coolant pumps are known. The failure behaviour is mapped and interpreted with a distribution model. For parameter estimation the methods Gumbel, regression, Dubey and the maximum likelihood estimator are applied.

Starting point
During the analysis of field data of a vehicle fleet in the service life phase, an accumulation of coolant pump failures—in total n = 16 pump failures—is detected. In addition to specific vehicle data, the failure data include the time of occurrence of the damage on the basis of the runtime variable operating hours [h]; cf. Table 12.2 (column measured values x_i).

Objective
A reliability analysis is carried out for the focus pump damage. The failure behaviour is mapped in a statistical model and key indicators for product reliability are determined. Finally, the results of the reliability analysis are interpreted.

Procedure

a) Determination of the probability of failure via the cumulative frequencies based on the damage data.

Table 12.2 Ranking, measured values, failure probability related to the respective damage case

Rank i	Measured values x_i [h]	Failure probability $F(x_i)$
1	1,500	0.04268
2	2,300	0.10366
3	2,800	0.16463
4	3,400	0.22561
5	3,900	0.28659
6	4,200	0.34756
7	4,800	0.40854
8	5,090	0.46951
9	5,300	0.53049
10	5,500	0.59146
11	6,200	0.65244
12	7,000	0.71341
13	7,600	0.77439
14	8,000	0.83537
15	9,000	0.89634
16	11,000	0.95732

b) Mapping of the failure behaviour with the aid of a two-parameter Weibull distribution model and estimation of the parameters according to Gumbel.
c) Determination of the confidence intervals of the Weibull parameters.
d) Determination of reliability parameters: B_3, B_{10}, B_{90}, mean, variance, range.
e) Mapping of the failure behaviour with the aid of a two-parameter Weibull distribution model; estimation of the parameters on the basis of regression analysis.
f) Mapping of the failure behaviour using a two-parameter Weibull distribution model; estimation of the parameters based on the maximum likelihood estimator.
g) Mapping of the failure behaviour using a three-parameter Weibull distribution model; estimation of the parameters based on Dubey / regression analysis.
h) Mapping of default behavior using a three-parameter Weibull distribution model; estimation of parameters using the maximum likelihood estimator.
i) Comparison and interpretation of the estimation of the Weibull model parameters in the two- and three-parameter case, respectively, as well as visualization of the failure probability $F(x)$ in the two- and three-parameter case.

Solution:
Ad a) For the determination of the failure probability via the cumulative frequencies the approximation via Eq. 12.27 is used. Ranking, measured values and failure probability are shown in Table 12.2.

12.2 Case Study: Analysis of Failure Behaviour ...

$$F(x) = \frac{i - 0.3}{n + 0.4} \qquad (12.27)$$

Ad b) The two-parameter Weibull distribution model is shown in Eq. 12.28. The approach for determining the parameters T and b according to Gumbel is carried out according to Eqs. 12.29 and 12.30. According to Table A.9 (Appendix), the parameters for N = 16 are obtained according to Eqs. 12.31 and 12.32. The scatter s_{log} is determined according to Eq. 12.33. Inserting Eq. 12.29 leads to the shape parameter b (Eq. 12.34). Equation 12.30 is transformed and yields the parameter T after substitution (Eq. 12.35).

$$F(x) = 1 - e^{-\left(\frac{x}{T}\right)^b} \qquad (12.28)$$

$$b = \frac{\sigma_N}{2.30258 \cdot s_{log}} \qquad (12.29)$$

$$\log T = \overline{\log x} + \frac{\bar{y}_N}{2.30258 \cdot b} \qquad (12.30)$$

$$\sigma_N = 1.0316 \qquad (12.31)$$

$$\bar{y}_N = 0.5157 \qquad (12.32)$$

$$s_{log} = \sqrt{\frac{1}{n-1}\left[\sum_{i=1}^{n}(\log x_i)^2 - \frac{1}{n}\left(\sum_{i=1}^{n}\log x_i\right)^2\right]} = 0.2286 \qquad (12.33)$$

$$b = 1.9598 \qquad (12.34)$$

$$T = 10^{\left[\frac{\left(\sum_{i=1}^{n}\log x_i\right)}{n} + \frac{0.224}{b}\right]} = 6336.102 \qquad (12.35)$$

Ad c) The confidence intervals for the parameters of the Weibull distribution model are determined for a confidence level of $\gamma = 0.95$ according to Eq. 12.36 (shape parameter b; here with k = 2) and Eq. 12.38 (characteristic lifetime T). Substitution leads directly to the confidence interval of the shape parameter b; Eq. 12.37. To determine the confidence interval for parameter T, the value of the χ^2-distribution at 2n = 32 is needed at $\gamma = 0.025$ and $\gamma = 0.975$; the standardized tabulations of the χ^2-distribution (Appendix Table A.5) provide values for 2n = 30 as well as 2n = 40. A simple interpolation (Eq. 12.39) at $\gamma = 0.025$, with the aid of Eq. 12.40 and Eq. 12.41, yields Eq. 12.42 (here: Approximation using only linear interpolation is sufficient). The same procedure directly yields Eq. 12.45 for $\gamma = 0.025$ via Eqs. 12.43, 12.44. Thus, by substituting in Eq. 12.38, the confidence interval for the parameter T can be determined; Eq. 12.46.

$$\widehat{b}\frac{1}{1+\sqrt{\frac{k}{n}}} \leq b \leq \widehat{b}\left(1+\sqrt{\frac{k}{n}}\right) \tag{12.36}$$

$$1.4479 \leq b \leq 2.6527 \tag{12.37}$$

$$\widehat{T}\left(\frac{2n}{\chi^2_{2n,1-\alpha/2}}\right)^{\frac{1}{b}} \leq T \leq \widehat{T}\left(\frac{2n}{\chi^2_{2n,\alpha/2}}\right)^{\frac{1}{b}} \tag{12.38}$$

$$f(x) = ax + b_g \tag{12.39}$$

$$f_{0.025}(30) = a \cdot 30 + b_g = 16.8 \text{ with } a = 0.76 \text{ and } b = -6 \tag{12.40}$$

$$f_{0.025}(40) = a \cdot 40 + b_g = 24.4 \tag{12.41}$$

$$f_{0.025}(32) = 18.32 \tag{12.42}$$

$$f_{0.975}(30) = a \cdot 30 + b_g = 47.0 \text{ with } a = 1.23 \text{ and } b = 10.1 \tag{12.43}$$

$$f_{0.975}(40) = a \cdot 40 + b_g = 59.3 \tag{12.44}$$

$$f_{0.975}(32) = 49.46 \tag{12.45}$$

$$5,073.77 \leq T \leq 8,422.07 \tag{12.46}$$

Ad d) The reliability parameters can be determined directly from the failure data: Arithmetic mean (Eq. 12.47), standard deviation (Eq. 12.48) and range (Eq. 12.49). The B_z parameters are obtained from Eq. 12.50, followed by Eqs. 12.51, 12.52, 12.53 for B_3, B_{10}, B_{90}.

$$\bar{x} = \frac{1}{n}\sum_{i=1}^{n} x_i = 5,474.375 \tag{12.47}$$

$$s = \sqrt{\frac{1}{n-1}\sum_{i=1}^{n}(x_i - \bar{x})^2} = 2,566.16 \tag{12.48}$$

$$R = x_{max} - x_{min} = 9,500 \tag{12.49}$$

$$B_z = T(-\ln(1-z))^{\frac{1}{b}} \tag{12.50}$$

$$B_3 = 1,066.916 \tag{12.51}$$

$$B_{10} = 2,009.73 \tag{12.52}$$

12.2 Case Study: Analysis of Failure Behaviour …

$$B_{90} = 9{,}697.69 \tag{12.53}$$

Ad e) Mapping of the failure behaviour using a two-parameter Weibull distribution model; estimation of the parameters based on regression analysis.

The estimation of the shape parameter b as well as the characteristic lifetime T is performed according to Eqs. 12.54 and 12.55; cf. in detail Chap. 8. It should be noted that here the failure times x_i as well as the cumulative failure probability $F(x_i)$ must be transformed according to Eqs. 12.56 and 12.57. The results of the estimation are summarized in Table 12.3.

$$\widehat{b} = \frac{\sum_{i=1}^{n}(x_i - \bar{x}) \cdot (y_i - \bar{y})}{\sum_{i=1}^{n}(x_i - \bar{x})^2} = \frac{\sum_{i=1}^{n} x_i y_i - n\bar{x}\bar{y}}{\sum_{i=1}^{n} x_i^2 - n\bar{x}^2} \tag{12.54}$$

$$\widehat{T} = e^{-\left(\frac{\bar{y} - \widehat{b} \cdot \bar{x}}{\widehat{b}}\right)} \tag{12.55}$$

$$x_i = \ln x_i \tag{12.56}$$

$$y_i = \ln(-\ln(1 - F(x_i))) \tag{12.57}$$

Ad f) Mapping of the failure behaviour using a two-parameter Weibull distribution model; estimation of the parameters on the basis of the maximum likelihood estimator.

Table 12.3 Estimation of Weibull distribution parameters according to the Gumbel approach, regression method and Maximum Likelihood Estimator (MLE). For the data set see Table 12.2

Weibull model two-parameter			
Parameters/AI	Gumbel	Regression	MLE
b	1.96	2.19	2.36
T	6,336 h	6,226 h	6,191 h
KI [$b_{0.025}$; $b_{0.975}$]	$1.45 \leq b \leq 2.65$	$1.62 \leq b \leq 2.96$	$1.74 \leq b \leq 3.19$
KI [$T_{0.025}$; $T_{0.975}$]	$5{,}074 \leq T \leq 8{,}422$	$5{,}103 \leq T \leq 8{,}032$	$5{,}148 \leq T \leq 7{,}841$
B_3	1,067 h	1,264 h	1,410 h
B_{10}	2,010 h	2,228 h	2,386 h
B_{90}	9,697 h	9,113 h	8,815 h
Weibull model three-parameter			
	Dubey	**Regression**	**MLE**
t_0	274.47 h	- - -	122.50 h
b	- - -	2.03	2.31
T	- - -	5,936 h	6,176 h
CI[$b_{0.025}$; $b_{0.975}$]	- - -	$1.50 \leq b \leq 2.75$	$1.71 \leq b \leq 3.13$
CI [$T_{0.025}$; $T_{0.975}$]	- - -	$4{,}790 \leq T \leq 7{,}813$	$5{,}115 \leq T \leq 7{,}863$

For a two-parameter Weibull distribution model, the maximum likelihood estimator for shape parameter b and the characteristic lifetime T according to Eqs. 12.58 and 12.59 are used and solved numerically, respectively; cf. in detail Chap. 8. The results of the estimation are summarized in Table 12.3.

$$T = \left[\frac{1}{q} \cdot \sum_{i=1}^{q} x_i^b \right]^{1/b} \tag{12.58}$$

$$\frac{q}{b} + \sum_{i=1}^{q} \ln x_i - \frac{q \cdot \sum_{i=1}^{q} \left(x_i^b \cdot (\ln x_i) \right)}{\sum_{i=1}^{q} x_i^b} = 0 \tag{12.59}$$

Ad g) Mapping of the failure behaviour with the aid of a three-parameter Weibull distribution model. Estimation of parameters based on Dubey/regression analysis.

The three-parameter Weibull distribution model (Eq. 12.60) additionally includes the parameter failure-free time (threshold) t_0 in comparison to the two-parameter Weibull distribution model. First, the failure-free time is estimated on the basis of the Dubey approach (cf. Eq. 12.61; for more details—in particular nomenclature t_1, t, t_3—Chap. 8). The times t_x are assigned as follows: t_1 refers to the first failure, t_2 to the middle failure, t_3 to the last failure with regard to the available damage data (cf. Table 12.2). The failure probability $F(t_2)$ is estimated (Eq. 12.62; for $F(t_1)$ and $F(t_2)$ cf. Table 12.2; Eq. 12.63 is obtained for $F(t_2)$, so that $t_2 = 3{,}900$ is chosen. Substituting in Eq. 12.61 gives the estimate of the threshold t_0 (failure-free time); Eq. 12.64.

$$F(x) = 1 - e^{-\left(\frac{x-t_0}{T-t_0}\right)^b} \tag{12.60}$$

$$t_0 = \frac{t_1 t_3 - t_2^2}{t_1 + t_3 - 2t_2} \tag{12.61}$$

$$F(t_2) = 1 - e^{-\sqrt{(-\ln(1-F(t_1))(-\ln(1-F(t_3))))}} \tag{12.62}$$

$$F(t_2) = 0.310 \tag{12.63}$$

$$t_0 = 274{,}47 \tag{12.64}$$

The regression analysis for estimating the shape parameter b as well as the characteristic lifetime T is now carried out analogously to the procedure for step (e); Eqs. 12.54 and 12.55 (cf. in detail Chap. 8). It should be noted that the shift of the model by the estimated failure-free time t_0 must be taken into account in the

12.2 Case Study: Analysis of Failure Behaviour ...

transformation of the failure times x_i and the cumulative failure probabilities $F(x_i - t_0)$; Eqs. 12.65, 12.66, 12.67. The results of the estimation are summarised in Table 12.3.

$$F(x_i - t_0) = \frac{i - 0.3}{n + 0.4} \quad (12.65)$$

$$x_i = \ln(x_i - t_0) \quad (12.66)$$

$$y_i = \ln(-\ln(1 - F(x_i - t_0))) \quad (12.67)$$

Ad h) Mapping of default behaviour using a three-parameter Weibull distribution model; estimation of parameters using maximum likelihood estimators.

For a three-parameter Weibull distribution model, the maximum likelihood estimators for the failure-free time (threshold) t_0, shape parameter b as well as the characteristic lifetime T according to Eqs. 8.13–8.18 (Chap. 8) are used or solved numerically. The results of the estimation are summarized in Table 12.3.

Ad i) Comparison and interpretation of the Weibull model parameter estimates
An overview of the results of the different estimation methods for the fitting of a two- or three-parameter Weibull distribution model is shown in Table 12.3. Deviations can be seen in the estimation of the shape parameter b and the characteristic lifetime T. In the present case, however, the deviations should not be serious with regard to the interpretation of the failure behaviour: A shape parameter b > 1 indicates a failure behaviour related to the running time (e.g.: failure behaviour due to wear); cf. Sect. 12.1 (step 6). Since the confidence interval (CI) in each case (independent of the estimation method) covers a range above the value 1, it can indeed be assumed that the failure behaviour is related to the running time ($\alpha = 0.05$). Furthermore, the comparison of the confidence intervals shows a clear overlap, so that no estimator yields a significantly different result when comparing the estimators with each other. This is true for the shape parameter b as well as for the characteristic lifetime T. A comparison of the two-parameter Weibull distribution models based on Table 12.3 is shown in Fig. 12.5, a comparison of the three-parameter Weibull distribution models is shown in Fig. 12.6.

Another indicator for a runtime-related failure behaviour is the existence of a failure-free time t_0, whereby the estimation of t_0 naturally results in a time that lies before the first failure time t_1. This succeeds if the first failure time t_1 (first damage) is not immediately after the time of commissioning of the product fleet or the start of the utilization phase. Both the Dubey method and the maximum likelihood estimator provide an estimate of the failure-free time (threshold) parameter $t_0 > 0$.

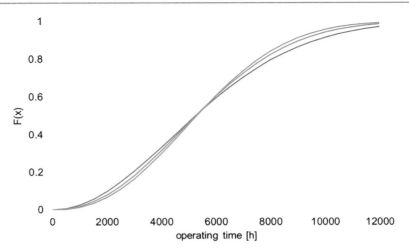

Fig. 12.5 Case study failure behaviour of coolant pump; comparison of the two-parameter Weibull distribution models based on the estimators according to Gumbel (blue), MLE (grey) as well as the regression method (orange); data set cf. Table 12.2, parameters cf. Table 12.3

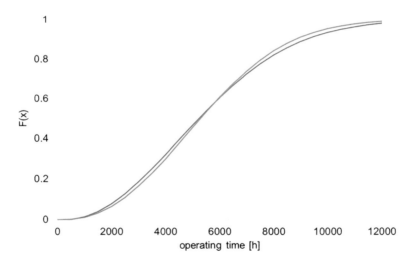

Fig. 12.6 Case study failure behaviour of coolant pump; comparison of the three-parameter Weibull distribution models based on the estimators according to MLE (orange) as well as the combination Dubey/regression method (blue); data set cf. Table 12.2, parameters cf. Table 12.3

12.3 Candidate Prognosis in Testing and Field

In the context of the present chapter, data analysis in the presence of censored data is discussed with regard to the testing of products as well as in the product use phase. In testing, not all components of the testing sample are tested to failure in every case. In this case, the sudden death test is one possible option (Sect. 12.3.1). However, in this case the units which did not fail during the test must be taken into account when mapping the failure behaviour. Possible correction methods are the approaches according to *Johnson* (Sect. 12.3.2) and *Nelson* (Sect. 12.3.3). The application of the sudden death test principle is also possible for the analysis of damage cases within a product fleet in the field (service life phase). In this case, the usage profile can be used as a correction on the basis of empirical distribution models; cf. the procedures according to *Kaplan-Meier* (Sect. 12.3.5) and *Eckel* (Sect. 12.3.6). The correction procedures according to Johnson, Nelson, Kaplan–Meier and Eckel were compared by Bracke and Neupert (2020). This comparison (Bracke and Neupert 2020) forms the basis of Sects. 12.3.2–12.3.6.

The term projection is also often used as a synonym for the term candidate prognosis by correction procedures – explicitly in the case of claims in the field – because ultimately the expected failures in the population (the product fleet) are inferred from a sample (concretely existing claims). Strictly speaking, the units that do not fail (andidates) are taken into account in the analysis of the data of known damaged products. Thus, in terms of mapping the failure behaviour, a damage data analysis can include two outcomes:

a) Mapping of the failure behaviour on the basis of a sample that includes the existing, (known) damage cases.
b) Mapping of the failure behaviour related to the product fleet on the basis of existing damage cases and empirically obtained utilisation profiles or information on non-failed units (candidate prognosis or projection).

The use of usage profiles on the basis of empirical distribution models is always recommended if there is no knowledge of possible load spectra with regard to the units that do not fail. If the load spectrum of the non-defaulted units (candidates) were known within the scope of operational data acquisition, this can be taken into account directly in the correction procedures, e.g. according to Kaplan-Meier or Eckel. The *RAPP procedure* (Sect. 12.4) also considers the utilisation profile of the fleet, but in contrast to the procedures according to Kaplan-Meier or Eckel, it also takes into account that not every unit of the fleet has to be a candidate with regard to a failure.

12.3.1 Sudden Death Test and Field Data Analytics

In the context of a classical component test (note: not sudden death test), a sample with scope n_{ges} is tested. The failure times of all tested components $n_1, n_2, \ldots, n_{ges}$ are used to model the failure behavior. The advantage of the method is that all failure time points are used to build the model; the disadvantage is that the duration of the test can be high, since the failure of all components is waited for. The goal of *sudden death testing* is to shorten the duration of the test time span as it would result from the classical approach. For this purpose, the sample size n_{ges} is divided into groups of equal size. Each group is tested until the first failure; thereafter, the non-failed units (candidates) in the affected group are not tested further. The failure behaviour of the components is now mapped considering the candidates. First, the failure times are brought in ascending order and a rank number i is assigned. The ranks are corrected depending on the candidates; there are different methods for this (cf. Sect. 12.3.2–12.3.6). Figure 12.7 shows the principle of the sudden death test in an example. The test volume $n_{ges} = 30$ components is divided into 6 groups of equal size. The first failure in each group after k cycles leads to a stop of the testing of the respective group. Accordingly, there are four candidates per group, which must be taken into account in the ranking.

If, within the framework of *component testing* according to the Sudden Death Test, the assumption is made that all components are subjected to the same load profile, the candidates are often considered according to the Johnson or Nelson procedure (Sect. 12.3.2 and 12.3.3). The principle of the sudden death test can also be applied to *data analysis in the field*. The known field damages correspond to the failed units in the component test; the non-failed units correspond to the candidates. This approach is valid if the failed units in the field have experienced the same load as the (not yet) failed units (candidates) in the field. In many cases, however, it must be assumed that the components are used differently, in contrast to component testing. The usage profile is usually represented by an empirical distribution model. The correction of the ranks to take into account the candidates

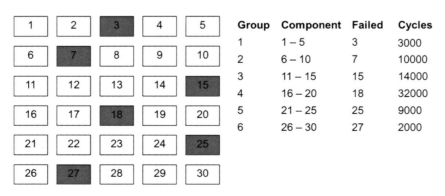

Fig. 12.7 Principle of the Sudden Death Test; division of the test volume into groups

(non-failed units) is done depending on the usage profile, which is described by an empirical distribution function. Frequently used correction methods, in which a utilisation profile can be taken into account (Sect. 12.3.4), are the methods according to Kaplan-Meier or Eckel (cf. Sect. 12.3.5 and 12.3.6). If the usage data of each non-defaulted unit are known, they can also be used directly; in this case, an empirical usage profile is not necessary.

Of course, the correction procedures for the consideration of usage profiles in the field can also be used in the context of prototype testing, e.g. by using an empirical usage profile of the previous product generation. However, this is only useful if no information is known about the units that did not fail (which is the rarest case within a prototype test).

12.3.2 Johnson Ranking

The method according to Johnson (1964) („Johnson ranking") takes into account the non-defaulted units (potential candidates) by assigning mean, hypothetical rank numbers in relation to the failed units; cf. also (Bracke and Neupert 2020).

The starting point is a sample (series of measurements) with damage data of failed units (e.g. failed products) from a population. The failure data (time of failure in testing or in the field) are sorted in ascending order and the rank i (ordinal number) is assigned starting with the first failure. However, if a test object/product – i.e. a candidate – is removed (testing) or cannot be evaluated due to information (field), an averaged rank number is assigned for the next following failure. The next failure could therefore be assigned rank j or rank j+1 if the candidate (the component that did not fail) had been assigned rank j+1 or rank j.

The averaged rank r_j is determined according to Eq. 12.68, where r_{j-1} is the previous averaged rank, x_j is the number of failures at the time of failure t_j, and n_j is the number of potential candidates no longer available at the time of failure t_j. This yields via approximate formula for the cumulative failure probability $F(t_j)$ based on the assigned rank j; Eq. 12.69.

$$r_j = r_{j-1} + x_j \cdot \frac{1 + n - r_{j-1}}{1 + n - n_j} \quad (12.68)$$

$$F(t_j) = \frac{r_j - 0.3}{n + 0.4} \quad (12.69)$$

If the Johnson approach were applied to the sudden death test, the number n_j of potential candidates no longer available at the time of failure j would be calculated according to Eq. 12.70, taking into account the k test units per group. In relation to a product fleet in the field, the candidates would be estimated, for example, with the aid of an empirical, logarithmic distribution model $F_{ln}(t_j; \mu_{ln}, \sigma_{ln})$ (cf. Chap.

12.3.4) and Eq. 12.71 is obtained, where X_{j-1} represents the cumulative number of failures up to the time of failure t_{j-1} and X_g the total number of failures.

$$n_j = (j-1) \cdot k \tag{12.70}$$

$$n_j = X_{j-1} + (n - X_g) \cdot F_{ln}(t_j; \mu_{ln}, \sigma_{ln}) \tag{12.71}$$

12.3.3 Procedure According to Nelson

The starting point of the procedure according to Nelson (1982) („Nelson procedure") is the failure rate $\lambda(t)$, Eq. 12.72; see also (Bracke and Neupert 2020). Integration of the failure rate yields the accumulation $H(t_j)$, Eq. 12.73. Integration yields Eq. 12.74 and transformation yields Eq. 12.75. This allows the failure behavior to be represented by any distribution model.

$$\lambda(x) = \frac{f(x)}{R(x)} = \frac{f(x)}{1 - F(x)} \tag{12.72}$$

$$H(t_j) = \int_0^{t_j} \lambda(t) \cdot dt \tag{12.73}$$

$$H(t_j) = -\ln(1 - F(t_j)) \tag{12.74}$$

$$F(t_j) = 1 - \exp(-H(t_j)) \tag{12.75}$$

If the failure behaviour is analysed after a certain period of prototype testing or field use, the candidates (non-failed units) are considered as follows: The time points of the failed units are sorted in ascending order and an inverse rank is assigned considering the non-failed units, Eq. 12.76. The stepwise reduction of the testing volume and the fleet volume, respectively, is then performed analogously to the Johnson procedure; the contender consideration is performed according to Eq. 12.70 for testing, and according to Eq. 12.71 for fleet analysis.

$$r_j = n - n_j \tag{12.76}$$

The failure rate $\lambda(t)$ can be estimated in terms of the time interval between two failures i and i-1, Eq. 12.77. Thus, Eq. 12.79 follows for $H(t_j)$ (Eq. 12.78) and by merging Eqs. 12.75 and 12.78 as an approximation for $F(t_j)$.

$$\widehat{\lambda}_i \cdot t_i = \widehat{\lambda}_i(t_i - t_{i-1}) = \frac{x_i}{n - n_i} \tag{12.77}$$

$$\widehat{H}_j = \sum_{i=1}^{j} \widehat{\lambda}_i \cdot \Delta t_i = \sum_{i=1}^{j} \frac{x_j}{n - n_j} \tag{12.78}$$

$$\widehat{F}(t_j) = 1 - \exp(-\widehat{H}(t_j)) \tag{12.79}$$

12.3.4 Product Use Profile: Empirical Distribution Function

Within the scope of the utilization phase of a product fleet, a different utilization intensity of the products can be assumed. The usage profile is usually represented by an empirical distribution model. When analyzing a damage event in the field, on the one hand the damage data (point of time) are known, and on the other hand often no information is available about the units that did not fail (candidates). Only the number of candidates is known (example: products sold).

The correction of the ranks for the consideration of the candidates can be done depending on the usage profile, which is described by an empirical distribution function. The variable X of the empirical distribution function corresponds, of course, to the runtime-related variable with which the damage events are documented (e.g.: switching cycles, travel distance, operating time).

For many technical products, the logarithmic normal distribution model is suitable $F_{ln}(t_j; \mu_{ln}, \sigma_{ln})$ to represent a usage profile; cf. Eq. 12.80 and 12.81 (and also Sect. 7.2.6).

$$f_{ln}(x) = \frac{1}{\sqrt{2\pi} \cdot \sigma_{ln} \cdot x} \cdot exp\left(-\frac{1}{2}\left(\frac{\ln(x) - \mu_{ln}}{\sigma_{ln}}\right)^2\right); x > 0 \quad (12.80)$$

$$F_{ln}(x) = \int_0^x f_{ln}(t; \mu_{ln}, \sigma_{ln}) \cdot dt \quad (12.81)$$

Expected value E(X) and variance Var(X) are determined by Eqs. 12.82 and 12.83. A transformation directly yields Eqs. 12.84 and 12.85, so that the mean μ_{ln} and standard deviation σ_{ln} of the log distribution model can be determined from the estimated values μ and σ of an empirically determined usage profile. The run-time related variable is t_e (index e = usage).

$$E(X) = \mu = exp\left(\mu_{ln} + \frac{\sigma_{ln}^2}{2}\right) \quad (12.82)$$

$$Var(X) = \sigma^2 = exp(2\mu_{ln} + \sigma_{ln}^2) \cdot \left(exp(\sigma_{ln}^2) - 1\right) \quad (12.83)$$

$$\mu_{ln} = \frac{1}{2} \cdot \ln\left(\frac{\mu^4(t_e)}{\mu^2(t_e) + \sigma^2(t_e)}\right) \quad (12.84)$$

$$\sigma_{ln} = \sqrt{\ln\left(\frac{\mu^2(t_e) + \sigma^2(t_e)}{\mu^2(t_e)}\right)} \quad (12.85)$$

If the assumption is made that the estimated values for mean μ and standard deviation σ are linearly dependent on the field deployment time t_e, the empirical distribution model for arbitrary field operation times can be determined from a defined field observation period.

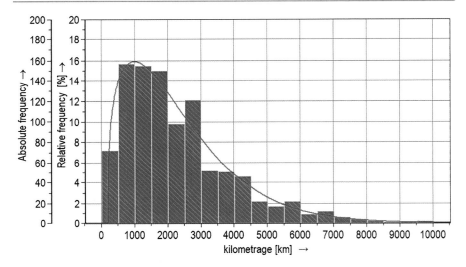

Fig. 12.8 Empirical distribution model for mapping a usage profile of an automobile fleet, the normalization period is one month in field use

Example of a usage profile

Figure 12.8 shows a normalized usage profile of a vehicle fleet (automobile) based on the travel time variable [km], which is in the usage phase. The normalization refers to the measured values regarding the distance travelled during the field runtime of one month. Furthermore, a logarithmic distribution model $f_{\ln}(t_j; \mu_{\ln}, \sigma_{\ln})$ was fitted to the normalized measurement data determined by empirical observation; cf. Eq. 12.80. The visualization of the density function and the histogram respectively clearly shows a typical usage profile of a vehicle fleet: few vehicles achieve a high mileage/month, many vehicles complete a lower mileage/month. In field observations, vehicle mileage data are often recorded with an annual reference (12 months) and the parameters μ_{12} and σ_{12} are estimated. Thus, if the assumption is made that mean μ and standard deviation σ are linearly dependent on field deployment time t_e, distribution models can be generated for arbitrary operating times t_e based on a pair of estimators for μ and σ (e.g. using estimators for μ_{12} and σ_{12}); cf. Eqs. 12.86 and 12.87.

$$\widehat{\mu}(t_e) = \frac{\widehat{\mu}_{12}}{12} \cdot t_e \tag{12.86}$$

$$\widehat{\sigma}(t_e) = \frac{\widehat{\sigma}_{12}}{12} \cdot t_e \tag{12.87}$$

12.3.5 Kaplan and Meier Method

The starting point of the approach according to Kaplan and Meier (1958) („Kaplan-Meier estimator") is the multiplication law of the probability calculus (Chap. 5); see also (Bracke and Neupert 2020). In terms of the survival probability $R(t_j)$, this means the product of the time intervals i without failure from the first time interval to time j; Eq. 12.88. At time i, the survival probability p_i is estimated with the quotient of the number of still functioning units n_i minus the failed units x_i at this time and the still functioning units n_i, Eq. 12.89.

$$R(t_j) = \prod_{i=1}^{j} p_i \qquad (12.88)$$

$$\widehat{p}_i = \frac{n_i - x_i}{n_i} \qquad (12.89)$$

The units still functional at time i results in the sudden death test in the test with k units per group according to Eq. 12.90. In the context of the application to field data, where a log-normal distribution model would be used $F_{ln}(t_j; \mu_{ln}, \sigma_{ln})$ would be used to represent the runtime variables, the determination is made according to Eq. 12.91, where n represents the sample size, X_{j-1} represents the cumulative number of failures up to the failure time t_{j-1}, and X_g represents the total number of failures. The failure probability is given by the complement to the survival probability according to Eq. 12.92.

$$n_i = n - (i-1) \cdot k \qquad (12.90)$$

$$n_i = n - X_{i-1} + (n - X_g) \cdot F_{ln}(t_j | \mu_{ln}, \sigma_{ln}) \qquad (12.91)$$

$$\widehat{F}(t_i) = 1 - \widehat{R}(t_i) \qquad (12.92)$$

12.3.6 Eckel Method

The method according to Eckel (1977) („Eckel method") takes the candidates (non-failed units) into account via a previously empirically determined distribution function with respect to the expected usage profile; cf. also (Bracke and Neupert 2020). In classical sudden-death testing and the use of Johnson's correction method, the non-failed units are assigned an averaged test time relative to the group; Sect. 12.3.2). In contrast, within the procedure according to Eckel the utilization profile of the component is mapped with a distribution function. The distribution function indicates the probability that products of a population have reached a certain value with respect to the runtime-related variable – which is used to describe the failure behavior – at a point in time i. In contrast to classical sudden death testing, this takes into account the fact that there can be different numbers of candidates at different points in time i depending on a product-specific usage profile.

First, the known failures are sorted in ascending order and a rank j is assigned to each. For each rank j, the number of candidates N_{anj} (index „an " ≙ candidates) is determined, Eq. 12.93. Using the log-normal distribution model F_{ln} (parameters μ_{ln} and σ_{ln}), the usage profile outlined earlier is included. X_j denotes the cumulative failures of failure j at time t_j.

Thereafter, the sum frequency related to the potential candidates, Eq. 12.94 is obtained by considering Eqs. 12.95 and 12.96. If the procedure is applied in the context of a sudden death test, then the estimation of the failure probability is done by Eq. 12.97, where k is the number of test objects per group.

$$N_{anj} = n \cdot F_{ln}(t_j | \mu_{ln}, \sigma_{ln}) - X_j - \sum_{i=1}^{j-1} N_{ani} \qquad (12.93)$$

$$F_{anj}(t_j) = F_{an}(t_{j-1}) + \frac{N_{anj}}{(n+1) \cdot (1 - F_k(t_{j-1}))} \qquad (12.94)$$

$$F_{an}(t_0) = 0 \qquad (12.95)$$

$$F_k(t_0) = 0 \qquad (12.96)$$

$$F_k(t_j) = F_k(t_{j-1}) + \frac{X_j}{(n+1) \cdot (1 - F_{ank}(t_j))} \qquad (12.97)$$

Note: The method according to Eckel (1977) was developed using the example of vehicle technology. Eckel uses the distance covered by a vehicle or the vehicle fleet in the field as the running time-related variable. The usage profile outlined refers to the vehicle fleet in the field and is therefore determined empirically. In principle, however, the procedure can be applied to many technically complex products; the running time-related variable can be, for example, switching cycles or operating time of the product; it must simply not be identical to the field time.

12.4 RAPP – Analysis of Serial Damages

The procedure "Risk Analysis and Prognosis of complex Products (RAPP)" focuses on the analysis of a damage focus within a product fleet in the use phase; cf. (Bracke 2013) and (Bracke and Sochacki 2015). The starting point is the damage data of product failures as well as time-related usage profiles of the product fleet. The method is particularly suitable for the analysis of serial damage, i.e. when there is a high incidence of damage events within the product fleet. The aim of the procedure is to determine the following parameters:

12.4 RAPP – Analysis of Serial Damages

a) Determination of the risk probability P_{RPF} to the damage focus within the product fleet N_{PF} at the field observation time i.
b) Candidates within the product fleet and at the field observation time i.
c) Determination of the damage peak (damage appearance) depending on the usage profile of the fleet and failure behaviour.
d) Detection of a saturation limit at which no increase of damage cases within the fleet is to be expected.

12.4.1 Risk Analysis and Prognosis of complex Products (RAPP)

First of all, the *base of operations for the RAPP procedure* is outlined: On the one hand, there is knowledge about the focus of the damage, on the other hand information about the usage profile of the product fleet itself. The knowledge of the damage focus includes the damage symptom or cause of damage as well as the damage data on the failed units. The damage data includes failure times related to the lifetime variable (e.g. distance, switching cycles) as well as the field runtime. Regarding the units that did not fail, a distribution model related to the lifetime variable is fitted to the usage data. Table 12.4 shows an overview of the required data.

RAPP procedure
Step A: Mapping and forecasting of product failure behaviour
The failure behaviour of the failed units is mapped using a service life distribution model, which is adapted to the damage data. A two- or three-parameter Weibull distribution model f_{FM} is well suited; cf. Eqs. 12.98 and 12.99 (Index FM: Failure Mode), respectively. The model parameters are determined by means of a suitable estimation procedure, e.g. via the maximum likelihood estimator (cf. Chap. 8).

Table 12.4 Overview of the data output basis of the RAPP procedure

1	Product fleet data	Parameter / Information
1.1	Production batch (Product fleet (PF))	N_{PF}
1.2	Period of field operation (field observation)	T
1.3	Number of non-failed units (candidates)	$n_{R1}, n_{R2}, ..., n_{Rn}$
1.4	Utilisation profile of the fleet	Distribution model; Example: Log-normal distribution (μ, σ)
2	**damage data**	
2.1	Damage focus	symptom/cause
2.2	Number of failed units (Failure F)	n_F
2.3	Failure point of time	$t_{nF1}, t_{nF2}, ..., t_{nFn}$

$$f_{FM}(x) = \frac{b}{T} \cdot \left(\frac{x}{T}\right)^{b-1} \cdot e^{-\left(\frac{x}{T}\right)^b} \qquad (12.98)$$

$$f_{FM}(x) = \frac{b}{T - t_o} \cdot \left(\frac{x - t_o}{T - t_o}\right)^{b-1} \cdot e^{-\left(\frac{x - t_o}{T - t_o}\right)^b} \qquad (12.99)$$

The field observation period t is divided into suitable observation phases $t_1 - t_n$. If, for example, a field observation period of 12 MIS (Months in Service) is available, twelve observation phases could be generated depending on the number of failed units. Accordingly, twelve distribution models $f_{FM,i}$ are adjusted (i = 1, …, 12), increasing the amount of damage data $n_{F,i}$ (cumulative). As a result, n parameter sets (e.g.: b_1, T_1; …; b_n, T_n) are determined at $t_1 - t_n$ observation phases. The development of the parameter set can be mapped and interpreted via a regression analysis. If the shape parameter b does not change any more compared to the previous observation phase (cf. Sect. 12.1), it can be assumed that the failure behaviour does not change any more. Furthermore, the parameters can be extrapolated over a certain period of time. Extrapolation should therefore only be carried out in a meaningful way for manageable periods of time, as potential further damage foci can have an impact on the modelling of the failure behaviour.

Step B: Mapping and forecasting the usage profile
With regard to the units that have not failed, a distribution model is adapted to the usage data of the lifetime variable in order to represent the usage profile of the product fleet (PF). For many technical products the log-normal distribution model is suitable to represent a usage profile; cf. Eq. 12.100 (in particular Chap. 7 and Sect. 12.3.4). The utilisation profile (distribution model) is also adapted with respect to the suitable observation phases $t_1 - t_n$ in the field observation period t.

$$f_{PF}(t) = \frac{1}{\sqrt{2\pi} \cdot \sigma_{ln} \cdot t} \cdot exp\left(-\frac{1}{2}\left(\frac{\ln(t) - \mu_{ln}}{\sigma_{ln}}\right)^2\right) \qquad (12.100)$$

Via a regression analysis, the evolution of the parameter set—here: μ_i; σ_i—can be mapped and interpreted. Furthermore, the parameters can be extrapolated over a certain period of time to forecast future usage profiles. Extrapolation can only be performed in a meaningful way for manageable periods of time: Market influences, such as the transfer of commercially used product fleets to the private consumer sector, can significantly affect the usage profile. After a long period of field observation, after the main product use phase has been passed, the parameter set μ_i; σ_i no longer changes significantly.

Step C: Risk analysis and key indicators
In the field observation period t, the probability density functions of the failure behaviour $f_{FM,i}$ are superimposed with the probability density function of the utilisation profile $f_{PF,i}$ for the selected observation phases $t_1 - t_n$ at each field runtime t_i.

12.4 RAPP – Analysis of Serial Damages

If the failure behavior is mapped by means of two-parameter Weibull distribution model and the utilization profile by means of logarithmic normal distribution, Eq. 12.98 and Eq. 12.100 directly yield Eq. 12.101. Furthermore, quantiles related to failure behavior [$Q_{FM,LL}$, $Q_{FM,UL}$] and utilization profile [$Q_{PF,LL}$, $Q_{PF,UL}$] are defined (index LL = lower limit; UL = upper limit). The schematic diagram in Fig. 12.9 shows the superposition of two probability density functions. The resulting function $f_{DA}(t_i)$ is exaggerated for illustration. The function $f_{DA}(t_i)$ represents the probability of the damage focus (index DA = Damage Appearance) within the fleet in relation to the lifetime variable t.

$$f_{DA}(t_i) = f_{FM}(t_i) \cdot f_{PF}(t_i) \quad (12.101)$$

The overlay of the probability density functions of the failure behavior $f_{FM,i}$ and the utilization profile $f_{PF,i}$ as well as the resulting $f_{DA}(t_i)$ leads to the following key indicators that quantify the risk in the product fleet.

a) Critical area C, in which failures are expected,
b) Damage appearance: Damage occurrence with a high probability,
c) Risk probability P_{RPF} to the damage focus within the product fleet at time t_i,
d) Number of candidates N_{Can} with respect to a failure at time point i,
e) Saturation limit S_F, at which no increase of damages within the fleet is to be expected.

The *critical range C* with the limits C_{LL} ($= Q_{PF,LL}$) and C_{UL} ($= Q_{FM,UL}$) results from the limits with respect to the previously selected quantile limits of the

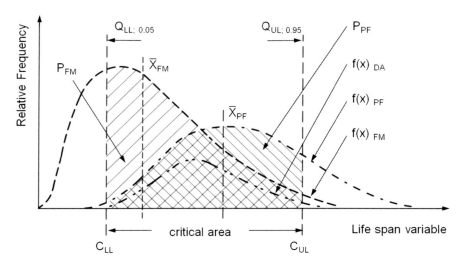

Fig. 12.9 Schematic diagram show the overlay of the probability density function of the failure behaviour $f_{FM,i}$ and the utilisation profile $f_{PF,i}$ as well as the resulting function $f_{DA}(t_i)$ in an exaggerated representation

distribution functions for the failure behaviour and utilisation profile of the product fleet; cf. Fig. 12.9. The critical range C is thus the interval with respect to the lifetime variable t in which non-failed units can be found which are to be regarded as potential candidates.

The *damage peak (Damage appearance)* is to be expected at the time of the maximum of the function $f_{DA}(t_i)$ (function for the probability of the damage peak within the fleet related to the lifetime variable t). The maximum results from the first derivative of the function $f_{DA}(t_i)$, which is set equal to zero (extreme value determination); Eq. 12.102.

$$\frac{df_{DA}(t_i)}{dt_i} = 0 \tag{12.102}$$

The *risk probability* P_{RPF} *to the damage centroid within the product fleet* at time t_i is obtained on the basis of the multiplication law from the probability of the occurrence of a damage event P_{FM} (integral of $f_{FM}(x)$) and the probability P_{PF}, in each case with the limits analogous to the critical area C; Eq. 12.103. P_{PF} denotes the probability related to the *critical product fleet proportion* in exactly this area (integral of $f_{PF}(x)$ with the limits of the critical area C).

$$P_{RPF,i} = P_{FM,i} \cdot P_{PF,i} = \int_{C_{LL}}^{C_{UL}} f_{FM}(t_i)dt_i \cdot \int_{C_{LL}}^{C_{UL}} f_{PF}(t_i)dt_i \tag{12.103}$$

The number of *candidates* N_{Can} (index Can = Candidates) with respect to a failure at time i is easily calculated according to Eq. 12.104.

$$N_{Can} = N_{PF} \cdot P_{PF,i} = N_{PF} \int_{C_{LL}}^{C_{UL}} f_{PF}(t_i)dt_i \tag{12.104}$$

The *saturation limit* S_F, at which no further failures are to be expected within the fleet, can be determined on the basis of the Verhulst model (cf. Chap. 7); Eq. 12.105. Where as parameters a represents the intercept (number of known damage occurrences at time t_0), S the growth limit (no further increases in failures to be expected) and k the proportionality factor. The saturation model can also be applied to the time of the damage peak (damage appearance). In this case, the time-related variable t is the field observation time; a detailed application is shown in (Bracke and Sochacki 2019).

$$f(t) = \frac{a \cdot S}{a + (S - a) \cdot e^{-Skt}} \tag{12.105}$$

Damage scenarios and risk probability $P_{RPF,i}$

A regression analysis of *risk probability* $P_{RPF,i}$ and field duration t_i (here: month in service) can provide information on the *risk trend*. In principle, three damage scenarios can be observed:

12.4 RAPP – Analysis of Serial Damages

a) *Scenario 1:* The focus of the documented failures is limited to the early field observation phase. The focus of the usage profile of the product fleet moves away from the focus of the failure model related to the lifetime variable. As a result, the risk probability P_{RPF} decreases and the critical area C becomes narrower. As a result, the damage peak (damage appearance) will not change significantly over the longer term observation. This principle is indicated in Fig. 12.9: From a time perspective, the fleet is already in higher value ranges compared to the failure model.

b) *Scenario 2:* The centroid of the failures in relation to the lifetime variable is in a higher value range compared to the centroid of the utilisation profile of the fleet. This scenario can occur if a part of the fleet is subject to intensive use, so that the relevant value range of the lifetime variable has not yet been reached by the remaining main part of the fleet. At the same time, this is usually a runtime-related failure behaviour; cf. Sect. 12.1. In this scenario, risk probability P_{RPF} increases and the critical area C increases. The expected damage peak (damage appearance) is still in the future from the point of view of the utilization profile.

c) *Scenario 3:* There is an (almost) random failure behaviour, in which the failure times extend over a wide observation period. In this case, the centroids of the utilization profile of the product fleet and the failure model can move in parallel as time progresses. This effect leads to a stagnating risk probability P_{RPF} as well as to a damage peak, which also moves continuously, corresponding to failure behavior and utilization profile.

Annotation:
Comparison of the RAPP (Bracke 2013) method versus method according to Eckel (1977) and Kaplan-Meier (1958)

The focus of the outlined methods according to Eckel (1977) and Kaplan and Meier (1958) is the determination of the probability of failure at time i related to the lifetime variable X (e.g.: switching cycles, distance travelled, operating time), usually using an empirical usage profile. Furthermore, the methods according to Eckel and Kaplan-Meier (as well as the methods according to Johnson (1964) and Nelson (1982)) are based on the assumption that each unit in the fleet is also a potential candidate with respect to the damage. In contrast to the methods of Eckel and Kaplan-Meier the RAPP method takes into account both the lifetime variable X (e.g. switching cycles, distance travelled, operating time) and the field lifetime (e.g. month in service). Based on the lifetime variable X, the failure behaviour is mapped in relation to the damage data and the usage profile of the fleet. The additional consideration of the field operating time enables the determination of the field risk with regard to risk probability, damage peak and saturation behaviour. A saturation behaviour cannot be represented by the methods according to Eckel and Kaplan and Meier due to the assumption outlined above. A saturation behavior can be observed if, for example, scatter effects in the manufacturing of different production batches lead to heterogeneity in the failure behaviour within the fleet. A comparison of the RAPP method with the approaches according to Kaplan-Meier, Johnson and Eckel is shown in Sochacki and Bracke (2017).

12.4.2 Case Study RAPP: Damage Focus in a Fleet

In this chapter, the application of the "Risk Analysis and Prognosis of complex Products (RAPP)" procedure for the analysis of damage foci within product fleets in the use phase is outlined on the basis of a case study.

Base of operations and objective of the analysis
Given a fleet of vehicles that has been in the field for nine months. Furthermore, there is a focus of damage of a mechanical component. Furthermore, the damage data of the affected vehicles as well as the data of the vehicle fleet according to the data structure in Table 12.4 based on the lifetime variable switching cycles x are given. The focus of the damage is characterised by a failure behaviour due to the operating time (cf. Fig. 3.2).

The goal ist the mapping of the development of the *risk probability* P_{RPF} to the damage focus within the product fleet as well as the *damage peak* $f'_{DA}(t)$ (Damage appearance); cf. 12.4.1.

Outline of the application of the RAPP procedure
Step A: Mapping and forecasting of product failure behaviour
The failure behaviour of the failed units is represented by a two-parameter Weibull distribution model (Eq. 12.98), which is fitted to the damage data. A field observation period of nine months (MIS = Month in Service) is available. Nine field observation points of time are chosen for risk quantification, accordingly Weibull distribution models $f_{FM,i}$ are fitted for t = 1, ..., 9 MIS (Eq. 12.98). Table 12.5 shows the corresponding parameters of the Weibull distribution models $f_{FM,i}$ with (i = 1, ..., 9 MIS) as well as mean and extreme values. On the basis of a regression analysis, an extrapolation of the parameters of the Weibull distribution models is carried out and thus the parameters are forecast for the period 10–12 MIS; cf. Table 12.6. The visualization of the temporal development of the model parameters (failure behaviour), both estimation on the basis of 9 MIS (available data) and extrapolation (forecast 10 to 12 MIS) is shown in Fig. 12.10 (left).

Step B: mapping and forecasting the usage profile
With regard to the vehicles that have not failed, a log-normal distribution model is fitted to the usage data with regard to the lifetime variable x (Eq. 12.100) in order to represent the usage profile of the product fleet (PF). The utilisation profile (distribution model) is also fitted with respect to the observation points of time 1, ..., 9 MIS in the field, ergo a observation period of nine MIS. An overview of the utilisation profiles for the field observation period of 9 MIS is shown in Table 12.7. On the basis of a regression analysis, an extrapolation of the parameters of the log-normal distribution models is carried out and thus the parameters are extrapolated for the period 10–12 MIS; cf. Table 12.8. The visualisation of the temporal development of the model parameters (utilisation profile), both estimation on the basis of 9 MIS (available data) and extrapolation (forecast up to 12 MIS) is shown in Fig. 12.10 (right).

12.4 RAPP – Analysis of Serial Damages

Table 12.5 Failure behaviour of the vehicles based on available damage data: Field observation period 1–9 MIS, estimated values for shape parameter b and characteristic life T of Weibull distribution models; mean value, minimum/maximum value

MIS	$\hat{b}$	$\hat{T}$	$\hat{\mu}$	x_{Min}	x_{Max}
1	0.91247719	3,030.22345	3171.39912	63.3272644	20,773.2379
2	1.15406527	3,332.27676	3170.80796	141.848585	8,945.35683
3	1.32064635	4,950.08442	4547.37981	490.839457	17,156.5301
4	1.28149686	5,452.04944	5043.72303	282.221955	15,397.4132
5	1.23007458	6,302.01278	5914.59464	94.3904772	18,544.3115
6	1.45426251	8,479.14505	7645.43846	699.402875	32,263.5134
7	1.44484201	8,476.64261	7696.08934	512.297961	21,933.7251
8	1.44808067	10,100.5571	9175.62261	326.255453	27,132.3361
9	1.50288512	9,105.10789	8181.56585	672.532065	24,572.9327

Table 12.6 Predicted failure behaviour of vehicles within the fleet: field observation period 10–12 MIS; estimated values for shape parameter b and characteristic lifetime T of the Weibull distribution models as well as mean value

MIS	$\hat{b}_{Prog}$	$\hat{T}_{Prog}$	$\hat{\mu}_{Prog}$
10	1.52358631	10,414.4321	9,425.03073
11	1.54721035	10,891.7316	9,878.46628
12	1.56877741	11,290.1021	10,272.059

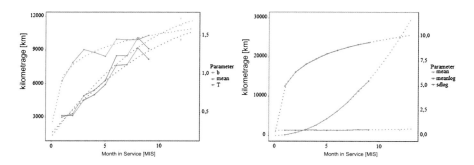

Fig. 12.10 Development of the parameters of the Weibull distribution model (failure behaviour) as well as the log-normal distribution model (utilisation profile); estimated parameters based on the available data set for the field duration 1–9 MIS as well as extrapolation up to the field duration 10–12 MIS (dashed curves)

Step C: Risk analysis and key figures
Within the field observation period $t = 9$ MIS, the probability density function of the failure behaviour $f_{FM,i}$ is overlayed with the probability density function of

Table 12.7 Product fleet: real existing usage profile; field observation period 1–9 MIS, estimated values for mean values and standard deviation of the log-normal distribution models; mean value, minimum/maximum value

MIS	$\hat{\mu}_{ln}$	$\hat{\sigma}_{ln}$	$\hat{\mu}$	x_{Min}	x_{Max}
1	4.98012415	0.51092698	165.672463	32.2803296	609.880982
2	6.38637883	0.50812684	674.937069	110.059027	2,943.87933
3	7.18219887	0.5322207	1,513.45945	211.070554	6,875.78679
4	7.7375198	0.53249595	2,649.62161	432.298451	15,505.9692
5	8.22240558	0.53060496	4,278.53759	706.653961	17,545.3847
6	8.58247707	0.53113601	6,146.98965	814.837557	33,479.3965
7	8.89862155	0.55241846	8,516.0784	1,316.25188	35,594.786
8	9.16514417	0.58556217	11,351.4684	1,471.17848	75,889.7792
9	9.3719232	0.58359181	13,927.4942	1,536.74862	69,527.8478

Table 12.8 Predicted parameters of the vehicle fleet usage profiles: field observation period 10–12 MIS; estimated values for mean values and standard deviation of the log-normal distribution models as well as mean value

MIS	$\hat{\mu}_{ln,Prog}$	$\hat{\sigma}_{ln,Prog}$	$\hat{\mu}_{Prog}$
10	9.60826989	0.59451517	17,766.437
11	9.78504247	0.59219319	21,127.3924
12	9.96457542	0.58939236	25,231.4055

the utilisation profile $f_{PF,i}$ at each field operation time t_i for the selected observation phases $t_1 - t_9$ (i = 1, ..., 9 MIS) (Eq. 12.101). Furthermore, the 0.95 quantiles related to failure behavior [$Q_{FM,LL}$, $Q_{FM,UL}$] and utilisation profile [$Q_{PF,LL}$, $Q_{PF,UL}$] are defined (index LL = lower limit; UL = upper limit).

As an example, Fig. 12.11 (left) shows the overlay of the density functions of failure behavior and utilisation profile at time t_9 after a field observation period of 9 MIS, which corresponds to the end time of the available database on damage events and utilisation data. The resulting function $f_{DA,i}$, which expresses the probability of the damage appearance (index DA) within the fleet in relation to the lifetime variable X (Eq. 12.101) is hardly recognizable, since the probabilities are correspondingly low due to the overlay (multiplication law). In contrast, Fig. 12.11 (right) shows the superposition of the failure behavior and utilisation profile functions based on the predicted distribution models for the field observation time 12 MIS. If the overlay is compared at field observation time 9 MIS (known database) and 12 MIS (forecast), it can be seen that the centroid of the documented failures develops in the early field observation phase; cf. also the outlined damage scenario 1 in Sect. 12.4.1. The centroid of the product fleet

12.4 RAPP – Analysis of Serial Damages

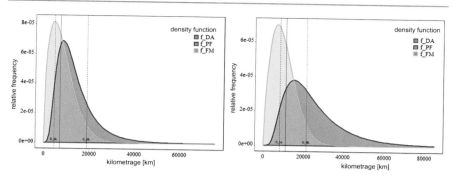

Fig. 12.11 Overlay of the probability density functions of the failure behaviour $f_{FM,i}$ and the utilisation profile $f_{PF,i}$ as well as the resulting function $f_{DA,i}$ at the field observation time 9 MIS (present database) as well as 12 MIS (forecast)

utilisation profile moves away from the centroid of the failure model as the field duration related to the lifetime variable X increases. Accordingly, the point in time at which the *damage peak (Damage appearance)* $f'_{DA,i}$ based on the function $f_{DA,i}$ is to be expected, moves forward (Eq. 12.102). However, the *risk probability* P_{RPF} *to the damage centroid within the product fleet* (Eq. 12.103) decreases, because the probability P_{PF} with respect to the *critical product fleet percentage* within the critical area declines.

The development of the resulting function $f_{DA,i}$, which depicts the damage centroid within the fleet in relation to the lifetime variable X (*damage appearance*), is shown in Fig. 12.12. As expected, the models $f_{DA,i}$ are characterised by an increase in scatter and a continuous increase in the time of the *damage peak* $f'_{DA,i}$ (*damage appearance;* maximum of the function $f_{DA,i}$).

Summary: Risk Probability P_{RPF} to the Damage Centroid within the Product Fleet and Damage Peak

The development (based on the available damage data) and forecast (extrapolation) of the *risk probability* P_{RPF} *to the damage peak within the product fleet* as well as the time of the *damage peak (damage appearance)* is shown in Fig. 12.13. Initially, the *risk probability* P_{RPF} *to the damage centroid within the product fleet* increases, since a large part of the fleet is within the critical area C. After field observation time point 7 MIS, a large part of the fleet has passed critical area C, ergo the *risk probability* P_{RPF} decreases in its trend (8 and 9 MIS). Therefore, the forecast (extrapolation) shows a decreasing *risk probability* P_{RPF} for field observation time points 10–12 MIS. Furthermore, the continuous increase in the time of the *damage peak* $f'_{DA,i}$ (*Damage appearance*) can be seen, due to the increasing mileage data of the product fleet.

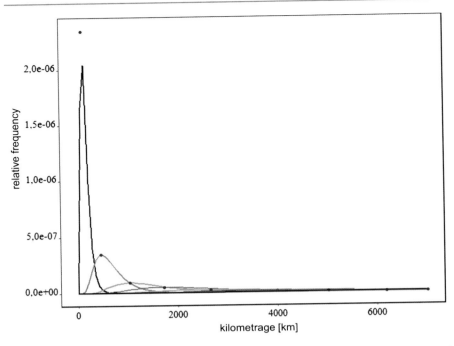

Fig. 12.12 Evolution of the resulting function $f_{DA,i}$, with (i = 1, ..., 12) based on the overlays of the failure behavior and utilisation profile models

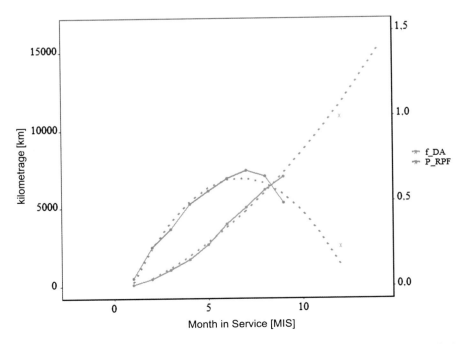

Fig. 12.13 The development (based on the available damage data) and forecast (extrapolation) of the *risk probability P_{RPF} to the damage peak within the product fleet* as well as the time of the *damage peak* (*damage appearance*)

12.5 Case Study Sensor: Failure Behaviour; Saturation Model Versus Weibull Model

The present chapter shows the analysis of a failure behavior using Verhulst saturation model and Weibull distribution model (state of the art) based on field data related to a vehicle fleet in the use phase. Based on a field observation period, failures of front-end sensors within the fleet are known. The failure behaviour is to be modelled, parameters interpreted and the models compared.

Base of operations
When analysing field data of a vehicle fleet in the use phase, an accumulation of sensor failures – in total n = 20 sensor failures; cf. Table 12.9 (column field runtime t, based on the life span variable operating months [m]) – is detected. The fleet itself comprises N = 1,000 vehicles. In addition to the time of occurrence of the damage, the failure data also include the number of damage events including the currently present damage event; cf. Table 12.9 (column Number of damage events N(t)).

Objective
A reliability analysis for the damage centroid of the sensor damage case is performed. The failure behaviour is analysed with the help of a Verhulst saturation model and a Weibull distribution model. Finally, the results of the reliability analysis are interpreted.

Table 12.9 Damage data regarding the sensor damages within a product fleet as well as required logarithms

Field running time t [m]	Number of claims N(t)	ln(1/N(t)-1/S)
1.1714	1	−0.0010
1.9125	2	−0.6951
1.9346	3	−1.1016
1.9452	4	−1.3903
1.9591	5	−1.6145
1.9786	6	−1.7978
2.3740	7	−1.9529
3.0582	8	−2.0875
3.4821	9	−2.2063
3.5447	10	−2.3126
3.5523	11	−2.4090
3.7471	12	−2.4970
4.3773	13	−2.5780
4.7362	14	−2.6532
4.7600	15	−2.7232
4.8016	16	−2.7887
4.9541	17	−2.8504
4.9923	18	−2.9085
5.2217	19	−2.9636
5.4922	20	−3.0159

Procedure

a) Mapping of the failure behaviour on the basis of a saturation model according to Verhulst related to the product fleet.
b) Mapping of the failure behaviour by means of a Weibull distribution model on the basis of the available damage data (sample) with the aid of a logarithmic regression.
c) Illustration of the failure behaviour with the aid of a Weibull distribution model on the basis of the available damage data related to the product fleet (basic population; cf. rank correction according to Johnson, Sect. 12.3.2).
d) Visualization of failure behavior on the topic of sensor malfunction within the product fleet: saturation model versus Weibull distribution model.
e) Comparison and interpretation of Verhulst and Weibull modelling.

Solution
First, the failure behavior of the sensor is modeled based on a Verhulst saturation model; cf. Eq. 12.106 (also Sect. 7.5.1). The model is fitted using the following steps:

a) Assumption: every vehicle in the fleet is a candidate, ergo the saturation limit $S = 1{,}000$ is set (rather than estimated) because the fleet contains 1,000 vehicles.
b) Preparation of the logarithmic regression analysis: From Eq. 12.106 a linear function is obtained (cf. Eq. 12.107) if the reciprocal of the saturation limit is subtracted from the reciprocal of the number of damage cases N(t) and the result is logarithmized, Table 12.5 column ln(1/N(t)-1/S).
c) Performance of a bivariate regression analysis with the variables field operation time t and the logarithmised number of damage events, taking into account the saturation limit ln(1/N(t)−1/S); determination of the parameters of the linear relationship; cf. Eq. 12.108 (gradient m), 12.109 (intercept n) as well as Fig. 12.14.
d) Derive the parameters of the saturation model from the logarithmic regression ("Conversion of parameters"); Eqs. 12.110 and 12.111.
e) Completion of the Verhulst saturation model with parameters $a = 1.229638813$ and $k = 0.000548438$; $S = 1{,}000$ (given saturation limit; since the product fleet is 1,000 vehicles.

$$f(t) = \frac{a \cdot S}{a + (S - a) \cdot e^{-Skt}} \qquad (12.106)$$

$$y = mx + n \qquad (12.107)$$

$$m = -0.5484 \qquad (12.108)$$

$$n = -0.208 \qquad (12.109)$$

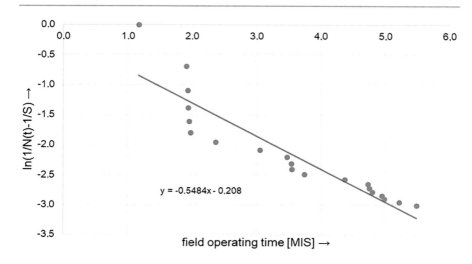

Fig. 12.14 Regression model based on the variables field operation time t as well as number of damage cases considering the saturation limit (logarithmized) to determine the parameters of the saturation model

$$k = -\frac{m}{S} = -\frac{-0.5484}{1000} = 0.000548438 \quad (12.110)$$

$$a = -\frac{S}{1 + Se^n} = \frac{1000}{1 + 1000e^{-0.208}} = 1.229638813 \quad (12.111)$$

Furthermore, the failure behaviour is modeled based on the two-parameter Weibull distribution model; cf. Eq. 12.112. The model is fitted based on the following steps:

a) Determine the cumulative probabilities related to the damage time points i; Eq. 12.113.
b) Estimation of Weibull distribution model parameters using the maximum likelihood estimator (MLE), initially based on the damage data, yields: Shape parameter b = 2.96; Characteristic life span T = 3.937 months.
c) Estimation of the Weibull distribution model parameters with the aid of the maximum likelihood estimator (MLE) on the basis of the damage data and using the candidate prediction by means of rank correction according to Johnson (cf. Sect. 12.3.2) yields: Shape parameter b = 2.96; Characteristic life T = 14.67 months.

$$F(t) = 1 - e^{-\left(\frac{t}{T}\right)^b} \quad (12.112)$$

$$F(t_i) = \frac{i - 0.3}{n + 0.4} \quad (12.113)$$

Discussion of the results
The illustration of the sensor failure behaviour by means of the Verhulst saturation model is shown in Fig. 12.15. The field observation period in which the damage cases were observed can be seen enlarged on the right in Fig. 12.15. The model was fitted based on damage data from a field observation period of 5.5 months, in that it should be noted that the extrapolation is based on the model assumption of exponential progression with limited growth (saturation limit). Given the extent of extrapolation shown, verification of the model at a later field observation period is advised. Nevertheless, a first estimate of the damage events still to be expected can be made on the basis of the known failure behaviour at time $t = 5.49$ months (last known damage event). It should be noted that a saturation model (growth model) always includes an initial inventory (intercept a). In the context of mapping the sensor failure behavior, the initial stock is $a = 1.23$; interpretable as the first defective sensor.

The illustration of the sensor failure behaviour with the aid of the Weibull distribution model is shown in Fig. 12.16: On the left is the fitting of the Weibull distribution model based on the known damage cases, on the right the candidate prognosis including the units that did not fail (note: product fleet 1,000 vehicles) based on the rank correction method according to Johnson. The shape parameter $b = 2.96$ indicates an operation time-related failure behaviour. It is equally true here, as already discussed for the saturation model, that the extrapolation covers a wide field observation period and should therefore be used with caution when estimating the number of failures still to be expected. In contrast to the saturation model, the Weibull model does not include an intercept which could be interpreted as an initial percentage.

The courses of the two functions are similar, both are based on exponential growth. The saturation model is a function (hence: it is of course not a probability density) and shows the expected failures in absolute frequency. In contrast, the Weibull model is a distribution function based on a probability density. Therefore, Figs. 12.15 and 12.16 represent the sensor failure behavior with respect to the fleet in two different models: The Verhulst saturation model (Fig. 12.15) with saturation limit ($S = 1,000$ vehicles) and the Weibull distribution model as well as candidate prognosis (Fig. 12.16) with limit $F(t) = P = 1$ (precisely: the limit at t towards infinity is one). Both exponential-based models show similar sensor failure behaviour.

Risk prognosis using the B_{90} parameter
A comparison of the B_{90} parameter (Sect. 12.1.4 and Eqs. 12.114 and 12.115) of the Weibull model (parameter $b = 2.96$ and $T = 14.67$ months) with the analogue of the percentage damage rate of 90% based on the Verhulst model substantiate the similar sensor failure behaviour in both models. The analogue is obtained based on rearranging the saturation model (Eq. 12.106) after t and substituting the parameters ($S = 1,000$; $a = 1.229638813$; $k = 0.000548438$) and the function value $f(t) = 900$ (damage rate 90%, based on a fleet of 1,000 vehicles) to Eq. 12.116. According to both models, similar field observation times are obtained at which 90% of the fleet may be affected by a failure. Estimating the B_{90} parameter of the Weibull model yields the probability $P = 0.9$ of a failure at 19.44 months field

12.5 Case Study Sensor: Failure Behaviour ...

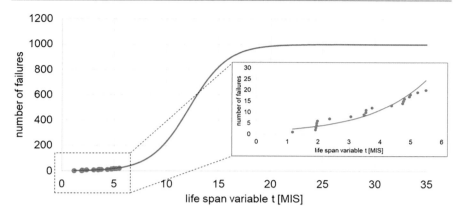

Fig. 12.15 Saturation model according to Verhulst for mapping the sensor failure behaviour in the vehicle fleet (S = 1000); optical enlargement of the field observation period in which sensor failures were observed

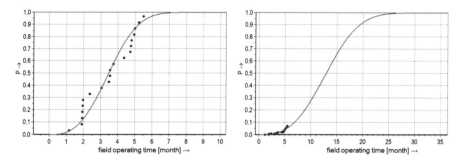

Fig. 12.16 Illustration of the failure behaviour with the aid of a Weibull distribution model (basis: damage data; Fig. left) and with consideration of the candidates (rank correction according to Johnson; basis: damage data, population; Fig. right)

duration. In contrast, according to the Verhulst model, n = 900 vehicles (90% of the fleet) failed at 16.22 months of field runtime.

$$B_Z = T[-\ln(1-Z)]^{\frac{1}{b}} \quad (12.114)$$

$$B_{90,Weibull} = T[-\ln(0.1)]^{\frac{1}{b}} = 19.44 \quad (12.115)$$

$$t_{90,Verhulst} = \frac{\ln\left(\frac{a \cdot S}{f(t)} - a\right) - \ln(S - a)}{-Sk} = 16.22 \quad (12.116)$$

Significance Tests 13

Chapter 13 focuses on data analysis on the basis of significance tests (hypothesis tests). Section 13.1 introduces the topic of significance testing and explains possible objectives in the context of an application example. The fundamentals of significance testing as well as the associated key indicators and terminology are the focus of Sect. 13.2. Significance tests for the investigation of an existing series of measurements—the one-sample case—are presented in Sect. 13.3. If analyses are performed on the basis of observations within two series of measurements (two-sample case), Sect. 13.4 shows suitable significance tests. If more than two samples are available, investigations with significance tests for the multiple sample case can be used; Sect. 13.5.

13.1 Introduction and Application Example

The focus of this chapter is an introduction to data analytics with the aid of significance tests. First, a hypothesis regarding an issue is defined. Subsequently, series of measurements are taken with the aim of substantiating the hypothesis through observations. The application of a significance test serves to examine these observations with regard to a significance statement.

Observations can be, for example:

- Trend in a series of measured values,
- Periodicities in a measured value series,
- Outlier within a value pattern,
- Different characteristics in different prototypes due to different stages of development,

- Different characteristics in two or more series of measurements with reference to different production batches,
- Different failure behaviour of products in the field based on failure data.

The statistical test – also called hypothesis test or significance test – belongs to the discipline of mathematical statistics. First, a null hypothesis is formulated on the basis of an observation, e.g. "The centroid of two series of measurements are not different". The alternative hypothesis is "The centroids of two series of measurements are different". A significance test can be used to examine whether the null hypothesis is valid or invalid on the basis of the available measured values or observations. If the null hypothesis is rejected, a decision is made in favour of the alternative hypothesis.

Thus, the statistical test can be used when the data analysis does not provide a clear, quantitative result and does not allow a decision to be made regarding an issue. The statistical significance test can support decisions when there is uncertainty due to a lack of knowledge about the population and only knowledge about samples is available. In principle, the analysis is performed on the basis of single samples (one-sample case), two samples (two-sample case) or of several samples (multiple sample case).

Example: Assessment of performance differences between prototype engines and series engines
Within the various phases of engine development and production, test bench trials are carried out (see also Chap. 2). During testing, test characteristics such as power, torque, oil consumption, oil pressure and fuel consumption are measured. Within the scope of an analysis, the measurement data of the development engines (prototypes) are compared with the engines from the production start-up (series engines). Figures 13.1 and 13.2 show the value patterns and frequency distributions of the corresponding measured values of the test characteristic power [kW] from testing and series production, as well as an adapted normal distribution model in each case. Table 13.1 shows the sample size as well as the estimated sample parameters arithmetic mean and standard deviation of both samples.

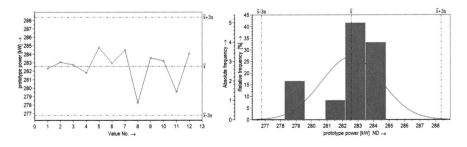

Fig. 13.1 Value pattern and frequency distribution of the engine prototype measurement series

13.1 Introduction and Application Example

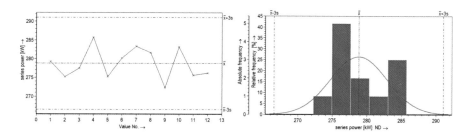

Fig. 13.2 Value pattern and frequency distribution of the motor series production measurement series

Table 13.1 Sample parameters and size based on measured performance data (prototypes versus series engines)

	Prototypes	Series engines
Arithmetic mean	286.57 kW	279.4 kW
Standard deviation	1.93 kW	4.10 kW
Sample size	12	12

When comparing the sample parameters, it is noticeable that the performance of the prototypes is higher compared to the series engines. Furthermore, the scatter of the performance data of the series engines seems to be higher compared to the prototypes. From a technical point of view, this seems plausible, since in prototype construction there are fewer variables influencing the manufacturing process (higher reproducibility; e.g.: no worker shift system, higher measurement accuracy) and thus manufactured engines have a lower performance scatter. However, due to the few measured values on the basis of which the parameters mean value and standard deviation were estimated, the question of whether these differences are significant would be open. The research question for the two-sample case presented here is: "Do prototype engines have significantly higher performance (mean) compared to production engines and is this performance reproducible with significantly lower scatter?" Possible procedure for investigating the series of measured values using significance tests:

(a) Distribution fit test (here: normal distribution) for both measurement series for the decision whether parametric or non-parameteric significance tests should be used in the further analysis. Also due to the small sample sizes, the decision would be made in favour of the non-parameteric significance tests.
(b) Separate examination of the samples: significance test with regard to the randomness of the sequence of measured values (cf. Sect. 13.3.1).
(c) Comparison of the centroids of both measurement series via significance test according to Mann–Whitney (cf. Sect. 13.4).

(d) Comparison of the dispersion of both series of measurements by means of a significance test according to Siegel-Tukey (cf. Sect. 13.4).

The assessment of the performance data according to the procedure described above provides indications as to whether the prototypes differ significantly from the series engines with regard to their performance data in terms of centre of gravity and dispersion.

13.2 Basics of Significance Tests

This chapter outlines the basics of a significance test. First, the general procedure of a significance test and the meaning of the p-value are presented (Sect. 13.2.1). Furthermore, the errors of the 1st and 2nd kind as well as the test efficiency are discussed (Sect. 13.2.2). Before carrying out a two- or multi-sample test, it must be known whether dependent or independent samples are involved; the basics of this are given in Sect. 13.2.3. Furthermore, parametric or non-parameteric tests can be used for the analysis of concrete samples: Background and selection options are discussed in Sect. 13.2.4. Section 13.2.5 shows an overview of the common significance tests for the one, two and multiple sample cases, most of which are also explained in the following Sects. 13.3–13.5.

13.2.1 Performance of Significance Tests, p-Value

The performance of a statistical significance test often follows a broadly similar scheme, which is outlined below.

Step 1: Formulation of hypotheses and determination of significance level
Formulation of a null hypothesis H_0 and an alternative hypothesis H_1. The hypothesis can be two-sided (e.g. according to Eqs. 13.1 and 13.2) or one-sided formulated. Subsequently, the significance level α is determined, on the basis of which the test decision is to be made (e.g.: $\alpha = 0.01$; $\alpha = 0.05$; $\alpha = 0.1$).

$$H_0 : U = U_0 \quad (13.1)$$

$$H_1 : U \neq U_0 \quad (13.2)$$

Step 2: Calculation of a test statistic T
The core of a statistical test is the calculation of a test statistic T with the variables X_i; Eq. 13.3. This is calculated on the basis of (an) existing sample(s) with the realizations x_i: The result is the estimate t; Eq. 13.4.

$$T = g(X_1; X_2; \ldots ; X_n) \quad (13.3)$$

$$\widehat{t} = g(x_1; x_2; \ldots ; x_n) \quad (13.4)$$

13.2 Basics of Significance Tests

Step 3: Determination of critical area
The test statistic T follows a distribution. On the basis of this distribution, a critical range is obtained (null hypothesis is rejected; alternative hypothesis is accepted) and a non-critical range (null hypothesis is accepted; alternative hypothesis is rejected); cf. Fig. 13.3. The limits of the critical or non-critical range (for two-sided test performance c_o and c_u; for one-sided test performance c) are dependent on the significance level α. In the case of two-sided tests, the test statistic T is related to the (non-)critical range according to Eq. 13.5. Equation 13.6 applies to the probability P that the test statistic T lies within the non-critical range.

$$c_u \leq T \leq c_o \tag{13.5}$$

$$P(c_u \leq T \leq c_o)_{H_0} = \gamma = 1 - \alpha \tag{13.6}$$

Step 4: Test decision
For the test decision the test statistic T (or the estimated value t based on the available sample) is compared with the critical range; cf. Fig. 13.3. If the estimated value t lies in the non-critical range, the null hypothesis is accepted. If the estimated value t is in the critical range, the null hypothesis is rejected in favor of the alternative hypothesis. The test decision is made with significance level α.

The p-value
When performing the significance test, the p-value is often documented. The p-value is the probability of making a statement regarding the null hypothesis with respect to the present sample (realization). Furthermore, the probability (the p-value) also reflects observing the statement regarding the null hypothesis on the basis of another sample; cf. Fig. 13.4. The p-value, which is always calculated on the basis of the present sample, thus always refers to the information content of

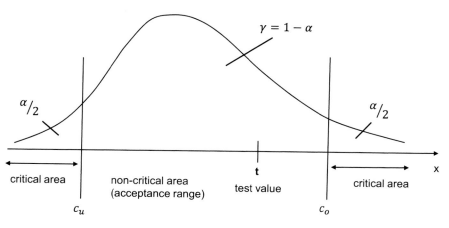

Fig. 13.3 Schematic diagram for the application of a statistical significance test (here: two-sided execution)

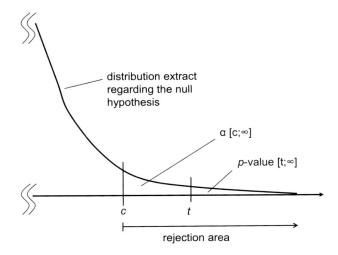

Fig. 13.4 The p-value on the basis of an existing sample (realisation) in relation to the critical range or the significance level α

this concrete realization. The smaller the p-value, the more likely the null hypothesis is rejected. The p-value is compared with the significance level α when the test is performed. If the p-value is below the significance level α, the null hypothesis is rejected. Thus, it is also true that in this case the test statistic T is in the critical region. The rejection of the null hypothesis is then called "statistically significant". Under no circumstances should a decision be made solely on the basis of the p-value, as it only applies to the sample at hand and can thus reflect a false certainty of decision.

If the significance level were to correspond exactly to the p-value, the p-value would be the smallest level at which the observation – realized by the present sample—could still be judged as "statistically significant" via hypothesis testing.

13.2.2 Error 1st Kind, Error 2nd Kind and Test Efficiency

Furthermore, it should be noted that when performing a significance test, an error of the 1st kind or an error of the 2nd kind can occur (cf. Table 13.2). The error of

Table 13.2 Test decision and truth related to errors of 1st and 2nd kind

		The test decision	
		H_0 is accepted	H_1 is accepted
The truth	H_0 is correct	No error	Error 1st kind (α)
	H_1 is correct	Error 2nd kind (β)	No error

13.2 Basics of Significance Tests

the first kind (also referred to as α) means that the null hypothesis H_0 is rejected, although H_0 is correct. The second type of error (also referred to as β) means that the null hypothesis H_0 is accepted although H_0 is false.

Figure 13.5 shows the relation of the error of the 1st type (α-error) and the error of the 2nd kind (β-error) to each other. Visualized are the assumed distribution model for parameter v_0 as well as the real distribution model for parameter v_1 and critical limit c. If the error of 1st kind is chosen small, the probability of error of 2nd kind is large (lower part of the figure). However, if the probability of the 2nd kind error is smaller, the probability of the 1st kind error increases (upper figure part).

Note on error 1st kind and error 2nd kind

1. The size of the 1st kind error is called the significance level α of the statistical test (cf. also performing a statistical significance test: Determining the significance level before performing the test).
2. The significance level α is chosen low (e.g.: $α=0.01$; $α=0.05$; $α=0.1$).
3. In a siginificance test, the error of the 2nd kind cannot be calculated because there are several alternatives (probability of H_1) to the assumed model (probability H_0) with regard to the null hypothesis. If, however, only one alternative is accepted (probability is different from H_0), this is referred to as an alternative test (instead of a siginificance test). Here, the calculation of the error of the 2nd kind is feasible: The non-critical area of the null hypothesis is in the center under the condition that the alternative hypothesis is correct. The error

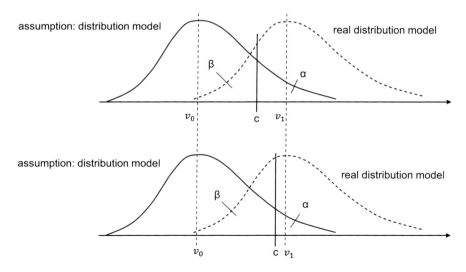

Fig. 13.5 Schematic diagram for error of the 1st kind (α) and error of the 2nd kind (β). If the possible error 1st kind is large, the probability of error of 2nd kind is small (upper part of the figure). If the error 1st kind is chosen lower, the probability of error 2nd kind is higher

of the 2nd kind is then exactly the probability that the test result lies in the non-critical range of the null hypothesis H_0, although H_1 holds.
4. 1st kind error and 2nd kind error are not independent of each other: a small 1st kind error leads to larger 2nd kind error (and vice versa). However, a 2nd kind error (β) is not the counter probability to the 1st kind error (α). Ergo, $\alpha + \beta = 1$ does not hold. If this equation is satisfied, it is a coincidence.

Criteria for evaluating the efficiency of a significance test
The efficiency of a significance test can be assessed using the following criteria: Power (also: goodness), Unbiasedness, consistency and robustness.

1. Power (also: goodness): The higher the probability of actually making a decision in favour of the alternative hypothesis H_1 in the presence of alternative hypothesis H_1, the higher the test power ($= 1 - \beta$). In this case, the error of the 2nd kind is all the smaller.
2. Unbiasedness: A test decision at significance level α is called unbiased if the probability of rejecting H_0, where H_0 is false, is at least as great as the probability of rejecting H_0 if the null hypothesis H_0 is true. In this case, $\beta \leq 1 - \alpha$ applies.
3. Consistency: A series of tests at a significance level α is performed. Consistency exists if the power of the tests converges to 1 with increasing sample size.
4. Robustness: Many statistical tests require assumptions to be made (e.g. population follows a normal distribution model). If the test power does not change or changes only slightly when an assumption is incorrect, the test is said to be robust.

13.2.3 Dependent and Independent Samples

Before analyzing multiple samples, it is necessary to examine whether the samples at hand are dependent or independent. If the observations/measured values of one sample are related to the observations/measured values of another sample, these samples are dependent on each other. However, if the observations/measured values of one sample do not contain information related to the observations/measured values of the other sample, the samples are independent of each other.

Definition
Dependent samples contain linked observations/measured values for a group of objects.

Independent samples are observations/measurements collected for two or more different, independent groups of objects.

13.2 Basics of Significance Tests

Example of fuel consumption measurement for vehicles
An analysis of the fuel consumption of vehicles of a given vehicle type X is carried out. Two studies are carried out.

Study A: The fuel consumption of $n=20$ vehicles is measured. The fuel consumption of 20 vehicles is measured on a test bench. Subsequently, the fuel consumption of the same 20 vehicles in real operation in road traffic is analysed. Both series of measurements (test bench versus operation in field) are compared with each other. It should be noted that the measured values are linked, since the two samples are based on the same objects (vehicles). Therefore, the samples are dependent on each other.

Study B: The fuel consumption of $n=20$ vehicles is measured. The fuel consumption of 10 vehicles is measured on a test bench. For a further 10 vehicles, the fuel consumption is measured in real operation. Subsequently, the fuel consumption is compared based on these two samples. The measured values were determined for groups of objects that have no binding. The vehicles in the two samples are not the same vehicles. Therefore, the samples are independent of each other.

13.2.4 Parametric and Parameter-Free Significance Tests

Basically, a difference can be made between parametric and non-parametric significance tests. Parametric significance tests refer to a parameter of the population, such as arithmetic mean or standard deviation. In the application of the significance test, the parameter estimates are analysed on the basis of an existing sample. However, knowledge or assumption of the distribution model (e.g. normal distribution for t-test or F-test) with respect to the population from which the sample is derived is assumed. If the distribution model of the population is not known, the analysis of a distribution model on the basis of the concrete sample is an important indicator. However, explicitly in the case of small sample sizes, it is often not possible to verify a specific distribution model: the data can be normally distributed or also follow a different (known or unknown) distribution model.

Non-parametric significance tests (also called "parameter-free significance tests" or "distribution-free significance tests") do not require any assumptions regarding the distribution of the population or sample under investigation. Parameter-free statistical significance tests are rank-based, i.e. ranks in one or

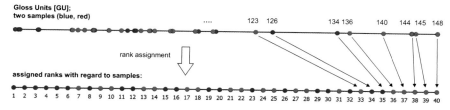

Fig. 13.6 Transformation of measured values (two samples; marked red/blue) into an ordinal rank scaling

more scales are assigned to the available data. The parameter-free test then refers to these ranks, not to the actual measured values.

Example:
The results of two different grinding processes are to be investigated. For this purpose, the gloss of the ground surfaces is analysed on the basis of two different random samples. The gloss measurement is carried out on 40 parts (two random samples of twenty parts each from two different production processes). The determined gloss values GU (Gloss Units) are plotted on a scale (cf. Fig. 13.6). Values of sample 1 (blue) and sample 2 (red) with different distances to each other can be seen. The measured values are then assigned ranks (ordinal transformation) so that the distance is equalized. The object of the significance test is now the ranks. The position or mixing of the values (blue/red) is an indicator for possible differences regarding to centroid and dispersion of the values.

Notes on non-parametric significance tests

1. Parameter-free significance tests are applicable regardless of the distribution model of the population or sample.
2. A low measurement level is required (nominal or ordinal measurement data).
3. Continuous/discrete characteristics may have to be transformed into a rank-based order (cf. Fig. 13.6); therefore, the following applies: The minimum level of the measurement data is nominal or ordinal, respectively.
4. Parameter-free tests leads to probabilities independent of the distribution of the population from which they are drawn.
5. Due to (1), non-parametric test procedures are universally applicable (compared to parametric procedures).
6. Suitable for small sample sizes and unknown populations (where no distribution model can be assumed or is known).
7. It should be noted that the use of ranks (transformation of data of continuous/discrete characteristics) involves a loss of information. The distance information is lost, empirically it is assumed that 10–15% of the actual information content is not taken into account. The example Fig. 13.6 shows this relationship, e.g. for measured value 126 [GU] and 134 [GU]. The clear distance between the measured values is no longer taken into account after the ranks have been assigned; the distance between the ranks is identical.

13.2.5 Overview: Statistical Tests for the One, Two, Multiple Sample Case

Chapter 13 contains the basics as well as the most important areas of application with regard to parametric and parameter-free statistical test procedures. Table 13.3 shows a selection of statistical test procedures; for most of them, both theoretical procedures and application examples are presented in Sects. 13.3 to 13.5.

13.2 Basics of Significance Tests

Table 13.3 Selection of statistical significance tests for the one, two and multiple sample case

Name	Test application	Parametric/Non-parametric	Note
Single sample case			
Wallis and Moore	Test for randomness of a measured value sequence	Parameter-free	Sect. 13.3.1
Kolmogorov–Smirnov Test	Test of goodness of fit	Parameter-free	Sect. 13.3.2; Here: Goodness of fit test, test for normal distribution
χ^2–Test	Test of goodness of fit	Parametric	Test for uniform distribution; test for distribution model; (Hedderich and Sachs 2020)
Anderson–Darling Test	Test of goodness of fit	Parameter-free	(Anderson and Darling 1952)
Swed-Eisenhart-Test / Wald Wolfowitz Runs Test	Randomization test	Parameter-free	(Büning and Trenkler 1994)
Sign-Test	Conformity of a position parameter with the default value	Parametric	Binomial test; Sect. 13.3.3
Outlier-Test	Test for outliers	Parametric	Population follows normal distribution, single-tailed skewed or arbitrary distribution; Sect. 13.3.4
t-Test	Comparison of mean value of a sample with reference value	Parametric	Sect. 13.3.5
Cox and Stuart Test	Test for trend	Parameter-free	Sect. 13.3.6
Two sample case			
Mann–Whitney-U Test	Comparison of two centroids	Parameter-free	Alternative to t-Test; Sect. 13.4.1
Siegel-Tukey Test	Comparison of two dispersions	Parameter-free	Alternative to the F-Test Sect. 13.4.2
Levene-Test	Comparison of two dispersions	Parameter-free	Alternative to the F-Test (Levene 1960)
t-Test	Comparing two mean values	Parametric	Assumes normally distributed populations; Sect. 13.4.3
F-Test	Comparison of two variances	Parametric	Assumes normally distributed populations; Sect. 13.4.4

(continued)

Table 13.3 (continued)

Name	Test application	Parametric/Non-parametric	Note
Multi-sample case			
Kruskal and Wallis Test	Comparison of several centroids	Parameter-free	Sect. 13.5.1
Post hoc analysis according to Conover	Detection of the centroid, which deviates in a multi-sample comparison	Parameter-free	Application after performing a Kruskal and Wallis test; Sect. 13.5.1
Bartlett-Test	Multiple variance comparison	Parametric	Assumes normally distributed populations; Sect. 13.5.2
Jonckheere and Terpstra-Test	Comparison of several samples with regard to a trend behaviour	Parameter-free	(Heddrich and Sachs 2020)
Meyer-Bahlburg-Test	Multiple variance comparison	Parameter-free	Based on Siegel-Tukey test; Sect. 13.5.3

For basic analyses of observations or series of measurements, the statistical tests shown provide a solid base of operations. Depending on the application, the well-known parametric test procedures (e.g.: t-test, F-test) or the respective parameter-free counterpart (e.g.: U-test, Siegel-Tukey test) can be used.

13.3 Single Sample Case

The present chapter shows a selection of significance tests for the investigation of a given sample (one-sample case) with regard to randomness of the measured value sequence, distribution model, reference value (location parameter), outlier behavior and trend behavior.

13.3.1 Wallis-Moore Test for Randomness

The test according to Wallis and Moore (1941) is a simple test procedure for investigating the randomness of a series of observations or measured values (Wallis-Moore test for randomness). The randomness refers to the temporal order in which the measured values were recorded. The analysis is focusing on the differences of the successive measured values. If the measured values are randomly distributed, the distribution of the differences should follow the normal distribution model (null hypothesis). In the literature, this test is also referred to as the difference sign iteration test.

13.3 Single Sample Case

Base of operations

- A sample or series of measured values with size n is available.
- The measured values $x_1, x_2, x_3, \ldots, x_n$ are present in the order of temporal acquisition, reflected by the index i.
- The size of the measured value series should be $n > 10$

Principle and goal
The principle ist the estimation of differences of the successive measured values; a "+" is assigned for a positive difference, a "−" is assigned for a negative difference. A succession of identical signs is referred to as a phase. Each time the sign changes, a new phase begins. The goal is to analyse the randomness of the phase distribution.

Procedure of the Wallis-Moore Test
Step 1: Formulation of the hypotheses and determination of the significance level
The hypotheses are stated as follows:
H_0: Random sequence of phases.
H_1: Non-random sequence of phases.
The significance level α is determined (e.g.: $\alpha = 0.01$ or $\alpha = 0.05$).

Step 2: Determination of differences, phases and number of phases
The measurement series with the measured values $x_1, x_2, x_3, \ldots, x_n$ is available. The differences $(x_{i+1} - x_i)$ are formed and checked according to Eqs. 13.7 and 13.8. If Eq. 13.7 is satisfied, the difference is considered positive. If Eq. 13.8 is satisfied, the difference is evaluated as negative. A succession of equal signs is called a phase. Then the total number h of phases is determined. The first phase and last phase are not taken into account.

$$x_{i+1} - x_i > 0 \qquad (13.7)$$

$$x_{i+1} - x_i < 0 \qquad (13.8)$$

Step 3: Calculation of the test statistics
The test statistic z is calculated according to Eq. 13.9 for an observation size (sample size) of $10 < n \leq 30$. For an observation size (sample size) of $n > 30$, Eq. 13.10 can be used. In addition to the sample size n, the total number of phases h is required (cf. Step 2).

$$\widehat{z} = \frac{\left| h - \frac{2n-7}{3} \right| - 0.5}{\sqrt{\frac{16n-29}{90}}} \qquad (13.9)$$

$$\widehat{z} = \frac{\left| h - \frac{2n-7}{3} \right|}{\sqrt{\frac{16n-29}{90}}} \qquad (13.10)$$

Step 4: Determination of critical value and test decision
The test statistic z is compared with the quantiles of the standard normal distribution cf. Appendix, Table A.2b. The test decision is made according to Eq. 13.11 with significance level α. If the condition is fulfilled, the null hypothesis H_0 is rejected in favour of the alternative hypothesis H_1. If H_0 is rejected, the phases (and hence the observed data) would not follow a random ordering.

$$\hat{z} > z_\alpha \qquad (13.11)$$

Note: Zero-Differences
If the formation of the differences $(x_{i+1} - x_i)$ results in one or more zeros (cf. step 2), the following procedure is used according to Kühlmeyer (2001:220): First, the difference is set equal to zero; thereafter, the following procedure is applied:

(a) Zeros between different signs are ignored.
(b) Zeros between equal signs are replaced by the adjacent sign and the number Z(equal) is determined (see step 3). On the other hand, zeros between the same signs are replaced by the opposite sign (i.e. the non-adjacent sign) and the number Z(different) phases is determined (cf. step 3).

The null hypothesis "The sample is random." is finally rejected if the null hypothesis is rejected with both test statistic Z(equal) and test statistic Z(different) (see step 4).

Application example for the Wallis and Moore test
Within the scope of a test equipment capability analysis (C_g-, C_{gk}-study) of a complex measuring system, two scanners A and B are examined. The C_g-, C_{gk}-study assumes measurement series whose measured values are normally distributed and not subject to any trend. Repeat measurements ($n_A = n_B = 24$) are performed on a normal with both scanners (Tables 13.4 and 13.5). In a first step, the random ranking is to be analysed before further tests on trend and distribution model (test for normal distribution) are carried out. With the aid of the Wallis and Moore test, it is to be examined whether the measured values follow a random sequence.

Solution
Step 1: Formulation of the hypotheses and determination of the significance level
H_0: Random sequence of phases; H_1: Non-random sequence of phases.
    Significance level $\alpha = 0.05$.

Step 2: Determination of differences, phases and number of phases
Tables 13.4 and 13.5 show the order of the measured values of scanner A and B as well as the assignment of the sign based on the measured value difference $(x_{i+1} - x_i)$. In the case of scanner A, counting the phases excluding the first and last phase gives $h_{TA} = 9$. In the case of scanner B, the total number of phases gives $h_{TB} = 15$.

13.3 Single Sample Case

Table 13.4 Sequence of measured values of scanner A and sign assignment in the Wallis and Moore test

0.010	0.020	0.016	0.020	0.018	0.090	0.020	0.030	0.040	0.020	0.045	0.050
↑ +	↑ −	↑ +	↑ −	↑ +	↑ −	↑ +	↑ +	↑ −	↑ +	↑ +	↑ +
0.049	0.037	0.055	0.058	0.070	0.071	0.072	0.074	0.075	0.080	0.081	0.087
↑ −	↑ +	↑ +	↑ +	↑ +	↑ +	↑ +	↑ +	↑ +	↑ +	↑ +	

Table 13.5 Sequence of measured values of scanner B and sign assignment in the Wallis and Moore test

0.040	0.077	0.028	0.024	0.060	0.015	0.022	0.034	0.004	0.050	0.056	0.047
↑ +	↑ −	↑ −	↑ +	↑ −	↑ +	↑ +	↑ −	↑ +	↑ +	↑ −	↑ −
0.034	0.041	0.094	0.039	0.046	0.044	0.010	0.025	0.020	0.070	0.043	0.060
↑ +	↑ +	↑ −	↑ +	↑ −	↑ −	↑ +	↑ −	↑ +	↑ +	↑ −	↑ +

13.3 Single Sample Case

Step 3: Calculation of the test statistics

$$\widehat{z}_{TA} = \frac{\left|h - \frac{2n-7}{3}\right| - 0.5}{\sqrt{\frac{16n-29}{90}}} = 2.098 \qquad (13.12)$$

$$\overline{z}_{TB} = 0.42 \qquad (13.13)$$

Step 4: Determination of critical value and test decision
According to the standard normal distribution (cf. Appendix, Sect. Table A.2a.), $\alpha = 0.05$ (two-sided) yields the critical value, Eq. 13.14. Since the test condition according to Eq. 13.15 is fulfilled for scanner A, H_0 is discarded in favor of H_1. The ordering of the measurements of scanner A is not random ($\alpha = 0.05$). On the other hand, it is found that the test condition in Eq. 13.16 is not satisfied for scanner B. Thus, for scanner B the null hypothesis H_0 is not rejected: The order of the measured values of scanner B may well be random ($\alpha = 0.05$).

$$z_{\alpha=0.05} = 1.96 \qquad (13.14)$$

$$\widehat{z}_{TA} = 2.098 > 1.96 = z_\alpha \qquad (13.15)$$

$$\widehat{z}_{TB} = 0.42 > 1.96 = z_\alpha \qquad (13.16)$$

13.3.2 Kolmogorov–Smirnov Test: Test of Goodness of Fit

The statistical test according to Kolmogorov and Smirnov (Kolmogorov 1933) (Smirnov 1948) can be used for two types of analyses. In the single sample case, it is examined whether a random variable (respectively the known observations) follows an assumed distribution model. In the two sample case, two independent series of observations (independent series of measurements) can be analyzed for agreement with respect to the assumed probability model. In the following, the focus is on the single sample case: When testing for a particular distribution model, the test is generally referred to as the Kolmogorov–Smirnov goodness-of-fit test (abbreviated to KS goodness-of-fit test). The random variable under test need not follow a normal distribution. The observations, realized in the form of a sample, can be used with respect to any distribution model. In the context of the present explanations, the KS test for normal distribution is outlined, the notation follows (Hartung 2009).

The Kolmogorov–Smirnov goodness-of-fit test is also well suited for small sample sizes (unlike the also well-known chi-square goodness-of-fit test) and is applicable for continuously distributed characteristics, but also for discrete or rank-scaled characteristics.

Base of operations

- The random variable X is examined.
- A realization in the form of a sample with size n_i and the observation values x_1 to x_n is available.
- The assumption is made that the random variable X follows a normal distribution model.

Procedure of the KS goodness-of-fit test
Step 1: Formulation of the hypotheses and determination of the significance level
The general formulation of the null hypothesis and the alternative hypothesis of the KS goodness-of-fit test are two-sided according to Eqs. 13.17 and 13.18. The test is to determine whether the observed values of the present measurement series follow a distribution $F_0(x)$ (in the following, $F_0(x)$ corresponds to the normal distribution model). The test is to be performed with a significance level α.

$$H_0 : F(x) = F_0(x) \tag{13.17}$$

$$H_0 : F(x) \neq F_0(x) \tag{13.18}$$

Step 2: Preparation of the test and carrying out of the KS work table
First, the observed values x_i are sorted in ascending order, where i is the rank number. Furthermore, the estimates for arithmetic mean (Eq. 13.19) and standard

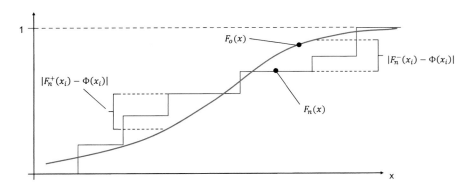

Fig. 13.7 Empirical distribution function based on rank numbers of the observed values, as well as hypothetical distribution function

Table 13.6 Structure of the work table for performing a KS goodness-of-fit test

i	x_i	$\phi(x_i)$	$F_n^-(x_i)$	$F_n^+(x_i)$	$d_- = \vert F_n^-(x_i) - \phi(x_i) \vert$	$d_+ = \vert F_n^+(x_i) - \phi(x_i) \vert$
1	...	...	...	...	...	...
...	...	...	...	...	...	...
n	...	...	...	...	...	...

13.3 Single Sample Case

deviation are obtained based on the available sample (Eq. 13.20). Subsequently, hypothetical distribution function (here: based on the determination of the quantiles $\phi(x)$ of the standard normal distribution (Eq. 13.21) as well as the empirical distribution function (Eqs. 13.22 and 13.23), cf. Fig. 13.7. Finally, the differences between limit value and hypothetical distribution function are formed (Eqs. 13.24 and 13.25). The KS work table shows a summary of the steps outlined (cf. Table 13.6).

$$\mu \approx \hat{\mu} = \bar{x} = \frac{1}{n}\sum_{i=1}^{n} x_i \qquad (13.19)$$

$$\sigma \approx \hat{\sigma} = s = \sqrt{\frac{1}{n-1}\sum_{i=1}^{n}(x_i - \bar{x})^2} \qquad (13.20)$$

$$F_0(x) = \phi(x_i) = \left(\frac{x_i - \bar{x}}{s}\right) \qquad (13.21)$$

$$F_n^-(x_i) = \left(\frac{i-1}{n}\right) \qquad (13.22)$$

$$F_n^+(x_i) = \left(\frac{i}{n}\right) \qquad (13.23)$$

$$d_- = \left|F_n^-(x_i) - \phi(x_i)\right| \qquad (13.24)$$

$$d_+ = \left|F_n^+(x_i) - \phi(x_i)\right| \qquad (13.25)$$

Step 3: Calculation of the test statistic T and determination of the critical value $K_{1-\alpha}$
The supremum of the differences between hypothetical distribution function and empirical distribution function corresponds to the test statistic T; cf. Eq. 13.26. The supremum is part of the set of real numbers. The test statistic T is compared with the critical value $K_{1-\alpha}$, which is taken from the KS table (critical values depending on the significance level and the present, concrete sample size); cf. Appendix Table A.7; quantiles of the KS test statistic as well as (Miller 1956).

$$T = \sup|F_0(x) - F_n(x)| \qquad (13.26)$$

Step 4: Test decision
The null hypothesis H_0 is retained if Eq. 13.27 holds: The present sample could originate from a population that follows the hypothetical distribution model $F_0(x)$. If Eq. 13.28 holds, H_0 is discarded and H_1 is assumed. That is, the present sample does not originate from a population that follows the hypothetical distribution model $F_0(x)$. The test decision with significance level α is made.

$$K_n \leq K_{1-\alpha} \qquad (13.27)$$

$$K_n > K_{1-\alpha} \qquad (13.28)$$

13 Significance Tests

Note on literature/table values KS goodness-of-fit test
Depending on the literature source used, Eq. 13.29 or 13.30 is used for test statistic T. Depending on the published table, the critical values of the KS table must either be multiplied by √n or divided by √n.

$$K_n = sup|F_0(x) - F_n(x)| \tag{13.29}$$

$$\sqrt{n}K_n = sup|F_0(x) - F_n(x)| \tag{13.30}$$

Note on further goodness-of-fit tests
In addition to the KS test, there are a number of other common adaptation tests: χ^2 goodness-of-fit test, Anderson–Darling test or Swed-Eisenhart test; sources for these can be found in Table 13.3.

Application example KS goodness-of-fit test
As part of the new development of a power train, a series of powertrain prototypes is built. During bench testing, the power [kW] of 15 engines is measured. An analysis of the determined performance data is to clarify whether the performance data follow a normal distribution model. The analysis is performed by means of KS goodness of fit test with a significance level of $\alpha = 0.05$.

Measured data Power [kW]: 258.63; 262.96; 263.67; 264.81; 268.34; 264.38; 257.03; 268.15; 268.52; 269.82; 266.73; 254.68; 260.72; 275.12; 270.49.

Solution
Step 1: Formulation of the hypotheses and determination of the significance level

$$H_0 : F(x) = F_0(x) \tag{13.31}$$

$$H_0 : F(x) \neq F_0(x) \tag{13.32}$$

Note: $F_0(x)$ corresponds to normal distribution model.
Significance level $\alpha = 0.05$.

Step 2: Preparation of the test and carrying out the KS work table
Estimation of arithmetic mean (Eq. 13.33) and standard deviation based on available sample (Eq. 13.34).

$$\bar{x} = 264.937 \text{ kW} \tag{13.33}$$

$$s = 5.526 \text{ kW} \tag{13.34}$$

The KS work table is developed according to the template Table 13.6: The application in the present case study can be seen in Table 13.7. In the following, the calculation path is now outlined. First, the observation values x_i are sorted in ascending order. Then the grid points of the hypothetical distribution function and the empirical distribution function are determined. As an example, the calculations are carried out here for the measured values with the rank $i = 1$ and $i = 2$.

The measured value $x_1 = 254.68$ kW is assigned to the rank $i = 1$. From this follows for the hypothetical distribution function Eq. 13.35. Where:

13.3 Single Sample Case

Table 13.7 Working table for carrying out a KS goodness-of-fit test on the basis of $n=15$ performance measurements as part of a prototype powertrain testing

i	x_i	$\phi(x_i)$	$F_n^-(x_i)$	$F_n^+(x_i)$	$d_- = \lvert F_n^-(x_i) - \phi(x_i)\rvert$	$d_+ = \lvert F_n^+(x_i) - \phi(x_i)\rvert$
1	254.68	0.0317	0.0000	0.0667	0.0317	0.0350
2	257.03	0.0762	0.0667	0.1333	0.0096	0.0571
3	258.63	0.1269	0.1333	0.2000	0.0065	0.0731
4	260.72	0.2227	0.2000	0.2667	0.0227	0.0440
5	262.96	0.3603	0.2667	0.3333	0.0936	0.0269
6	263.67	0.4093	0.3333	0.4000	0.0760	0.0093
7	264.38	0.4599	0.4000	0.4667	0.0599	0.0068
8	264.81	0.4909	0.4667	0.5333	0.0242	0.0425
9	266.73	0.6272	0.5333	0.6000	0.0939	0.0272
10	268.15	0.7196	0.6000	0.6667	**0.1196**	0.0529
11	268.34	0.7310	0.6667	0.7333	0.0644	0.0023
12	268.52	0.7417	0.7333	0.8000	0.0083	0.0583
13	269.82	0.8116	0.8000	0.8667	0.0116	0.0551
14	270.49	0.8426	0.8667	0.9333	0.0241	0.0908
15	275.12	0.9673	0.9333	1.0000	0.0340	0.0327

$1 - \phi(1.856)$ can be taken from the set of tables for the standard NV (cf. Appendix). Equations 13.36 and 13.37 apply to the empirical distribution function. Forming the differences between the limit and the hypothetical distribution function yields Eqs. 13.38 and 13.39.

$$F_0(x) = \phi(x_1) = \phi\left(\frac{x_1 - \bar{x}}{s}\right) = \phi(-1.856) = 1 - \phi(1.856) = 0.0317 \quad (13.35)$$

$$F_n^-(x_1) = \left(\frac{i-1}{n}\right) = \frac{0}{15} = 0 \quad (13.36)$$

$$F_n^+(x_1) = \left(\frac{i}{n}\right) = \frac{1}{15} = 0.0667 \quad (13.37)$$

$$d_- = \lvert F_n^-(x_1) - \phi(x_1)\rvert = 0.0317 \quad (13.38)$$

$$d_+ = \lvert F_n^+(x_1) - \phi(x_1)\rvert = 0.0350 \quad (13.39)$$

The measured value $x_2 = 257.03$ kW is assigned to rank $i = 2$. It follows for the hypothetical distribution function Eq. 13.40 as well as for the empirical distribution function 13.41 and 13.42. The differences between limit value and hypothetical distribution function result in Eq. 13.43 and 13.44.

$$F_0(x) = \phi(x_2) = \phi\left(\frac{x_2 - \bar{x}}{s}\right) = \phi(-1.4309) = 0.0762 \qquad (13.40)$$

$$F_n^-(x_2) = \frac{i-1}{n} = \frac{1}{15} = 0.0667 \qquad (13.41)$$

$$F_n^+(x_2) = \frac{i}{n} = \frac{2}{15} = 0.1333 \qquad (13.42)$$

$$\left|F_n^-(x_2) - \phi(x_2)\right| = 0.0096 \qquad (13.43)$$

$$\left|F_n^+(x_2) - \phi(x_2)\right| = 0.0571 \qquad (13.44)$$

According to the exemplary procedure for the first two ranks, the complete working table for the KS adjustment test can be developed for all sorted measured values with the assigned ranks $i = 1, \ldots, 15$; cf. Table 13.7.

Step 3: Calculation of the test statistic T and determination of the critical value $K_{1-\alpha}$

The supremum between hypothetical distribution function and empirical distribution function is shown by Eq. 13.45, cf. also Table 13.7, row for rank $i = 10$. The critical value $K_{1-\alpha}$ from the KS table for $n = 15$ and significance level $\alpha = 0.05$ is $K_{1-\alpha} = 0.338$ (cf. Appendix Table A.7).

$$T = \sup|F_0(x) - F_n(x)| = 0.1196 \qquad (13.45)$$

Step 4: Test decision

Step 3 is immediately followed by the test decision based on Eq. 13.46.

$$K_n = 0.1196 \leq 0.338 = K_{1-\alpha} \qquad (13.46)$$

The null hypothesis H_0 is retained, the distribution of the measured data with respect to the power of the powertrain could follow a normal distribution model.

13.3.3 Sign Test: Test for Location Parameters

The sign test is a distribution-free procedure for investigating whether an observed location parameter Θ (median) based on an existing sample / series of measurements agrees with a previously assumed (or specified) location parameter Θ_0 (Conover 1999:157–176).

The sign test is a special case of the general binomial test, with which random variables or the present realization can be examined with regard to dichotomy (two expressions). The possible hypotheses include testing whether the probability P of an observation is equal to a previously assumed probability P_0 (two-sided test), or

13.3 Single Sample Case

can be equal to/below or equal to/above a previously assumed probability P_0 (one-sided test). Ergo, the test statistic T is binomially distributed.

The sign test also refers to the binomial distributed test statistic. The starting point is the present sample or series of measured values. In the focus of the sign test is the assumption that approx. half of the observations lie below and the other half of the observations above the assumed position parameter Θ_0. The measurement series is thus considered dichotomized, the scatter of these alternatives is binomially distributed.

Performing the Sign Test
Step 1: Formulation of the hypotheses and determination of the significance level
In the case of a two-sided test, the hypotheses are stated according to Eqs. 13.47 and 13.48. In case of the application of a one-sided test with investigation of possible exceeding of the assumed position parameter Θ_0 the hypotheses are according to Eqs. 13.49 and 13.50. In case of the application of a one-sided test, with investigation of possible falling short of the assumed position parameter Θ_0 Eqs. 13.51 and 13.52 are used.

$$H_{01} : \theta = \theta_0 \tag{13.47}$$

$$H_{11} : \theta \neq \theta_0 \tag{13.48}$$

$$H_{02} : \theta = \theta_0 \tag{13.49}$$

$$H_{12} : \theta > \theta_0 \tag{13.50}$$

$$H_{03} : \theta = \theta_0 \tag{13.51}$$

$$H_{13} : \theta < \theta_0 \tag{13.52}$$

Step 2: Carrying out the work table for the sign test
First, the observed values (the i values x_i of the measurement series with scope n) are transformed (cf. Eq. 13.53). Furthermore, the variable Z_i is assigned according to the conditions formulated in Eqs. 13.54–13.56 (ties: Compare also notes on the sign test in the present chapter). Table 13.8 shows a possible structure of the working table for the sign test.

$$x'_i = x_i - \theta_0 \tag{13.53}$$

Table 13.8 Template for the work table when performing a sign test

Rank i	x_i	$x_i - \Theta_0$	Z_i
1			
...			
n			

$$Z_i = 1; \text{ if } x'_i > 0 \tag{13.54}$$

$$Z_i = 0.5; \text{ if } x'_i = 0 \left(\text{tie}: x'_i = x_i - \theta_0 = 0\right) \tag{13.55}$$

$$Z_i = 0; \text{ if } x'_i < 0 \tag{13.56}$$

Step 3: Calculation of the test statistic T and calculation of the critical limits

$$T = \sum_{i=1}^{n} g(i) \cdot Z_i = \sum_{i=1}^{n} Z_i \tag{13.57}$$

Note: A weighting of the variable assignment Z_i can be carried out via the function g(i). If no weighting of the variable assignment Z_i is to take place, g(i)=1 is set; cf. Eg. 13.57.

The critical limits are calculated with the aid of the binomial distribution (cf. Eq. 13.58). Since the assumption is made that the observation is equally likely to lie below or above the default value θ_0 if it coincides with the median, the parameter p=0.5 is set.

$$f(x) = \binom{n}{x} p^x (1-p)^{n-x} \tag{13.58}$$

The determination of the critical values $t_{\alpha/2}$ as well as $t_{1-\alpha/2}$ (in the two-sided case) is carried out by means of cumulative probability values on the basis of Eqs. 13.59 and 13.60 (parameter p=0.5; n=extent series of measurements).

$$\Pr(T \le t_{\alpha/2}) = \sum_{i=0}^{n} \binom{n}{i} p^i (1-p)^{n-i} \ge \alpha/2 \tag{13.59}$$

$$\Pr(T \le t_{1-\alpha/2}) = \sum_{i=0}^{n} \binom{n}{i} p^i (1-p)^{n-i} \ge 1 - \alpha/2 \tag{13.60}$$

Step 4: Test decision

The test decision is made by comparing the test statistic T and the critical values $t_{\alpha/2}$ and $t_{1-\alpha/2}$ (in the two-sided case) (cf. Eqs. 13.61 and 13.62). In case one of the conditions is true, the null hypothesis H_0 is rejected and the alternative hypothesis H_1 is accepted.

$$T = \sum_{i=1}^{n} Z_i < t_{\alpha/2} \tag{13.61}$$

$$T = \sum_{i=1}^{n} Z_i > t_{1-\alpha/2} \tag{13.62}$$

13.3 Single Sample Case

Notes on the Sign Test:

1. If the sample size is $n \geq 20$, the binomial distribution can be approximated by a normal distribution model and thus the standard NV can be used. In this case, Eq. 13.63 applies to the test statistic T.

$$T = \frac{\sum_{i=1}^{n} Z_i - \frac{n}{2}}{1/2 \cdot \sqrt{n}} \qquad (13.63)$$

2. In the event that ties occur in the transformation of the observed values occur (cf. Eq. 13.64), one of the following options for taking ties into account can be selected, as is also the case with other statistical tests.

1. The observations are not taken into account.
2. If there is an even number of ties, an equal division is made; if there is an odd number, the last tie is eliminated.
3. Observations are scored 0.5 within variable Z_i.
4. Ties are considered with the less frequent sign, which corresponds to a conservative consideration.

$$x'_i = x_i - \Theta_0 = 0 \qquad (13.64)$$

Application example sign-test

The failure centroid of two control units of different generations A and B is to be analysed and compared after a comparable period of use in the field. There is considerably more empirical information available on generation A, and the failure centroid can be safely assumed to be $\Theta_0 = 660$ h (operating hours). For generation B, the completed operating times [h] until failure are known from $n = 15$ control units: 650, 159, 39, 25, 895, 358, 1450, 532, 596, 456, 785, 565, 715, 378, 485. The analysis via sign test should provide clarity as to whether the failure centroid of generation B is significantly different compared to generation A (centroid: $\Theta_0 = 660$ h).

Solution

Step 1: Formulation of the hypotheses and determination of the significance level

The test is two-sided, and the hypotheses show Eqs. 13.65 and 13.66. The significance level is set at $\alpha = 0.05$.

$$H_{01} : \theta = \theta_0 \qquad (13.65)$$

$$H_{11} : \theta \neq \theta_0 \qquad (13.66)$$

Step 2: Developing of the working table for the sign test

The observations are assigned ranks and then the observation values are transformed. Then the variable Z_i is calculated according to Eq. 13.54 to 13.56, here

Table 13.9 Work table for performing the sign test based on the failure data of a control unit type

Rank i	x_i	$x_i - \Theta_0$	Z_i
1	650	−10	0
2	159	−501	0
3	39	−621	0
4	25	−635	0
5	895	235	1
6	358	−302	0
7	1450	790	1
8	532	−128	0
9	596	−64	0
10	456	−204	0
11	785	125	1
12	565	−95	0
13	715	55	1
14	378	−282	0
15	485	−175	0

without weighting, i.e. g(i) = 1, and the results are written down in the working table according to Table 13.9.

Step 3: Calculation of the test statistic T and calculation of the critical limits
Calculate the test statistic T without weighting (g(i) = 1) according to Eq. 13.67.

$$T = \sum_{i=1}^{n} g(i) \cdot Z_i = \sum_{i=1}^{n} Z_i = 4 \tag{13.67}$$

Calculation of the critical limit values
The density function of the binomial distribution according to Eq. 13.58 (here: parameter p = 0.5 and n = 15) is shown in Fig. 13.8.

The critical values are determined using cumulative probability values based on Eqs. 13.59 and 13.60 and are presented in Table 13.10 for clarity. The critical values are obtained according to Eqs. 13.68 and 13.69.

$$t_{\alpha/2} = 4 \tag{13.68}$$

$$t_{1-\alpha/2} = 11 \tag{13.69}$$

Step 4: Test decision
On the basis of Eqs. 13.61 and 13.62, Eqs. 13.70 and 13.71 respectively are obtained, from which it follows that the null hypothesis H_0 cannot be rejected: The centroid of the failure data concerning electronic control unit generation A ($\Theta_0 = 660$ h) can be identical with the centroid of the known failures of the electronic control unit generation B.

$$T = \sum_{i=1}^{n} Z_i = 4 \not< t_{\alpha/2} = 4 \tag{13.70}$$

13.3 Single Sample Case

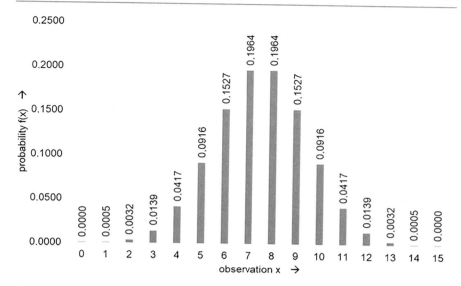

Fig. 13.8 Binomial distribution (parameter: p = 0.5; n = 15)

Table 13.10 Cumulated probabilities based on the binomial distribution (p = 0.5; n = 15)

T	Pr	Critical boundaries Pr
0	3.0518E-05	
1	0.00048828	
2	0.00369263	
3	0.01757813	
4	0.05923462	$Pr(T \leq t_{\alpha/2}) \geq \alpha/2$
5	0.15087891	
6	0.30361938	
7	0.5	
8	0.69638062	
9	0.84912109	
10	0.94076538	
11	0.98242188	$Pr(T \leq t_{1-\alpha/2}) \geq 1-\alpha/2$
12	0.99630737	
13	0.99951172	
14	0.99996948	
15	1	

$$T = \sum_{i=1}^{n} Z_i \not> 4 \not> t_{1-\alpha/2} = 11 \qquad (13.71)$$

13.3.4 Test for Outliers Within an Observation Series

Let the starting point be a measurement series with scope n and measured values $x_1, x_2, \ldots, x_n$. If a value x_i differs significantly compared to the other measured values, this measured value is referred to as an outlier. It should not be forgotten that the observer – directly or indirectly or consciously or unconsciously – has assumed a model of the value distribution when detecting the outlier. The observed measured value (outlier) thus deviates considerably from the observer's model assumption. Within the framework of the present chapter, a simple test for outliers is with the goal of testing whether the deviation is significant with respect to the model assumption.

Even if the measured value is recognized as an outlier, a general elimination of outliers must be questioned before analyzing the measured value series: An outlier usually has a technical background, e.g. measurement error, calculation error or measured value processing error. In this respect, an analysis of the outlier before its potential elimination is meaningful.

Base of operations
A measurement series with the values $x_1, x_2, \ldots, x_n$ is available, consisting of $n \geq 10$ values (better: $n \geq 25$ values). One (or more) observed value(s) seems to deviate from the other observed values, whereby the observer has made a certain model assumption regarding the distribution of the measured values.

Performing the outlier test
Step 1: Formulation of the hypotheses
The observed value x_i deviates significantly from the range of the other known observed values. The range is defined as the 4σ-range ("4-sigma range"); note that the "non-critical" range is thus $\mu \pm 4\sigma$. The critical range is outside the 4-sigma range. The hypotheses are according to Eqs. 13.72 and 13.73.

$$H_0 : x_i \in \pm 4\sigma \qquad (13.72)$$

$$H_1 : x_i \notin \pm 4\sigma \qquad (13.73)$$

Step 2: Preparation of the test decision
The determination of the 4-sigma-range related to the measurement series (of course without the observation value to be examined) is carried out with the help of the estimators according to Eqs. 13.74a and 13.74b. The limits of the 4-sigma range are obtained according to Eqs. 13.75a and 13.75b.

$$\mu \approx \widehat{\mu} = \bar{x} = \frac{1}{n} \sum_{i=1}^{n} x_i \qquad (13.74a)$$

13.3 Single Sample Case

Table 13.11 Model assumption and probability related to 4-sigma range

Model assumption	Probability to the 4-sigma range
Normal distribution	$P(|X - \mu| \leq 4\sigma) = 0.9999$
Symmetrical single-peaked distribution	$P(|X - \mu| \leq 4\sigma) \geq 0.97$
Any distribution	$P(|X - \mu| \leq 4\sigma) \geq 0.94$

$$\sigma \approx \hat{\sigma} = s = \sqrt{\frac{1}{n-1} \sum_{i=1}^{n} (x_i - \bar{x})^2} \tag{13.74b}$$

$$c_{LL} = \bar{x} - 4s \tag{13.75a}$$

$$c_{UL} = \bar{x} + 4s \tag{13.75b}$$

Step 3: Test decision
Depending on the model assumption, the probability of a value x_i lying within the 4-sigma range (non-critical range) can be estimated (cf. H_0 (Eq. 13.72) and Table 13.11). If the observation value in focus lies outside the estimated 4-sigma range (outside the critical limits according to Eqs. 13.75a and 13.75b), the null hypothesis is rejected and a decision is made in favor of the alternative hypothesis (outlier assumption).

Notes on model assumption and 4-sigma range
The probability of a $k\sigma$ range can be determined with reference to a model assumption. If a normal distribution is assumed, the probability can be estimated using the standard normal distribution, cf. Chap. 7. Table 13.12 shows the probabilities for relevant scatter ranges.

Assuming a symmetrical single-peaked distribution Eq. 13.76 is valid according to Gauss (1823); where k should follow the condition according to Eq. 13.77. Assuming an arbitrary distribution, Eq. 13.78 after Bienaymé (1853) and Chebyshev (1874) applies to the probability; under the condition Eq. 13.79.

$$P(|X - \mu| \geq k\sigma) \leq \frac{4}{9k^2} \tag{13.76}$$

$$k \geq \frac{2}{\sqrt{3}} \approx 1.155 \tag{13.77}$$

Table 13.12 Model assumption normal distribution; correlation between scatter range and probability

Scatter range	Probability
$\pm 2\sigma$	$P = 0.9545$
$\pm 3\sigma$	$P = 0.9973$
$\pm 4\sigma$	$P = 0.9999$

Table 13.13 Model assumption as well as scatter range kσ and associated probability P

kσ – range	Distribution model	Probability P
$\mu \pm 1.960\sigma$	Arbitrary Symmetrical-one-peaked Normal distribution	≥ 0.74 ≥ 0.884 $= 0.95$
$\mu \pm 2\sigma$	Arbitrary Symmetrical-one-peaked Normal distribution	≥ 0.75 ≥ 0.889 $= 0.9545$
$\mu \pm 3\sigma$	Arbitrary Symmetrical-one-peaked Normal distribution	≥ 0.889 ≥ 0.951 $= 0.997$
$\mu \pm 4\sigma$	Arbitrary Symmetrical-one-peaked Normal distribution	≥ 0.94 ≥ 0.97 $= 0.9999$

$$P(|X - \mu| \geq k\sigma) \leq \frac{1}{k^2} \quad (13.78)$$

$$k > 0 \quad (13.79)$$

Summary:
Table 13.13 shows the probabilities with respect to important kσ-ranges related to a model assumption.

Application example outlier test
The Hawaiian archipelago in the Pacific protects itself with a defence system against missiles from a potential attacker. If an approaching missile is detected by the defence system, a warning is sent to the population via short message (SMS) on the mobile phone. In 2018, an alarm was inadvertently triggered by a supposed incident and a corresponding warning was sent via short message. In the aftermath, the operator of an independent website noticed that at the moment of the warning, the number of hits dropped. After it was determined that a false alarm had been triggered, the number of website hits rose again and for a short time clearly exceeded the average value. The access numbers related to the Internet pages are to be examined (cf. Table 13.14). In this context, both a possible outlier at the time of the warning due to visits to corresponding protective facilities and a

Table 13.14 Average proportion deviation of access numbers in relation to a comparable working day. The penultimate value shows the drop in the number of accesses after the short message (SMS) was sent. The last observation value shows the number of accesses after the all-clear

10	4	12	−1	−13	−8	−12	−11	9	−1
−4	−7	−13	−8	−7	9	10	12	−4	9
−2	5	4	6	−1	1	−1	6	0	−7
6	1	−1	−5	2	11	8	−77	48	

13.3 Single Sample Case

possible outlier at the time of the all-clear due to an increased need for information are to be examined.

Solution: Perform an outlier test
Step 1: Formulation of the null hypothesis
The observed values $x_{38} = -77$ and $x_{39} = -48$ deviates significantly from the other known observed values. Thus the hypotheses are according to Eqs. 13.80a and 13.80b.

$$H_0 : x_{38} \text{ ist } \in \pm 4\sigma \quad \text{and} \quad H_1 : x_{38} \notin \pm 4\sigma \qquad (13.80a)$$

$$H_0 : x_{39} \text{ ist } \in \pm 4\sigma \quad \text{and} \quad H_1 : x_{39} \notin \pm 4\sigma \qquad (13.80b)$$

Step 2: Preparation of the test decision
Determination of the 4-sigma range related to the concrete measurement series at hand (deviation of the website accesses). Carrying out the estimation of the parameters μ and σ; Eqs. 13.80 and 13.81 without using the possible outliers $x_{38} = -77$ and $x_{39} = 48$. This results in the estimation of the fourfold standard deviation (Eq. 13.82) as well as, with the inclusion of the mean, the limits of the non-critical range c_{LL} and c_{UL}; Eqs. 13.83 and 13.84.

$$\mu \approx \hat{\mu} = \bar{x} = 0.1891 \qquad (13.80)$$

$$\sigma \approx \hat{\sigma} = s = 7.5087 \qquad (13.81)$$

$$4s = 30.03 \qquad (13.82)$$

$$c_{LL} = \bar{x} - 4s = -29.85 \qquad (13.83)$$

$$c_{UL} = \bar{x} + 4s = 30.22 \qquad (13.84)$$

Step 3: Test decision
With regard to the present series of measurements (with the exception of the two deviating measurement values 38 and 39), a normal distribution model is assumed, so that the statement with $P = 0.9999$ is made. Equations 13.85 and 13.86 apply. Both measured values – apparently deviating from the series of measurements – do not lie within the estimated 4-sigma range after performing the outlier test, and are thus interpreted as outliers. Obviously, both the missile warning and the all-clear signal had a significant impact on the number of hits on the website.

$$x_{38} = -77 \notin 4s \qquad (13.85)$$

$$x_{39} = 48 \notin 4s \qquad (13.86)$$

13.3.5 t-Test: Comparison of Sample Mean Against Reference Value

The t-test can be applied in the single-sample case as well as in the two-sample case. In the single-sample case, it is tested whether a mean value of a sample representing a population can correspond to a specified reference value. In the two-sample case (cf. Sect. 13.4.3), the test is carried out on the basis of the mean values of two samples with regard to whether the mean values of the represented populations are significantly distinguishable or indistinguishable.

The following procedure outlines the application of the t-test in the single-sample case.

Base of operations
A sample with n observed values for the random variable X is available. It is assumed that the population is normally distributed, furthermore the sample size should be $n \geq 30$. Furthermore, $\mu \in \mathrm{IR}$ and the standard deviation $\sigma > 0$. The test is to be performed with respect to a reference value μ_0, which is given.

Procedure of the t-test
Step 1: Formulation of the hypotheses and determination of the significance level
In the case of a two-sided test, the hypotheses are formulated according to Eq. 13.87. In the case of a one-sided test, Eq. 13.88 (right-hand side) and Eq. 13.89 (left-hand side) apply to the hypotheses. The test to be performed should be carried out with a significance level α (e.g.: $\alpha = 0.05$).

$$H_0 : \mu = \mu_0 \quad H_1 : \mu \neq \mu_0 \tag{13.87}$$

$$H_0 : \mu \leq \mu_0 \quad H_1 : \mu > \mu_0 \tag{13.88}$$

$$H_0 : \mu \geq \mu_0 \quad H_1 : \mu < \mu_0 \tag{13.89}$$

Step 2: Calculation of the test statistic T
The test statistic T is determined with Eq. 13.90. The estimators for mean and standard deviation are estimated with Eqs. 13.91 and 13.92 on the basis of the available sample.

$$T = \sqrt{n} \cdot \frac{\bar{x} - \mu_0}{s} \tag{13.90}$$

$$\bar{x} = \frac{1}{n} \sum_{i=1}^{n} x_i \tag{13.91}$$

$$s = \sqrt{\frac{\sum_{i=1}^{n} (x_i - \bar{x})^2}{n - 1}} \tag{13.92}$$

13.3 Single Sample Case

Step 3: Determination of critical value and test decision
In the case of a two-sided test procedure, the test decision is made according to Eqs. 13.93; the critical value is determined on the basis of the quantiles of the t-distribution (cf. Appendix Table A.3). If the equation is not satisfied, the null hypothesis is rejected in favor of the alternative hypothesis. If the test is one-sided, the test decision is made according to Eq. 13.94 or 13.95, depending on the test design. If the null hypothesis is chosen according to Eq. 13.88 (right-sided), the null hypothesis is rejected if Eq. 13.94 is fulfilled. If the null hypothesis is chosen according to Eq. 13.89 (left-hand side), the null hypothesis is rejected if Eq. 13.95 is satisfied.

$$-t_{(1-\alpha/2;n-1)} \leq T \leq t_{(1-\alpha/2;n-1)} \tag{13.93}$$

$$T > t_{(1-\alpha;n-1)} \tag{13.94}$$

$$T < -t_{(1-\alpha;n-1)} \tag{13.95}$$

13.3.6 Cox and Stuart Test: Trend Analysis

Testing for trend according to Cox and Stuart (1955) is a parameter-free procedure with the goal of examining measured values – which were recorded time-dependently – with regard to a trend. There are $x_1, x_2, ..., x_n$ observed values are available in the form of a sample with size n.

Procedure of the Cox and Stuart trend test
Step 1: Formulation of the hypotheses and determination of the significance level
The hypotheses relate to the differences y_i formed in pairs with respect to the end of the measured value series x_{i+m} (second half) and the respective measured value x_i (first half of the measured value series); cf. step 2. For one-sided observation, possibility one is: null hypothesis "no upward trend" (Eq. 13.96) against alternative hypothesis: "Upward trend" (Eq. 13.97). Furthermore, possibility two is: the null hypothesis "no downward trend" (Eq. 13.98) is tested against the alternative hypothesis "downward trend" (Eq. 13.99). Two-sided consideration yields possibility three: the null hypothesis is "no trend" (Eq. 13.100); on the other hand, the alternative hypothesis is "trend exists" (Eq. 13.101). The test to be performed should be carried out with a significance level α (e.g.: $\alpha = 0.05$).

$$H_{01} : P(y_i > 0) \leq P(y_i < 0) \tag{13.96}$$

$$H_{11} : P(y_i > 0) > P(y_i < 0) \tag{13.97}$$

$$H_{02} : P(y_i > 0) \geq P(y_i < 0) \tag{13.98}$$

$$H_{12} : P(y_i > 0) < P(y_i < 0) \tag{13.99}$$

$$H_{03} : P(y_i > 0) = P(y_i < 0) \tag{13.100}$$

$$H_{13} : P(y_i > 0) \neq P(y_i < 0) \qquad (13.101)$$

Step 2: Preparation of the test, formation of the division parameter m and the differences y_i

First, the sample is divided into a front and a back part (halved or approximately halved sample size) by determining the division parameter m according to Eq. 13.102 (if sample size is even) and Eq. 13.103 (if sample size is odd). Subsequently, the pairwise differences (cf. Eq. 13.104) are formed with respect to the front or back part of the sample on the basis of the division size m, whereby the following applies: if n is even, Eq. 13.105 applies to index i; if n is not even, Eq. 13.106 applies to index i.

$$m = \frac{n}{2}; \text{ if n even} \qquad (13.102)$$

$$m = \frac{n+1}{2}; \text{ if n odd} \qquad (13.103)$$

$$y_i = (x_{i+m} - x_i) \qquad (13.104)$$

$$i = 1, \ldots, m; \text{ if n even} \qquad (13.105)$$

$$i = 1, \ldots, m-1; \text{ if n odd} \qquad (13.106)$$

Step 3: Test statistic T and auxiliary variable L

The test statistic T is given by the number of *positive* differences y_i, where pairwise differences with value zero ($y_i = 0$) are neglected. The auxiliary variable L corresponds to the number of *all* differences for which $y_i \neq 0$ applies.

Step 4: Test decision

The hypothesis H_0 is rejected if Eq. 13.107 (regarding lower bound) or Eq. 13.108 (regarding upper bound) is satisfied. Where, for two-sided questions, Eq. 13.109 holds for r^*. If the question is one-sided, Eq. 13.110 applies. The value $u_{1-\alpha/2}$ or $u_{1-\alpha}$ is determined from the table for the standard normal distribution according to Gauss (cf. Appendix Table A.2). The use of the standard normal distribution is a simplification: it would be more accurate to use the binomial distribution model (cf. Cox and Stuart 1955).

$$T < r^* \qquad (13.107)$$

Table 13.15 Measurement of the pressure in a pipe system within a time interval at time i

Time i	1	2	3	4	5	6	7	8	9
x_i [$1 \cdot 10^5$ Pa]	4	4.8	4.6	4.3	5	4.0	3.8	3.9	3.8
Time i	10	11	12	13	14	15	16	17	18
x_i [$1 \cdot 10^5$ Pa]	3.7	3.5	3.0	3.5	3.4	3.3	3.2	3.0	4.0

13.3 Single Sample Case

$$T > L - r^* \quad (13.108)$$

$$r^* = \frac{1}{2}\left(L - u_{1-\alpha/2}\sqrt{12}\right) \quad (13.109)$$

$$r^* = \frac{1}{2}\left(L - u_{1-\alpha}\sqrt{12}\right) \quad (13.110)$$

Application example Cox-Stuart trend test
A filter system is integrated within a process plant for the purpose of separating particles from a fluid. As the filter system is continuously enriched with particles, the pressure in the downstream pipe system changes. The pressure is continuously monitored. On the basis of a representative series of measurements (see Table 13.15), it is investigated whether a significant drop in pressure can be detected in a time interval. The series of measurements is analysed with regard to whether there is a trend. Furthermore, it is to be clarified whether a downward or upward trend is present.

Solution: Two-sided performance of the Cox and Stuart Test
First, an analysis is carried out with regard to a possible trend in the measured values.

Step 1: Formulation of the hypotheses and determination of the significance level
The following null hypothesis is stated: There is no trend (Eq. 13.111; index chosen analogous to Eq. 13.100, two-sided test design). The alternative hypothesis is: There is a trend (Eq. 13.112; index analogous to Eq. 13.101).

$$H_{03} : P(y_i > 0) = P(y_i < 0) \quad (13.111)$$

$$H_{13} : P(y_i > 0) \neq P(y_i < 0) \quad (13.112)$$

The significance level is set at $\alpha = 0.1$.

Step 2: Preparation of the test, formation of the division parameter m and the differences y_i
The division size m is obtained according to Eq. 13.113. The formation of the pairwise differences considering $i = 1, \ldots, 9$ (cf. Eq. 13.114) is shown in Table 13.16.

$$m = \frac{n}{2} = 9 \quad (13.113)$$

Table 13.16 Formation of the pairwise differences in the Cox and Stuart test on the basis of the sample values according to Table 13.15

I	1	2	3	4	5	6	7	8	9
y_i	−0.3	−1.3	−1.6	−0.8	−1.6	−0.7	−0.6	−0.9	0.2

$$y_i = (x_{i+m} - x_i) \tag{13.114}$$

Step 3: Test statistic T and auxiliary variable L
The calculation yields L=9 (all differences) for auxiliary variables and T=1 (positive differences) for the test statistic. For $u_{1-\alpha/2}$, Eq. 13.115 results (for $\alpha = 0.1$). From this follows for r* Eq. 13.116.

$$u_{1-\alpha/2} = 1.645 \tag{13.115}$$

$$r^* = \frac{1}{2}\left(L - u_{1-\alpha/2}\sqrt{12}\right) = 1.651 \tag{13.116}$$

Step 4: Test decision
The hypothesis H_0 is rejected if Eq. 13.107 or Eq. 13.108 is satisfied. In the present example, Eqs. 13.117 and 13.118 result.

$$T = 1 < r^* = 1.651 \tag{13.117}$$

$$T = 1 \not> L - r^* = 7.349 \tag{13.118}$$

The condition of Eq. 13.117 is fulfilled, the hypothesis H_0 is rejected in favor of H_1, the trend regarding the pressure measurement is evaluated as significant.

The Cox-Stuart test was performed two-sided. Since the pairwise differences are negative, it is reasonable to conclude that the trend must be downward. The Cox-Stuart test can also be performed one-sided with the aim of directly examining the significance of the downward trend.

Solution: One-sided performance of the Cox and Stuart Test
Step 1: Formulation of the hypotheses and determination of the significance level
The following null hypothesis is established: There is no downward trend (Eq. 13.119; index chosen analogous to Eq. 13.98, one-sided test procedure). The alternative hypothesis is: There is a downward trend (Eq. 13.120; index analogous to Eq. 13.99).

$$H_{02} : P(y_i > 0) \geq P(y_i < 0) \tag{13.119}$$

$$H_{12} : P(y_i > 0) < P(y_i < 0) \tag{13.120}$$

The significance level is set at $\alpha = 0.1$.

Step 2: Preparation of the test, formation of the division parameter m and the differences y_i
The test is carried out in the same way as a two-sided test.

Step 3: Calculation of the test statistic T and the auxiliary variable L
The calculation yields L=9 (all differences) for auxiliary variables and T=1 (positive differences) for the test statistic. For $u_{1-\alpha}$, Eq. 13.121 results (for $\alpha = 0.1$). From this follows directly for r* the result according to Eq. 13.122.

13.3 Single Sample Case

$$u_{1-\alpha} = 1.282 \tag{13.121}$$

$$r^* = 2.280 \tag{13.122}$$

Step 4: Test decision
Hypothesis H_0 is rejected if Eq. 13.107 or Eq. 13.108 is satisfied. In the present example, the one-sided test procedure yields Eq. 13.123.

$$T = 1 < r^* = 2.280 \tag{13.123}$$

The condition of Eq. 13.123 is satisfied, the hypothesis H_0 is rejected in favor of H_1, and the downward trend with respect to the pressure measurement is considered significant.

13.4 Two Sample Case

The focus of the present chapter is the investigation of two existing samples (two-sample case) with respect to significant differences in mean (centroid) and dispersion. First, the following parameter-free procedures are presented: Mann–Whitney U test (comparison of two centroids; Sect. 13.4.1) and Siegel-Tukey test (comparison of two dispersions; Sect. 13.4.2). This is followed by the parametric test procedures t-test (comparison of two means; Sect. 13.4.3) and F-test (comparison of two variances; Sect. 13.4.4).

13.4.1 Mann–Whitney U Test: Comparison of Two Centroids

The Rank Test from Mann and Whitney (1947) is based on the so-called Wilcoxon test (Wilcoxon 1945) and can be carried out independently of an assumption of a concrete distribution model with respect to the measured or observed values. It is thus a parameter-free procedure; the analysis is on the investigation of two expected values of continuous distributions with regard to a significant difference. The test statistic U also gives the test its name: Mann–Whitney U-Test or just U-test for short is the term used in the literature.

Base of Operations

- Two concrete, independent samples with the sizes n_1 and n_2 are available.
- Metric measurement data or at least rank data are available.
- The variables analyzed follow similar or the same continuous distribution functions.

Goal
The null hypothesis is tested whether the probability of an observation in a first population compared to an observation in a second population could be P = 0.5.

The observations refer to the centroids of the measurement series. The test is sensitive to different medians but insensitive to different variances.

Procedure of the U-Test
Step 1: Formulation of the hypotheses and determination of the significance level
In general, Eqs. 13.124 and 13.125 apply to null and alternative hypotheses for two-sided questions, and Eqs. 13.126 and 13.127 or Eqs. 13.128 and 13.129 for one-sided questions, respectively. Since the analysis is on the investigation of centroids, Eqs. 13.130 to 13.135 can also be formulated as alternatives.

$$H_{01} : P(X_1 > X_2) = \frac{1}{2} \tag{13.124}$$

$$H_{11} : P(X_1 > X_2) \neq \frac{1}{2} \tag{13.125}$$

$$H_{02} : P(X_1 > X_2) \geq \frac{1}{2} \tag{13.126}$$

$$H_{12} : P(X_1 > X_2) < \frac{1}{2} \tag{13.127}$$

$$H_{03} : P(X_1 > X_2) \leq \frac{1}{2} \tag{13.128}$$

$$H_{13} : P(X_1 > X_2) > \frac{1}{2} \tag{13.129}$$

$$H_{01} : \mu_1 = \mu_2 \tag{13.130}$$

$$H_{11} : \mu_1 \neq \mu_2 \tag{13.131}$$

$$H_{02} : \mu_1 \geq \mu_2 \tag{13.132}$$

$$H_{12} : \mu_1 < \mu_2 \tag{13.133}$$

$$H_{03} : \mu_1 \leq \mu_2 \tag{13.134}$$

$$H_{13} : \mu_1 > \mu_2 \tag{13.135}$$

The null hypothesis should be chosen in a meaningful way if the *question* is *one-sided*. The preliminary examination of the centroids via the formation of medians provides an indication of which of the two centroids is potentially the larger. Accordingly, this hint (suspicion) should be recorded in the alternative hypothesis. Ergo, an attempt would be made to refute the null hypothesis. If there is a significant difference in the centroids which confirms the suspicion, the null

13.4 Two Sample Case

hypothesis would be rejected and proof would be provided by accepting the alternative hypothesis. With this procedure, the minimum of the potential test variables results directly as the test variable U_{pruef} and the test decision criterion (cf. step 4) is identical for both one-sided and two-sided questions.

Exemplary is the examination of two existing samples A and B; the median of sample B is greater than the median of sample A. This relationship should be the subject of the alternative hypothesis. The null hypothesis, on the other hand, should express that the centroid of sample A is greater than the centroid of sample B. Accordingly, in the following nomenclature, sample A would be sample 1 and sample B would correspond to sample 2; the corresponding hypotheses are shown in Eqs. 13.132 and 13.133.

Step 2: Rank assignment and determination of rank sums
The observations are in the form of two samples with sizes n_1 and n_2. For the assignment of ranks, the $n = n_1 + n_2$ observations are sorted by size on a scale. A rank is assigned to each observation value. These ranks are summed up for the respective sample, cf. Eqs. 13.136 and 13.137. If, for example, observation values of one sample were assigned rather low ranks compared to the ranks of the other sample, the rank sums of both samples would differ; and finally also the centroids.

$$R_1 = \sum R_{1;i} \qquad (13.136)$$

$$R_2 = \sum R_{2;i} \qquad (13.137)$$

Example of rank assignment and determination of rank sums:
Table 13.17 shows the observation values of two samples with size $n = n_1 + n_2 = 8$, the sorting in an ascending ranking order and the assignment of the ranks. Furthermore, the difference of the formed rank sums can be clearly seen.

Step 3: Calculation of the test variable U
To calculate the test variable U, the sample sizes n_1, n_2 and the rank sums according to Eqs. 13.136 and 13.137 are required. First, the potential test variables are

Table 13.17 Example of rank assignment and determination of rank sums

Sample	Observed values	Sample size
1	$x_{1,i} = 2; 8; 12; 15$	$n_1 = 4$
2	$x_{2,i} = 17; 21; 16; 19$	$n_2 = 4$
Ranking of the unified sample:	$2 < 8 < 12 < 15 < 16 < 17 < 19 < 21$	
Assignment of ranks	1 2 3 4 5 6 7 8	
Allocation of observation to sample	1 1 1 1 2 2 2 2	
Rank sums $\Sigma R_{1;i}$ and $\Sigma R_{2;i}$	$\Sigma R_{1;i} = 10$; $\Sigma R_{2;i} = 26$	

determined according to Eqs. 13.138 and 13.139. Equation 13.140 provides the check. The final test variable $U_{prüf}$ results from the minimum of U_1 and U_2; Eq. 13.141.

$$U_1 = n_1 n_2 + \frac{n_1(n_1 + 1)}{2} - R_1 \tag{13.138}$$

$$U_2 = n_1 n_2 + \frac{n_2(n_2 + 1)}{2} - R_2 \tag{13.139}$$

$$U_1 + U_2 = n_1 n_2 \tag{13.140}$$

$$U_{pruef} = min(U_1, U_2) \tag{13.141}$$

Step 4: Determination of critical value and test decision
First, the critical value for assessing the test variable is determined. For this purpose, the table of Milton (1964) is used, it includes critical values depending on the significance level up to a sample size of $n = n_1 + n_2 \leq 60$ (cf. Appendix Tab A.6); Eq. 13.142 applies.

The test decision is made as follows for two-sided questions: If Eq. 13.143 is satisfied, then the null hypothesis is rejected according to Eq. 13.124 or Eq. 13.130, and the alternative hypothesis Eq. 13.125 or 13.131 is accepted.

The test decision for one-sided questions is made according to the same principle, but it must be ensured that the null hypothesis and the alternative hypothesis are chosen in a meaningful way; cf. notes within step 1.

$$U_{krit} = U\left(n_{1;}; n_2; \alpha\right) \tag{13.142}$$

$$U_{pruef} \leq U_{krit} \tag{13.143}$$

Notes on the application of the Mann–Whitney U-Test
1. approximation for large sample sizes
For sample sizes with $n = n_1 + n_2 > 60$, an approximation can be performed so that Eq. 13.144 applies to U_{krit}. The parameter z follows the normal distribution and can be taken from the table on the quantiles of the standard normal distribution; cf. appendix Table 18.2. When determining the parameter z, the significance level and the one-sided or two-sided test procedure must be taken into account.

$$U_{krit} = U(n_1, n_2; \alpha) = \frac{n_1 n_2}{2} - \hat{z} \cdot \sqrt{\frac{n_1 n_2 (n_1 + n_2 + 1)}{12}} \tag{13.144}$$

2. approximation in the absence of tables
If no tables are available, e.g. according to Milton (1964), the U-Test can also be carried out by means of approximation according to Eq. 13.145. The sample sizes n_1 and n_2 should be greater than eight. The test value $\hat{z}$ is then compared directly with the quantiles of the standard normal distribution (cf. Eq. 13.146); Appendix

13.4 Two Sample Case

Table 13.18 Documentation of the number of cycles in case of pump failure (Generation A and Generation B; sorted data) within the scope of a condition monitoring system

Pump i	Generation A: Cycles on failure	Generation B: Cycles on failure
1	12,000	11,500
2	12,500	13,200
3	13,200	17,500
4	15,500	19,000
5	17,000	20,000
6	17,000	21,500
7	17,100	21,500
Arithmetic mean	14,900	17,742.86
Median	15,500	19,000
Standard deviation	2,275.96	3,970.25

Table 13.19 Failure data of pump generations A and B and ranking according to Mann–Whitney

Index i	Pump A	Pump B	R_A	R_B
1		11,500		1
2	12,000		2	
3	12,500		3	
4	13,200		[4] 4.5	
5		13,200		[5] 4.5
6	15,500		6	
7	17,000		[7] 7.5	
8	17,000		[8] 7.5	
9	17,100		9	
10		17,500		10
11		19,000		11
12		20,000		12
13		21,500		[13] 13.5
14		21,500		[14] 13.5

Tables A.2a and A.2b. If the test value $\hat{z}$ is within the quantiles, the null hypothesis is retained.

$$\hat{z} = \frac{\left|U_{pruef} - \frac{n_1 n_2}{2}\right|}{\sqrt{\frac{n_1 n_2 (n_1 + n_2 + 1)}{12}}} \tag{13.145}$$

$$U(\phi_{\alpha/2}) \leq \hat{z} \leq U(\phi_{1-\alpha/2}) \tag{13.146}$$

3. rank ties
For the assignment of ranks, the $n = n_1 + n_2$ observations are sorted by size on a scale. Each observation value is assigned a rank. If an observation value or measured value occurs more than once, a tie exists. No different ranks are assigned for the same observations, as it would influence the formation of the rank sums. Ergo, numerically equal observations receive the mean rank number, calculated from the arithmetic mean of all ranks assigned to this observation value, cf. also the application example in this chapter, here in particular Tables 13.18 and 13.19. However, the tie only influences the rank sum if the same observation is present in different samples. The rank sums are unaffected if the binding occurs within a sample.

Application example: Mann–Whitney-U-test
Two process plants (A and B) for pumping wastewater are monitored as part of a condition monitoring system. The design of both conveying systems is identical, but the construction states of the pumps are from two different generations (A and B). The pumps are operated until failure. The cycles at failure are documented for seven pumps of each generation (cf. Table 13.18). After the same observation period, failures of pump generation A are compared with those of pump generation B. By means of a U-test, it is to be investigated whether the failure centroid of pump generation A is significantly lower compared to generation B.

Solution: Carry out the U-Test
Step 1: Formulation of the hypotheses and determination of the significance level
The comparison of the estimators for the arithmetic mean shows that the centroid of the failures of generation A is lower in comparison to generation B. Ergo, the null hypothesis is chosen according to Eqs. 13.147 and 13.148, i.e.: the observation (centroid A lower B) is part of the alternative hypothesis. Thus, sample A is noted as 1 and sample B is noted as 2. If the U-Test were to yield a rejection of the null hypothesis, the alternative hypothesis would be accepted and generation A would fail significantly earlier than generation B, at chosen significance level $\alpha = 0.05$.

$$H_{01} : \mu_1 \geq \mu_2 \tag{13.147}$$

$$H_{11} : \mu_1 < \mu_2 \tag{13.148}$$

Step 2: Rank assignment and determination of rank totals
The rank assignment can be seen in Table 13.19: As ties are observed within the measurement series, they are assigned medium rank numbers.
Equations 13.149 and 13.150 show the calculation of rank sums.

$$R_1 = R_A = \sum R_{A;i} = 39.5 \tag{13.149}$$

$$R_2 = R_B = \sum R_{B;i} = 65.5 \tag{13.150}$$

13.4 Two Sample Case

Step 3: Calculation of the test variable U
Equations 13.151 to 13.154 show the calculation of the test quantities, their verification (Eq. 13.153), and the determination of the test quantity U_{pruef}.

$$U_1 = n_1 n_2 + \frac{n_1(n_1+1)}{2} - R_1 = 37.5 \qquad (13.151)$$

$$U_2 = n_1 n_2 + \frac{n_2(n_2+1)}{2} - R_2 = 11.5 \qquad (13.152)$$

$$U_1 + U_2 = n_1 n_2 = 49 \qquad (13.153)$$

$$U_{pruef} = min(U_1, U_2) = 11.5 \qquad (13.154)$$

Step 4: Determination of critical value and test decision
With the aid of Milton's (1964) table, the critical limit (cf. Appendix Tab A.6; Note: One-sided test; $\alpha = 0.05$) is determined (Eq. 13.155) and the test decision is made directly on the basis of Eq. 13.156.

$$U_{krit} = U_{n_1; n_2; \alpha} = U_{7;7;0.05} = 11 \qquad (13.155)$$

$$U_{pruef} \not\leq U_{krit} \qquad (13.156)$$

The test statistic is not less than critical value (Eq. 13.156), thus the null hypothesis is not rejected. The failure centroids of generations A and B are not significantly different, respectively the centroid A is not significantly below B, as the alternative hypothesis would imply (at a significance level of $\alpha = 0.05$).

Note
The test decision can also be made by approximation according to Eq. 13.157; e.g., the test here is two-sided according to hypotheses Eqs. 13.130 and 13.131. The test statistic $\hat{z}$ does not violate the test condition (Eqs. 13.158 and 13.159); ergo, the null hypothesis is retained: The foci are not significantly different.

$$\hat{z} = \frac{\left|U_{pruef} - \frac{n_1 n_2}{2}\right|}{\sqrt{\frac{n_1 n_2 (n_1 + n_2 + 1)}{12}}} = 1.66 \qquad (13.157)$$

$$U(\phi_{\alpha/2}) \leq \hat{z} \leq U(\phi_{1-\alpha/2}) \qquad (13.158)$$

$$-1.96 \leq 1.66 \leq 1.96 \qquad (13.159)$$

13.4.2 Siegel-Tukey Test: Dispersion Comparison

The rank test according to Siegel and Tukey (1960) is based on the Wilcoxon test (Wilcoxon 1945) and can be carried out independently of an assumption of a concrete distribution model with regard to the measured or observed values (parameter-free procedure). The focus is on the examination of two dispersions (scatterings) of continuous distributions with regard to a significant difference. Generally formulated, the null hypothesis would be: Both samples are part of the same population. The alternative hypothesis is: Both samples are not part of the same population. Note: As the difference between the means of the populations increases, the probability of rejecting H_0 (dispersions are equal) decreases, even though dispersion differences actually exist. Ergo, the probability of a 2nd kind error is high. The test power drops to zero if the populations do not show any intersections. As a consequence, this may mean that, given the samples at hand, the measurement series must be aligned. If the mean values are (almost) equal, the Siegel-Tukey test is very sensitive to dispersion differences. Furthermore, it should be noted that several steps of the test procedure are identical to the Mann–Whitney U-Test. Central differences are the formulations of the hypotheses, the alignment of the measurement series, if necessary, and the rank assignment.

Base of operations

- The realization of two independent samples with sizes n_1 and n_2 is available.
- The observations or measurement data are at least rank-based.
- The variables under study follow similar or the same continuous distribution functions.
- Same or similar median for both populations.

Goal
Generally formulated, the null hypothesis is: Both samples originate from the same population with regard to dispersion. The alternative hypothesis is: Both samples do not originate from a common population with regard to dispersion.

Performance of the Siegel-Tukey Test
Step 1: Formulation of the hypotheses and determination of the significance level
In general, Eqs. 13.160 and 13.161 apply to null and alternative hypotheses for two-sided questions, and Eqs. 13.162 to 13.165 apply accordingly for one-sided questions. Since the focus is on the study of dispersions, Eqs. 13.166 and 13.167 can also be formulated alternatively for two-sided studies, and Eqs. 13.168 to 13.171 for one-sided studies; parameter Θ (multiple) is used to express the diversity of dispersions.

Two-sided or one-sided questioning:

$$H_{01} : \sigma_1 = \sigma_2 \tag{13.160}$$

$$H_{11} : \sigma_1 \neq \sigma_2 \tag{13.161}$$

$$H_{02} : \sigma_1 \leq \sigma_2 \tag{13.162}$$

$$H_{12} : \sigma_1 > \sigma_2 \tag{13.163}$$

$$H_{03} : \sigma_1 \geq \sigma_2 \qquad (13.164)$$

$$H_{13} : \sigma_1 < \sigma_2 \qquad (13.165)$$

Alternative formulation of the hypotheses, both sides with $\Theta \neq 1$ as well as $\Theta > 1$:

$$H_{01} : G(z) = F(z) \qquad (13.166)$$

$$H_{11} : G(z) = F(\theta z) \qquad (13.167)$$

Alternative formulation of the hypotheses, one-sided with $0 < \Theta < 1$ (cf. Eqs. 13.162 and 13.163):

$$H_{02} : G(z) = F(z) \qquad (13.168)$$

$$H_{12} : G(z) = F(\theta z) \qquad (13.169)$$

Alternative formulation of the hypotheses, one-sided with $\Theta > 1$ (cf. Eqs. 13.164 and 13.165):

$$H_{03} : G(z) = F(z) \qquad (13.170)$$

$$H_{13} : G(z) = F(\theta z) \qquad (13.171)$$

The null hypothesis should be chosen to be meaningful when the question is one-sided. From the preliminary examination of the dispersions via estimation of the standard deviation, the indication arises which of the two dispersions is potentially the larger. Accordingly, this hint (suspicion) is to be recorded in the alternative hypothesis. Ergo, an attempt would be made to prove the opposite via the null hypothesis. If there is a significant difference between the dispersions which confirms the suspicion, the null hypothesis would be rejected and the proof would be provided by accepting the alternative hypothesis. With this approach, the minimum from the potential test variables results directly as the test variable U_{pruef}. The test decision criterion (cf. step 4) is identical for both one-sided and two-sided questions.

Step 2: Alignment of the measurement series (mean values) in case of large mean value difference

The power of the Siegel-Tukey test decreases with increasing differences in the centroids of the populations. It is recommended to align the measurement series in case of large mean differences. There are two possibilities for this:

Option 1: The mean of the corresponding sample is subtracted from each observation. Thus, both samples are transformed with respect to their measured values; Eqs. 13.172 and 13.173.

$$\widehat{x}_i = x_i - \overline{x} \qquad (13.172)$$

$$\widehat{y}_i = y_i - \overline{y} \qquad (13.173)$$

Option 2: The difference between the mean values of the two samples is subtracted from the measured values of the sample with the larger mean value. Example: Let x be the sample with the larger mean value. The sample with the transformed measured values is determined according to Eq. 13.174.

$$\widehat{x}_i = x_i - (\bar{x} - \bar{y}) \tag{13.174}$$

Step 3: Rank assignment and determination of rank totals
The observations are in the form of two samples with sizes n_1 and n_2. For the assignment of the ranks, the $n = n_1 + n_2$ observations are sorted according to size on a scale. Before forming the ranks, it is necessary to check whether the size of the unified sample with $n = n_1 + n_2$ observations is even. If the unified sample with $n = n_1 + n_2$ observations forms an odd number of observations, the middle observation is dropped and the sample size is reduced by one. Each observation value is assigned a rank g(i). However, unlike the U-test, the assignment is made according to the regularities given in Eq. 13.175, with extreme observation values receiving low ranks and central observation values receiving high ranks. This means that the highest rank is an even number. Then, rank sums are formed based on the assigned ranks, Eqs. 13.176 and 13.177. Rank sum R_1 refers to all ranks of observation values from sample 1, the same applies to rank sum R_2 with respect to sample 2. In consequence, this means: The more different the dispersion of sample 1 and 2, the more different the rank sums (cf. step 4).

$$g(i) = \begin{cases} 2i & \text{if } i \text{ equal and } 1 < i \leq N/2 \\ 2(N - i) + 2 & \text{if } i \text{ equal and } N/2 < i \leq N \\ 2i - 1 & \text{if } i \text{ unequal and } 1 \leq i \leq N/2 \\ 2(N - i) + 1 & \text{if } i \text{ equal and } N/2 < i < N \end{cases} \tag{13.175}$$

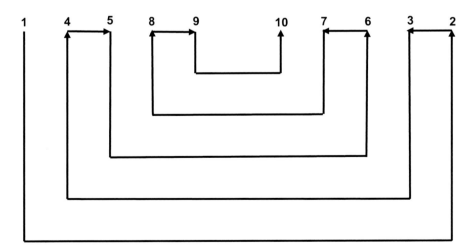

Fig. 13.9 Rank assignment within a Siegel-Tukey Test using the example of ten available observations

13.4 Two Sample Case

Table 13.20 Transformation of the failure data with respect to generation B

Measured value i	Transformed value $\widehat{b}_i$
1	8,700
2	10,400
3	14,700
4	16,200
5	17,200
6	18,700
7	18,700

Table 13.21 Failure data of pump generations A and B as well as rank assignment according to Siegel-Tukey and correction for ties within the sample (not mandatory)

Index i	Pump A	Pump B	Rank A	Rank B
1		8,700		1
2		10,400		4
3	12,000		5	
4	12,500		8	
5	13,200		9	
6		14,700		12
7	15,500		13	
8		16,200		14
9	17,000		[11] 10.5	
10	17,000		[10] 10.5	
11	17,100		7	
12		17,200		6
13		18,700		[3] 2.5
14		18,700		[2] 2.5

The scheme of rank assignment based on 10 measured values is shown in Fig. 13.9. The smallest value is assigned to rank 1, the two largest values are assigned to rank 2 and 3, respectively. Rank 4 and 5 are assigned the next smallest values. The next largest values are assigned rank 6 and 7, then the following small values are assigned rank 8 and 9. Finally, the last value is assigned rank 10.

$$R_1 = \sum R_{1;i} \qquad (13.176)$$

$$R_2 = \sum R_{2;i} \qquad (13.177)$$

If the rank sums R_1 and R_2 have been correctly formed, then Eq. 13.178 holds.

$$R_1 + R_2 = \frac{(n_1 + n_2)(n_1 + n_2 + 1)}{2} \qquad (13.178)$$

Note on rank ties:
If ties occur in the realizations to be examined, these must be taken into account when carrying out the Siegel-Tukey Test. The procedure is the same as for the U-Test (Sect. 13.4.1). To assign the ranks, the $n = n_1 + n_2$ observations are sorted by size on a scale. Each observation value is assigned a rank. If an observation value or measured value occurs more than once, a tie is present. Different ranks cannot be assigned to the same observations, as this would influence the formation of the rank sums. Ergo, numerically equal observations receive the mean rank, calculated from the arithmetic mean of all ranks assigned to this observation value, cf. also the application example regarding the Siegel-Tukey-Test in the present chapter; Table 13.20 and 13.21. However, the tie influences the rank sum only if the same observation is present in different samples. The rank sums are unaffected if the binding is present within one sample.

Step 4: Calculation of the test variable U (carried out in the same way as the U-Test)
The test statistic U is determined in the same way as the test statistic U of the U-Test (Sect. 13.4.1). The sample sizes n_1, n_2 and the rank sums according to Eqs. 13.176 and 13.177 are required for the calculation of the test variable U. First, the potential test variables are determined according to Eqs. 13.179 and 13.180. Equation 13.181 provides the check. The final test quantity U_{pruef} results from the minimum of U_1 and U_2; Eq. 13.182.

$$U_1 = n_1 n_2 + \frac{n_1(n_1 + 1)}{2} - R_1 \qquad (13.179)$$

$$U_2 = n_1 n_2 + \frac{n_2(n_2 + 1)}{2} - R_2 \qquad (13.180)$$

$$U_1 + U_2 = n_1 n_2 \qquad (13.181)$$

$$U_{pruef} = min(U_1, U_2) \qquad (13.182)$$

Step 5: Test decision (analogous to U-Test)
First, the critical value for assessing the test variable is determined. For this purpose, the table work of Milton (1964) is used, it includes critical values up to a sample size of $n = n_1 + n_2 \leq 60$ (cf. Appendix Tab A.6); Eq. 13.183 applies.

The test decision for two-sided questions is as follows: If Eq. 13.184 is satisfied, then the null hypothesis is rejected according to Eq. 13.160 or 13.166, and the alternative hypothesis Eq. 13.161 or 13.167 is accepted.

The test decision for one-sided questions is made according to the same principle, but it must be ensured that the null hypothesis and alternative hypothesis are chosen in a meaningful way; cf. notes within step 1.

13.4 Two Sample Case

$$U_{krit} = U_{n_1;n_2;\alpha} \qquad (13.183)$$

$$U_{pruef} \leq U_{krit} \qquad (13.184)$$

Note on the performance of the Siegel-Tukey Test
If large sample sizes are available with $n = n_1 + n_2 > 60$, an approximation can be made in determining the critical value U_{krit}, where Eq. 13.185 holds. The parameter $\hat{z}$ follows the normal distribution and can be taken from the table on the quantiles of the standard normal distribution (Appendix Tables A.2a and A.2b). When determining the parameter $\hat{z}$ the significance level as well as the one-sided or two-sided test procedure must be taken into account.

$$U_{krit} = U(n_1, n_2; \alpha) = \frac{n_1 n_2}{2} - \hat{z} \cdot \sqrt{\frac{n_1 n_2 (n_1 + n_2 + 1)}{12}} \qquad (13.185)$$

Application example: Siegel-Tukey Test
The database of failure data of two different pump generations A and B for pumping wastewater is examined; cf. application example Mann–Whitney U-Test; Table 13.18. After the same observation period, failures of pump generation A are compared with pump generation B. The test is performed by means of the Siegel-Tukey Test. Using the Siegel-Tukey Test, it is investigated whether the dispersion of failures of pump generation B is significantly higher in comparison to generation A.

Step 1: Formulation of the hypotheses and determination of the significance level
Comparison of the standard deviation estimators shows that the dispersion of failures generation A ($\sigma_A = 2275.96$) is lower compared to generation B ($\sigma_B = 3970.25$). Ergo, the null hypothesis is chosen according to Eq. 13.186. If a rejection of the null hypothesis were to result from the performance of the Siegel-Tukey Test, the alternative hypothesis would be accepted and Generation A would show a significantly lower dispersion in failures than Generation B (at chosen significance level $\alpha = 0.05$).

$$H_0 : \sigma_A \geq \sigma_B \qquad (13.186)$$

$$H_1 : \sigma_A < \sigma_B \qquad (13.187)$$

Step 2: Alignment of the measurement series (mean values) in case of large mean value difference
It is recommended to adjust the measurement series due to the large difference in mean values, whereby the arithmetic mean values are used here. The transformation is carried out by means of Eq. 13.188. The transformed values of generation B are shown in Table 13.20.

$$\hat{b}_i = b_i - (\overline{b} - \overline{a}) = b_i - 2800 \qquad (13.188)$$

Step 3: Rank assignment and determination of rank totals
The rank assignment is performed with respect to the observations of both samples; cf. Table 13.21. After the transformation, there are no ties between the samples; furthermore, the number of observations is even and thus no value elimination is necessary. The ties within the sample are marked, but has no influence on the rank sum formation. The rank sums are calculated according to Eqs. 13.189 and 13.190.

$$R_1 = R_A = \sum R_{A;i} = 63 \tag{13.189}$$

$$R_2 = R_B = \sum R_{B;i} = 42 \tag{13.190}$$

Step 4: Calculation of the test variable U (carried out in the same way as the U-Test)
Determine the test variables (Eqs. 13.191 and 13.192) and the final test variable (Eq. 13.193).

$$U_1 = n_1 n_2 + \frac{n_1(n_1+1)}{2} - R_1 = 14 \tag{13.191}$$

$$U_2 = n_1 n_2 + \frac{n_2(n_2+1)}{2} - R_2 = 35 \tag{13.192}$$

$$U_{pruef} = min(U_1, U_2) = 14 \tag{13.193}$$

Step 5: Determination of critical value and test decision
The condition according to Eq. 13.195 is not fulfilled, thus the null hypothesis is not rejected. The dispersions of generations A and B are not significantly different, respectively the dispersion of the failures of generation A is not significantly above the level of generation B (significance level $\alpha = 0.05$).

$$U_{krit} = U_{n_1;n_2;\alpha} = U_{7;7;0.05} = 11 \tag{13.194}$$

$$U_{pruef} \nleq U_{krit} \tag{13.195}$$

13.4.3 t-Test: Comparison of Two Mean Values

The t-test already outlined in Sect. 13.3.5 can also be applied in the two-sample case. The goal is to investigate the discriminability of the mean values of two normally distributed populations; it is a parametric test. Both two independent and two dependent samples can be present as realizations of a random variable. The variances and sample sizes of the two samples should be approximately equal. If the populations deviate from the assumption of normal distribution, the t-test is robust, within limits. For different sample sizes or variances, reference is made here to the Behrens-Fisher problem: The t-statistic is adjusted, but then does not follow a t-distribution, but the Behrens-Fisher distribution (Hedderich and Sachs 2020: 513).

Procedure of the two independent samples t-test
Step 1: Formulation of the hypothesis and determination of the significance level
The null hypothesis as well as the alternative hypothesis are established according to Eqs. 13.196 and 13.197 for two-sided testing. In the case of right-sided testing, the hypotheses result in Eqs. 13.198 and 13.199. In the case of left-sided testing, the hypotheses result in Eqs. 13.200 and 13.201. The significance level α is determined (e.g.: α = 0.01; α = 0.05; α = 0.1).

$$H_{01} : \mu_x = \mu_y \tag{13.196}$$

$$H_{11} : \mu_x \neq \mu_y \tag{13.197}$$

$$H_{02} : \mu_x \leq \mu_y \tag{13.198}$$

$$H_{12} : \mu_x > \mu_y \tag{13.199}$$

$$H_{03} : \mu_x \geq \mu_y \tag{13.200}$$

$$H_{13} : \mu_x < \mu_y \tag{13.201}$$

Step 2: Calculation of the test statistic T
The test statistic T is calculated according to Eq. 13.202. Note that the parameter s here is the weighted mean of the sample variances s_x^2 and s_y^2 (Eq. 13.203). For this, the arithmetic means as well as sample standard deviations s_x (sample x; size n) and s_y (sample y; size m) are required.

$$T = \frac{\bar{x} - \bar{y}}{s\sqrt{\frac{1}{n} + \frac{1}{m}}} \tag{13.202}$$

$$s^2 = \frac{(n-1)s_x^2 + (m-1)s_y^2}{n + m - 2} \tag{13.203}$$

Step 3: Determination of critical values and test decision
The critical value t results from the t-distribution depending on the sample size and the significance level; cf. quantiles of the t-distribution (cf. Appendix, Table A.3). If the condition according to Eq. 13.204 is fulfilled (and the test is two-sided), the null hypothesis H_0 is not rejected. Otherwise, H_0 is rejected and decided in favor of the alternative hypothesis H_1. The critical values t are determined using the quantiles of the t-distribution (Appendix, Table A.3) as a function of the degrees of freedom. The degree of freedom f corresponds to the sample size n minus 1, f = n − 1. Accordingly, the sample size are added here and minus 2 results in the degree of freedom, f = n + m − 2. For right-sided questions, the condition must be checked according to Eq. 13.205, for left-sided questions according

to Eq. 13.206: If the respective equation is satisfied, the null hypothesis is not rejected.

$$-t_{(1-\frac{\alpha}{2};n+m-2)} \leq T \leq t_{(1-\frac{\alpha}{2};n+m-2)} \tag{13.204}$$

$$T < t_{(1-\alpha;n+m-2)} \tag{13.205}$$

$$T > -t_{(1-\alpha;n+m-2)} \tag{13.206}$$

Application example: t-test for independent samples
A manufacturer of laptops receives two batches of microprocessors, which are to be installed in the current series. According to the manufacturer, the maximum frequency is 2.16 GHz and is subject to a certain degree of variation due to production. With regard to the distribution of the frequency, a normal distribution model may be assumed. Both batches comprise 4000 microprocessors each. A full test of all processors is not possible. The two batches are to be compared with regard to the question of whether the frequency differs on average. For this purpose, two samples (sample sizes: $n = m = 10$) are taken from the batches and the frequency is measured (cf. Table 13.22). It is assumed that both batches are independent. It is examined whether the batches differ with regard to the mean value of the clock frequency with an error probability of $\alpha = 0.05$.

Solution
Step 1: Formulation of the hypothesis and determination of the significance level
The null hypothesis as well as alternative hypothesis are established according to Eqs. 13.207 and 13.208, and the significance level is set at $\alpha = 0.05$.

$$H_0 : \mu_x = \mu_y \tag{13.207}$$

$$H_1 : \mu_x \neq \mu_y \tag{13.208}$$

Table 13.22 Measured values for the frequency [GHz] of two representative samples of processors from batch x and y

Sample from batch x	Sample from batch y
2.095	1.826
2.103	1.909
2.11	1.977
2.123	2.025
2.134	2.059
2.151	2.077
2.226	2.108
2.247	2.125
2.294	2.158
2.418	2.256

13.4 Two Sample Case

Step 2: Calculation of the test statistic T
First, the characteristics for the test statistic T are determined (Eqs. 13.209 and 13.210); for the sample means and standard deviations, Eqs. 13.211–13.214 are obtained. Thus, with n=m=10, the weighted mean of the sample variances is determined; Eq. 13.215. Hence, for the test statistic T, Eq. 13.216 is obtained.

$$\bar{x} = \frac{1}{n}\sum_{i=1}^{n} x_i \qquad (13.209)$$

$$s = \sqrt{\frac{\sum_{i=1}^{n}(x_i - \bar{x})^2}{n-1}} \qquad (13.210)$$

$$\bar{x} = 2.1901 \qquad (13.211)$$

$$\bar{y} = 2.0520 \qquad (13.212)$$

$$s_x = 0.1051 \qquad (13.213)$$

$$s_y = 0.1247 \qquad (13.214)$$

$$s^2 = \frac{(n-1)s_x^2 + (m-1)s_y^2}{n+m-2} = \frac{(9)0.1051^2 + (9)0.1247^2}{18} = 0.0133 \qquad (13.215)$$

$$T = \frac{\bar{x} - \bar{y}}{s\sqrt{\frac{1}{n} + \frac{1}{m}}} = 2.6776 \qquad (13.216)$$

Step 3: Determination of critical values and test decision
The critical limits for $\alpha = 0.05$ and m=n=10 are obtained based on the quantiles of the t-distribution (Appendix, Table A.3); substituting in Eq. 13.217 yields Eq. 13.218.

$$-t_{(1-\frac{\alpha}{2};n+m-2)} \leqslant T \leqslant t_{(1-\frac{\alpha}{2};n+m-2)} \qquad (13.217)$$

$$-2.101 \leq 2.6776 \not\leq 2.101 \qquad (13.218)$$

The condition is not fulfilled according to Eq. 13.218, the null hypothesis H_0 is rejected and it is decided in favor of the alternative hypothesis H_1. The arithmetic means of the frequencies of both samples examined differ significantly from each other with a probability of error of $\alpha = 0.05$.

The t-test for two dependent samples
Dependent samples exist if observations of the samples (e.g.: sample X and sample Y with values x_i and y_i) are tied in pairs. The following examples are given:

(a) Repeat measurement of a characteristic with respect to an object using the same or different measurement methods.
(b) Position tolerance of a hole (natural pairs).
(c) Measurement of two different surface parameters (e.g. roughness parameters Ra and Rz) for the same surface.

Step 1: Formulation of the hypothesis and determination of the significance level
The null hypothesis as well as the alternative hypothesis are established according to Eqs. 13.219 and 13.220 for two-sided testing. In the case of right-sided testing, the hypotheses result in Eqs. 13.221 and 13.222. In the case of left-sided testing, the hypotheses result in Eqs. 13.223–13.224. The significance level α is determined (e.g.: α = 0.01; α = 0.05; α = 0.1).

$$H_{01} : \mu_x = \mu_y \tag{13.219}$$

$$H_{11} : \mu_x \neq \mu_y \tag{13.220}$$

$$H_{02} : \mu_x \leq \mu_y \tag{13.221}$$

$$H_{12} : \mu_x > \mu_y \tag{13.222}$$

$$H_{03} : \mu_x \geq \mu_y \tag{13.223}$$

$$H_{13} : \mu_x < \mu_y \tag{13.224}$$

Step 2: Calculation of the test statistic T
The test statistic T (Eq. 13.225) is essentially based on the mean (cf. Eq. 13.227) and standard deviation s_d (cf. Eq. 13.228) of the differences D_i (cf. Eq. 13.226) of the paired sample values. The parameter n represents the sample sizes (which are equal) and, logically, the number of differences D_i. In principle, examining the differences converts the two-sample test into a one-sample test, since the significant deviation of the differences from the value 0 is examined.

$$T = \sqrt{n}\frac{\overline{D}}{s_d} \tag{13.225}$$

$$D_i = x_i - y_i \tag{13.226}$$

$$\overline{D} = \frac{1}{n}\sum_{i=1}^{n} D_i \tag{13.227}$$

13.4 Two Sample Case

$$s_d = \sqrt{\frac{1}{n-1} \sum_{i=1}^{n} (D_i - \overline{D})^2} \qquad (13.228)$$

Step 3: Determination of critical values and test decision
If the condition according to Eq. 13.229 is fulfilled when the test is two-sided, the null hypothesis H_0 is not rejected. Otherwise, H_0 is rejected and decided in favor of the alternative hypothesis H_1. The critical values t are determined using the quantiles of the t-distribution (Appendix, Table A.3) based on the degree of freedom $f = n-1$. For right-sided questions, the condition must be checked according to Eq. 13.230, for left-sided questions according to Eq. 13.231: If the respective equation is satisfied, the null hypothesis is retained.

$$-t_{(1-\frac{\alpha}{2};n-1)} \leq T \leq t_{(1-\frac{\alpha}{2};n-1)} \qquad (13.229)$$

$$T < t_{(1-\alpha;n-1)} \qquad (13.230)$$

$$T > -t_{(1-\alpha;n-1)} \qquad (13.231)$$

Application example for the t-test with dependent samples
As an application example, the batch problem of the laptop manufacturer is discussed again; cf. application example t-test with independent samples (Table 13.22). The assumption is made that the two batches are dependent on each other. The two batches are to be compared with regard to the question of whether, on average, the frequency is different. For this purpose, two samples (sample sizes: $n = m = 10$) are taken from the batches and the cycle frequency is measured (Table 13.22). The application of the t-test should allow a statement as to whether the series of measurements are distinguishable with regard to the mean value of the

Table 13.23 Measured values for the clock frequency [GHz] of two representative samples of processors from batch x and y; furthermore pair differences D_i

Sample from batch x	Sample from batch y	Pair differences D_i
2.095	1.826	0.269
2.103	1.909	0.194
2.11	1.977	0.133
2.123	2.025	0.098
2.134	2.059	0.075
2.151	2.077	0.074
2.226	2.108	0.118
2.247	2.125	0.122
2.294	2.158	0.136
2.418	2.256	0.162

cycle frequency under the assumption that the samples are dependent (selected level of significance $\alpha = 0.05$).

Solution
Step 1: Formulation of the hypothesis and determination of the significance level
The null hypothesis and alternative hypothesis are set up according to Eqs. 13.232 and 13.233, and the significance level is set at $\alpha = 0.05$.

$$H_0 : \mu_x = \mu_y \tag{13.232}$$

$$H_1 : \mu_x \neq \mu_y \tag{13.233}$$

Step 2: Calculation of the test statistic T
First, the pair differences D_i of the measured values are formed according to Eq. 13.226 (Table 13.23).
The mean and standard deviation of the pair differences are obtained according to Eqs. 13.234 and 13.235. Thus, the test statistic T can be calculated according to Eq. 13.236.

$$\overline{D} = \frac{1}{n}\sum_{i=1}^{n} D_i = 0.1381 \tag{13.234}$$

$$s_d = \sqrt{\frac{1}{n-1}\sum_{i=1}^{n}(D_i - \overline{D})^2} = 0.0589 \tag{13.235}$$

$$T = \sqrt{n}\frac{\overline{D}}{s_d} = 7.4144 \tag{13.236}$$

Step 3: Determination of critical values and test decision
The critical limits for $\alpha = 0.05$ as well as $n = 10$ are obtained on the basis of the quantiles of the t-distribution (cf. Appendix, Table A.3) and are inserted into Eq. 13.237; Eq. 13.238 follows.

$$-t_{(1-\frac{\alpha}{2};n-1)} \leq T \leq t_{(1-\frac{\alpha}{2};n-1)} \tag{13.237}$$

$$-2.262 \leq 7.414 \nleq 2.262 \tag{13.238}$$

The condition according to Eq. 13.238 is not fulfilled, the null hypothesis H_0 is rejected and decided in favor of the alternative hypothesis H_1. The arithmetic means of the frequencies of both samples examined – assuming sample dependence – differ significantly from each other at a significance level of $\alpha = 0.05$.

13.4.4 F-Test: Comparison of Two Variances

The comparison of the variances of two samples can be done using the F-Test according to Ronald Aylmer Fisher. This is a parametric analysis of variance with the goal of testing whether two samples come from different, normally distributed populations.

Base of operations
There are two different populations with mean and dispersion (μ_A and σ_A respectively μ_B and σ_B), from which there are two samples A and B with size n_A and n_B. The samples are independent of each other.

Procedure of the F-Test
Step 1: Formulation of the hypothesis; determination of the significance level
The null hypothesis H_0 is formulated according to Eq. 13.239 when the test is two-sided, and the alternative hypothesis H_1 is formulated according to Eq. 13.240. When the test is one-sided, Eqs. 13.241–13.244 apply. The significance level α is specified (e.g., $\alpha = 0.01$; $\alpha = 0.05$; $\alpha = 0.1$).

$$H_{01}: \sigma_A^2 = \sigma_B^2 \tag{13.239}$$

$$H_{11}: \sigma_A^2 \neq \sigma_B^2 \tag{13.240}$$

$$H_{02}: \sigma_A^2 \leq \sigma_B^2 \tag{13.241}$$

$$H_{12}: \sigma_A^2 > \sigma_B^2 \tag{13.242}$$

$$H_{03}: \sigma_A^2 \geq \sigma_B^2 \tag{13.243}$$

$$H_{13}: \sigma_A^2 < \sigma_B^2 \tag{13.244}$$

Step 2: Determiation of test statistic T
First, the variances are estimated on the basis of the available samples. Then the variances are inserted into the quotient of the test statistic T according to Eq. 13.245: The larger variance is to be inserted in the numerator (is given index 1), the smaller variance is to be inserted in the denominator (given index 2).

$$T = \frac{\max\left(s_A^2; s_B^2\right)}{\min\left(s_A^2; s_B^2\right)} = \frac{s_1^2}{s_2^2} = \frac{\frac{1}{n_1-1}\sum_{i=1}^{n_1}(x_{1i}-\bar{x}_1)^2}{\frac{1}{n_2-1}\sum_{i=1}^{n_2}(x_{2i}-\bar{x}_2)^2} \tag{13.245}$$

Step 3: Determination of critical value and test decision
The test statistic T follows Fisher's F-distribution; cf. Appendix Table A.4 a–l). To determine the critical value, the number of degrees of freedom according to Eqs. 13.246 and 13.247 is required. The indexing corresponds to the determination in step 2. In the case of a two-sided test procedure the critical value is determined

according to Eq. 13.248, in the case of one-sided test procedure according to Eq. 13.249. Using the critical value, the test decision can be made: If the test is two-sided, the null hypothesis is rejected and the alternative hypothesis is accepted if Eq. 13.250 is fulfilled.

$$f_1 = n_1 - 1 \tag{13.246}$$

$$f_2 = n_2 - 1 \tag{13.247}$$

$$P_{krit} = P(F_{f1;f2;1-\alpha/2}) \tag{13.248}$$

$$P_{krit} = P(F_{f1;f2;1-\alpha}) \tag{13.249}$$

$$T > P_{krit} \tag{13.250}$$

If the test is one-sided, note: If the hypotheses were chosen according to Eqs. 13.251 and 13.252, the test decision is made according to Eq. 13.253. The null hypothesis is rejected and the alternative hypothesis is accepted if Eq. 13.253 is fulfilled.

$$H_{02} : \sigma_1^2 \leq \sigma_2^2 \tag{13.251}$$

$$H_{12} : \sigma_1^2 > \sigma_2^2 \tag{13.252}$$

$$T > P_{krit} \tag{13.253}$$

If the hypotheses were chosen according to Eqs. 13.254 and 13.255, the test decision is made as follows: The null hypothesis H_0 is retained, since $s_1 > s_2$. Equation 13.256 is thus satisfied, and separate testing is not required.

$$H_{03} : \sigma_1^2 \geq \sigma_2^2 \tag{13.254}$$

$$H_{13} : \sigma_1^2 < \sigma_2^2 \tag{13.255}$$

$$T > P_{krit} \tag{13.256}$$

Application example for the F-test
As an application example, the batch problem of the laptop manufacturer (cf. application example for the t-test) is discussed again; cf. Table 13.22. The assumption is made that the two batches are independent of each other and the populations follow the normal distribution model. The two batches are compared with regard to the question of whether the variances of the frequency related to the available measurement series for batch x and y are significantly different.

Solution
Step 1: Formulation of the null hypothesis; determine of significance level

$$H_0 : \sigma_A^2 = \sigma_B^2 \tag{13.257}$$

13.4 Two Sample Case

$$H_1 : \sigma_A^2 \neq \sigma_B^2 \tag{13.258}$$

$$\alpha = 0.05 \tag{13.259}$$

Step 2: Determination of test statistic T
Estimating the sample variances (Eq. 13.260) yields the variances for batch x (Eq. 13.261) and batch y (Eq. 13.262). The test statistic T is obtained according to Eq. 13.263.

$$s^2 = \frac{\sum_{i=1}^{n}(x_i - \bar{x})^2}{n-1} \tag{13.260}$$

$$s_x^2 = 0.0111 \tag{13.261}$$

$$s_y^2 = 0.0155 \tag{13.262}$$

$$T = \frac{\max(s_A^2; s_B^2)}{\min(s_A^2; s_B^2)} = \frac{s_y^2}{s_x^2} = \frac{0.0155}{0.0111} = 1.3964 \tag{13.263}$$

Step 3: Determination of critical value and test decision
Using the set of tables for the F-distribution (cf. Appendix Table A.4), the critical value is obtained according to Eq. 13.264. The condition in Eq. 13.265 is not fulfilled, therefore the null hypothesis is retained: The samples are not significantly different with respect to the estimated values of their dispersions.

$$P_{krit} = P(F_{f1;f2;1-\alpha/2}) = P(F_{9;9;0.975}) = 4.026 \tag{13.264}$$

$$1.3964 = T \not> P_{krit} = 4.026 \tag{13.265}$$

13.5 Multi-Sample Case

If the database to be analysed includes more than two observation series or samples, this is referred to as a multiple sample case. In this case, standard investigations focus on the analysis of centroid and the dispersion of the samples in relation to each other. The following procedures belong to the known hypothesis tests:

(a) Comparison of several centroids using analysis of variance (ANOVA) with normally distributed populations; cf. (Hedderich and Sachs 2020:628ff.).
(b) Comparison of several centroids according to Kruskal and Wallis (parameter-free method).
(c) Comparison of several dispersions according to Bartlett for normally distributed populations.
(d) Comparison of several dispersions according to Meyer-Bahlburg (parameter-free method).

Furthermore, trend analysis (cf. Jonckheere 1954) or multiple pairwise comparisons (cf. Rao and Swarupchand 2009), for example, can also be carried out with regard to the centroids of the samples. The following explanations focus on the procedures according to Kruskal (1952) as well as Kruskal and Wallis (1952), Bartlett (1937) and Meyer-Bahlburg (1970).

13.5.1 Kruskal–Wallis H Test and Conover Post hoc Analysis: Centroids

The comparison of several centroids of different samples can be done by means of the H-test according to Kruskal and Wallis (1952) and Kruskal (1952) respectively. The H-test is a generalization of the U-test and, compared to the optimal (!) analysis of variance in the case of normal distribution, has an asymptotic efficiency of $100 \cdot 3/\pi$ ($\approx 95\%$); cf. (Hedderich and Sachs 2020). This is a parameter-free procedure. It does not require that the populations from which the samples under investigation are drawn follow a normal distribution model. The H-test is sensitive to median differences and insensitive to dispersion differences.

Starting Point

1. There are $k \geq 3$ independent samples with sizes $n_1, n_2, ..., n_k$.
2. The observations or measured values have at least rank level.
3. The distributional forms of the populations from which the samples are drawn should be similar or the same.

Procedure of the H-test according to Kruskal and Wallis
Step 1: Formulation of hypotheses and determination of the significance level
The object of the null hypothesis is that the centroids of the present k samples do not differ (Eq. 13.266). The alternative hypothesis of the statement is that at least one pair of the medians (index i and j related to all present samples 1, ..., k) differs (Eq. 13.267). Since the focus is on divergence, this option of hypothesis formulation is the only one (thus the test is always two-sided). The significance level α is set (usually $\alpha = 0.01$; $\alpha = 0.05$; $\alpha = 0.10$).

$$H_0 : \Theta_1 = \Theta_2 = \cdots = \Theta_k \tag{13.266}$$

$$H_1 : \Theta_i \neq \Theta_j \quad \text{for} \quad i \neq j \tag{13.267}$$

Step 2: Preparation of the test
First, all values of all samples with size n_i are sorted by size and a corresponding rank is assigned; for the size n_{Ges} of all samples $i = 1, ..., k$, Eq. 13.268 applies. At the same time, it is noted from which sample j the respective observation value originates. In the case that observations have the same rank (i.e. a tie exists), the mean value of the assigned rank is determined for the assigned rank and this mean rank is assigned. If ties occur in the same samples, no mean rank needs to be

13.5 Multi-Sample Case

formed. Then the rank sums R_i are formed for each of the k samples; Eq. 13.269. The summation can be checked for correctness according to Eq. 13.270.

$$n_{Ges} = \sum_{i=1}^{k} n_i \tag{13.268}$$

$$R_i = \sum_{j=1}^{n_i} R_{ij} \tag{13.269}$$

$$\sum_{i=1}^{k} R_i = \frac{n_{Ges}(n_{Ges} + 1)}{2} \tag{13.270}$$

Step 3: Test statistics H
The calculation of the test statistic H is generally done according to Eq. 13.271. In the case that all available samples k have the same size n_i, Eq. 13.272 can be used, since $n_i = n_{Ges}/k$. If more than ¼ of all observed values x_{ij} have ties, then Eq. 13.273 is used for the test statistic H_{corr}. The parameter C refers to the ties and is determined using Eq. 13.274 (Hedderich and Sachs 2020), where t_i is the number of equal values in the ith tie, from which the mean rank was calculated in the same tie.

$$H = \left[\frac{12}{n_{Ges}(n_{Ges}+1)} \sum_{i=1}^{k} \frac{R_i^2}{n_i} \right] - 3(n_{Ges}+1) \tag{13.271}$$

$$H = \left[\frac{12}{n_{Ges}(n_{Ges}+1)} \cdot \frac{k}{n_{Ges}} \sum_{i=1}^{k} R_i^2 \right] - 3(n_{Ges}+1) \tag{13.272}$$

$$H_{corr} = \left\{ \left[\frac{12}{n_{Ges}(n_{Ges}+1)} \sum_{i=1}^{k} \frac{R_i^2}{n_i} \right] - 3(n_{Ges}+1) \right\} \cdot \frac{1}{C} \tag{13.273}$$

$$C = 1 - \frac{\sum_{i=1}^{k}(t_i^3 - t_i)}{n_{Ges}^3 - n_{Ges}} \tag{13.274}$$

$$H_{corr} > H \tag{13.275}$$

Note on Test Statistics H
If a correction of the test statistic has been carried out due to the tie problem, the test decision is to be made using the test statistic H_{corr}. If the test decision using

the test statistic H already yields a significant result (rejection of the null hypothesis), the determination and a test decision with test statistic H_{corr} are unnecessary, since Eq. 13.275 applies.

Step 4: Determination of critical value and test decision
If $n_i \leq 5$ and $k = 3$, the critical value in the form of the exceedance probability can be taken from the table work (Kruskall and Wallis 1952); cf. Appendix Tab A.8. The test decision is made according to Eq. 13.276: if the test statistic H exceeds the critical value, the null hypothesis is rejected.

If larger, summed sample sizes n_{Ges} are available ($n_i \geq 5$ and sample number $k \geq 4$), the test statistic H approximately follows the χ^2-distribution. With the help of the degrees of freedom (Eq. 13.277), the critical value can be taken based on the quantiles of the χ^2-distribution (Appendix Table A.8). Thereafter, the test decision is made according to Eq. 13.278. It should be noted that the approximation via χ^2-distribution should be sufficient in most cases.

$$H > H_{crit} \tag{13.276}$$

$$f = k - 1 \tag{13.277}$$

$$H > \chi^2_{k-1; 1-\alpha} \tag{13.278}$$

Post hoc analysis according to Conover (1999)
The Kruskal and Wallis procedure provides an indication of whether the centroids of different samples differ significantly. However, no statement can be made about which of the samples are different with regard to the centroid.

The method of Conover can be used to determine which sample is significantly different from the other samples available by means of a pairwise comparison; cf. (Conover 1999) and also (Conover and Iman 1979). This parameter-free procedure compares the differences in rank numbers – based on the Kruskal–Wallis test statistic – of the samples according to Eq. 13.279. The differences refer to the mean ranks R_i/n_i. The null hypothesis H_0 is "equality of expected mean ranks". The null hypothesis can be rejected as soon as Eq. 13.279 holds (noting that $n_i \geq 6$). In the case where there are no ties, Eq. 13.280 holds for the total mean square s^2 and, consequently, Eq. 13.281 holds for the test statistic of difference.

$$\left| \overline{R}_i - \overline{R}_j \right| > t_{1-\frac{\alpha}{2}; n_{Ges}-k} \sqrt{s^2 \left[\frac{n_{Ges} - 1 - H}{n_{Ges} - k} \right] \left[\frac{1}{n_i} + \frac{1}{n_j} \right]} \tag{13.279}$$

$$s^2 = \frac{n_{Ges}(n_{Ges} + 1)}{12} \tag{13.280}$$

$$\left| \overline{R}_i - \overline{R}_j \right| > t_{1-\frac{\alpha}{2}; n_{Ges}-k} \sqrt{\left[\frac{n_{Ges}(n_{Ges} + 1)}{12} \right] \left[\frac{n_{Ges} - 1 - H}{n_{Ges} - k} \right] \left[\frac{1}{n_i} + \frac{1}{n_j} \right]} \tag{13.281}$$

13.5.2 Bartlett-Test: Comparison of Variances

The parametric test according to Bartlett (1937) focuses on the equality of variances in the multiple sample case and is based on the F-test. In essence, it is a parametric analysis of variance. The prerequisite for carrying out the Bartlett-test is the assumption that the populations from which the samples are taken follow the normal distribution model (parametric test). The Bartlett-test is sensitive to differences in variances and insensitive to differences in location.

Base of operations

1. There are $k \geq 3$ independent samples with sizes $n_1, n_2, \ldots, n_k$.
2. The normal distribution model is assumed for the distribution forms of the populations from which the samples are taken. When examining the sample characteristics, the focus is on the medians.
3. For sample sizes, $n_i \geq 5$ for all samples i.

Procedure of the Bartlett-Test
Step 1: Formulation of hypotheses and determination of significance level
The focus of the test is to examine the variances related to the samples at hand with regard to significant differences. Accordingly, the null hypothesis is H_0 according to Eq. 13.282 and the alternative hypothesis is H_1 (at least two variances are unequal) according to Eq. 13.283. Alternatively, the hypotheses can also be related to the standard deviations, whereby the statement is comparable in essence as it relates to dispersion; cf. Eqs. 13.284 and 13.285.

$$H_0 : \sigma_1^2 = \cdots = \sigma_k^2 \tag{13.282}$$

$$H_1 : \sigma_i^2 \neq \sigma_j^2 \quad \text{with } i \neq j \tag{13.283}$$

$$H_0^a : \sigma_1 = \cdots = \sigma_k \tag{13.284}$$

$$H_1^a : \sigma_i \neq \sigma_j \quad \text{with } i \neq j \tag{13.285}$$

Step 2: Test preparation
First, the standard deviations are calculated for all available samples; Eq. 13.286. The values x_{ij} correspond to the values of sample i, the same applies to the arithmetic mean. To estimate the standard deviation s_i, the jth value of sample i is considered in each case (in the sum), from which the mean value of sample i is subtracted, Eq. 13.286. This yields the standard deviation of sample i. Furthermore, with the goal of the degrees of freedom f_i (Eq. 13.287) and the size of all samples n_{Ges} (Eq. 13.288), the variance s_p^2 is determined in relation to all sample variances s_i, cf. Eq. 13.289.

$$s_i = \sqrt{\frac{1}{n_i - 1} \sum_{j=1}^{n_i} (x_{ij} - \bar{x}_i)^2} \qquad (13.286)$$

$$f_i = n_i - 1 \qquad (13.287)$$

$$n_{Ges} = \sum_{i=1}^{k} n_i \qquad (13.288)$$

$$s_p^2 = \frac{\sum_{i=1}^{k}(n_i - 1)s_i^2}{\sum_{i=1}^{k}(n_i - 1)} = \frac{\sum_{i=1}^{k} f_i s_i^2}{\sum_{i=1}^{k} f_i} = \frac{\sum_{i=1}^{k} f_i s_i^2}{f_{Ges}} = \frac{1}{n_{Ges} - k} \sum_{i=1}^{k} (n_i - 1)s_i^2 \qquad (13.289)$$

Step 3: Calculation of the test statistic T
The test statistic T is determined according to Eq. 13.290, where the quantity c is obtained from Eq. 13.291. If the number of degrees of freedom f_i is high (rule of thumb: $f_i \geq 10$), then $c \approx 1$ may be set and thus neglected in Eq. 13.290. If rather small degrees of freedom f_i (rule of thumb: $f_i < 10$) are assumed with respect to the context under investigation, Eq. 13.292 can also be used directly—without the separate calculation of parameter c.

$$T = \frac{1}{c}(n_{Ges} - k)\ln(s_p^2) - \sum_{i=1}^{k}(n_i - 1)\ln(s_i^2) \qquad (13.290)$$

$$c = 1 + \frac{1}{3(k-1)} \left(\sum_{i=1}^{k} \frac{1}{f_i} - \frac{1}{f_{Ges}} \right) = 1 + \frac{1}{3(k-1)} \left(\sum_{i=1}^{k} \left(\frac{1}{n_i - 1} \right) - \frac{1}{n_{Ges} - k} \right) \qquad (13.291)$$

$$T = \frac{(n_{Ges} - k)\ln(s_p^2) - \sum_{i=1}^{k}(n_i - 1)\ln(s_i^2)}{1 + \frac{1}{3(k-1)}\left(\sum_{i=1}^{k} \left(\frac{1}{n_i - 1} \right) - \frac{1}{n_{Ges} - k} \right)} \qquad (13.292)$$

Step 4: Determination of the critical value and test decision
The test statistic T follows the χ^2-distribution model; cf. Sect. 7.3.3. Thus, the critical value can be taken from the quantiles of the χ^2-distribution, using the degrees of freedom $f_{Ges} = k - 1$ as well as the significance level α (cf. Appendix Table A.5). The test decision is made according to Eq. 13.293: The null hypothesis is rejected if the condition according to Eq. 13.293 is fulfilled.

$$T > \chi^2_{k-1; 1-\alpha} \qquad (13.293)$$

13.5.3 Meyer-Bahlburg Test: Comparison of Variances

The aim of the Meyer-Bahlburg Test is the comparison of several variances related to m available samples. In principle, the test according to Meyer-Bahlburg (1970) test is based on the Siegel-Tukey (1960) test, which can be used to detect possible dispersion differences in two independent samples, cf. Sect. 13.4.2. Ergo, the Meyer-Bahlburg test is also a parameter-free procedure for investigating m independent samples, so that no preconditions need to be met with regard to a specific distribution model related to the populations.

Base of Operations

1. The populations from which the present samples are drawn follow a continuous distribution model $F_i(x)$.
2. There are m independent samples with size n_i (i = 1, ..., k).
3. For the total sample size, $n_{Ges} = 4$ m; where $n_{ges} = n_1 + n_2 + ... + n_k$ is an integer and thus n_{Ges} is divisible by four.

Procedure of the Meyer-Bahlburg test
Step 1: Formulation of hypotheses and determination of significance level
The test focuses on the examination of the distribution functions $F_i(x)$ with regard to significant differences. Accordingly, the null hypothesis is H_0 according to Eq. 13.294 and the alternative hypothesis is H_1 (at least two distribution functions are different) according to Eq. 13.295.

$$H_0 : F_1(x) = \cdots = F_k(x) \tag{13.294}$$

$$H_1 : F_i(x) = F_j(\theta x) \quad \text{with } \theta > 0 \text{ and } i \neq j \tag{13.295}$$

Step 2: Test preparation
The observations are in the form of k samples with sizes $n_1, ... n_k$. For the assignment of the ranks, the observations in the combined sample with size n_{Ges} are sorted according to size. Afterwards, the rank assignment is carried out on the basis of the Siegel-Tukey method (cf. in detail Sect. 13.4.2); the respective rank j is assigned to each observation value. Subsequently, the rank sums are calculated with respect to each sample i = 1, ..., k; Eq. 13.296. Equation 13.297 may be taken as a control calculation with respect to the rank sum.

$$R_i = \sum R_{i;j} \tag{13.296}$$

$$\sum_{i=1}^{k} R_i = \frac{n_{Ges}(n_{Ges} + 1)}{2} \tag{13.297}$$

Step 3: Calculation of the test statistic T
The test statistic T is determined according to Eq. 13.298: The test statistic T corresponds to the test statistic of the significance test according to Kruskal and Wallis (1952); see in detail Sect. 13.5.1. If all available sample sizes n_i have the same size and the condition $n_{Ges} = 4 \cdot n_i$ is fulfilled, Eq. 13.299 can be applied directly.

$$T = \frac{12}{n_{Ges}(n_{Ges}+1)} \sum_{i=1}^{k} \frac{R_i^2}{n_i} - 3(n_{Ges}+1) \tag{13.298}$$

$$T = \frac{12}{n_{Ges}(n_{Ges}+1)} \cdot \frac{k}{n_{Ges}} \sum_{i=1}^{k} R_i^2 - 3(n_{Ges}+1) \tag{13.299}$$

Step 4: Determination of the critical value and test decision
The test statistic T follows the χ^2-distribution model; cf. Sect. 7.3.3. Thus, the critical value can be taken from the quantiles of the χ^2-distribution, using the degrees of freedom $f_{Ges} = k - 1$ and the significance level α (cf. Appendix Table A.5). The test decision is made according to Eq. 13.300: The null hypothesis is rejected if the condition according to Eq. 13.300 is fulfilled.

$$T > \chi^2_{k-1;1-\alpha} \tag{13.300}$$

Notes Regarding the Meyer-Bahlburg Test
A) If the condition $n_{Ges} = 4 \cdot m$ is not satisfied, then the following procedure can be followed:

1. For $n_{Ges} = 4m+1$: deletion of the median of the total sample.
2. For $n_{Ges} = 4m+2$: delete quartiles Q_1 and Q_3.
3. For $n_{Ges} = 4m+3$: deletion of the median and the quartiles Q_1 and Q_3.

The quartile Q_1 corresponds to the sample element for which one quarter of the sample is less than Q_1. Q_3 corresponds to the sample element for which three quarters of the values are smaller than Q_3.

B) In the case of ties related to observed values that originate from different samples, the final rank assignment is based on an averaging of the initially assigned ranks (cf. also Siegel-Tukey Sect. 13.4.2).

Prototype Testing and Accelerated Testing 14

As part of the series development of a product emergence process, prototype testing plays a decisive role in analyzing the technical reliability of a product at a certain stage of development. The present chapter is first devoted to the fundamentals of accelerated testing (Sect. 14.1). Afterwards, important models are presented which can be used in the planning and analysis of accelerated testing activities, e.g. Arrhenius model, Wöhler test, Coffin-Manson model; Sect. 14.2. Furthermore, the number of prototypes (sample size) to be tested is of central importance in the planning of a test; Sect. 14.3 shows two approaches for determining a minimum sample size. A case study of accelerated testing using the Weibull distribution model and corresponding data analysis is shown in Sect. 14.4. Qualitative accelerated testing using the Highly Accelerated Life Test (HALT) and Highly Accelerated Stress Screening (HASS) methods is the focus of Sect. 14.5.

14.1 Accelerated Testing

Within the scope of series development, a series of prototype tests are carried out in order to obtain a statement about the technical reliability of a product. The statement refers to the current design status (design maturity) of the product (cf. Chap. 2). The test object can be, for example, a component (e.g. engine connecting rod), assembly (e.g. engine), system (e.g. power train consisting of engine and transmission) or the entire product (e.g. vehicle). The difficulty, however, is to obtain a statement about the reliability of the product in relation to the planned service life (useful life). For example, the series development phase of a vehicle can last three to four years, whereas the use phase in the field can last an average of 12 years; with a warranty period of two to four years. The problem is obvious: If a product is tested under real conditions, as they exist in the use phase, the required test time significantly exceeds the series development time.

© Springer-Verlag GmbH Germany, part of Springer Nature 2024
S. Bracke, *Reliability Engineering*, https://doi.org/10.1007/978-3-662-67446-8_14

This chapter outlines the main features as well as different procedures of accelerated testing and is based on the explanations of Hinz et al. (2014).

The goal of Accelerated Testing of products is to gain knowledge about the technical reliability in a shorter period of time compared to the planned service life. In accelerated testing, one variable or a combination of several variables is used to analyze and determine the aging process or degradation of a product. The variable(s) can be, for example, the rate of use, temperature, voltage or humidity.

A model can descibe the relationship between the different product degradations: accelerated failure behaviour under test conditions and real failure behaviour within the usage phase. The challenge is to activate or detect the same damage causality at an earlier point in time by accelerating the degradation and at the same time not to provoke further damage causalities by the acceleration itself. Furthermore, the model must be chosen correctly in order to be able to represent the assumed or existing degradation behaviour. With the aid of this model, an acceleration factor can be determined: The acceleration factor describes the relationship between test time and utilization time.

Quantitative Accelerated Test
The quantitative accelerated test is carried out with the aid of an influencing variable or a combination of several influencing variables (e.g. temperature, humidity) with the goal of determining information about the distribution of the failure points or the degradation behaviour in relation to the time under increased loads. On this basis, predictions can be made for operation under normal conditions. Mathematical models used in this test are, for example, the "cumulative exposure model" and the "Khamis-Higgins model" (Khames and Higgins 1999). Frequently, the relevant cause-effect relationships between damage causality and damage symptom are already known in advance, as is the knowledge of the relationship between damage and influence quantity. This is based on either known physical or chemical relationships or simply empirical data and information. Based on the gained findings within the test procedure, the product design as well as the manufacturing process can be optimized with regard to technical reliability in addition to the service life prognosis.

Qualitative Accelerated Test
The qualitative accelerated test—like the quantitative accelerated test—is also carried out with the aid of an influencing variable or a combination of influencing variables (e.g. temperature, humidity, vibration). Common names for procedures of this type are HALT (Highly Accelerated Life Test), STRIFE (Stress-Life), and EST (Environmental Stress Testing); (Chan and Englert 2001). The aim of the qualitative accelerated test is to detect product weaknesses with regard to product design and/or the manufacturing process. It is a targeted damage provocation under operating conditions outside the specification. The provoked damage is analysed with the goal of gaining knowledge about the extent to which the observed defect could also occur in real operation, and also to detect defective products ("weak products") in the manufacturing process (Highly Accelerated Stress Screening

14.1 Accelerated Testing

(HASS)). A sound knowledge of the physical fundamentals of the cause-and-effect relationship between damage causality and damage symptom is important here. Since the results of a qualitative accelerated test are mainly used to optimize the product design and the manufacturing process, it is not easy (or at least risky or bold) to map a time-related failure behavior for real operation on the basis of these data. Ergo, the data are generally not suitable for a service life prognosis.

Accelerated testing by increasing the usage rate and/or load intensity
Two strategies can be used for both qualitative and quantitative accelerated tests:

1. Increasing the usage rate,
2. Increasing the load intensity.

Increasing the usage rate in the test cycle is applied to products that are not in continuous operation. For example, the assumed life of a bearing for a washing machine may be twelve years based on a usage rate of eight wash cycles per week. If this machine is tested at 112 cycles per week (16 a day), the expected life is reduced to ten months. This is subject to the assumption that the increased usage rate (actuation cycles performed at shorter intervals) does not affect the failure behavior. An illustrative case study is shown by Yang and Zaghati (2006).

The second strategy for shortening the test time is to increase the load intensity. An increased load intensity leads to intensified wear. For example, loading a product with increased force leads to more rapid failures (cf. also Sect. 14.4.2) compared to the failure behaviour in normal operation. A combination of these accelerating factors can also be applied. In some cases, physical modelling can become very difficult. Empirical models may be helpful in these cases to infer the service life under normal conditions of use.

Finally, various forms of accelerated testing are presented in an overview based on the result that can be obtained after an accelerated test has been carried out; Table 14.1.

Table 14.1 Overview of accelerated testing procedures depending on the test result

Test procedure	Abbreviation	Test result
Accelerated Binary Test	ABT	Failure or no failure of the device under test (DUT)
Accelerated Life Test	ALT	Test result is related to the lifetime of the test object Right-censored or interval-censored for test collective
Accelerated Repeated Measures Degradation Test	ARMDT	Degradation is detected by means of one or more variables at specific points in time with non-destructive measurement
Accelerated Destructive Degradation Test	ADDT	Degradation is measured by means of one or more variables at specific points in time with destructive measurement

14.2 Quantitative Accelerated Testing: Models

An important characteristic of all quantitative accelerated tests is the relationship between product operation under real conditions and under test conditions with the inclusion of an acceleration factor.

The understanding of quantitative accelerated tests requires models that relate the influence of acceleration related variables to the time of use. The difficulty in implementation lies in the correlation between the test and the subsequent field use. The so called acceleration factor describes this connection. The determination of the acceleration factor is the actual challenge in the planning of reliability tests under increased loads. In the approach, the cause-effect relationship between damage causality and damage symptom in the field is stimulated by the loads in the test operation. If this is not the case, extremely high test loads will lead to product damage, but this information is not suitable for determining potential field life.

Models for the acceleration of a degradation behavior are distinguished between two types:

1. Physical and chemical acceleration models,
2. Empirical acceleration models.

For known cause-effect relationships between damage causality and damage symptom, physical or chemical theories may exist that can describe the fault-causing process with the aid of the available data, usually with reference to time. Furthermore, these physical and/or chemical acceleration models allow the extrapolation of the effects from test operation to real operation. Although the relationship between influencing variables and damage causality is complex in many cases, simple models often exist that can adequately describe the degradation behavior.

If the chemical or physical processes leading to damage are unknown, it is not possible to make a sound prediction based on a physical or chemical model. An empirical acceleration model is then the only alternative. A model of this form is fitted to the available data and thus describes the failure behaviour with sufficient precision. It should be noted, however, that extrapolation to areas without prior knowledge may well be highly error-prone.

The present Sect. 14.2 shows a compilation of important acceleration models as well as their fields of application.

14.2.1 Arrhenius Model

The law according to Arrhenius focuses on the relationship between the reaction rate k and the temperature T (Arrhenius 1889) and is essentially based on the rule according to van't Hoff (1884). Thus, the rate of a chemical reaction is directly related to the absolute temperature. The Arrhenius law is used for the

14.2 Quantitative Accelerated Testing: Models

phenomenological, macroscopic illustration of chemical as well as physical processes. The Arrhenius law is very similar to the Eyring law, which focuses on the microscopic level; see Eyring (1935). Note that the Arrhenius law may involve various constants. Equation 14.1 shows the Arrhenius law involving the gas constant R, as it is used in chemistry. On the other hand, Eq. 14.2 shows the Arrhenius law including the Boltzmann constant k_B, as used in physics.

$$k = A \cdot e^{-\left(\frac{E_A}{RT}\right)} \tag{14.1}$$

$$k = A \cdot e^{-\left(\frac{E_A}{k_B T}\right)} \tag{14.2}$$

Here, the parameter k means the reaction speed, or collision frequency, which causes a reaction. The parameter T represents the temperature unit [K]. A is a pre-exponential factor; constant with respect to the chemical reaction under consideration. R is the general gas constant ($8.314 \text{ J} \cdot \text{mol}^{-1} \cdot \text{K}^{-1}$) in the chemical model and k_B is the Boltzmann constant ($1.381 \cdot 10^{-23} \text{ J} \cdot \text{K}^{-1}$) in the physical model. The parameter E_A corresponds to the activation energy, for Eq. 14.1 in the unit corresponding to the product $R \cdot T$ or for Eq. 14.2 in the unit corresponding to the product $k_B \cdot T$. In the following, the variant with the Boltzmann constant is used.

In electronic components, many failure mechanisms are temperature dependent. This thermally induced aging can be mapped using the following approach: The two reaction rates—or, in the context of technical reliability, better known as failure rate λ_1 and λ_2 (unit [ppm/a])—are put into the ratio and converted to λ_1; Eq. 14.3. Equations 14.4 and 14.5 should be noted.

$$\lambda_1 = \lambda_2 e^{-\frac{E_A}{k_B}\left(\frac{1}{T_1} - \frac{1}{T_2}\right)} \tag{14.3}$$

$$T_1 > T_2 \tag{14.4}$$

$$\lambda_1 > \lambda_2 \tag{14.5}$$

The acceleration factor α is given by the ratio of λ_1 (failure rate 1, associated with temperature T_1) and λ_2 (failure rate 2, associated with temperature T_2), cf. Eq. 14.6.

$$\alpha = \frac{\lambda_1}{\lambda_2} = e^{-\frac{E_A}{k_B}\left(\frac{1}{T_1} - \frac{1}{T_2}\right)} \tag{14.6}$$

The use of Arrhenius' law to design an accelerated test procedure requires knowledge of the activation energy E_A. However, this is unlikely to be known in most cases when testing technical components. Two approaches are conceivable to estimate the activation energy approximately: The application of the rule according to van't Hoff (1884) or the use of empirically determined failure rates.

Rule van't Hoff (1884)
An increase in temperature of 10 K leads to at least a doubling of the reaction rate. The factor Q_{10} maps the increase in the reaction rate for a temperature increase of T = 10 K, i.e. as a function of the reaction rates at temperature T_1 and T_2; cf. Eq. 14.7. In this case, $Q_{10} = \alpha$; cf. Eq. 14.6.

$$Q_{10} = \left(\frac{k_2}{k_1}\right)^{\frac{10K}{T_1-T_2}} \tag{14.7}$$

A transfer to the testing of components is conceivable: An increase of the temperature by 10 K leads to a doubling of the reaction speed, thus to a doubling of the failure rate as well as to a halving of the average service life. It should be noted that the van't Hoff rule has a higher uncertainty with larger temperature differences, ergo the approximation is only permitted to a limited extent.

Empirically determined failure rates
Failure rates can be determined for components at different temperatures with sufficiently large sample sizes. From the thus known failure rate with respect to two known temperatures, the activation energy can be calculated (cf. Eq. 14.8 based on conversion of Eqs. 14.3 or 14.6). The calculated activation energy can be transferred to other temperature differences and thus the acceleration factor α and thus test duration t_{test} can be determined as a function of the specification t_{spec}; Eq. 14.9.

$$E_A = -\frac{k_B \ln\left(\frac{\lambda_1}{\lambda_2}\right)}{\frac{1}{T_1} - \frac{1}{T_2}} \tag{14.8}$$

$$t_{test} \approx \frac{t_{spec}}{\alpha} \tag{14.9}$$

Notes on the Arrhenius law
The Arrhenius law is often used to estimate a test time for a test under increased temperature conditions (temperature as a stress factor). If a technical realization is possible, a validation of the test result should be carried out, since the outlined application is based on the following assumptions:

a) The assumption is made that the activation energy relates to a specific chemical or physical cause-effect relationship to damage causality and damage symptom (failure mechanism). However, a more complex cause-effect structure could be based on several, different activation energies, which are not represented by the approach shown.

14.2 Quantitative Accelerated Testing: Models

b) The transfer of a calculated activation energy for known failure rates at known temperatures also assumes the outlined regularity. However, inaccuracies can occur with larger temperature differences, so that the transfer of the activation energy E_A to other temperature fields is always associated with the assumption that the activation energy is constant.
c) The transfer of the Arrhenius law (chemical or physical processes) to technical components follows a phenomenological approach. The more complex the component, the more uncertain the approximation is likely to be.
d) The uncertainty concerning the approximation by using the van't Hoff rule, which formed the initial basis for the Arrhenius law, is already known for higher temperature differences with respect to chemical and physical processes.

Conclusion: The use of the Arrhenius law for the determination of test times as well as acceleration factors is to be classified as an estimation. If possible, validation should be carried out with regard to observed ageing processes or damage patterns. The uncertainty increases with increasing temperature difference and complexity of the test object.

Application Example: Accelerated Test for an Electronic Control Unit

The engine room of a ship is planned as the installation location of an electronic control unit. A service life of $t_{spec} = 4000$ h is planned. The temperature at the installation location in the engine room is approximately $T = 60\ °C$. Furthermore, the failure rates at different temperatures are known from preliminary tests and field experience (carry over part); cf. Table 14.2. In the laboratory, the control unit is tested under elevated temperature at $T = 95\ °C$ ($\triangleq 368$ K). The estimation of the required test time t_{test} – at which the control unit is to be tested – is the goal.

Solution: Determination of the test time t_{test}
The acceleration factor $\alpha_{60/70}$ is determined with the aid of Eq. 14.10 for the temperature levels $T_2 = 60\ °C$ and $T_1 = 70\ °C$. Furthermore, the activation energy E_A can be determined directly with the aid of the known failure rates at $T_2 = 60\ °C$ and $T_2 = 70\ °C$; cf. Eq. 14.11. Testing takes place at $T_3 = 95\ °C$ ($\triangleq 368$ K), so that the acceleration factor $\alpha_{60/95}$ is obtained for the temperature levels $T_2 = 60\ °C$ and $T_3 = 95\ °C$ according to Eq. 14.12, and thus the test time t_{test} for testing can be estimated; Eq. 14.13.

$$\alpha_{60/70} = \frac{\lambda_1}{\lambda_2} = 1.3 \qquad (14.10)$$

Table 14.2 Failure rates of a control unit as a function of installation location and temperature

	Temperature [°C] and [K]	Failure rate [ppm/year]
Control unit installation location: Normal operation; moderate climate zone	$T_2 = 60\ °C$ ($\triangleq 333$ K)	$\lambda_2 = 1.000$ ppm/year
Control unit installation location: Hot and humid earth zone	$T_1 = 70\ °C$ ($\triangleq 343$ K)	$\lambda_1 = 1.300$ ppm/year

$$E_A = -\frac{k_B \ln\left(\frac{\lambda_1}{\lambda_2}\right)}{\frac{1}{T_1} - \frac{1}{T_2}} = -\frac{1.381 \cdot 10^{-23} JK^{-1} \cdot \ln(1.3)}{\frac{1}{343K} - \frac{1}{333K}} = 4.1384 \cdot 10^{-20} J \quad (14.11)$$

$$\alpha_{60/95} = e^{-\frac{E_A}{k_B}\left(\frac{1}{T_3} - \frac{1}{T_2}\right)} = e^{-\frac{4.1384 \cdot 10^{-20} J}{1.381 \cdot 10^{-23} JK^{-1}}\left(\frac{1}{368K} - \frac{1}{333K}\right)} = 2.3535 \quad (14.12)$$

$$t_{test} \approx \frac{t_{spec}}{\alpha_{60/95}} \approx 1699.60\,\text{h} \quad (14.13)$$

14.2.2 Wöhler Test and Damage Accumulation Hypotheses

During operation, components products and systems are generally subject to dynamic loads and rather less static loads. The design and testing of components is therefore based on changing loads. August Wöhler (1819, Soltau—1914, Hanover; Germany) developed a test in which the Wöhler line can be determined under different loads and stress cycles. The trigger for the development of the Wöhler test was the breakage of a railway wheel tyre of the Amstetten locomotive (1875) near the Timelkam railway station on the Linz to Salzburg line (Fig. 14.1). The damage was caused by the alternating stress, i.e. a dynamic load, on the wheel tyre. However, the design of the wheel tyre was based on static loads, as other, dynamic models were not yet known at that time.

Fig. 14.1 Amstetten locomotive (1875) crashed near Timelkam station on the Linz – Salzburg line due to a broken wheel tyre (photo: author unknown)

Figure 14.2 shows a typical sinusoidal stress curve for an alternating load (t_0 = period of oscillation, σ_0 = amplitude); following Hornbogen (1994:144), see also Haibach (2006) and (DIN 50,100:2016-12). If a static tensile or compressive stress is applied to the component in addition to the alternating load is applied to the component, it results in a shift accordingly in the vertical direction, Fig. 14.3 (Hornbogen 1994:144). Based on the performance of several tests with different amount of load and stress amplitude the Wöhler curve can be determined: The amount of load is continuously increased to a certain stress amplitude until the tested specimen breaks at N load cycles. As an example, Fig. 14.4 shows the determination of Wöhler curves for steel and grey cast iron with spheroidal graphite on the basis of tests carried out. The scattering of specimen fractures at different numbers of load cycles with respect to the same stress amplitude can be seen. This scatter is often due to heterogeneities in the material, rather than inconsistencies in the test procedure. Therefore, Fig. 14.4 shows three different Wöhler curves as a function of the survival probability R. For this purpose, an analysis of the distribution of damages (fractures) related to the number of load cycles is performed and a distribution model is fitted. From this model, the B_{10}, B_{50}, B_{90} parameters are determined (cf. Chap. 12) and the Wöhler curves for the survival probabilities $R = 0.1$, $R = 0.5$ and $R = 0.9$ are estimated accordingly on the basis of several supporting points (different load cycles). If the Wöhler test is repeated several times, deviations from the determined Wöhler curves may occur due to heterogeneities in the material.

A schematic diagram of Wöhler lines with the three characteristic strength ranges – short-term strength range, fatigue strength range, endurance range – is shown in Fig. 14.5 (abscissa and ordinate: logarithmic scaling):

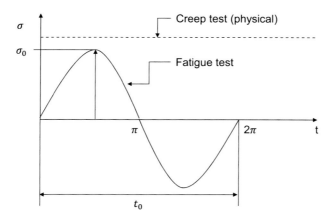

Fig. 14.2 Sinusoidal stress curve with alternating load in the fatigue test versus creep test with constant stress

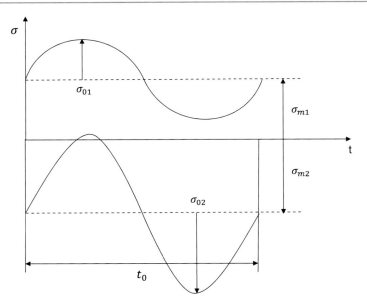

Fig. 14.3 Sinusoidal stress curve for alternating load in fatigue test with additional alternating load due to a static tensile or compressive stress (level shift by σ_{m1} and σ_{m2})

a) Short-term strength range: Small number of oscillating cycles N with high stress amplitude σ. The stress amplitude $\sigma_{a,1}$ corresponds to the tensile strength R_m or the yield strength R_e.
b) Fatigue strength range: The possible stress amplitudes σ decrease with increasing number of oscillating cycles. Micro-cracks can occur in the stressed component, which expand to macro-cracks in the course of further loading and finally lead to component fracture.
c) Endurance range: If the stress amplitude is below a certain stress amplitude σ, the component will not fail even if the number of oscillating cycles N is arbitrarily high.

Note: The endurance strength range can be described in many technical applications; the example of steel and grey cast iron in Fig. 14.4 shows this range well. However, there are also some applications where a endurance range cannot be represented. Due to material heterogeneities or environmental conditions, the growth of component damage can progress more rapidly than outlined in Fig. 14.5; cf. in particular the following section on damage accumulation and the damage accumulation hypotheses.

Linear damage accumulation
The loading of a component is often based on different stress amplitudes. Each partial load with stress amplitude $\sigma_{a,i}$ causes a partial damage d_i; for this purpose,

14.2 Quantitative Accelerated Testing: Models

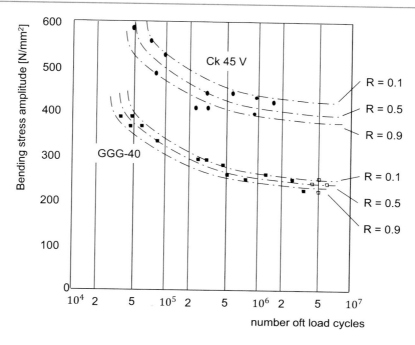

Fig. 14.4 Schematic diagram of experimentally determined Wöhler curves related to different survival probabilities R for steel and grey cast iron, respectively, according to Hornbogen (1996:390)

the number of occurring oscillating cycles n_i is related to the maximum possible vibration cycles N_i; Eq. 14.14. The maximum possible oscillating cycles N_i are determined from the Wöhler curve, which is based on experimental data. It should be noted that Wöhler curves can be determined both for incipient component damage (incipient cracks) and for complete component failure: The schematic diagram in Fig. 14.7 shows this relationship as well as the Wöhler curve for the case of component failure. Furthermore, the assumption is made that stress amplitudes that lie within the endurance range (cf. Fig. 14.5) do not lead to any damage.

$$d_i = \frac{n_i}{N_i} \qquad (14.14)$$

Damage accumulation hypothesis according to Miner (1945); Miner rule, Miner original

The sum of all partial damages from the amplitude collective with stress amplitudes $\sigma_{a,i}$ (i = 1, ..., j) results in the total damage of the component, cf. Eq. 14.15 as well as Fig. 14.6. In the case that the total damage is D = 1, failure of the component occurs. According to the hypothesis of Miner (1945), it is irrelevant in

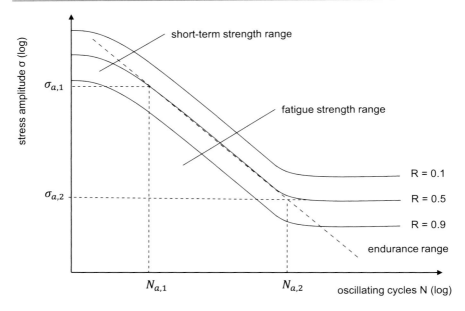

Fig. 14.5 Wöhler lines (related to survival probability R) and characteristic strength ranges as a function of the number of oscillating cycles N and stress amplitude σ

which combination or sequence the load levels or sub-collectives reach the damage (and thus the maximum number of oscillations). Thus, the sub-collectives can be converted into each other; Eq. 14.16.

$$D = \sum_{i=1}^{j} \frac{n_i}{N_i} \qquad (14.15)$$

$$\frac{n_i}{N_i} = \frac{n_j}{N_j} \qquad (14.16)$$

Assumptions in the Miner (1945) approach:

Miner's damage accumulation hypothesis makes the following assumptions:

a) The stress amplitudes $\sigma_{a,i}$ in the endurance range below the stressamplitude $\sigma_{a,D}$ do not lead to partial damage.
b) The progress of component damage is independent of the sequence of oscillating cycles n_i.
c) The elastic–plastic deformation behavior of the material is mapped with a hysteresis in the stress–strain diagram. The deformation energy corresponds to the damage and is constant over the number of cyclic loads.

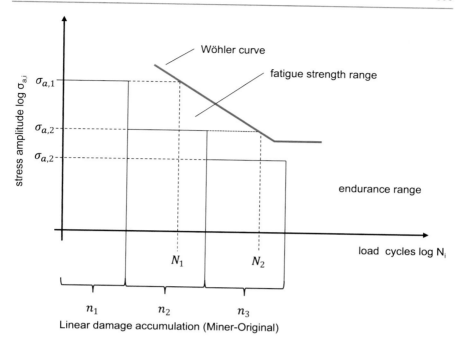

Fig. 14.6 Miner's damage accumulation hypothesis (Miner rule; Miner original)

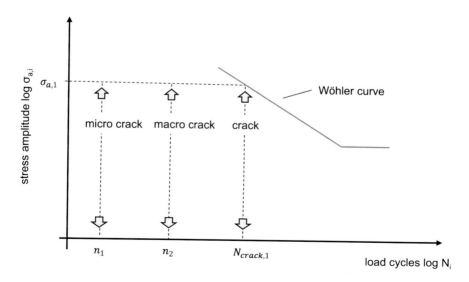

Fig. 14.7 Component damage during growth (microcrack, macrocrack, fracture) and Wöhler line

Based on assumption a), Eq. 14.17 applies to all stress amplitudes σ_a above the stress amplitude $\sigma_{a,D}$, which characterizes the endurance range. However, it can be shown experimentally that stress amplitudes within the endurance range can also lead to component damage. Therefore, a component design based on Miner's approach (it should be emphasized again: Miner original) can lead to undersizing. As a result, a modified Miner approach is often used; cf. in particular the modification of Miner according to Palmgren (1924).

$$\sigma_a > \sigma_{a,D} : N = N_D \cdot \left(\frac{\sigma_a}{\sigma_{a,D}}\right)^{-k} \qquad (14.17)$$

Modifications of the approach according to miner
Figure 14.8 shows a comparison of the different damage accumulation hypotheses Miner (Miner-Original), Miner-Palmgren, Miner-Haibach, Miner Eurocode (Haibach) and Miner-Zenner-Liu. The modifications of the Miner damage accumulation hypothesis are differentiated from each other by naming the authors and are explained below.

Miner-Palmgren damage accumulation hypothesis.
The modification of the Miner approach according to Palmgren (1924) assumes a continuation of the decreasing course in the fatigue-strength range with the same gradient; therefore, in addition to Eqs. 14.17 und 14.18 also applies. The

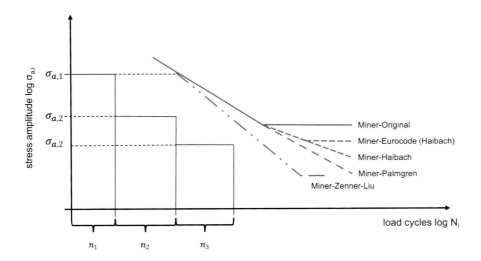

Fig. 14.8 Schematic diagram comparing the damage accumulation hypotheses according to Miner (Miner original) and the modifications according to Haibach, Eurocode, Palmgren and Zenner-Liu

14.2 Quantitative Accelerated Testing: Models

hypothesis according to the Miner-Palmgren approach is considered a conservative approach. Palmgren developed and published the damage hypothesis as early as 1924; however, it did not become widely known until the publication of Miner (1945), which may cause some confusion due to the non-chronological order of names in the Miner-Palmgren approach.

$$\sigma_a \leq \sigma_{a,D} : N = N_D \cdot \left(\frac{\sigma_a}{\sigma_{a,D}}\right)^{-k} \quad (14.18)$$

Damage accumulation hypothesis according to Miner-Haibach
The modification of the Miner-Original rule according to Haibach (2006) assumes a continuation of the sloping course in the fatigue-strength range with a modified, lower gradient, therefore Eq. 14.19 also applies in addition to Eq. 14.17. The hypothesis according to Miner-Haibach is often used in an industrial context.

$$\sigma_a \leq \sigma_{a,D} : N = N_D \cdot \left(\frac{\sigma_a}{\sigma_{a,D}}\right)^{-(2k-1)} \quad (14.19)$$

Damage accumulation hypothesis according to Miner Eurocode
The modification of the rule Miner-Original according to Ungermann et al. (DIN 2015) is known as Eurocode 3: Design of steel structures; DIN EN 1993-1-8 (DIN and BAUFORUMSTAHL 2015). Ungermann et al. assume a continuation of the decreasing progression in the fatigue strength range with a modified, lower gradient according to Miner-Haibach, but only up to a further lower limit, which characterizes the endurance range with stress amplitude $\sigma_{a,D2}$. The assumption is that no damage to the component occurs below the stress amplitude $\sigma_{a,D2}$. Therefore, Eq. 14.20 (as Eq. 14.17) and Eq. 14.21 (building on Eq. 14.19) apply.

$$\sigma_a > \sigma_{a,D} : N = N_D \cdot \left(\frac{\sigma_a}{\sigma_{a,D}}\right)^{-k} \quad (14.20)$$

$$\sigma_{a,D2} \leq \sigma_a \leq \sigma_{a,D} : N = N_D \cdot \left(\frac{\sigma_a}{\sigma_{a,D}}\right)^{-(2k-1)} \quad (14.21)$$

Damage accumulation hypothesis according to Miner-Liu-Zenner
The modification of the Miner-Original rule according to Liu and Zenner (1995) provides a rotation at the highest stress amplitude in the collective, the gradient is also modified, Eq. 14.22 holds. Furthermore, the endurance range is modified compared to the Miner-Original approach according to Eq. 14.23. The Liu-Zenner approach is less frequently used in an industrial context.

$$\sigma_a > \sigma_{a,D}^* : N = N_D \cdot \left(\frac{\sigma_a}{\sigma_{a,D}}\right)^{-\frac{k+m}{2}} \qquad (14.22)$$

$$\sigma_{a,D}^* = \frac{\sigma_{a,D}}{2} \qquad (14.23)$$

14.2.3 Inverse Power Law: Introduction

The Inverse Power Law generally describes the relationship between the service life of a component and the defined service life in relation to accelerated testing. The term Inverse Power Law, which is commonly used in the literature, is deliberately used in this book. Strictly speaking, however, it is not (!) **one** singular law with validity for different forms of stress (e.g. temperature, pressure, humidity). Rather, the Inverse Power Law generally describes the relationship between decreasing service life of a component under increased stress. However, since the stress to which a component is subjected can be of different types, there are a number of laws for specific forms of stress, all of which are based on the basic ideas of the Inverse Power Law mentioned above. A very good overview of various forms of the Inverse Power Law can be found in Nelson (1990:85) and (MIL 1998:8-160). The nomenclature according to Nelson (1990) will be adopted in this chapter, as it allows a comparability of the different forms of the Inverse Power Law.

The general formulation of the Inverse Power Law can be found in Eq. 14.24: The stress variable V must assume a positive value. The change in the stress variable determines the acceleration in testing. The value τ is the value of a function depending on the stress V and refers to the nominal product lifetime. Parameters A and γ_1 are used to describe characteristics of the product, manufacturing process or testing. γ_1 describes the power of the accelerated testing. Other common forms of the inverse power law are Eqs. 14.25 and 14.26. The parameter V_0 represents a defined standard stress (e.g. stress under real operation of the component or the product), the parameter V represents the stress in testing increased to the parameter V_0.

$$\tau(V) = \frac{A}{V^{\gamma_1}} \qquad (14.24)$$

$$\tau(V) = \left(\frac{A'}{V}\right)^{\gamma_1} \qquad (14.25)$$

$$\tau(V) = A'' \left(\frac{V_0}{V}\right)^{\gamma_1} \qquad (14.26)$$

14.2.4 Coffin-Manson Model

The focus of the Coffin-Manson model is to represent the relationship between component damage and thermal cycling; see Coffin (1954 and 1974) and Manson (1953 and 1965). The Coffin-Manson model represents a function in which the variable N describes the number of cycles up to component damage depending on the range of the temperature cycles ΔT, cf. Eq. 14.27. The parameters A and B are constants which can describe the test procedure and the material properties of the component respectively. The base of operations for the Coffin-Manson model is also the assumption that a higher stress on the component in a shorter time of testing leads to a comparable damage at lower stress and longer time of testing.

$$N = \frac{A}{(\Delta T)^B} \qquad (14.27)$$

The Coffin-Manson model is often used in the analysis of mechanical and electrical components. For example, the parameter B takes the value $B \sim 2$ for metallic components. For electronic components based on a material mix of predominantly plastics and metallic components, the parameter B is $B \sim 5$.

14.2.5 Inverse Power Weibull Model

The Inverse Power Weibull Model is based on the Weibull distribution model (cf. in detail Chap. 7 and Weibull (1951)). Equation 14.28 shows the two-parameter Weibull distribution model with the parameters T (characteristic life) and b (shape parameter). The Inverse Power Weibull model represents the characteristic life T as a function of stress V; cf. Eq. 14.29. The product properties are described by the parameters γ_0 and γ_1, and the shape parameter b refers to the failure behavior property. Thus, for the probability of failure F(x) at a given stress level V, based on Eqs. 14.28 and 14.29 yields Eq. 14.30.

$$F(x) = 1 - e^{-\left(\frac{x}{T}\right)^b} \qquad (14.28)$$

$$T(V) = \frac{e^{\gamma_0}}{V^{\gamma_1}} \qquad (14.29)$$

$$F(x) = 1 - e^{-\left(\frac{x}{T(V)}\right)^b} = 1 - e^{-\left(x \cdot e^{-\gamma_0} \cdot V^{\gamma_1}\right)^b} \qquad (14.30)$$

Assumptions and comments on the inverse power Weibull model:

1. The Weibull distribution model is used to represent the product lifetime at a defined stress level V.
2. The shape parameter b of the Weibull distribution model is constant and thus independent of stress level V. This means that the failure characteristic remains the same with increased stress and no new cause-effect relationship to damage causality and damage symptom occurs due to the increased stress. This is an elementary precondition for the implementation of accelerated testing; cf. also in detail Sect. 14.1.
3. Using a mathematical paper (here double logarithmic), the probability function for a given stress level V appears as a straight line.
4. The application range of the Inverse Power Weibull model is broad: The failure behavior, caused by mechanical or electrical stress, can be mapped for mechanical, electronic as well as mechatronic components.
5. Thus, when developing a method for accelerated testing, a mechanical or electrical stress can form the basis for test acceleration.

14.2.6 Lundberg-Palmgren Equation

One of the best known forms of the inverse power law is the Lundberg-Palmgren equation which refers to the failure behaviour of rolling bearings; cf. Lundberg and Palmgren (1947) and (1953). Both Nils Arvid Palmgren (1890, Falun – 1971, Lerum) and Gustav Lundberg (1901, Ingatorp – 1961, Lerum) were research engineers at Svenska Kullagerfabriken (SKF) and developed, among other things, lifetime models for the reliability of rolling bearing (SKF 2014). The life of bearings can be calculated for point contact (rolling element: ball) using Eq. 14.31 and for line contact (ex: cylindrical rolling element) using Eq. 14.32. The Palmgren equation represents the relationship between B_{10} parameter (note: failure probability $P = 0.1$ referred to a bearing collective; cf. in detail Chap. 12), the bearing-dependent constant C_x (x = r for radial load or x = a for axial load) and the radial load or axial load P_x (x = r for radial load or x = a for axial load). The parameter c represents the stress in relation to the service life. The parameter h represents the depth of the maximum shear stress. The parameters b_1 (rolling elements with point contact) and b_2 (rolling elements with line contact) represent the shape parameters of the Weibull distribution model (cf. Eq. 14.28 as well as Chap. 7), which describe the wear mechanism; Lundberg and Palmgren (1947) and (1953) as well as (DIN ISO 281:1994).

Palmgren and Lundberg (1947) determined the values $c = 31/3$ and $h = 7/3$ for the parameters c and h on the basis of experimental investigations. The shape parameters of the Weibull distribution were determined experimentally to be $b_1 = 10/9$ and $b_2 = 9/8$.

Thus Eqs. 14.31 and 14.32 can also be given in the form according to Eq. 14.33, where the exponent p is called power (or in German: Lebensdauerexponent). For Power p, the empirically obtained parameters c, h, b_1

and b_2 result in the value $p=3$ for the case of point contact and the value $p=4$ for the case of line contact; (DIN ISO 281:1994,39-40) or (DIN ISO 281:2010).

In the industrial application context of rolling element bearings, values of $p=3$ for ball bearings (point contact) and $p=10/3$ for roller bearings (line contact) were determined empirically. The difference to the previously calculated values of the service life exponent p results from empirical observations of damage to bearings; (DIN 281:1994,39-40) and (SKF 2014:64).

$$B_{10} = \left(\frac{C_x}{P_x}\right)^{\frac{c-h+2}{3b_1}} \quad (14.31)$$

$$B_{10} = \left(\frac{C_x}{P_x}\right)^{\frac{c-h+1}{2b_2}} \quad (14.32)$$

$$B_{10} = \left(\frac{C_x}{P_x}\right)^{p} \quad (14.33)$$

14.2.7 Taylor Equation

Frederick Winslow Taylor (1856, Germantown – 1915, Philadelphia; U.S.A.) is among other things the inventor of the high-speed steel as well as the founder of the work science ("Taylorism"). The Taylor equation named after him refers to the mean service life τ ("tool life T") of a cutting tool, Eq. 14.34; quoted in notation after Nelson (1990). The exponent m refers to the gradient of the tool life curve which represents the empirically determined relationship between cutting speed and tool life. The tool life curve is determined on the basis of test series, an ablation within a mathematical paper with double logarithmic scaling results approximately in a descending straight line with gradient m: The higher the cutting speed (abscissa), the lower the tool life (ordinate). The parameter T is the theoretical tool life at a cutting speed $V=1$ m/min; for details see (Westkämper and Warnecke 2010). The parameters T and m are directly dependent on the material and geometry of the cutting tool and are determined empirically on the basis of test series.

$$\tau = \frac{T}{V^m} \quad (14.34)$$

Note: The tool life in real operation can deviate from the calculated tool life due to heterogeneities in the material to be cut and in the material of the cutting tool itself. Furthermore, influences of the cutting tool geometry and the workpiece to be cut can cause a deviation of the actual tool life.

14.2.8 Eyring Model

The acceleration of testing of components is often done by varying the temperature conditions under which the component are investigated in a testing program. Increasing the temperature can, for example, cause an acceleration of the aging process, so that the damage symptom is detectable at an earlier point in time compared to testing under standard environmental conditions (usually: T = 20 °C). For the design of accelerated tests, the Arrhenius law is often used, which is used for the phenomenological, macroscopic illustration of chemical as well as physical processes and also includes a temperature component (Sect. 14.2.1). An alternative to the Arrhenius approach is the model according to Eyring (1935): The Eyring model was derived from quantum mechanical effects and takes into account the reaction rate for chemical degradation processes; cf. Eq. 14.35. In essence, the Eyring equation represents the dependence of the mean lifetime τ on the temperature T. The parameters A and B are constants related to the component to be tested and the test method used. The parameter k_B is the Boltzmann constant ($1.381 \cdot 10^{23}$ JK^{-1}). For small temperature deltas, the quotient A/T may be assumed to be constant, whereby the Eyring model approximates Arrhenius' law.

$$\tau = \frac{A}{T} exp\left[\frac{B}{k_B T}\right] \tag{14.35}$$

14.2.9 Power Acceleration Factor

If a component test is carried out under accelerating conditions, the power accelerating factor K (German: Beschleunigungsfaktor R; cf. also application examples in Sects. 14.2.1 and 14.4) defines the acceleration; cf. Eq. 14.36 (Nelson 1990). This is basically a rearrangement of Eq. 14.24. The variable τ represents the lifetime related to the stress V. The reference base is the lifetime τ' related to the stress V'.

$$K = \frac{\tau}{\tau'} = \left(\frac{V'}{V}\right)^{\gamma_1} \tag{14.36}$$

14.3 Minimum Test Scope in a Prototype Test

When designing a prototype test, one of the central questions is the determination of the number of components to be tested. This chapter outlines the basic principles for calculating the number of test objects (e.g. within a prototype testing) to be tested as a function of the product reliability per component and the probability of failure related to the total test scope.

14.3.1 Test Scope on the Basis of Survival Probability

A simple procedure for determining the number of test items is to use the multiplication law from probability theory (cf. Chap. 5). The preconditions for applying the multiplication law to the determination of the testing minimum are as follows:

1. All test specimens are of the same design and construction.
2. All examinees are tested within the same testing program.

If the prerequisites are fulfilled, then Eq. 14.37 applies with regard to the survival probability R(t) of the test items (i = 1, 2, …, n) with test scope n (test item collective). From this, Eq. 14.38 follows as a complement to the survival probability for the failure probability $F_A(t)$. Simply rearranging Eq. 14.38 leads to the survival probability of a single test item $R_i(t)$; Eq. 14.39. If both a determination for the survival probability of a single test specimen $R_i(t)$, as well as a failure probability $F_A(t)$ of the entire test scope (at least one failure within the test scope), the number of test items n (test scope) can be easily determined with Eq. 14.40 based on the transformation of Eq. 14.39.

$$R_1(t) = R_2(t) = \cdots = R_n(t) \tag{14.37}$$

$$F_A(t) = 1 - R_i^n(t) \tag{14.38}$$

$$R_i(t) = (1 - F_A(t))^{\frac{1}{n}} \tag{14.39}$$

$$n = \frac{\ln(1 - F_A(t))}{\ln(R_i(t))} \tag{14.40}$$

Note: When determining the required number of test specimens, the failure probability of the test specimen collective F_A would be set high (e.g. $F_A = 0.99$). The background is that this would provide a confidence (certainty) that at least one component is defective after passing through the test program and thus the potential defect(s) would be detected. Furthermore, with reference to an expected service life (component design), the survival probability of a product (device under test (DUT)) would also be chosen to be high (e.g.: $R_i(t) = 0.95$). The background to this would be that the high survival probability of a test specimen is desired with regard to the mapped service life cycle. Of course, due to Eq. 14.40, the correlation between the number of test specimens and the probability of detection is not linear. Doubling of the number of test specimens cannot mean a doubling of the certainty of detection. Figure 14.9 shows the functional correlation between the quantities test size n, failure probability of the test sample collective $F_A(t)$ and the survival probability of a product (test sample) $R_i(t)$.

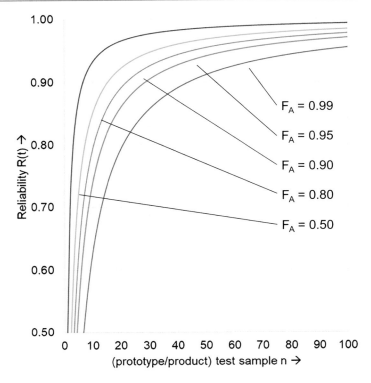

Fig. 14.9 Functional relationship between the extent of testing n, the failure probability of the test sample $F_A(t)$ and the survival probability of a test sample $R_i(t)$

Application example: Determination of a test scope
A function-critical component of a vehicle is to be tested within the scope of a test program and a possible malfunction is to be detected with a high probability. The test programme covers a total distance of $t_{ges} = 250{,}000$ km. The following requirements are placed on the component and the test program: The component survival probability shall be $R(t_{ges}) = 0.95$. Furthermore, the probability of component failure in the whole test scope (test sample) during testing shall be $F_A(t) = 0.99$. Using Eq. 14.41, the testing scope n is obtained.

$$n = \frac{\ln(1 - F_A(t))}{\ln(R_i(t))} = \frac{\ln(1 - 0.99)}{\ln(0.95)} = 89.78 \approx 90 \tag{14.41}$$

14.3.2 Test Scope Based on Failure Behaviour and Test Duration

The determination of the number of device under test (DUT) can be done with the help of assumptions regarding the expected failure behaviour on the basis of a

14.3 Minimum Test Scope in a Prototype Test

Weibull distribution model as well as assumptions for the test duration. The preconditions for the determination of the testing minimum are as follows:

1. All test specimens are of the same design and construction.
2. All prototypes are tested within the same testing program.
3. An assumption is made about the failure behavior based on prior knowledge.

The testing scope n can be changed by varying the scheduled testing time t_1. If the test time is shortened to the time t_2, the testing scope n would increase, and if the test time were extended to a time t_2, the number of components to be tested would decrease.

First, it is assumed that the failure behavior of the component under test can be represented by a two-parameter Weibull distribution model; therefore, the survival probability is given by Eq. 14.42. If the testing time were varied to a test time t_2 different from test time t_1, a Weibull distribution model would also be applied with Eq. 14.43. The shape parameter b is identical ($b_1 = b_2$), since the failure behavior must be identical. Thus, the distribution models can be set in terms of a lifetime ratio L, Eq. 14.44. By transforming (Eq. 14.45), the following follows for the lifetime ratio L, Eq. 14.46. Substituting Eq. 14.46 into Eq. 14.39 yields Eq. 14.47. After transforming, the extent of testing n according to Eq. 14.48 or the test time t_2 can be determined via the service life ratio L (Eq. 14.46), Eq. 14.49.

$$R_1(t_1) = e^{(-(t_1/T)^b)} \tag{14.42}$$

$$R_2(t_2) = e^{(-(t_2/T)^b)} \tag{14.43}$$

$$\frac{ln(R_2(t_2))}{ln(R_1(t_1))} = \left(\frac{t_2}{t_1}\right)^b = L^b \tag{14.44}$$

$$R_1(t_1)^{L^b} = R_2(t_2) \tag{14.45}$$

$$L = \frac{t_2}{t_1} \tag{14.46}$$

$$R_i(t) = (1 - F_A(t))^{\frac{1}{L^b \cdot n}} \tag{14.47}$$

$$n = \frac{1}{L^b}\left(\frac{ln(1 - F_A(t))}{ln(R_i(t))}\right) \tag{14.48}$$

$$L = \left(\frac{1}{n}\left(\frac{ln(1 - F_A(t))}{ln(R_i(t))}\right)\right)^{1/b} \tag{14.49}$$

In addition, the assumption can be made that there is a failure-free time of t_0 until the first failure. At this point, reference is made to the three-parameter Weibull distribution model (cf. Chap. 7). In this case, the lifetime ratio is given by Eq. 14.50. The test time t_2 is thus consistent with Eq. 14.51.

$$L = \frac{t_2 - t_o}{t_1 - t_o} \tag{14.50}$$

$$t_2 = L(t_1 - t_o) + t_o \tag{14.51}$$

If the test times related to the tested components are of different durations, the reliability $R_i(t)$ (Eq. 14.47) is extended by the summation of the lifetime ratios L_i related to each component j of the k components; Eq. 14.52.

$$R_i(t) = (1 - F_A(t))^{\left(\sum_{j=1}^{k} L_j^b n_j\right)^{-1}} \tag{14.52}$$

Summary: The test scope can be determined using the following procedure based on Eq. 14.48:

a) Definition of a requirement for reliability $R_i(t)$ of the component under test,
b) Definition of the service life ratio L (test time t_1 and t_2),
c) Assumption of failure behaviour with the aid of a Weibull distribution model; estimation of the shape parameter b of the Weibull distribution model as a function of the expected failure characteristic on the basis of prior knowledge (cf. interpretation of the shape parameter b in Chap. 12),
d) A potentially expected failure-free time t_0 is taken into account in the service life ratio; cf. Eqs. 14.50 and 14.51.
e) For different test times, the service life ratios L_i of the components must be included; Eq. 14.52.

Exemplary results for the determination of a testing scope n on the basis of the reliability probability $R_i(t)$, service life ratio L, shape parameter b of the Weibull distribution model (model assumption) and failure probability $F_A(t)$ of the test specimen collective (testing) are outlined in the following application example.

Application Example: Prototype Test Scope in an Engine Testing Phase
Within the scope of engine testing, a operation time-related failure behavior is expected, which can be represented by a Weibull distribution model. The expected damage symptom is an oil leakage at the sealing complex crankshaft/engine housing. Based on empirical values, the shape parameter b=2 is estimated. For different reliabilities in the range of $R_i(t) = 0.8$ to 0.99, service life ratios L in the range of 0.5 (test time reduction) to 3.0 (test time extension) as well as failure probabilities $F_A(t)$ in the range of 0.8 to 0.99, the required test volumes n are determined according to Eq. 14.48, the result is shown in Table 14.3.

14.3 Minimum Test Scope in a Prototype Test

Table 14.3 Extent of testing n as a function of reliability $R_i(t)$, service life ratio L, shape parameter $b = 2.0$ of the Weibull distribution model (model assumption) and failure probability F_A of the test specimen collective (testing)

$R_i(t)$	L	Sample size for $F_A = 0.80$	Sample size for $F_A = 0.90$	Sample size for $F_A = 0.95$	Sample size for $F_A = 0.99$
0.80	0.50	29	42	54	83
0.90	0.50	62	88	114	175
0.95	0.50	126	180	234	360
0.99	0.50	641	917	1193	1833
0.80	0.75	13	19	24	37
0.90	0.75	28	39	51	78
0.95	0.75	56	80	104	160
0.99	0.75	285	408	530	815
0.80	1.00	8	11	14	21
0.90	1.00	16	22	29	44
0.95	1.00	32	45	59	90
0.99	1.00	161	230	299	459
0.80	1.50	4	5	6	10
0.90	1.50	7	10	13	20
0.95	1.50	14	20	26	40
0.99	1.50	72	102	133	204
0.80	2.00	2	3	4	6
0.90	2.00	4	6	8	11
0.95	2.00	8	12	15	23
0.99	2.00	41	58	75	115
0.80	2.50	2	2	3	4
0.90	2.50	3	4	5	7
0.95	2.50	6	8	10	15
0.99	2.50	26	37	48	74
0.80	3.00	1	2	2	3
0.90	3.00	2	3	4	5
0.95	3.00	4	5	7	10
0.99	3.00	18	26	34	51

14.4 Case Study: Accelerated Testing and Weibull Distribution Model

The present chapter outlines the design of an accelerated testing and the analysis of damage data based on the Weibull distribution model. Section 14.4.1 introduces the topic of accelerated testing and the Weibull distribution model, while Sect. 14.4.2 illustrates the application of the model to the paper clip bending test.

14.4.1 Introduction

In the context of product testing, accelerated testing procedures are often used in order to achieve accelerated ageing of the component to be tested and thus to obtain information about the product reliability at an earlier stage (cf. Sects. 14.1 and 14.2). Accelerated testing is based on two possible principles:

1. The usage rate of the product is increased during test operation. This method is used for products that are not continuously in operation in the field. The increased usage rate shifts the failure behavior to an earlier point in time.
2. The test operation is carried out at a higher load intensity compared to the field use phase; e.g. increased temperature amplitude during thermal cycling. An increased load intensity leads to intensified degradation and thus to an earlier failure.

It should be noted that combinations of the above procedures can also be used. The failures of the tested components are documented and a distribution model is adapted to the failure data. Usually a Weibull distribution model is chosen. It is crucial that the failure symptom as well as the failure cause do not change when accelerated testing is performed. The goal of testing is to map the field failure spectrum, but not to provoke causes of damage that are not field relevant. In terms of fitting a Weibull distribution model to failure data, this means the following: The shape parameter b does not differ in an accelerated test compared to a test under real conditions. Only the values of the localization parameter T (characteristic lifetime) as well as the threshold parameter t_0 (if a three-parameter model is used) will be reduced if the accelerated test procedure is carried out correctly compared to the test under normal conditions; cf. Fig. 14.10. With double logarithmic ablation of the failure data as well as fitted distribution models, a shift in the failure behavior can be seen. The shift can be expressed in terms of the acceleration factor: The acceleration factor R is the quotient formed from t_{Ref} and t_{Acc}; cf. also Sect. 14.2.9.

14.4 Case Study: Accelerated Testing and Weibull Distribution Model

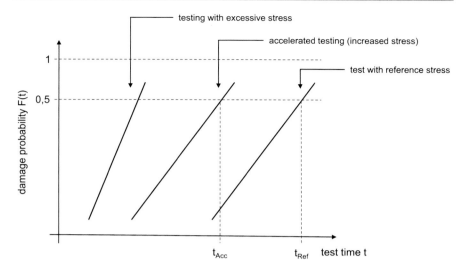

Fig. 14.10 Schematic diagram of accelerated testing compared to testing under reference conditions and testing with excessive stress (double logarithmic scaling)

14.4.2 Accelerated Testing: Example Paper Clip Bending

A simple example illustrates the principle of accelerated testing: In an experiment, the life span of paper clips is tested, which are exposed to a cyclic bending stress. Note: The illustration of accelerated testing, which is predestined from a didactic point of view, using the paper clip bending experiment was outlined by Pantelis Vassiliou as part of the Reliability and Maintainability Symposium (RAMS) Tutorials 2001. This involves bending a paper clip open at a defined angle α and bending it back to its original position; cf. Fig. 14.11. The procedure is repeated until a break occurs. The test procedure in the normal method provides for a bending angle of $\alpha = 45°$. In order to achieve an acceleration of the test procedure, a test series with bending angle $\alpha = 90°$ is carried out in an accelerated testing. By widening the bending angle, the damage is achieved in a smaller number of bending operations. Table 14.4 shows the results of the test performance of the bending test under standard condition with bending angle $\alpha = 45°$ and under increased load by a bending angle of $\alpha = 90°$.

The failure behaviour of the paper clips under both test conditions (normal procedure and accelerated procedure) is mapped using a Weibull distribution model and the parameters of the models are interpreted. Furthermore, a statement on the significance with regard to the parameter interpretation and the failure behaviour is made. Finally, a statement about the acceleration factor will be made.

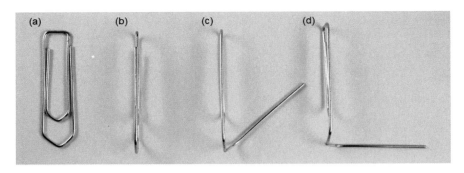

Fig. 14.11 Life test paper clip bending; initial position in top view (a), side view (b), bending with angle $\alpha = 45°$ (c) and $\alpha = 90°$ (d)

Table 14.4 Results of the bending test of the paper clip (test object) under different bending angles under standard condition and in accelerated testing

Test under standard condition; bending angle $\alpha = 45°$; cycles to failure					
Test object	1	2	3	4	5
Bending cycles	36	40	32	34	28
Accelerated testing with increased stress; bending angle $\alpha = 90°$; cycles to failure					
Test object	1	2	3	4	5
Bending cycles	8	13	10	11	14

Damage Data Analysis: Paperclip Bending Test—Standard Condition versus Accelerated Testing.

A three-parameter Weibull distribution model is chosen to represent the failure behavior; cf. Eq. 14.53 (Chap. 7); the parameters are shape parameter b, position parameter T, and the failure-free time t_0 (threshold). The number of bending cycles is used as the lifetime variable t; therefore, the damage data represent the cycles to failure. Due to the simple, manual traceability, the parameters of the Weibull model are determined via regression analysis as well as the approach according to Dubey (1967) (cf. Chap. 8).

$$F(x) = 1 - e^{-\left(\frac{t-t_0}{T-t_0}\right)^b} \tag{14.53}$$

The calculation is outlined in detail using the example of the damage data based on the test performance under standard conditions (bending angle $\alpha = 45°$). The evaluation of the measurement series for the test procedure with increased stress (accelerated testing; bending angle $\alpha = 90°$) is carried out analogously, therefore only the calculation results are presented in this case.

First, the rank-based cumulative probabilities are formed; cf. Eq. 14.54 and Table 14.5. Subsequently, the parameter t_0 (failure-free time) is estimated

14.4 Case Study: Accelerated Testing and Weibull Distribution Model

according to Dubey (1967), where $t_1 = 28$ cycles (first failure) and $t_3 = 40$ cycles (last failure) is set; cf. Eq. 14.56. To determine t_2, the failure probability $F(t_2) = 0.413$ is determined beforehand with Eq. 14.55. This results in $t_2 = 33$ (Dubey) from averaging t_2 (second failure time!) and t_3 (third failure time!). It should be noted that the nomenclature Dubey refers to the first, last and an intermediate failure, whereby the increases related to the probability should be approximately equal (detailed explanation: cf. Chap. 8) and not to the ordinal numbers of Table 14.5. Subsequently, the simplified regression analysis is carried out related to the logarithmized values; cf. Table 14.6 (detailed description of procedure in Chap. 8). The failure data are corrected by t_0.

$$F(t) = \frac{i - 0.3}{n + 0.4} \tag{14.54}$$

$$\begin{aligned} F(t_2) &= 1 - e^{-\sqrt{-\ln(1-F(t_1))(-\ln(1-F(t_3)))}} \\ &= 1 - e^{-\sqrt{-\ln(1-0.130)(-\ln(1-0.870))}} \\ &\approx 0.413 \end{aligned} \tag{14.55}$$

$$t_0 = \frac{t_1 t_3 - t_2^2}{t_1 + t_3 - 2t_2} = \frac{28 \cdot 40 - 33^2}{28 + 40 - 2 \cdot 33} = t_{0[45]} = 15.50 \tag{14.56}$$

Using the logarithmized values, the parameters of the line of best fit are determined; cf. Eqs. 14.57 and 14.58. Since the regression analysis can be applied to the two-parameter Weibull distribution model (and not directly to the three-parameter model), the characteristic life T must be corrected by t_0; cf. Eq. 14.59. In addition, the confidence interval for the parameters b (Eqs. 14.60, 14.61 with k = 1.4 for confidence level 0.9; cf. also Chap. 8) and T (cf. Eqs. 14.62, 14.63, 14.64 and 14.65; with χ^2-quantile for confidence level 0.9 (Appendix Table A.5); cf. Sect. 9.4) are calculated. Thus, the estimated values of the Weibull parameters b and T of both measurement series can be compared, taking into account the high scatter effects (confidence intervals Eqs. 14.66 and 14.67), due to the small paper clip test size n = 5.

$$\widehat{b}_{45} = \frac{\sum_{i=1}^{n} x_i y_i - n\overline{x}\overline{y}}{\sum_{i=1}^{n} x_i^2 - n\overline{x}^2} = \frac{-6.08 - 5 \cdot 2.89 \cdot (-0.49)}{42.12 - 5 \cdot 2.89^2} \approx 2.78 \tag{14.57}$$

Table 14.5 Test under standard condition; bending angle $\alpha = 45°$; damage data: Actuation cycles to failure and sum probabilities

Rank i	Bending cycles t_i until breakage	Summed probability $F(t_i)$
1	28	0.13
2	32	0.31
3	34	0.50
4	36	0.69
5	40	0.87

Table 14.6 Test under standard condition; bending angle α = 45°; damage data: Correction for failure-free time and double-logarithmic transformation for regression analysis

i	t_i	t_i-t_0	$\ln(t_i-t_0)$	$\ln(t_i-t_0)^2$	$F(t_i-t_0)$	$\ln[-\ln(1-F(t_i-t_0))]$	$\ln(t_i-t_0)*\ln[-\ln(1-F(t_i-t_0))]$
1	28	12.50	2.53	6.40	0.13	−1.97	−4.98
2	32	16.50	2.80	7.84	0.31	−0.99	−2.77
3	34	18.50	2.92	8.53	0.50	−0.37	−1.08
4	36	20.50	3.02	9.12	0.69	0.16	0.48
5	40	24.50	3.20	10.23	0.87	0.71	2.27
Σ				42.12			−6.08
$\bar{x}$			2.89			−0.49	

$$\widehat{T}_{45} = e^{-\left(\frac{\bar{y}-\hat{b}\cdot\bar{x}}{b}\right)} = e^{-\left(\frac{-0.49-2.78\cdot 2.89}{2.78}\right)} \approx 21.46 \qquad (14.58)$$

$$\widehat{T}_{45} = \widehat{T} + t_0 \approx 36.96 \qquad (14.59)$$

$$b_{45} \geq \widehat{b}_{45}\frac{1}{1+\sqrt{\frac{k}{n}}} = 2.78 \cdot \frac{1}{1+\sqrt{\frac{1.4}{5}}} \approx 1.82 \qquad (14.60)$$

$$b_{45} \leq \widehat{b}_{45}\left(1+\sqrt{\frac{k}{n}}\right) = 2.78 \cdot \left(1+\sqrt{\frac{1.4}{5}}\right) \approx 4.25 \qquad (14.61)$$

$$T_{45} \geq \widehat{T}_{45}\left(\frac{2n}{X^2_{2n,1-\frac{\alpha}{2}}}\right)^{\frac{1}{b_{45}}} = 36.96 \cdot \left(\frac{10}{18.31}\right)^{\frac{1}{2.78}} \approx 29.73 \qquad (14.62)$$

$$\chi^2_{10,1-\frac{0.1}{2}} = 18.31 \qquad (14.63)$$

$$T_{45} \leq \widehat{T}_{45}\left(\frac{2n}{X^2_{2n,\frac{\alpha}{2}}}\right)^{\frac{1}{b_{45}}} = 36.96 \cdot \left(\frac{10}{3.94}\right)^{\frac{1}{2.78}} \approx 51.69 \qquad (14.64)$$

$$\chi^2_{10,\frac{0.1}{2}} = 3.94 \qquad (14.65)$$

As a first intermediate summary, the model parameters for testing under standard condition (bending angle α = 45°) are:

Estimator for shape parameter b_{45} = 2.78; confidence interval, cf. Eq. 14.66

14.4 Case Study: Accelerated Testing and Weibull Distribution Model

$$1.82 \leq b_{45} \leq 4.25 \qquad (14.66)$$

Estimator for shape parameter $T_{45} = 36.96$; confidence interval, cf. Eq. 14.67

$$29.73 \leq T_{45} \leq 51.69 \qquad (14.67)$$

The evaluation is carried out analogously for the series of measurements of the test performance with increased stress (accelerated testing) and yields the estimate of the parameters t_0, T_{90} and b_{90} according to Eqs. as well as the confidence intervals according to Eqs. 14.71 and 14.72.

$$t_{0[90]} = 1.75 \qquad (14.68)$$

$$\widehat{b}_{90} \approx 4.22 \qquad (14.69)$$

$$\widehat{T}_{90} = 12.09 \left(\widehat{T}_{90} \approx 10.34 \right); \text{ correction with } t_0 \qquad (14.70)$$

$$2.76 \leq b_{90} \leq 6.45 \qquad (14.71)$$

$$10.48 \leq T_{90} \leq 15.08 \qquad (14.72)$$

Summary: Interpretation of results on paper clip bending test under standard conditions versus accelerated testing conditions.

The visualization of the failure behavior in Fig. 14.12 illustrates that the damage in the test collective within the accelerated testing obviously occurred at earlier points in time compared to the test under standard condition and thus the desired acceleration related to the test cycle occurred. In the following, the results are compared quantitatively with regard to the acceleration.

The acceleration of the failure behavior due to the increased stress can be determined from the parameters of the Weibull distribution models. On the one hand, the failure-free time t_0 is reduced by 13.75 cycles ($t_{0[45]}$ versus $t_{0[90]}$), and on the other hand, the characteristic lifetime T is reduced by 24.87 cycles. The shape parameter b_{45} differs from the shape parameter b_{90}, but it is determined on the basis of only five test specimens in each case, so that the different values can be justified by scattering effects: The confidence intervals of the shape parameters b_{45} and b_{90} overlap, so that the difference is not significant. The reduction of the characteristic life, on the other hand, is significant because the confidence intervals of T_{45} and T_{90} do not overlap. The quotient of both characteristic lifetimes forms the acceleration factor RF_{test}. The increased stress in the form of the increased bending angle reduces the test time by a factor of 3, cf. Eq. 14.73.

$$RF_{test} = \frac{\widehat{T}_{45°}}{\widehat{T}_{90°}} = \frac{36.96}{12.09} \approx 3.06 \qquad (14.73)$$

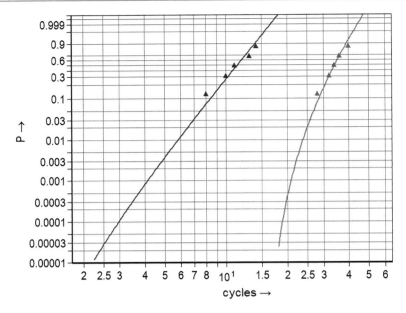

Fig. 14.12 Test paper clip bending; three-parameter Weibull distribution models based on damage data (cycles to failure); test performed under standard condition (bending angle $\alpha = 45°$, ▲ right) and with increased stress (accelerated testing; bending angle $\alpha = 90°$, ▲ left)

14.5 Qualitative Accelerated Testing: HALT and HASS

The Highly Accelerated Life Test (HALT) is used to detect design and process weaknesses. A provocation of defects is carried out by a targeted overload outside the product specification. The aim is to detect product weaknesses, caused by the design or by the manufacturing process. It is an iterative procedure to determine the extreme load capacity of the product. The goal is not to prove that the product specifications have been met, as is the case with conventional testing technology or product testing. The conventional testing is usually based on load spectra and usage profiles which cover the loads to be expected in field use. Furthermore, the misuse is also a component of conventional testing.

On the other hand, the following ranges are determined by the Highly Accelerated Life Test; cf. Fig. 14.13: In the range of the destruction margin to the destruction limit (Lower/Upper Destruction Level [LDL, UDL]) of the product, product defects can be provoked. In the range of the operating span up to the operating load limit (Lower/Upper Operating Level [LOL, UOL]), the provocation of faults is in principle possible. However, faultless operation of the product should be feasible. The operating span as well as the destruction span are outside the product specification.

14.5 Qualitative Accelerated Testing: HALT and HASS

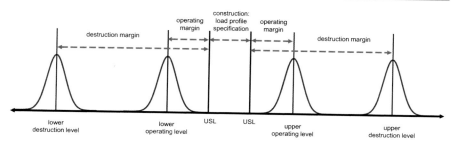

Fig. 14.13 Specification, operation and destruction of technical products

The Highly Accelerated Life Test (HALT) is also used to define the test conditions for the Highly Accelerated Stress Screening (HASS). In Highly Accelerated Stress Screening, the product is also subjected to a short-term overload outside the product specification. However, the product is not destroyed during this test; the aim of this test is rather to detect weak points in the product, which are likely to lead to failure in the subsequent use phase in the field.

The loading of the product in the Highly Accelerated Life Test is carried out iteratively, mainly by increasing/decreasing the temperature and/or increasing the load by vibration. The iterative loading is carried out in a test facility in which temperature ranges in the interval from −100 °C to +180 °C (example), e.g. with a temperature delta of 50 °C/min, can be realized. The applied vibration covers a frequency range from 5 to 200 Hz (example) over six possible degrees of freedom (3 linear axes, three rotational axes). The test execution can also be carried out with a combination of temperature change and vibration.

Summary:

Highly Accelerated Life Test (HALT)
The Highly Accelerated Life Test enables the early identification of potential weak points (focus: product design and layout errors) during the development process through extreme stressing of the system (assembly/component/product). The iterative stressing takes place outside the specification to determine the extreme load capacity up to destruction.

Highly Accelerated Stress Screening (HASS)
Non-destructive reliability test based on the Highly Accelerated Life Test, which focuses on the detection of defective products within series production. The load profile is in the range of the load limit outside the specification limit.

Elementary aspects of the *test bench technology of the Highly Accelerated Life Test*:

1. Through cold stage/heat stage test, thermal cycling test, vibration test outside the specification limits, the limit values of the product functionality are up to the product destruction achieved.
2. The product is tested in the operating condition, both in the operating span and in the destruction span.
3. The provoked damage symptom forms the starting point for development and/or production process optimization.

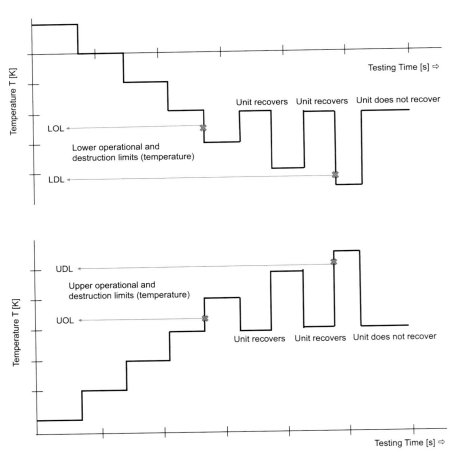

Fig. 14.14 Schematic diagram of an iterative stress loading by cold steps (Fig. above) as well as by heat steps (Fig. below) in the Highly Accelerated Life Test

4. The damage symptom is analysed in order to understand the damage causalities and to be able to carry out a comparison with the known defect spectra which were determined during the testing within the product specification.
5. Use of special test facilities for environmental impact tests to perform combined stress testing. By overlay vibration and thermal cycling testing, extreme stress profiles are realized against the background that the failure behavior of many products changes at different temperatures.

Figure 14.14 shows the schematic diagrams of iterative stress loading by cold steps (Fig. 14.14 left) and by heat steps (Fig. 14.14 right) in the Highly Accelerated Life Test. When the component has failed for the first time, but recovers after the failure, the operational stress limit has been reached. If the component does not recover after further iterations, the destruction limit has been reached. The Highly Accelerated Life Test can also be performed with rapid temperature changes. Furthermore, a vibration load can be introduced step by step. Figure 14.15 shows the schematic diagram of a rapid temperature change in combination with the stepwise increase of the vibration frequency. As a further possible variation in the design of a Highly Accelerated Life Test, Fig. 14.16 shows an iterative stress load by temperature change with gradually increasing amplitude in combination with a stepwise increase of the vibration load. The vibration is represented as random oscillations; the acceleration spectral density (ASD) is measured to represent the random oscillations. The mean square acceleration (Grms) is the square root of the area under the ASD curve in the frequency domain.

A combination and superposition of rapid temperature change and vibration loads (structured or random) is also called Rapid Highly Accelerated Life Test.

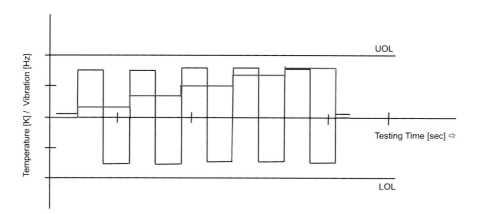

Fig. 14.15 Schematic diagram of an iterative stress loading by temperature change (black) in combination with stepwise increase of vibration load (blue) in the Highly Accelerated Life Test

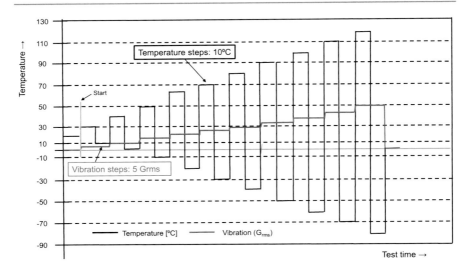

Fig. 14.16 Schematic diagram of an iterative stress loading by temperature change with stepwise increasing amplitude in combination with stepwise increase of the vibration load in the (Rapid) Highly Accelerated Life Test

Test Process Suitability 15

The execution of measurements and the use of measuring equipment takes place within test processes in all phases of the product life cycle: development, production and field support. The analysis of measurement and test processes takes place within the framework of test planning. Test planning includes the planning of test procedures, test equipment, test aids, test scope and frequency, test personnel, test time, test location as well as the documentation of the test result with regard to the characteristics to be tested (cf. in detail Sects. 3.2 and 3.4 as well as VDI 1985). Furthermore, inspection planning includes the definition of decision rules with regard to the inspection result.

The present chapter first shows the fundamentals of the analysis of measurement and test processes, in particular on variables influencing the measurement result as well as the significance of measurement uncertainty (Sect. 15.1). Section 15.2 presents the most important standards of test process suitability analysis for analyzing the suitability or capability of a test process with regard to the given test object specifications. Thereafter, the analysis of the short-term test equipment capability (C_g-, C_{gk}-study, cf. Sect. 15.3), the repeatability and reproducibility (%GRR-study; Sect. 15.4) as well as the comprehensive analysis of the test process suitability (Q_{MS}-/Q_{MP}-study) with determination of the expanded measurement uncertainty U are explained, Sect. 15.5. The chapter concludes with remarks on the historically known "procedure X" (German language: „Verfahren X") for the analysis of test equipment capability and test process suitability; Sect. 15.6.

15.1 Basics of the Analysis of Measuring and Testing Processes

The test process suitability is the basic precondition for monitoring product or process characteristics in the manufacturing process and thus serves to ensure the proper functioning of a product in the use phase (cf. Sect. 2.1). Product characteristics are usually safety–critical characteristics or function-critical characteristics as well as general inspection characteristics (cf. definitions in Sect. 3.2). On the one hand, an inspection process suitability analysis can be performed when implementing a new test process or purchasing new test equipment. On the other hand, the inspection process suitability analysis is carried out within the scope of the planning of manufacturing processes; the point in time of the procedure of an inspection process suitability in the product development process is shown in Fig. 2.4, Sect. 2.1.

The suitability of a test process is thus, among other things, the precondition for carrying out the following capability tests with a view to series production (Bracke 2016):

1. Machine capability testing prior to start of production (SOP),
2. Preliminary process capability tests and long-term process capability tests before or after start of production (SOP),
3. Full inspection and sampling,
4. Statistical Process Control (SPC).

Furthermore, the test process suitability analysis is a precondition for the general analysis of measurement and test data (data analytics, cf. e.g. Chaps. 6 and 8–14), for example with regard to the detection of significant deviations, trend developments, jumps in the course of values or the comparison of several samples. Statistical hypothesis tests are frequently used for this purpose in the context of data analysis (cf. Chap. 13), ergo the recording of the observed values also plays a special role here: the proof of significance with regard to an observation on the basis of measurement data presupposes that possible influences due to the recording of the measurement data are known beforehand. This is to avoid that significance within the observed values is falsely detected (or not detected), which can be attributed to the measurement system used.

The term test process suitability includes the analysis of the measurement or test system. On the one hand, influences of the measuring system on the measurement result are analyzed, on the other hand, the measuring system is evaluated on the basis of defined specification values with regard to the test task. Consequently, when investigating the test process, a distinction should always be made between test equipment (influences of the test equipment on the measurement result) and the actual test process (influences of the test process on the measurement result; i.e.: application of the test equipment within a test process). Figure 15.1 shows an overview of potential influences on the uncertainty of a measurement result, differentiated according to the main influences: personnel, measuring process, data analysis, environment, measured object, measuring equipment, component part fixture and reference standard.

Furthermore, a distinction must be made within repetitive manufacturing as to the location at which a measurement is carried out. The environment of a measurement laboratory, in which measurements are frequently carried out in small samples during series production, differs considerably from an inline measurement,

15.1 Basics of the Analysis of Measuring and Testing Processes

which is integrated into the running production. Table 15.1 shows a comparison of the influences on the measurement uncertainty of a measurement result in a laboratory or production environment.

In addition to the place where a measurement is carried out (laboratory or inline (production)), the phase within the product development process is of importance. Figure 15.2 schematically shows the increasing measurement uncertainty in relation to the time between product development and product realization (production under series conditions). At the beginning of the product development the measurement uncertainty is zero, since only the technical drawing with the specification limits (lower specification limit (LSL) and upper specification limit (USL)) is available. In the course of prototype construction, a few components are manufactured, which are precisely measured with a view to testing: The measurement uncertainty tends to be at a low level at this stage, since precise measurement systems, mostly in laboratories (cf. Table 15.1), are used and therefore many influencing variables are small or almost zero (cf. Fig. 15.1). With the start of production, the number and respective characteristics of the influences on the measurement uncertainty are likely to have assumed a higher level. The production and environmental conditions are often major contributors with regard to the expected measurement uncertainty. In addition, Fig. 15.2 shows the three possibilities of a measurement result (cf. Eq. 15.1) in relation to a given specification and uncertainty range: Measurement result 1 allows a positive, measurement result 3 a negative assessment with regard to the specification limits. Measurement result 2 is not unambiguous, since a repeated measurement can also yield a result outside the specification.

The complete specification of a measurement result y on the basis of a measurement process (MP) in each phase of the product development process is carried out according to Eq. 15.1. In addition to the determined measurement value x, the expanded measurement uncertainty U_{MP} is given. Furthermore, the coverage factor

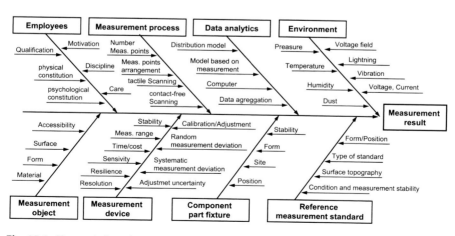

Fig. 15.1 Factors influencing the uncertainty of measurement results, based on VDA (2010:26)

Table 15.1 Comparison of potential influencing variables on a measurement within a laboratory or a production environment

Influencing variable	Laboratory environment	Production environment
Staff	Metrology technician: expertise in measurement-specific problems and testing tasks	Machine worker: As a rule, no metrology/skilled worker training
Measuring room	Climatised environment; conditioned, filtered air; free from vibrations and noise	High temperature and humidity differences; oily, dusty air; sometimes strong vibrations, noises or shocks
Measurement method	Defined measuring forces and points; exact test plans; measuring time is not in the foreground	partly operator-dependent; usually only simple test plans; economical measurement in a short time
Measurement standard	Factory measurement standards	Service standards, generally less accurate than factory measurement standards
Test object	Production standards of use; production parts cleaned and tempered	Production parts with relatively large dimensional and shape deviations; not tempered, partly not cleaned (e.g. machining residues, oil adhesions)
Measuring equipment	High quality and accurate; intensive maintenance, good state of care	Frequent use, e.g. 3 shifts/7 days; maintenance not intensive

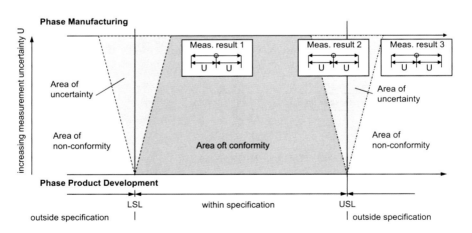

Fig. 15.2 Relationship between measurement uncertainty and specification in the product emergence process

15.1 Basics of the Analysis of Measuring and Testing Processes

for the calculation of the expanded measurement uncertainty U_{MP} should be specified (here: $k=2$; corresponds to probability of $P=0.9545$ with assumed normal distribution model); cf. also Sect. 7.2.2.

$$y = x \pm U_{MP}(k=2) \tag{15.1}$$

Notes on measurement uncertainty:

1. By specifying the measurement uncertainty U, it becomes transparent that no exact value is determined by a measurement.
2. The determination of a measured value x by a measurement process is subject to a number of influences which often cannot be quantified exactly.
3. The possible range $\pm U$ in which the measured value lies is specified with a probability, which is based on a probability distribution model.
4. The aim of the test process analysis is to record each individual influence i on the measurement result, to describe it with the aid of distribution models and to quantify it by means of standard uncertainty u_i ($\triangleq$ standard deviation).
5. Standard uncertainties u_i of individual influences i on the measurement result x are combined to an expanded measurement uncertainty U.

15.2 Test Equipment Capability and Test Process Suitability

First, the terms test process suitability and test equipment capability are differentiated. The investigation of the *test process suitability* is an analysis of the properties and handling instructions of a test equipment with regard to its suitability for performing the planned test task under the specified test conditions. The result is the proof of suitability for the planned or already existing test process; cf. also Pfeifer (2001), VDA (2010) and Bracke (2016).

The basic investigations – under ideal and typical conditions of use – within the suitability analysis of a test equipment as well as the test process generally refer to the following points (cf. Fig. 15.3): Accuracy, repeatability, reproducibility, linearity, stability.

The term *test equipment capability* focuses only on the capability of the test equipment to reproduce a measurement result under almost ideal conditions (cf. in detail Sect. 15.3, C_g-, C_{gk}-Study). In contrast, *test process suitability* includes both the suitability of the test equipment and the process in which the test equipment is used. An overview of the standards with regard to the performance of an inspection process suitability analysis or an investigation of the test equipment capability is shown in Table 15.2. The procedure according to VDA Volume 5 (VDA 2010) was derived from ISO/IEC Guide 98 (GUM:1995) (ISO 2008) or ISO 14253 (ISO 2017). However, VDA (2010) concentrates on the test process suitability analysis, taking into account some simplifications (example: neglecting the interdependencies of standard uncertainty components), in order to ensure a safe applicability;

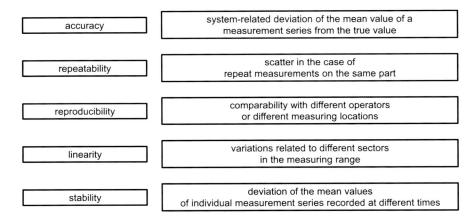

Fig. 15.3 Elementary points of the suitability test of a test equipment as well as of a test process

cf. detailed comments in Sect. 15.5. Furthermore, the classic C_g-/C_{gk}-Study as it is established in the industrial context, has been taken into account in the VDA (2010) standard. The Measurement System Analysis (MSA; cf. AIAG 2010) includes the %GRR study in addition to the significance analyses. However, it does not include a dedicated investigation of individual influences of test equipment and test process on the measurement result. The standards ISO/IEC Guide 98 (GUM) and ISO 14253 (ISO 2017), respectively, are industry-independent and complex in their complete application. In contrast, C_g-/C_{gk}-Study, %GRR-Study (MSA; AIAG 2010) as well as the VDA (2010) standard (Q_{MS}-/Q_{MP}-Study; based on (GUM)) were developed by the automotive industry and adopted by many other technical disciplines due to their very good applicability.
Against this background, this chapter focuses on the following procedures (Bracke 2016):

- Short-term test equipment capability: C_g-/C_{gk}-Study
- Repeat and comparative precision study: %GRR study (MSA)
- Test process suitability study: Q_{MS}-/Q_{MP}-Study (VDA 2010)

The short-term test equipment capability (C_g-/C_{gk}-study) basically assesses the repeatability of a test equipment in relation to the given specification of the test object at hand (cf. Sect. 15.3; Ford 1997; AIAG 2010). The C_g-/C_{gk}-study can be considered as a precondition to be fulfilled for the more comprehensive test process suitability studies, such as the %GRR study or the Q_{MS}-/Q_{MP}-study. In any case, the investigated scatter related to repeatability at the standard is part of the VDA (2010) standard (cf. Sect. 15.5). The repeatability and reproducibility study (%GRR study) assesses the suitability of the test process using a procedure that includes the main influencing variables from the test process (e.g.: worker, repeatability, test object). The %GRR study is described within the Measurement System

15.2 Test Equipment Capability and Test Process Suitability

Table 15.2 Standards for the analysis of the test process suitability as well as for the investigation of the test equipment capability, cf. also (Dietrich and Schulze 2014)

Standard	ISO/IEC Guide 98 (GUM); ISO 14253 Guide to the expression of uncertainty in measurement (GUM)	VDA*) Vol./Vol. 5 (2nd ed.) ISO 22514 Test process suitability	AIAG**) MSA 4th Edition Measurement System Analysis (MSA)	Industry standards "Procedure X", with $X = 1, \ldots, 5$
Contents	**ISO/IEC Guide 98 (GUM)** Combined standard measurement uncertainty, Degrees of freedom, Expanded uncertainty of measurement	**Suitability measuring system** Consideration C_g-/C_{gk}-Study (Procedure 1), Linearity, Consideration of prior knowledge/assumptions, **Suitability of measuring/testing process** Consideration of %GRR study (procedure 2/3), Consideration of further influences, e.g. temperature, measuring object, stability, Consideration of prior knowledge/assumptions, Consideration of attributive feature analysis	**Suitability measuring system** Test for significant bias (sample or QRK), Test for linearity (significant intercept and gradient), %GRR (Gage R&R), and/or ndc, Stability, *Annotation*: No separate consideration of the influences environment, temperature et cetera, No measurement uncertainty	**Capability of test equipment or test process** Procedure $1 - C_g/C_{gk}$, Procedure $2 - \%GRR$ (with operator influence), Procedure $3 - \%GRR$ (without operator influence), Procedure 4: Measurement stability, Procedure 5: Test process capability for attributive characteristics *Annotation*: Selection of the procedure depending on the test
Decision and limit value demand	**ISO 14253:** Decision rules for determining conformity or non-conformity with specifications	Limit value requirement and decision rules for determining conformity or non-conformity with specifications	Limit value requirement	Limit value requirement within the procedures

*) VDA: Verband der Automobilindustrie (engl.: German Association of the Automotive Industry); here: (VDA 2010)
**) AIAG: Automotive Industry Action Group; here (AIAG 2010)

Analysis (MSA) manual and is a recommendation of AIAG (2010). The third standard was developed by the German Association of the Automotive Industry (Verband der Automobilindustrie (VDA)) and is essentially based on the standard ISO/IEC Guide 98 (GUM) "Guide to the expression of uncertainty in measurement" (ISO 2008). The test process suitability study (Q_{MS}-/Q_{MP}-Study) comprises the systematic analysis of all influencing variables, differentiated according to test equipment influences and test process influences, on the measurement result (Sect. 15.5), (VDA 2010). Thus, both methods, %GRR study and Q_{MS}-/Q_{MP}-study, aim at the assessment of the test process suitability. The procedures of the C_g-/C_{gk}-study and Q_{MS}-/Q_{MP}-study are described in detail in the following. The explanations of the Q_{MS}-/Q_{MP}-study are more comprehensive compared to the %GRR study, as it considers the analysis of all potential influencing variables and is less limited from a methodological point of view. Therefore, the application of the %GRR study is only outlined in its main features. Finally, a brief reference to the industrial standards procedure "X" (with X = 1, ..., 5) is outlined in Sect. 15.6.

15.3 Short-Term Test Equipment Capability: C_g-/C_{gk}-Study

15.3.1 Basics of the C_g-/C_{gk}-Study

The study of the short-term test equipment capability C_g-/C_{gk}-Study is carried out to evaluate new and existing measuring systems prior to acceptance testing at the supplier's premises or at the final place of use at the user's premises. The C_g-/C_{gk}-study was first published in the standardization work (Ford 1997) and has established itself as the first standard. The following explanations of the C_g-/C_{gk}-study are a revised version of the explanations according to (Bracke 2016). The formula symbols of the capability indices C_{gk} mean: C = capability; g = gauge; k = katayori (= offset, presumably taken from the Japanese language; corresponding to bias). On the basis of the capability indices C_g or C_{gk} can be used to assess the capability of a test equipment. The focus here is on the assessment of the repeat measurement under constant boundary conditions.

The general procedure for the C_g-/C_{gk}-study includes the following steps:

a) Gauge resolution analysis,
b) Repeat measurement of the setting standard; number of measurement processes: $n \geq 25$,
c) Determination of the test equipment capability indices C_g and C_{gk},
d) Assessment of short-term test equipment capability by limit value requirement;
e) test equipment capability is given at: $C_g \geq 1.33$ and $C_{gk} \geq 1.33$.

The *evaluation of the resolution (RE = resolution) of* the measuring device (step a)) related to the test task is performed with the aid of Eq. 15.2. The background of the limit value is the requirement that the display accuracy is at least 20 times

15.3 Short-Term Test Equipment Capability: C_g-/C_{gk}-Study

higher in relation to the component specification in order to obtain a sufficiently high number of potential supporting points with regard to the adaptation of the distribution model. With a lower number of grid points, an unwanted classification of the measurement data can occur.

$$\%RE = \frac{RE}{Specification} \cdot 100\% \leq 5\% \tag{15.2}$$

Subsequently, the *short-time test eqipment capability* is assessed via the capability indices C_g and C_{gk} (Eqs. 15.3 and 15.4). The core of the C_g capability index (step c)) is the analysis of the measured value scatter of the repeat measurement (step b)) in relation to the given component specification. The C_{gk} capability index (step c)), on the other hand, includes not only the scatter but also the position of the repeat measurement in relation to the correct value (reference value).

$$C_g = \frac{0.2 \cdot Specification}{k \cdot \hat{\sigma}} \geq 1.33 \tag{15.3}$$

$$C_{gk} = \frac{0.1 \cdot Spec - |\hat{\mu} - x_m|}{\frac{1}{2}k \cdot \hat{\sigma}} = \frac{0.1 \cdot Spec - |Bi|}{\frac{1}{2}k \cdot \hat{\sigma}} \geq 1.33 \tag{15.4}$$

Meaning Parameter:

$\hat{\mu}$ = estimator for the mean value
x_m = normal/reference value of the standard or reference part
$\hat{\sigma}$ = estimator for the standard deviation
k = multiple of the standard deviation (reference value)
n = number of measurements/measured values
Bi = systematic measurement offset (bias)

The reference quantity of the multiple of the standard deviation in the denominator of the capability indices C_g and C_{gk} of the repeat measurement series is often 4σ and 6σ, respectively, in industrial applications. (The quantile 4σ ($\pm 2\sigma$, k=4) corresponds to approx. P=0.9545; the range 6σ ($\pm 3\sigma$, k=6) corresponds to approx. P=0.9973 related to a normal distribution model).
Interpretations and notes of the capability indices C_g and C_{gk}:

a) If a repeat measurement is performed on a standard, the C_g-/C_{gk}-study assumes a random scatter of measured values, which is represented by a normal distribution model. If systematic influences are present, the test equipment technique must be questioned. Systematic influences can be recognized, for example, in the value pattern or in the measurement offset Bi.

b) The numerator of the C_g and C_{gk} quotients contains the factor 0.2 and 0.1 respectively. This was chosen – for historical reasons – so that the limit value requirement is 1.33 and is the same as the known limit value requirement in process capability investigations (cf. Chap. 16). However, the statement is different. If Eqs. 15.3 and 15.4 are each divided by the value 1.33, it is easy to see the actual requirement: only ~15% of the specification area should be allowed to be stressed by the kσ measurement scatter. In the case of the process capability indices C_p and C_{pk}, however, the limit value requirement of 1.33, for example, refers to the ratio of specification to process scatter width, i.e. to the ratio 8 to 6 (Chap. 16).

c) In the presence of a bias Bi, the capability index C_{gk} is lower compared to the C_g, because the bias is considered as an absolute value. Nevertheless, the calculation of the C_g capability index is important, as it is used to assess the random scatter of the test equipment in relation to the specification. As a result, a test equipment may be perfectly capable in terms of repeat measurement dispersion, but if the measurement results are systematically offset from the nominal value, the capability index C_{gk} will be below the limit value requirement due to the bias. A technical correction of the measurement offset would be the consequence. Therefore, both characteristic values must always be interpreted.

15.3.2 Application Example C_g-/C_{gk}-Study: Cylinder Head Testing

The production planning department of an engine manufacturer would like to check function-critical dimensions of a cylinder head during ongoing production. One of the function-critical dimensions is the height of the cylinder head $h = 140 \pm 0.09$ mm. A new, suitable test system is to be procured with which the height can be accurately tested. A supplier offers a testing device: A C_g-/C_{gk}-study is used to assess the short-time test equipment capability. A series of measurements with a total of 25 measured values is recorded on site (cf. Table 15.3), a measurement standard is used as the test object, which represents an exact height of 140 mm (tolerance centre).

Solution
The value pattern shows no trend or other anomaly, a normal distribution model may be assumed. First, the resolution of the measurement system (RE = 0.001 mm; results from the decimal places, cf. Table 15.3) is judged to be suitable, since the limit value requirement according to Eq. 15.5 is fulfilled. By estimating the arithmetic mean (Eq. 15.6), the standard deviation (Eq. 15.7) and the bias (Eq. 15.8) the characteristic indices C_g and C_{gk} with k = 4 can be

Table 15.3 Measured values [mm] of a series of repeat measurements; test object normal (height h = 140 mm), sample scope n = 25

140.005	140.010	140.000	139.997	140.006
140.008	139.997	140.008	140.002	140.011
139.996	140.002	140.001	140.006	140.004
140.007	140.004	140.011	140.004	140.013
140.008	140.002	139.999	140.008	140.005

determined (Eqs. 15.9 and 15.10). Both characteristic values meet the capability requirement of $C_g \geq 1.33$ and $C_{gk} \geq 1.33$. By rearranging Eqs. 15.9 and 15.10, where $C_g = C_{gk} = 1.33$, the specification for which the measuring system is still capable can be calculated; cf. Eqs. 15.11 and 15.12. Due to the present systematic offset Bi, the minimum tolerance for which the measuring system could still be used would be T = 0.171 mm.

$$\%RE = \frac{RE}{Specification} \cdot 100\% = \frac{0.001 \text{ mm}}{2 \cdot 0.09 \text{ mm}} \cdot 100\% = 0.56\% \leq 5\% \quad (15.5)$$

$$\widehat{\mu} = \bar{x} = \frac{1}{n} \cdot \sum_{i=1}^{n} x_i = 140.00475 \text{ mm} \quad (15.6)$$

$$\widehat{\sigma} = s = \sqrt{\frac{1}{n} \cdot \sum_{i=1}^{n} (x_i - \bar{x})^2} = 0.00463 \text{ mm} \quad (15.7)$$

$$Bi = |\widehat{\mu} - x_m| = |140.00475 \text{ mm} - 140 \text{ mm}| = 0.00475 \text{ mm} \quad (15.8)$$

$$C_g = \frac{0.2 \cdot Specification}{k\widehat{\sigma}} = \frac{0.2 \cdot 2 \cdot 0.09 \text{ mm}}{4 \cdot 0.00463 \text{ mm}} = 1.94 > 1.33 \quad (15.9)$$

$$C_{gk} = \frac{0.1 \cdot Specification - |\widehat{\mu} - x_m|}{\frac{1}{2} k \cdot \widehat{\sigma}} = \frac{0.1 \cdot 2 \cdot 0.09 \text{ mm} - 0.00475 \text{ mm}}{\frac{1}{2} \cdot 4 \cdot 0.00463 \text{ mm}} = 1.43 > 1.33 \quad (15.10)$$

$$Specification_{min,C_g} = 0.123 \text{ mm} \quad (15.11)$$

$$Specification_{min,C_{gk}} = 0.171 \text{ mm} \quad (15.12)$$

15.4 Repeatability and Reproducibility: %GRR Study

15.4.1 Basics of the %GRR Study

The study of repeatability and reproducibility %GRR study (G = Gauge, R = Repeatability, R = Reproducibility) is carried out to evaluate new and existing measuring systems at the final point of use (e.g. within production). Therefore, the performance of a %GRR study requires a positive assessment of the resolution of the measuring system as well as a positive assessment of the short-term measurement system capability (C_g-/C_{gk}-study).

The assessment by means of the %GRR indicator is carried out under conditions that are as real as possible with regard to the place of use of the test system, e.g. the consideration of series conditions. The investigations are therefore ideally carried out with representative test objects and testers from the series production process. In the focus of a %GRR study, an overall scatter is determined on the basis of the scatter influences of the testers as well as of repeat measurements related to several components. This is compared in relation to the given component specification or a known process scattering with a limit value requirement (AIAG 2010). Illustration of this principle is provided later in this chapter using an application example. The general procedure for the %GRR study thus comprises the following preconditions and steps:

a) Precondition: Positive evaluation of the measurement system resolution.
b) Precondition: Positive assessment of the short-term measurement system capability (C_g-/C_{gk}-study).
c) Determination of the starting point: number of testers k (e.g.: k = 3), test objects n (e.g.: n = 5), number of repeat measurements r (e.g.: r = 2).

Equation 15.13 is used as a rule of thumb for establishing the baseline.

$$n \cdot k \cdot r > 30 \tag{15.13}$$

If operator influence can be excluded, k = 1 and, for example, the number of test objects n = 25 with r = 2 repeat measurements is selected.

d) Determination of repeatability EV (Equipment Variation), reproducibility AV (Appraiser Variation) and total dispersion %GRR
e) Assessment of the repeatability and reproducibility %GRR by means of a limit value requirement as suitable (Eq. 15.14), conditionally suitable (Eq. 15.15), unsuitable (Eq. 15.16).

$$\%GRR < 10\% \tag{15.14}$$

$$10\% \leq \%GRR \leq 30\% \tag{15.15}$$

$$\%GRR > 30\% \tag{15.16}$$

The calculation of the %GRR indicator can be calculated in two ways: The repeatability and reproducibility can be related to the given specification (cf. Eq. 15.17) or to the process variation (cf. Eq. 15.18). The characteristic value repeatability EV (Equipment Variation) refers to the scatter of the measuring system, the characteristic value appraiser precision AV (Appraiser Variation) refers to the variation caused by the different testers.

$$\%GRR = \frac{6 \cdot \sqrt{EV^2 + AV^2}}{Specification} \cdot 100\% \tag{15.17}$$

$$\%GRR = \frac{6 \cdot \sqrt{EV^2 + AV^2}}{Process\ scattering} \cdot 100\% \tag{15.18}$$

In the industrial-historical context, four procedures are mentioned to perform the dispersion analysis within a %GRR study:

15.4 Repeatability and Reproducibility: %GRR Study

a) Span method (Short Method/Short Range): Estimation of the total dispersion of a system, provides an overall impression of the measuring system. However, there is no differentiation of the total scatter into repeatability and reproducibility.

b) ARM = Average Range Method: Focus of the calculation are average values (Average) and ranges (Range) of the respective measured value series (evaluation of several measured value series possible).

c) ANOVA = Analysis of Variance: Focus of the calculation are mean values (Average) and variances (Variance) of the respective measurement series.

d) Difference method (mean-standard deviation method): Focus of the calculation are the differences from the 1st and 2nd measured value series of the respective testers; determination of the standard deviation s based on the differences.

The span method is a simple estimation of the total scatter of the system, a differentiation of the scatter of individual influences is not carried out, ergo a potential measurement process optimization is difficult. The difference method has not been accepted because it requires a limitation of two testers. The procedure according to the ARM has established itself from a historical point of view, since mean values and ranges can be easily determined from the measurement series and the scatter of the individual influences can be calculated – also manually – via corresponding correction factors (K1, K2, K3 from tables; see Duncan 1986). However, the use of ranges as a measure of dispersion is less precise than focusing directly on variances. As a result, the method of analysis of variance (ANOVA), which was mainly developed by Ronald Aylmer Fisher (among others also maximum likelihood estimator, F-distribution, discriminant function), finally prevailed. The procedure is complex and cannot be carried out manually with reasonable effort, but it is deposited in numerous libraries, so that it is easily applicable in industrial applications (ANOVA: see Hedderich and Sachs 2020). The scatter of the influences repeatability measuring system, tester, interactions, total scatter, scatter related to the test objects are calculated individually. This allows the targeted optimization of the testing process, since the scatter of the individual influences is known. The analysis of variance (ANOVA) is therefore the most accurate of the above-mentioned procedures within a measurement system analysis.

15.4.2 Application Example %GRR Study: Engine Piston Rod Testing

The manufacture of an engine piston rod is carried out, among other things, by means of a machining process (cf. case study engine piston rod; Sect. 4.1). After machining, function-critical characteristics are tested. The test equipment and the test process are examined with regard to its suitability within the framework of a %GRR study. The present example shows the results of a %GRR study with regard to the function-critical characteristic "large piston rod eye". The %GRR study is only outlined in the following explanations, since the manual calculation is associated with high effort. The calculations were performed using analysis of variance (ANOVA).

The specification covers the diameter Ø 64 mm with tolerance zone T = 0.02 mm. On n = 5 different piston rods by k = 3 testers (testers A, B, C) r = 3 repeat measurements (Test 1, 2, 3) are carried out for each component; see Table 15.4.

The resolution of the test system RE = 0.0001 mm is judged to be capable, since the limit value requirement is fulfilled; Eq. 15.19. According to the %GRR study, the test system is judged to be conditionally capable, since $10\% \leq \%GRR \leq 30\%$ applies, Eq. 15.20. Furthermore, the repeatability EV (Eq. 15.21) shows a higher scatter compared to the influence of the gaugers (Eq. 15.22). This is also reflected in the representation of the value patterns or histograms of the %GRR measurement series; cf. Figs. 15.4 and 15.5: The measurement results of the three gaugers are similar in relation to the five connecting rods. For example, the position of the measured values with respect to connecting rod A is almost identical for all three gaugers and corresponds approximately to the mean value of the entire series of measurements. The repeat measurement of each tester also shows a similar result in each case. Only the first measurement of tester A shows a higher deviation in comparison to two other repeat measurements of tester A as well as in comparison to testers B and C. The high scattering of the components in relation to the given specification (PV = 88.40%; the calculation basis can be taken from AIAG 2010) is striking.

$$\%RE = \frac{RE}{Specification} \cdot 100\% = 0.5\% \leq 5\% \tag{15.19}$$

$$\%GRR = \frac{6 \cdot \sqrt{EV^2 + AV^2}}{Specification} \cdot 100\% = 10.11\% \tag{15.20}$$

$$\%EV = 9.25\% \tag{15.21}$$

$$\%AV = 4.08\% \tag{15.22}$$

Table 15.4 Measurement series for testing a piston rod eye with Ø 64 mm; tolerance T = 0.02 mm

Gauger A			Gauger B			Gauger C		
Test 1	Test 2	Test 3	Test 1	Test 2	Test 3	Test 1	Test 2	Test 3
64.0060	64.0070	64.0070	64.0070	64.0070	64.0072	64.0068	64.0069	64.0065
64.0110	64.0100	64.0105	64.0100	64.0105	64.0108	64.0110	64.0107	64.0107
64.0070	64.0075	64.0078	64.0080	64.0075	64.0078	64.0076	64.0074	64.0073
64.0050	64.0060	64.0052	64.0050	64.0048	64.0050	64.0046	64.0049	64.0043
64.0030	64.0034	64.0032	64.0037	64.0036	64.0036	64.0030	64.0031	64.0028

15.4 Repeatability and Reproducibility: %GRR Study

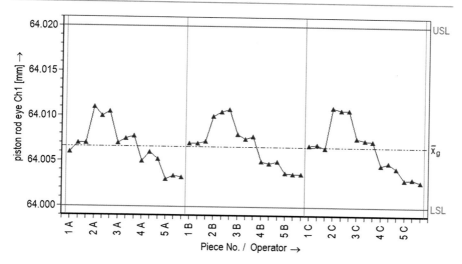

Fig. 15.4 Value patterns of the measurement series of a %GRR study; case study engine piston rod, characteristic "large piston rod eye" with 5 components, 3 repeat measurements each by gauger A, B and C

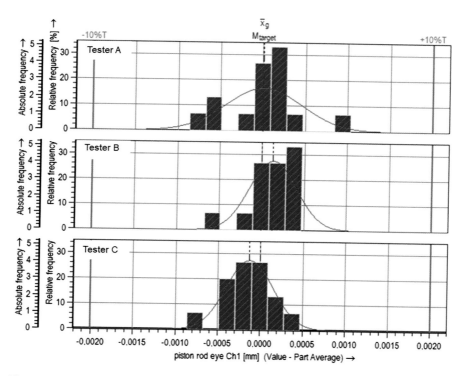

Fig. 15.5 Histograms of the measurement series of a %GRR study; case study engine piston rod, characteristic "large piston rod eye" with 5 components, 3 repeat measurements each by gauger A, B and C

15.5 Test Process Suitability: Analysis of Measurement Uncertainty (Q_{MS}-/Q_{MP}-Study)

The Test Process Suitability Study (Q_{MS}-/Q_{MP}-Study) according to (VDA 2010) comprises the evaluation of the entire test process on the basis of the systematic analysis of all influencing variables, differentiated according to test equipment influences and test process influences, on the measurement result. The flexible structure of the procedure allows individual adaptation with regard to the test task to be assessed, yet the relevant influencing variables on the test process can be comprehensively and traceably taken into account (in contrast to the %GRR study, which only allows a limited analysis of influencing variables). The Q_{MS}-/Q_{MP}-study according to (VDA 2010) is based in its approach on the Guide to the expression of uncertainty in measurement (ISO/IEC Guide 98 (GUM); (ISO/IEC 2008)) respectively ISO 14253 (ISO 2017). However, compared to ISO/IEC Guide 98 (GUM), some assumptions and simplifications have been made, so that the Q_{MS}-/Q_{MP}-study represents a practical procedure for the comprehensive performance of a measurement system assessment and an analysis of the test process suitability. Furthermore, the Q_{MS}-/Q_{MP}-study takes into account components of the C_g-/C_{gk}-study and %GRR study and is compatible with ISO/CD 22514-7 "Capability of Measurement Processes" (ISO 2021).

15.5.1 Analysis of the measurement uncertainty

The basic procedure for carrying out a Q_{MS}-/Q_{MP}-study is shown in Fig. 15.6. First, the relevant uncertainty components $x_1, x_2, \ldots, x_n$ are analysed; differentiated according to test equipment and test process. Subsequently, the standard uncertainties u_x ($x=1, \ldots, n$) are determined for each influence quantity. Here, test series can be carried out and analyzed in relation to each influence quantity (approach: "test"). Or empirical values are available for the ranges of the influence quantities, from which the standard uncertainties u_x are estimated (approach: "empiricism"). The standard uncertainties are combined, whereby the differentiation by measurement system (u_{MS}) and measurement process (u_{MP}) is maintained. Based on the combined standard uncertainties u_{MS} and u_{MP} the expanded measurement system uncertainty U_{MS} and the expanded measurement process uncertainty U_{MP} are calculated.

If no test series are carried out as a basis for estimating standard uncertainties u_x (approach: test), empirical values or estimates with regard to the expected range can be used (approach: empiricism). Here, an assumption must be made about the expected distribution model. The standard uncertainty u_x can be determined from the span and based on the distribution model. Figure 15.7 shows four distribution models (triangular, normal, rectangular, U-distribution) which can typically be assumed with regard to an influence quantity. The rectangular distribution is a conservative, common assumption, since within the expected range ±a the probability

15.5 Test Process Suitability: Analysis of Measurement Uncertainty ...

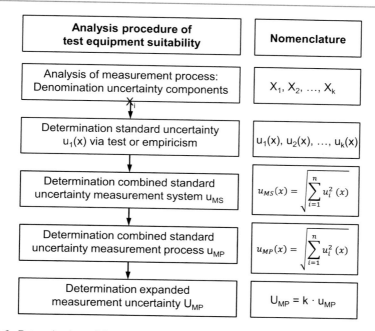

Fig. 15.6 Determination of the expanded measurement uncertainty U on the basis of standard uncertainties for the measurement system (MS) and measurement process (MP)

is equally distributed and accordingly the standard uncertainty u_x is higher compared to the normal or triangular distribution.

Table 15.5 shows a representative overview of the expected uncertainty components with respect to the measurement system and the test process. Depending on the measurement system or test process used, some of these standard uncertainties are taken into account or can be neglected. If necessary, further influencing variables or standard uncertainties must be added (summarised here under u_{other}). Finally, the selected and, if necessary, supplemented standard uncertainties are taken into account in the calculation of the standard uncertainties u_{MS} or u_{MP}.

The standard uncertainties for the measurement system u_{MS} and the test process u_{MP} can be determined with the aid of Eqs. 15.23 and 15.24. If the repeatability on the test object (not on the standard) can be neglected, $u_{EVO} = 0$ and Eq. 15.25 follows.

$$u_{MS} = \sqrt{u_{CAL}^2 + max\{u_{RE}^2, u_{EVR}^2\} + u_{LIN}^2 + u_{BI}^2 + u_{MS\_other,i}^2} \quad (15.23)$$

$$u_{MP} = \sqrt{\begin{array}{c} u_{CAL}^2 + max\{u_{RE}^2, u_{EVR}^2, u_{EVO}^2\} + u_{LIN}^2 + u_{BI}^2 + u_{MS\_other,i}^2 + u_{OBJ}^2 + u_{GV}^2 + u_{AV}^2 + u_{STAB}^2 + \\ \sum_i u_{IA,i}^2 + u_T^2 + u_{MP\_other,i}^2 \end{array}} \quad (15.24)$$

Distribution	Schematic draw density function Probability P regarding measurement value within range ± a	Standard uncertainty u(x)
Triangel distribution	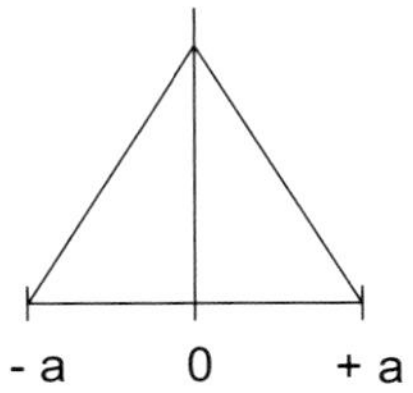 P = 1	$u(x_B) = \dfrac{2a}{\sqrt{24}} \approx 0{,}4 \cdot a$
Normal distribution	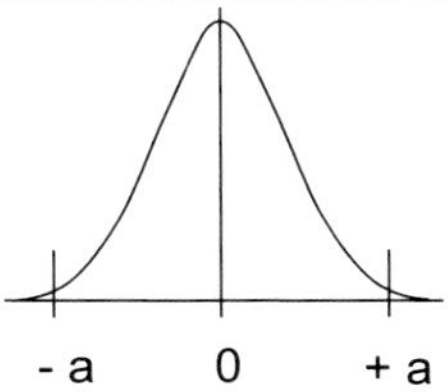 P = 0,945	$u(x_B) = \dfrac{2a}{\sqrt{16}} = 0{,}5 \cdot a$
Uniform distribution	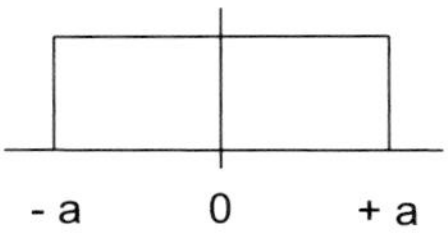 P = 1	$u(x_B) = \dfrac{2a}{\sqrt{12}} \approx 0{,}6 \cdot a$
U distribution	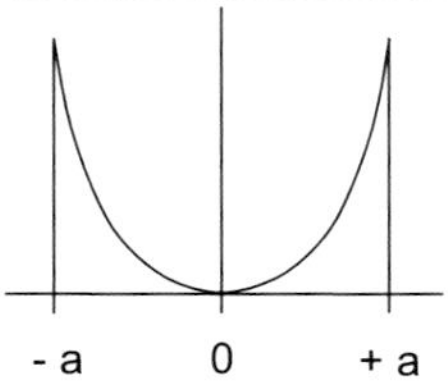 P = 1	$u(x_B) = \dfrac{a}{\sqrt{2}} \approx 0{,}7 \cdot a$

Fig. 15.7 Common distribution models as a base of operations for determining the standard uncertainty $u(x_i)$ when assuming a limit value or the range (via ratio a)

$$u_{MP} = \sqrt{u_{MS}^2 + u_{OBJ}^2 + u_{GV}^2 + u_{AV}^2 + u_{STAB}^2 + \sum_i u_{IA,i}^2 + u_T^2 + u_{MP\_other,i}^2} \tag{15.25}$$

Subsequently, the expanded measurement uncertainty U is determined for the measurement system and test process according to Eqs. 15.26 and 15.27. The factor k is selected depending on the probability P related to the distribution model which represents the measurement uncertainty. If a normal distribution model is assumed or proven (quite common), P = 0.9545 (corresponding to 4σ, ergo k = 2) or P = 0.9973 (corresponding to 6σ; ergo k = 3) can be chosen, for example.

15.5 Test Process Suitability: Analysis of Measurement Uncertainty ...

Table 15.5 Uncertainty components of measurement system (MS) and measurement process (MP); representative selection

u_x/U_x	Uncertainty component	MS or MP
u_{CAL}	Calibration of the standard/reference	MS
u_{LIN}	Linearity deviation	MS
u_{EVR}	Repeatability (on measurement standard/on reference)	MS
u_{EVO}	Repeatability (on the test object)	MP
u_{AV}	Comparability of operators (operator influence)	MP
u_{GV}	Comparability of the measuring equipment (measuring points)	MP
u_{STAB}	Comparability at different points in time	MP
$u_{IA,i}$	Interaction(s)	MP
u_{OBJ}	Inhomogeneity of the test object	MP
u_{RE}	Resolution of the display/the reading	MS
u_T	Temperature	MP
u_{other}	other influences with index i	MS/MP
u_{MS}	combined standard uncertainty (measuring system)	MS
U_{MS}	expanded uncertainty of measurement (measuring system)	MS
u_{MP}	combined standard uncertainty (measurement process)	MP
U_{MP}	expanded measurement uncertainty (measurement process)	MP

Finally, the measurement system and test process are evaluated via the capability characteristics Q_{MS} and Q_{MP}: The quotient of expanded measurement uncertainty and given specification (here: tolerance T from existing test requirement) forms the basis of a comparison with a limit value requirement; Eqs. 15.28 and 15.29. The measuring system is considered suitable if the limit value requirement according to Eq. 15.30 is fulfilled; the test process is evaluated as suitable if the limit value requirement according to Eq. 15.31 is fulfilled.

$$U_{MS} = k \cdot u_{MS} \tag{15.26}$$

$$U_{MP} = k \cdot u_{MP} \tag{15.27}$$

$$Q_{MS} = \frac{2\,U_{MS}}{Specification} \cdot 100\% \tag{15.28}$$

$$Q_{MP} = \frac{2\,U_{MP}}{Specification} \cdot 100\% \tag{15.29}$$

$$Q_{MS} \leq Q_{MS\_max} = 15\% \tag{15.30}$$

$$Q_{MP} \leq Q_{MP\_max} = 30\% \tag{15.31}$$

If the measurement system and test process are classified as suitable, future measurement results are specified in accordance with Eq. 15.32. The specification refers, of course, only to the actual test process, see Note 4 in Sect. 15.5.2. Furthermore, the coverage factor for the calculation of the expanded measurement uncertainty U_{MP} should be given (here: $k=2$; corresponds to probability of $P=0.9545$ if a normal distribution model is assumed or proven).

$$y = x \pm U_{MP}(k=2) \tag{15.32}$$

15.5.2 Notes on the Analysis of Measurement Uncertainty

The procedure shown for performing an inspection process suitability analysis (Q_{MS}-/Q_{MP}-Study) is based on the approach on the Guide to the expression of uncertainty in measurement (ISO/IEC Guide 98 (GUM)) (ISO 2008) and ISO 14253 (ISO 2017), respectively. However, compared to ISO/IEC Guide 98 (GUM)), some assumptions or simplifications were made, which are briefly explained in the following notes.

Note 1: GUM assumptions and simplification
According to standard ISO/IEC Guide 98 (GUM)), the combined standard uncertainty $u_c(y)$ for the measurement result y is determined from the standard uncertainties and covariances associated with the input estimates. A Taylor series is used to map the function at one point with a power series. The power series is the limit of the Taylor polynomials. The series expansion is called Taylor series expansion. Thus, if several influencing variables are present, the interactions of these can also be taken into account by considering the covariances. The combined standard uncertainty $u_c(y)$ is thus obtained on the basis of the Taylor series expansion, which is terminated after the first element – due to the sufficient accuracy, cf. Eq. 15.33. The covariances are taken into account here. The measure of correlation between x_i and x_j is expressed by the correlation coefficient $r(x_i,x_j)$, Eq. 15.34. Transforming Eq. 15.34 and substituting it into Eq. 15.33 yields Eq. 15.35, which is easier to handle because the correlation coefficient is easier to interpret than the covariance between the quantities x_i and x_j.

$$\begin{aligned} u_c^2(y) &= \sum_{i=1}^{N} \sum_{j=1}^{N} \frac{\partial f}{\partial x_i} \frac{\partial f}{\partial x_j} u(x_i, x_j) \\ &= \sum_{i=1}^{N} \left(\frac{\partial f}{\partial x_i} \right)^2 u^2(x_i) + 2 \sum_{i=1}^{N-1} \sum_{j=i+1}^{N} \frac{\partial f}{\partial x_i} \frac{\partial f}{\partial x_j} u(x_i, x_j) \end{aligned} \tag{15.33}$$

$$r(x_i, x_j) = \frac{u(x_i, x_j)}{u(x_i)u(x_j)} \tag{15.34}$$

15.5 Test Process Suitability: Analysis of Measurement Uncertainty ...

$$u_c^2(y) = \sum_{i=1}^{N}\left(\frac{\partial f}{\partial x_i}\right)^2 u^2(x_i) + 2\sum_{i=1}^{N-1}\sum_{j=i+1}^{N}\frac{\partial f}{\partial x_1}\frac{\partial f}{\partial x_j}u(x_i)u(x_j)r(x_i,x_j) \tag{15.35}$$

The following assumptions are now made:
The influences are represented purely phenomenologically, there are no interactions, so that Eq. 15.36 applies to the covariances and these can be neglected in Eq. 15.33, from which Eq. 15.37 follows directly for the combined standard uncertainty $u_c(y)$.

$$\sum_{i=1}^{N-1}\sum_{j=i+1}^{N}\frac{\partial f}{\partial x_i}\frac{\partial f}{\partial x_j}u(x_i,x_j) = 0 \tag{15.36}$$

$$u_c^2(y) = \left(\sum_{i=1}^{N}\frac{\partial f}{\partial x_i}u(x_i)\right)^2 = \left(\sum_{i=1}^{N}c_i u(x_i)\right)^2 \tag{15.37}$$

The partial derivatives c_i in Eq. 15.37 are generally also called sensitivity coefficients, since they describe the change in the output variable y as a function of a change in the input variables $x_1, x_2, \ldots, x_n$. The partial derivatives thus describe the expected value related to the influence variable x_i.

If the sensitivity coefficients are neglected, since the influencing variables are, for example, geometric parameters, Eq. 15.38 applies. From this, Eq. 15.39 follows directly for the standard uncertainty $u_c(y)$. Thus, the detailed modelling according to GUM (ISO 2008) is not necessary and the standard uncertainty u_c can be determined with Eq. 15.39, or in a simple representation with Eq. 15.40. Equation 15.39 (or Eq. 15.40) is also called "quadratic addition", the index c is replaced by MS in the context of the measurement system analysis and by MP in the measurement process analysis. The quadratic addition of the n standard uncertainties is easy to perform after careful determination of the same, by experiment or empiricism.

$$c_i = \frac{\partial f}{\partial x_i} = 1 \tag{15.38}$$

$$u_c^2(y) = \left(\sum_{i=1}^{N}u(x_i)\right)^2 \tag{15.39}$$

$$u_c^2 = u_1^2 + u_2^2 + u_3^2 + u_4^2 + \cdots + u_N^2 \tag{15.40}$$

The quadratic addition can also be seen in the %GRR study (Sect. 15.4, Eq. 15.20): Here, equipment variation and appraiser variation are added to the total variance and a potential covariance of equipment and appraiser is neglected.

Note 2: Systematic measurement offset
The systematic measurement bias within the measurement system is usually known or has already been determined within a C_g-/C_{gk}-study. For the determination of the associated standard uncertainty u_{Bi} a rectangular distribution model (cf. Fig. 15.7) is assumed; thus Eq. 15.41 results.

$$u_{BI} = \frac{BI}{\sqrt{3}} = \frac{|\hat{\mu} - x_m|}{\sqrt{3}} \qquad (15.41)$$

Note 3: Calibration uncertainty
The calibration uncertainty, a component of the standard uncertainty u_{MS}, can be easily determined in many cases (Eq. 15.42), since the expanded calibration uncertainty U_{CAL} of the measuring system is known and documented in the calibration certificate of a measuring system. Division by the associated coverage factor k_{CAL} is performed.

$$u_{CAL} = \frac{U_{CAL}}{k_{CAL}} \qquad (15.42)$$

Note 4: Differentiation of expanded measurement uncertainty U_{MP} and calibration uncertainty U_{CAL} or uncertainty U of the measuring system manufacturer
The expanded measurement uncertainty U_{MP} is (of course) not identical with the measurement uncertainty U_{CAL} or any other measurement uncertainty U specified by the measuring system manufacturer, which is stated in the calibration certificate of the measuring system. This measurement uncertainty refers to the measurement system uncertainty under laboratory conditions.

Note 5: Correlation and multiple assessment
In principle, the standard uncertainty u_{EVR} (repetition scatter) is already included in the standard uncertainty u_{RE}, thus u_{MS} is determined without considering u_{RE} (Eq. 15.43). On the other hand, the standard uncertainty u_{RE} dominates at high resolution, so u_{EVR} can be neglected (Eq. 15.44). This results in the recommendation to select the maximum of the standard uncertainties u_{RE}, u_{EVR} and u_{EVO}, respectively, in order to avoid multiple evaluation, cf. Eq. 15.45. Note: The influence u_{EVO} is only included in the analysis of the measurement process, so it is missing in Eq. 15.45. Alternatively, the resolution can also be evaluated separately beforehand (cf. Sect. 15.3, Eq. 15.2) and the influence can be neglected if the accuracy is sufficient.

$$u_{MS} = \sqrt{u_{CAL}^2 + u_{LIN}^2 + u_{EVR}^2 + u_{BI}^2 + u_{MS\_other,i}^2} \qquad (15.43)$$

$$u_{MS} = \sqrt{u_{CAL}^2 + u_{RE}^2 + u_{LIN}^2 + u_{BI}^2 + u_{MS\_other,i}^2} \qquad (15.44)$$

$$u_{MS} = \sqrt{u_{CAL}^2 + max\{u_{RE}^2, u_{EVR}^2\} + u_{LIN}^2 + \ldots + u_{MS\_other,i}^2} \quad (15.45)$$

15.5.3 Application Example Q_{MS}-/Q_{MP}-Study: Measurement of Body-in-White

Within a automotive body shop it is planned to determine the position of underbody set screws (for fastening plastic add-on parts) with the aid of a coordinate measuring machine with mechanical probe. Only the deviation is determined; i.e. the specification is: 0 mm ±0.4 mm. The suitability of the underlying test process is investigated: First, the short-term test equipment capability C_g-/C_{gk}-study is to be assessed. Subsequently, the expanded measurement uncertainty U is determined and the entire test process is evaluated by means of the Q_{MS}-/Q_{MP}-study.

Base of operations
A repeat measurement was carried out at a defined position; cf. measurement series with scope n = 25 parts in Table 15.6. The correct value (reference value position due to a slight offset of the body mount) is $x_m = 0.05$ mm.

The following facts are known:

a) From the calibration certificate of the coordinate measuring machine, calibration uncertainty $U_{CAL} = 0.02$ ($k_{CAL} = 2$) is read, so the standard calibration uncertainty $u_{CAL} = 10$ μm.
b) A rectangular distribution is assumed for the environmental effects, and the span is estimated to be 80 μm.
c) There is a test influence due to the manual clamping of the subfloor. The characteristic value %GRR = 12% is known from a %GRR study already carried out. The influence and interaction of different inspectors could be neglected in this study.
d) Due to material properties, the standard uncertainty due to test object influence is estimated on a range of 16.67 μm.
e) Other influences can be neglected.

Procedure of the C_g-/C_{gk}-Study
The value pattern shows no trend or other anomalies, a normal distribution model may be assumed. The resolution of the measurement system is RE = 0.001 mm; the resolution meets the limit value requirement according to Eq. 15.46. The mean and standard deviation are estimated directly (Eqs. 15.47 and 15.48). With specification T = 0.8 mm, $x_M = 0.05$ mm, the systematic measurement offset is Bi = 0.0065 mm. The coverage factor is set as k = 4 (P = 0.9545; normal distribution model). This yields the capability indices C_g (Eq. 15.49) and C_{gk} (Eq. 15.50), which satisfy the limit requirement of 1.33. The short-term test equipment capability of the measurement system is proven via C_g-/C_{gk}-Study.

Table 15.6 Measurement series for testing set screw positions (deviation measurement)

0.040	0.015	0.056	0.039	0.020
0.077	0.022	0.047	0.046	0.070
0.028	0.034	0.034	0.044	0.043
0.024	0.004	0.041	0.010	0.060
0.060	0.050	0.094	0.025	0.105

$$\%RE = \frac{RE}{Specification} \cdot 100\% = 0.125\% \leq 5\% \quad (15.46)$$

$$\widehat{\mu} = 0.0435 \, \text{mm} \quad (15.47)$$

$$\widehat{\sigma} = 0.0247 \, \text{mm} \quad (15.48)$$

$$C_g = \frac{0.2 \cdot T}{k\widehat{\sigma}} = 1.62 \geq 1.33 \quad (15.49)$$

$$C_{gk} = \frac{0.1 \cdot T - |\widehat{\mu} - x_m|}{\frac{1}{2} k \cdot \widehat{\sigma}} = 1.49 \geq 1.33 \quad (15.50)$$

Determination of expanded measurement uncertainty and performance of a Q_{MS}-/Q_{MP}-Study
First, the standard uncertainty components $u_i(x)$ are analyzed with respect to the measurement system. The calibration uncertainty is known (Eq. 15.51), the repeatability uncertainty can be taken from the previously performed C_g-/C_{gk}-Study (Eq. 15.52). The uncertainty component due to bias is determined based on a rectangular distribution, and the bias itself is known from the C_{gk}-Study (Eq. 15.53). Linearity error as well as other influences are neglected (Eq. 15.54). The standard uncertainty of the measurement system u_{MS} is given by Eq. 15.55, where standard uncertainty u_{EVR} is taken into account and u_{RE} is neglected as it has already been assessed in the C_g-/C_{gk}-Study. The measurement uncertainty U_{MS} is thus determined, Eq. 15.56 (with coverage factor k=2 (P=0.9545)). The limit value requirement with regard to the suitability of the measurement system is fulfilled, thus the measurement system is judged to be suitable, Eq. 15.57.

$$u_{CAL} = 0.01 \, \text{mm} \quad (15.51)$$

$$u_{EVR} = \widehat{\sigma} = 0.0247 \, \text{mm} \quad (15.52)$$

$$u_{BI} = \frac{BI}{\sqrt{3}} = 0.00375 \, \text{mm} \quad (15.53)$$

15.5 Test Process Suitability: Analysis of Measurement Uncertainty …

$$u_{LIN} = u_{MS\_other} = 0 \tag{15.54}$$

$$u_{MS} = \sqrt{u_{CAL}^2 + max\{u_{RE}^2, u_{EVR}^2\} + u_{LIN}^2 + u_{BI}^2 + u_{MS\_other}^2} = 0.0269 \text{ mm} \tag{15.55}$$

$$U_{MS} = 2u_{MS} = 0.0538 \text{ mm} \tag{15.56}$$

$$Q_{MS} = \frac{2 \cdot U_{MS}}{T} \cdot 100\% = 13.45\% \leq 15\% \tag{15.57}$$

For the determination of the expanded measurement uncertainty U_{MP} related to the measurement process, the associated standard uncertainty components $u_i(x)$ are estimated. The appraiser variation u_{AV} is determined from the already known %GRR study, Eq. 15.58. For the environmental influences as well as the test object influences the range is known in each case, ergo the respective standard uncertainty is determined via the assumption of a rectangular distribution model (Eqs. 15.60 and 15.61, cf. Fig. 15.7). Thus, the standard uncertainty of the measurement process u_{MP} (Eqs. 15.62 and 15.63) as well as the expanded measurement uncertainty of the test process U_{MP} can be determined (Eq. 15.64). This is immediately followed by the characteristic value Q_{MP}, which satisfies the limit value requirement (Eq. 15.65). The test process is thus judged to be suitable on the basis of the procedure according to (VDA 2010).

$$\%GRR = \frac{6 \cdot \sqrt{EV^2 + AV^2}}{Specification} \cdot 100\% = 12\% \tag{15.58}$$

$$u_{AV} = AV = 0.016 \text{ mm} \tag{15.59}$$

$$u_{Enviro} = \frac{2a}{\sqrt{12}} = \frac{80 \, \mu m}{\sqrt{12}} = 0.02309 \text{ mm} \tag{15.60}$$

$$u_{OBJ} = \frac{2a}{12} = \frac{16.67 \, \mu m}{\sqrt{12}} = 0.0048 \text{ mm} \tag{15.61}$$

$$u_{MP} = \sqrt{u_{MS}^2 + u_{GV}^2 + u_{AV}^2 + u_{STAB}^2 + \sum_i u_{IA,i}^2 + u_{EVO}^2 + u_{OBJ}^2 + u_T^2 + u_{MP\_other,i}^2} \tag{15.62}$$

$$u_{MP} = \sqrt{u_{MS}^2 + u_{AV}^2 + u_{OBJ}^2 + u_{Enviro}^2} = 0.0393 \text{ mm} \tag{15.63}$$

$$U_{MP} = 2 \cdot u_{MP} = 0.0795 \text{ mm} \tag{15.64}$$

$$Q_{MP} = \frac{2 \cdot U_{MP}}{Specification} \cdot 100\% = 19,63\% \leq 30\% \qquad (15.65)$$

15.6 Notes and References: Procedure X

From a historical point of view, the various methods for analyzing the capability of test equipment and the suitability of test processes were (and are) sometimes referred to as „procedure X" (X = 1, ..., 5; in german: „Verfahren X"). In the following, these terms are classified as they are often used in earlier as well as current sources; e.g. (Keferstein et al. 2018; Linß 2005).

Procedure 1: Proof of the capability of a measurement system
The term "Procedure 1" refers to the short-term test equipment capability and is also known under the term C_g-/C_{gk}-Study (cf. Sect. 15.3).

Procedure 2: Proof of the capability of a measurement process with operator influence
The performance of procedure 2 involves analysis of a measurement process taking into account the final location of use, therefore it is the Gage Repeatability & Reproducibility – study (%GRR study), cf. Sect. 15.4. Procedure 2 requires the successful evaluation of the test equipment with procedure 1.

Procedure 3: Proof of the capability of a measurement process without operator influence
Procedure 3 focuses on the analysis of automated measurement processes and is essentially carried out in the same way as Procedure 2 (%GRR study), but without operator influence (Sect. 15.4).

Procedure 4: Measurement stability of a measurement process
Procedure 4 is based on Procedure 1 and examines the test equipment on the basis of a standard at different points in time. Using a control chart, the stability of a measuring device is investigated over a longer period of time. In practice, a C_g-/C_{gk}-study (Sect. 15.3) can be carried out at different points in time. The quality control chart is an additional tool for detecting the potential overstepping of intervention limits at an early stage (cf. also Chap. 16).

Procedure 5: Ability of inspection processes for attributive characteristics
C_g-/C_{gk}-Study, %GRR-Study as well as Q_{MS}-/Q_{MP}-Study (expanded measurement uncertainty) refer to continuous characteristics (e.g.: component dimensions length, diameter) and can therefore not be applied to attributive tests, i.e.: test processes based on the use of gauges. Procedure 5 is suitable exclusively for these purposes and is also mentioned in (VDA 2010) and is not part of this book on data analysis.

Statistical Process Control (SPC) 16

The controllability and quality capability of production processes is a precondition for ensuring the reliable manufacture of technically complex products. The term "reliability" refers here – in deviation from the definition in Chap. 3 – to the probability of a defect-free proportion with respect to a monitored product characteristic within a defined time interval of product manufacture. If the monitoring refers to a production process characteristic, "reliability" refers to the probability that the production process parameters of the production process range within the specification limits within a time interval and thus the prerequisite for fault-free production is fulfilled.

Thus, the controllability and quality capability of a manufacturing process can relate to characteristics of the manufactured product or to characteristics of the manufacturing process (process parameters). With regard to the product, these are often function-critical or safety-relevant characteristics. With regard to process parameters, these are often those that are directly correlated with a specific product feature.

Statistical process control (SPC) is an established method for ensuring controlled and quality capable manufacturing processes in industrial practice. Statistical process control is a way of maintaining an optimized manufacturing process in an optimized state through continuous monitoring and, if necessary, corrections. The basics of Statistical Process Control were developed by Walter A. Shewhart (1891, New Canton (Illinois) 1967, Troy Hills (New Jersey)) and its impact on other disciplines, especially the natural sciences (Shewhart 1939).

Within Sect. 16.1, the basics of Statistical Process Control are first presented. Section 16.2 deals with the technical realisation of statistical process control (SPC) within manufacturing and presents the planning and realization in series preparation as well as the application in the production phase. The univariate process analysis by means of process capability indices as well as process visualizations within a statistical process control in the context of the different phases of the product development process is shown in Sect. 16.3. Approaches to

© Springer-Verlag GmbH Germany, part of Springer Nature 2024
S. Bracke, *Reliability Engineering*, https://doi.org/10.1007/978-3-662-67446-8_16

multivariate process analysis by means of process capability indices are outlined in Sect. 16.4. Section 16.5 focuses on control technology of production processes with its components process visualization, design of control charts and control chart analysis.

16.1 Basics and Definitions

Within this chapter, the basic terms and concepts of statistical process control are explained. The controllability and quality capability of production processes is a precondition for ensuring the realization of reliable manufacturing of technically complex products.

In a *controlled process,* the parameters of the distribution (e.g. mean value, standard deviation, range) of the measured values of a product characteristic (e.g. component diameter) practically do not change or only change in a known way or within known limits (Bracke 2016). A controlled production process thus delivers reproducible results, respectively products with a certain predictability regarding the product characteristics. However, a controlled process can deliver a high reject rate with respect to the part specifications (cf. Fig. 16.1, left: excess rate of the normal distribution model). The following example illustrates this: Due to a worn out tool guide, the scatter with respect to a part characteristic is too high relative to the part specification. Possibly, the distribution model does not change with respect to the measured data for a product characteristic (the mean value as well as the scatter of the measured values within the analysed samples remain approximately constant). As a consequence, the manufacturing process is controlled, but the reject rate is (constantly) too high because the specification limits are continuously exceeded (cf. also Sect. 16.3.4, Fig. 16.8 left). Therefore, the process would not be quality capable.

A *quality capable process* means the suitability of a delimited process for the realization of a component (Bracke 2016). With the help of the manufacturing process, the requirements for this component are fulfilled: After the manufacturing process and testing, the produced components meet the given specification limits (e.g. tolerance specifications) with regard to the function-critical characteristics. Quality capability can therefore also be ensured, for example, by a final inspection alone, if any rejects are sorted out (cf. Fig. 16.1, center). The distribution model,

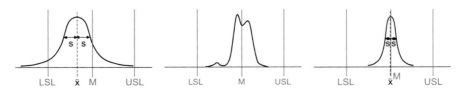

Fig. 16.1 Schematic diagram of different processes: controlled (left); quality capable (centre); robust (right)

16.1 Basics and Definitions

respectively the characteristic values mean value and dispersion based on the analysed samples, may well change in this case, cf. Sect. 16.3.4, Fig. 16.9: Process not controlled, but quality capable.

However, the goal of *robust manufacturing* is both: process control and quality capability (cf. Fig. 16.1, right); (Bracke 2016). The production process delivers reproducible results, the distribution of the measured values of a product characteristic of interest therefore does not change (or only in a known way). At the same time, the process is quality capable, the specification is fulfilled, because, for example, the process is centered and the scatter of measured values is low.

Random influence and systematic influence
During the production and subsequent measurement of manufactured parts, a scatter can usually be detected with regard to the (inspection) characteristic under consideration. The cause for the occurrence of scatter are systematic and/or random influences within the manufacturing process or the measuring process. *Random influences* can be described by a normal distribution model, which is adapted to the measurement data. If the manufacturing process is controlled, random influences have the following properties:

- constantly present,
- stable over time,
- predictable,
- constant sum of many individual influences.

Example: A slight backlash in the tool guide of a boring mill leads to a slight deviation of the hole position. The slight backlash – and thus the slight deviation in the realized result – is constantly present, stable in time, predictable and the sum of many, small influencing variables (backlash in the tool bearing, the holder, the traversing device, etc.).

Systemic influences in manufacturing destabilize or shift a manufacturing process. If a systematic influence on a production process is known, it is either eliminated or corrected by appropriate process measures. Example: The temperature influence caused by a warm-up of the hall after the start of production in the morning hours can affect a machining process. The temperature increase can lead to component expansion, which has an influence on the machining process. This influence is systematic, because with continuous temperature increase the process results (respectively the distribution model of the measured data) are continuously shifted. It can be eliminated by air-conditioning the hall. It can be corrected by fixing the workpieces with a preload that is opposite to the shifting. Systematic influences are analysed and eliminated during production planning, so that production can continue without interruption during the running series. However, if a systematic influence occurs during the running series production, the effects can be characterized as follows:

- Appearance follows no rule,
- Destabilization of the process,
- Prediction not possible,
- Causation by one or a few main influences.

Example: In the running manufacturing process (setting of bores) a forklift collides with a boring mill. After the collision, all holes produced have a similar offset to the nominal value or to the previously set holes. The collision of the forklift with the boring mill is a systematic influence, which rarely occurs (does not follow any rule), destabilizes the process, was not predictable and is not connected to other influences.

16.1.1 Process suitability analyses

For the complete assessment of a production process, four process suitability analyses are essentially carried out in the product emergence process:

- Test equipment capability or test process suitability,
- Machine capability,
- Preliminary process capability,
- Long-term process capability.

The chronological classification of the above analyses in relation to series preparation and the start of production (SOP) within the product development process is outlined in Fig. 16.2.

The precondition for the evaluation of machines and manufacturing processes on the basis of random sample inspections is the use of a capable test equipment

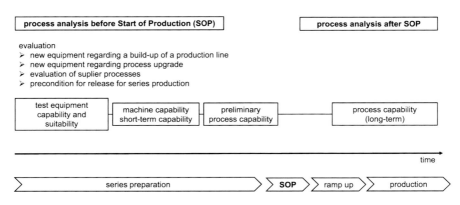

Fig. 16.2 Suitability and capability analyses in the chronological order and application in relation to the start of series production (SOP) and production in the product emergence process; cf. also Chap. 2

16.1 Basics and Definitions

or a suitable testing process. The investigation of the *test process suitability* is an analysis of the characteristics as well as the handling instructions of a test equipment. The focus is on the analysis of the suitability to fulfil the planned test task under the specified test conditions. The result is the proof of suitability for the planned or existing test process, see Chap. 15.

The basic investigations – under ideal and typical conditions of use – within the test suitability analysis regarding a test equipment as well as the test process generally refer to the following points: Accuracy, repeatability, reproducibility, linearity, stability (cf. Explanation Chap. 15.2, Fig. 15.3). In general, a distinction is made between the *short-term test equipment capability* (investigation of the test equipment with repeat measurements on the standard) and the *test process capability* (detailed investigation of influencing variables on the measurement result, based on test equipment and test process). The subject area of test equipment capability or test process suitability as well as the associated test equipment capability and test process suitability indices are presented in detail in Chap. 15.

The evaluation of a manufacturing process is carried out using so-called capability indices. *Machine capability indices* are used to examine the capability of a machine with regard to the realization of a technical feature, Sect. 16.3.1. The point of time regarding the procedure of a machine capability analysis is prior to the use of the same within series production. The *preliminary process capability indices* is used to investigate the realized manufacturing process within a pilot series before the time of Start of Production (SOP), i.e. the start of series production, but under near-series conditions, Sect. 16.3.1.

Within the long-term process capability study, the *process capability indices* C_p and C_{pk} are determined and interpreted (cf. Sects. 16.3.2 and 16.3.3). The manufacturing process can be assessed with the aid of these characteristic values. The state of the art is the calculation of univariate capability parameters which refer to a product characteristic (e.g. component length or diameter). Process characteristics can also be analysed which allow an indirect inference to be drawn about the product characteristic (e.g.: analysis of a welding process duration as a contribution to drawing conclusions about the quality of the welding spot). Thus, characteristic-related statements can be made on the basis of the associated manufacturing process. As a rule, a selection of characteristics of a product is examined within the scope of process capability analyses. This selection refers in particular to function-critical product or process characteristics in order to be able to ensure process capability while at the same time taking economic aspects into account.

The *long-term process capability study* focuses on the analysis of process control and quality capability in relation to the specification. It includes the analysis of all influencing factors affecting the manufacturing process. Essentially, the process capability investigation comprises the following elements:

1. Analysis of process parameters with regard to product realization,
2. Elimination of all systematic process influences,
3. Minimization of random process scatter,
4. Centering of the process position on the setpoint of the specification,
5. Minimization of testing expenses.

16.1.2 Inspection Strategy: Full Inspection versus Sampling Inspection Procedure

The statistical process control as well as the analyses for the process capability investigation are carried out in the current series in most cases on the basis of random samples. Depending on the number of products to be manufacutred and the type of product characteristic (function/safety-critical features), a full inspection (100% inspection) can also be carried out. In a full inspection, all units of a series production are inspected with regard to the inspection characteristics. In addition to the complete monitoring (and thus, if necessary, the complete sorting of rejects), a full inspection also allows an additional, statistical analysis of the process in order to obtain further knowledge about the manufacturing process – for example, about mean, dispersion, trends, jumps – in addition to the actual sorting inspection (full inspection). This allows conclusions to be drawn about possible systematic, organisational or technical influences and thus a process correction. For example, tool wear can be identified by detecting a trend in relation to the measured values. If, in addition to the full inspection, the measured data are analyzed statistically, the trend can still be determined within the given specification limits and the process can be corrected before a specification limit is exceeded. Without statistical trend detection, the process would only be intervened in when the first rejects are sorted out.

A full inspection is often planned, especially for safety-critical features. However, this can only be performed with non-destructive testing methods. For example, the following safety-critical characteristics or processes from the automotive industry are monitored using a full inspection:

1. Bolting of a chassis,
2. Mounting of brake calipers,
3. Bolting of wheels,
4. Seam of an airbag bag,
5. Geometry of a brake disc.

Summary
A full inspection is applied within series production with regard to the following conditions:

1. Manufacturing processes in which there is neither process control nor quality capability; ergo, there is no robust process.
2. A full test may be provided for safety-critical features which may cause critical failures and failure symptoms.
3. A full inspection is useful for single part and small batch production (also in the context of prototype construction).
4. It is economically feasible to carry out a full audit and the added information value as opposed to a sample audit is desirable.
5. The use of non-destructive testing methods in a full-scale test is mandatory.

16.1 Basics and Definitions

If an inspection can be carried out on the basis of random samples within a sampling inspection strategy (if necessary also within a statistical process control), the following advantages usually result:

1. The use of destructive testing methods is possible; a supplementary information content in addition to non-destructive testing methods is also possible.
2. Application to components/assemblies in series production is possible, for which the application purpose is single use (example: airbag release device).
3. Inspection lots are usually directly available or can be controlled from production.
4. Use of fewer inspection personnel compared to a full inspection.
5. The inspection content of random inspections is often more complex. This results in higher motivation; the qualification of the personnel is more comprehensive.
6. As a rule, auditing on the basis of random samples is more economical than full auditing.

16.2 Planning and Realization of SPC

This chapter deals with the introduction of statistical process control (SPC) within manufacturing. Planning and implementation take place during series preparation, while application takes place during the production phase (cf. product emergence process; Chap. 2). The following elementary planning and implementation steps are explained in Sect. 16.2.1:

1. Analysis of components and assemblies as well as associated production processes with regard to the use of SPC as a test and control strategy.
2. Determination of the function-critical test characteristics for the identified components.
3. Planning and definition of SPC control loops within individual production processes.
4. Execution of test process, machine as well as process capability analyses.
5. Continuous monitoring and analysis of the production process using SPC.

16.2.1 Procedure

The following explanations describe the elementary planning and implementation steps with regard to SPC.

Step 1: Analysis of components and assemblies as well as associated production processes with regard to the use of SPC as a test and control strategy.
First, the entire scope of the component that is (is to be) produced is analyzed. This includes in particular the component design, functional characteristics, safety-critical characteristics, process characteristics and quality characteristics.

It should be noted that assemblies or systems can also be selected with regard to characteristics to be monitored (e.g. assembly dimensions or connection dimensions). Furthermore, the planned (or already implemented) manufacturing processes are to be analyzed, this includes in particular the process flow, the test strategy, test planning and control loops. Depending on the importance of the components, as well as the already planned measures for feature validation (example: Poka Yoke measures for failure avoidance), the selection of function-critical components and associated manufacturing processes is available after step one.

Step 2: Definiton of the function-critical test characteristics for the identified components.
Step two involves defining the function-critical test characteristics and the associated test strategy for the identified components. Basic test strategies can be: sampling, 100% or skip-lot test methods. It should be noted that process characteristics can also be selected to indirectly ensure the assurance of functionally critical component characteristics.

The result of step two is the definition of the characteristics to be monitored as well as the corresponding inspection strategies, cf. also VDI/VDE/DGQ 2619 (VDI 1985).

Step 3: Planning and definition of SPC control loops within individual production processes.
Step three comprises the planning and implementation of the inspection steps in machine-oriented and process chain-oriented control loops. The components of a SPC control loop include the controlled system (e.g., machine, interlinked machines, production line), the time at which the sample is taken, the controller including inspection equipment and inspection process (e.g., analysis of sample and key indicators), and the control and manipulated variable for process control (e.g., adjustment of tool or tool change).

Step 4: Carrying out test process suitability, machine and process capability investigations.
Within step four, the analysis and assessment of the manufacturing process within the planned control loop takes place. First of all, a test process suitability respectively capability study is carried out, as this is the precondition for the subsequent machine capability study. Capable test processes as well as capable machines are in turn the precondition for the subsequent process capability investigation. The following procedure is recommended in detail:

Step 4.1: Performing the test process capability study. Capable test processes (indices respectively indicators C_g; C_{gk}/ %GRR/ Q_{MS}, Q_{MP}; cf. Chap. 15) are a precondition for the application and assessment of measurement and test procedures, new equipment when setting up a production line, new equipment when upgrading existing processes, as well as for the decision on the final series release and assessment of process modifications.

Step 4.2: Decision on the use of test and measuring equipment. Test process suitability analyses are used as acceptance criteria for the procurement of workpiece-related test equipment. Furthermore, they form the basis for the use of universal measuring devices.

Step 4.3: Carrying out the machine capability study (german: Maschinenfähigkeitsuntersuchung (MFU)) for the release of the planned manufacturing process. The capability of machines is assessed with the characteristic values C_m and C_{mk}.

Step 4.4: Definition of the inspection plan depending on the capable measuring equipment and machines. The exact sequence of an inspection planning is carried out, for example, according to VDI/VDE/DGQ 2619 (VDI 1985).

Step 4.5: Conduct preliminary process capability studies (german: vorläufige Prozessfähigkeitsuntersuchung (PFU)). Under near-series conditions, the preliminary process capability test is carried out and the characteristic values for the short-term process capability P_p and P_{pk} are determined. The result is an assessment of the capability of the manufacturing process with regard to being able to produce components within the given specification under near-series conditions.

Step 4.6: After proving the preliminary process capability for all manufacturing processes required for the realization of the component, a final definition of the SPC control loops is made before the start of production (SOP); e.g.: machine-internal, machine-near, production line-oriented or production-internal control loop (cf. step 3).

Step 5: Continuous monitoring of the production process using SPC.
The monitoring of a production process is based on an analysis of the value patterns, statistical evaluations and on the basis of the capability indices C_p and C_{pk}. In the case of machine adaptations, the analysis is carried out by means of the characteristic values C_m and C_{mk} and P_p and P_{pk} respectively. After the start of series production and the achievement of the production line (planned number of pieces per time unit (e.g. hour, day, week)), long-term process capability tests (PFU) are carried out at regular intervals in the ongoing series production process. The regular intervals are defined on the basis of the number of pieces to be produced and the sample sizes to be analyzed. The results of step 5 are regular long-term process capability tests which, with the aid of the capability indices (C_p, C_{pk}), value patterns and statistical analyses, allow the current series to be assessed with regard to process control and quality capability.

In the case of adaptations of machines during the current series, it is planned to carry out machine capability tests (MFU) and preliminary process capability tests (PFU) during the implementation.

16.2.2 Summary: Scope of Analysis, Indices and Notes

Table 16.1 shows a summary of the different ability tests and guideline values for the underlying sample size(s) and number of ability indices; cf. also (Schulze 2014).

Comments on recommended values concerning the scope of investigations
The mentioned inspection scopes (sample size and number) are to be understood as guideline values and are based, among other things, on empirical values in the industrial context, e.g. (Dietrich and Schulze 2005) and (VDA 1996). Of course, the scope of testing depends on the component to be manufactured and the underlying technology, which is illustrated by the following three examples.

Example A: A manufacturer produces car body parts, the number of pieces is 400 parts per day. The recommended inspection size of n = 125 parts with sample sizes n = 5, distributed over 5 days (thus 5 samples/day) is suitable.

Example B: A screw manufacturer produces in a continuous production line at a rate of 2 screws/second (approximately 57,600 screws per day). It is recommended to increase the sample size and to take samples on different days so that all potential systematic influences can be detected. For example, 20 production days are recommended as a basis for evaluating process control and quality capability.

Example C: A manufacturer produces 5 interior parts per day in a small series. The earliest possible statement on the long-term capability of the production process can theoretically only be made after 5 weeks. The time for a process correction could be too late. If necessary, the statistical analysis should be carried out on smaller scopes (with ongoing full inspection) in order to obtain statements about possible process changes at an early stage. Furthermore, the use of parameter-free

Table 16.1 Overview of capability tests in series preparation and series production: standard values for the scope of tests and capability indices

Capability Assessment	Scope of analysis	Capability indices
Test process capability	Adjustment standard; different testers/workpieces	C_g, C_{gk}; %GRR; Q_{MS}, Q_{MP}
Machine capability	Minimum (total): ≥ 50 parts in a row	C_m, C_{mk}
Preliminary process capability	Minimum size (total): ≥ 100 parts; Minimum number of samples: ≥ 20	P_p, P_{pk}
Long-term process capability	Minimum size (total): ≥ 125 parts; Minimum sample size: n = 3, ..., 5 and/or: 20 days of production	C_p, C_{pk}

significance tests (cf. Chap. 13) is recommended in order to assess small sample sizes without distribution and to be able to detect changes in the centroid and dispersion of the measurement series at an early stage.

16.3 Machine and Process Analysis: Capability Indices

Every capability study of machines or production processes presupposes that the test equipment and test processes used have been assessed as capable (cf. Sect. 16.2; Step 1). The performance of test process capability investigations and the analysis of the capability indices C_g, C_{gk}; %GRR; Q_{MS}, Q_{MP}; are discussed in detail in Chap. 15.

The result of capability studies (machine capability, preliminary process capability and long-term process capability) is the assessment of machines and manufacturing processes. In addition to graphical and statistical analyses, the evaluation is based on the calculation of capability indices (cf. Sects. 16.3.1 and 16.3.2). The method for calculating the capability indices is identical for all capability studies (with the exception of test equipment capability and test process suitability). The difference is, on the one hand, the point of time of the capability investigation within the product emergence process (cf. Fig. 16.2) and, on the other hand, the sample size investigated and the number of samples taken.

16.3.1 Machine Capability and Preliminary Process Capability

A *machine capability analysis* (german: *Maschinenfähigkeitsuntersuchung (MFU)*) is a short-term investigation on machines and plants regarding the suitability related to the fulfilment of a given component specification; cf. also (Bracke 2016). Basis of the evaluation is a single sample (recommended sample size $n \geq 50$) from the manufactured components. The background for the investigation of only one sample is as follows: Since the evaluation of the machine is the main focus, all potential further influences (systematic or random) are avoided during the MFU: The manufacturing parameters should be constant during the sample production; e.g. machine at operating temperature, no temperature jumps, no shift changes. Ergo, the time phase in which the sample is produced should be short in order to avoid systematic influences due to daily processes and day/night changes.

The machine capability test is therefore the basis for the approval decision during machine acceptance. The machine capability test is planned and carried out before the manufacturing process is set up, i.e. at the time of series preparation in the product development process (see Fig. 16.2). The machine capability test can also be useful during the ongoing series production, e.g. for machine adaptations within the scope of maintenance measures (tool or control exchange, etc.) or when using a new machine.

The calculation of the *machine capability indices C_m and C_{mk}* is carried out in the same way as the determination of the process capability indices C_p and C_{pk} (cf.

Sect. 16.3.2). Ergo, please refer to Sects. 16.3.2 and 16.3.3 for a detailed explanation of the quotients, the selection of the C_{mk} minimum and the procedure in the case of non-normally distributed measured values.

$$C_m = \frac{USL - LSL}{6\hat{\sigma}} \tag{16.1}$$

$$C_{mk} = min\left\{\frac{USL - \hat{\mu}}{3\hat{\sigma}}; \frac{\hat{\mu} - LSL}{3\hat{\sigma}}\right\} \tag{16.2}$$

Parameters:

USL = Upper Specification Limit
LSL = Lower Specification Limit
$\hat{\sigma}$ = estimator for the standard deviation
$\hat{\mu}$ = estimator for the mean value

With the application of a *preliminary process capability study*, a manufacturing process is evaluated: The analysis presupposes the capability of the machines and, in contrast to the MFU, is carried out under near-series conditions. The expression "close to series" here refers to the consideration of the typical influencing variables from series production (e.g.: shift change, weekend, temperature change; cf. Chap. 15; Fig. 15.1). The aim is to assess the short-term process capability, among other things on the basis of the *preliminary process capability indices P_p and P_{pk}* (Eqs. 16.3 and 16.4). The calculation of the indices P_p and P_{pk} is analogous to the determination of the process capability indices C_p and C_{pk} (Sect. 16.3.2). The detailed explanation of the quotients, the selection of the P_{pk} minimum and the procedure in the case of non-normally distributed measured values are explained in Sects. 16.3.2 and 16.3.3. The result is the assessment of a manufacturing process with regard to its ability to meet the requirements of the component to be manufactured under near-series conditions.

$$P_p = \frac{USL - LSL}{6\hat{\sigma}} \tag{16.3}$$

$$P_{pk} = min\left\{\frac{USL - \hat{\mu}}{3\hat{\sigma}}; \frac{\hat{\mu} - LSL}{3\hat{\sigma}}\right\} \tag{16.4}$$

Parameters:

USL = Upper Specification Limit
LSL = Lower Specification Limit
$\hat{\sigma}$ = estimator for the standard deviation
$\hat{\mu}$ = estimator for the mean value

16.3.2 Process Capability in the Current Series

With the *long-term process capability analysis* (german: *Langzeit-Prozessfähigkeitsuntersuchung (PFU)*) manufacturing processes are assessed after the

16.3 Machine and Process Analysis: Capability Indices

start of production (SOP); Fig. 16.2 and (Bracke 2016). The calculation of the *process capability indices* C_p and C_{pk} is done in the same way as for the machine capability indices C_m and C_{mk} as well as the characteristic values P_p and P_{pk} of the preliminary process capability. Ergo, reference is made to the explanation of the quotients in the present chapter in the respective sections. The capability index C_p (c = capability; p = process) represents, in relation to a certain product characteristic, the ratio of the given product characteristic specification and the scatter of the manufacturing process, cf. Eq. 16.5. The manufacturing process scatter is defined as six times the standard deviation, assuming a normal distribution of the measured values. The characteristic value C_{pk} (Eq. 16.6); k = katayori (presumably taken from the Japanese language, corresponds to bias or offset) additionally takes into account the manufacturing process location (estimator mean) in comparison to the capability index C_p. The manufacturing process scatter is defined as the six-fold standard deviation.

$$C_p = \frac{USL - LSL}{6\hat{\sigma}} \tag{16.5}$$

$$C_{pk} = min\left\{\frac{USL - \hat{\mu}}{3\hat{\sigma}}; \frac{\hat{\mu} - LSL}{3\hat{\sigma}}\right\} \tag{16.6}$$

Parameters:

USL = Upper Specification Limit
LSL = Lower Specification Limit
$\hat{\sigma}$ = estimator for the standard deviation
$\hat{\mu}$ = estimator for the mean value

The process capability index C_{pk}, like the capability index C_p, relates the given product characteristic specification to the scatter of the manufacturing process, but includes both the upper and lower specification limits in relation to the mean value. The critical C_{pk} capability index is the smaller value (Eq. 16.6), since the manufacturing process is approaching this specification limit and a higher reject rate can possibly be expected here due to the manufacturing process scatter. Equations 16.5 and 16.6 inevitably show that the index C_{pk} is lower than the index C_p for features specified on two sides. This is not the case for unilaterally specified characteristics; a detailed explanation can be found in Sect. 16.3.3.

The term manufacturing process scatter (corresponds to six times the standard deviation) is based on the assumption that the measured values follow a normal distribution model (classical approach). The quotient of specification limits and manufacturing process dispersion is thus directly related to the expected reject probability, see Sect. 16.3.4 and Table 16.2. Even if the measured values follow an arbitrary distribution model (cf. Sect. 16.3.3), a direct relationship between capability index and expected reject proportion applies. Thus, a correlation to the expected reject rate can also be established in multivariate capability analyses.

Table 16.2 Characteristic values of capability indices, probabilities and potential reject rate on the basis of a capability study; precondition: normally distributed measured values

Capability index	Specification in relation to process dispersion	Quotient specification to process scatter width	Probability: Characteristic within specification	Potential reject rate [ppm]
0.667	±2σ	4/6	0.9545	45,500
1.000	±3σ	6/6	0.9973	2700
1.333	±4σ	8/6	0.9999(367)	63.3
1.667	±5σ	10/6	0.9999(995)	0.57
2.000	±6σ	12/6	0.9999(999)	0.002

Example process capability analysis within a case study engine connecting rod

A long-term process capability analysis (PFU) is carried out within the engine connecting rod case study; cf. Sect. 4.1. Two hundred connecting rods are manufactured and a long-term process capability analysis is carried out on the basis of the measured values. The focus is on the final machining process honing to realize the large connecting rod outer diameter (specification: LSL = 64.000 mm; USL = 64.020 mm; nominal value: 64.010 mm). The components were taken from the current series within different random samples, a normal distribution is assumed with regard to the measured values.

The representation of the measured values in value pattern and histogram is shown in Fig. 16.3. The classified measured values, specification limits and a fitted normal distribution model can be seen. Furthermore, the estimator for the location μ (arithmetic mean $\bar{x}$ = 64.00594 mm) as well as the multiple of the estimated standard deviations 6σ (manufacturing process dispersion) can be seen (with s = 3.185 · 10^{-3} mm). Within the 6σ range, assuming a normal distribution model, the manufacturing process measurements are expected to have probability P = 0.9973.

If the area below the normal distribution fit is determined taking into account the given specification limits (Chap. 7), the reject probability can be calculated. In

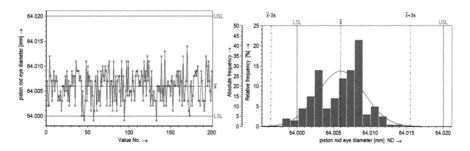

Fig. 16.3 Value pattern (left), histogram with distribution function (right) of a PFU (characteristic "diameter of large piston rod eye", case study engine piston rod)

16.3 Machine and Process Analysis: Capability Indices

the present example (Fig. 16.3), the statistically expected reject rate is P=0.0311 (≙ 3.11%). The calculation of the process capability characteristic values results in $C_p = 1.05$ and $C_{pk} = 0.62$. If $C_p \geq 1.67$ and $C_{pk} \geq 1.33$ were selected as limit values (meaning and interpretation: cf. Sect. 16.3.4, Tables 16.2 and 16.3), the process would be assessed as not capable of quality.

16.3.3 Capability Analysis with Arbitrary Distribution Model: Quantile Approach

Capability analyses are often performed with the assumption that the measured values follow a normal distribution model. However, if the assumption of normally distributed measured values does not apply, the manufacturing process dispersion is mapped via the difference of the quantiles $X_{0.99865}$ and $X_{0.00135}$, which has the same length as the reference length (regarding to P=0.9973) as six times the standard deviation in the case of normally distributed measured values. Therefore, the reference length (manufacturing process dispersion) is comparable, but in the approach using the determination of the quantile $Q_{0.9973}$ independent of the distribution model at hand, cf. Fig. 16.4. This procedure is called the quantile approach; cf. Eqs. 16.7–16.10, DIN ISO 22514-2 (DIN 2019), (ISO 2021) as well as (Schulze 2014). In some literature sources, the term percentile approach is also used for the outlined factual context, which is, however, strictly speaking incorrect, since percentiles (percentile ranks) means the decomposition of the present distribution into one hundred equal shares (1% segments). The quantile approach, however, refers to a quantile that comprises several decimal places.

$$C_p = \frac{USL - LSL}{X_{0.99865} - X_{0.00135}} \qquad (16.7)$$

$$C_{pkU} = \frac{USL - X_{0.5}}{X_{0.99865} - X_{0.5}} \qquad (16.8)$$

$$C_{pkL} = \frac{X_{0.5} - LSL}{X_{0.5} - X_{0.00135}} \qquad (16.9)$$

$$C_{pk} = min\{C_{pkU}; C_{pkL}\} \qquad (16.10)$$

Parameters and quantiles:
$X_{0.00135}$, $X_{0.50}$ and $X_{0.99865}$ = quantiles

Table 16.3 Typical limit value requirements for capability studies (Example)

Machine capability	$C_m \geq 2.0$ and $C_{mk} \geq 1.67$
Preliminary process capability	$P_p \geq 1.67$ and $P_{pk} \geq 1.67$ (1.33)
Long-term process capability	$C_p \geq 1.67$ and $C_{pk} \geq 1.33$

USL = Upper Specification Limit
LSL = Lower Specification Limit

With the aid of the quantile approach, it is also possible to assess manufacturing processes whose measured values follow a left-skewed or right-skewed distribution model, for example. Figure 16.5 shows the expected distribution models with regard to the most important characteristic types. For length measurements, a normal distribution model is generally expected, for form and position tolerances a folded normal distribution or a Rayleigh distribution is expected. Mixed distribution models can also be analyzed. Figure 16.4 shows the corresponding principle by means of a two-peaked distribution model and the corresponding quantile.

The critical process capability index C_{pk} is again the smaller of the two indices C_{pkU} and C_{pkL} in the case of a two-sided specification with a nominal value at the middle of the tolerance (Eq. 16.10).

In the case of non-normally distributed measurement data, and especially in the case of mixed distributions, the subsequently following principles of process control and quality capability must be considered:

When assessing process capability, care should be taken to ensure that the characteristic distribution model, the position and the dispersion do not change, change little or only change due to known causes. This can also apply to normally distributed as well as non-normally distributed measured values, so that in this case one can evaluate the process capability (cf. also explanations in Sect. 16.1). If the statistically expected reject rate is within the agreed limits (expressed by the capability parameters), a quality capability is assumed.

A large part of the technical specifications contains a tolerance, whereby, depending on the technical context, the optimum is not located in the tolerance

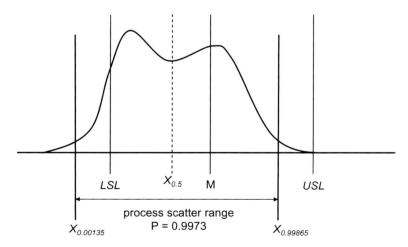

Fig. 16.4 Two-peaked mixture distribution model and manufacturing process dispersion (quantile $Q_{0.9973}$ with $P=0.9973$) versus given specification limits LSL and USL

16.3 Machine and Process Analysis: Capability Indices

centre: The optimum is realized with agreement of the lower or upper specification limit. As a result, the distribution of characteristics in these cases is skewed and/or usually follows the folded normal distribution or the Rayleigh distribution, cf. also Fig. 16.5. Note that for this type of specified characteristics, the capability index C_{pk} is not determined according to Eq. 16.10, but as follows: If the upper specification limit is the optimum, then the C_{pkL} is the critical index; if the lower specification limit is the optimum, then the C_{pkU} is the critical index. In most cases – when the maximum of the skewed distribution approaches the optimum – it will be the larger value of the two calculated indices C_{pkU} and C_{pkL}, since an approximation to the optimum is desired.

The most common case might be a specification where the lower specification limit is the natural limit with value zero and an upper specification limit is given. This is a one-sided tolerance. The optimum is the natural limit with value zero; cf. Fig. 16.5. Examples are the characteristics straightness, parallelism, flatness, concentricity.

Example turning process (machining)
A turned part is inspected with regard to concentricity. The characteristic concentricity has the zero point (ideal) as its natural specification limit, since the measured value only represents the deviation from the ideal. A right-skewed distribution is to be expected, usually a folded normal distribution model.

Example leakage test (assembly wheel rim and tire)
A wheel with pneumatic tyres is tested with regard to the leakage rate ("tyre pressure loss" characteristic). The characteristic leakage rate is specified unilaterally, the ideal is the value 0 ("no leakage"), a deviation up to an upper limit value is permitted. Ergo, it is to be expected that for the majority of the wheels produced with pneumatic tyres the leakage rate – assuming a controlled/quality process – is

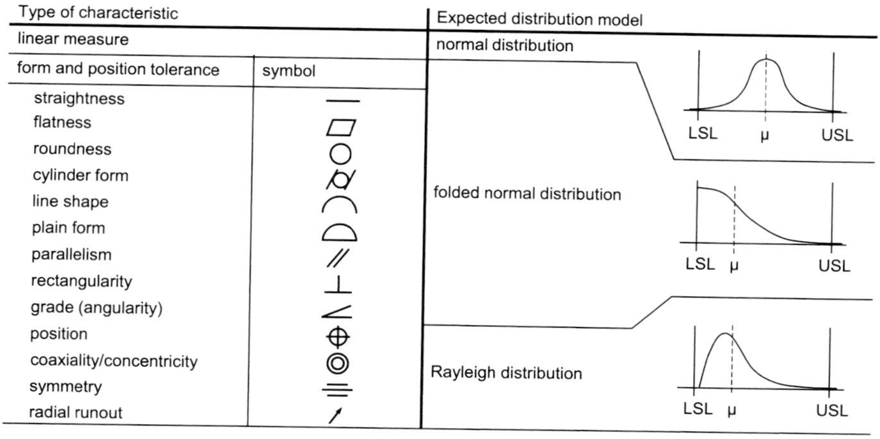

Type of characteristic		Expected distribution model
linear measure		normal distribution
form and position tolerance	symbol	
straightness	—	
flatness	▱	
roundness	○	folded normal distribution
cylinder form	⌭	
line shape	⌒	
plain form	⌓	
parallelism	∥	
rectangularity	⊥	
grade (angularity)	∠	
position	⊕	
coaxiality/concentricity	◎	Rayleigh distribution
symmetry	≡	
radial runout	↗	

Fig. 16.5 Expected distribution models related to different characteristic types

approximately zero, but for no pneumatic tyre is it completely zero. The distribution of the measured values will most likely follow a right skewed log normal distribution model. Note: If the leakage rate were zero, there would be complete tightness; it would be a pneumatic tyre which theoretically would only need to be inflated once.

Example spot welding
In welded constructions, the geometries of spot welds can be specified on one side: The specification refers, for example, to a minimum diameter of the spot weld. A specification of the maximum spot weld diameter is omitted, since practice shows that the manufactured spot welds achieve the minimum specification (and thus strength), but no realized spot weld has a diameter with an unreasonably high value. Here, too, the logarithmic distribution model can be expected with regard to the distribution of measured values, but by an offset with respect to the abscissa corresponding to the minimum diameter.

16.3.4 Capability Analysis and Interpretation of Process Images

Typical values of capability indices related to a process analysis are shown in Table 16.2. The reference to the 6σ process dispersion applies only to the case of normally distributed characteristics. Some capability indices also correspond to generally accepted limit value requirements in an industrial context (cf. also Table 16.3); the values are based on the quotient consisting of an integer multiple of the specification related to the manufacturing process dispersion and the process variation itself (six times the standard deviation); cf. column 3 of Table 16.2. Since the quotients for calculating the machine capability, preliminary and long-term process capability are identical, the statements of Table 16.2 are also the same for all capability analyses.

Table 16.3 (column 1) shows typical limit value requirements for capability analyses. It should be noted that the limit value requirements shown have grown historically over decades. In part, limit values related to a manufacturing process are defined independently by the user depending on the process capability requirement. If several manufacturers are involved, the limit values are agreed between the product manufacturer and the supplier. In the case of machine capability and long-term process capability, the capability index which includes the process distribution location parameter (C_{mk}, C_{pk}) is set lower than the pure assessment of process variation (C_m, C_p). Thus, a slight process offset with respect to the nominal value is taken into account, and the limit value requirement is correspondingly lower and thus the potentially expected reject rate is higher. This fact was not taken into account in the definition of the analysis of the preliminary process capability – for historical reasons – so that different limit value requirements exist depending on the technical discipline. In the industrial application context, the requirements are agreed upon individually (e.g. the quotient of specification to process spread is 9/6). In general, however, the limit value requirement is highest for machine capability, since only the machine is assessed here. As the number of expected influences increases, the limit value requirement is at a correspondingly lower level. The long-term process assessment refers to the most comprehensive period of time and therefore contains the most influences, consequently the limit value requirement is the lowest.

Process Analysis: Data Visualization and Interpretation of Patterns
The long-term evaluation of manufacturing processes is carried out on the one hand via the process capability indices and on the other hand via distribution models related to the analysis of successive samples. The combination of both evaluation possibilities is meaningful, since a sufficient process capability index does not necessarily mean that a manufacturing process is controlled. The present section shows principle possibilities of process evaluation. The actual limit values of the process capability indices to be complied with depend on the industrial discipline as well as the component under consideration and may deviate from each other.

When evaluating the long-term process capability of manufacturing processes, frequently recurring or common patterns emerge, which are briefly outlined below. Figures 16.6–16.9 show the visualization of different process sequences on the basis of successive samples. In each case, a distribution model was fitted to the measured values of the samples; the figures show the distribution models as well as the specification limits of the analyzed characteristic. Sample (A) in Fig. 16.6 shows a controlled and quality process: It is centered to the specification center, the feature distribution model does not change in centroid and dispersion. Pattern (B) in Fig. 16.6 and pattern (C) in Fig. 16.7 also show controlled and quality capable processes under the condition that the systematic influences are known. In

Fig. 16.6 Possible patterns (A) and (B) within a process capability analysis

Fig. 16.7 Possible patterns (C) and (D) within a process capability analysis

case of pattern (B), the process mean position is subject to minor jumps and the dispersion is constant. Since the distribution model statistically does not suggest any rejects, the process can be judged as quality capable. For pattern (C) a trend can be seen, the centroid shifts continuously, the dispersion remains constant. If the causing influence is known (e.g. tool wear), the process can be corrected accordingly (e.g. tool replacement and adjustment) so that the specification limit is not exceeded. The pattern (D) in Fig. 16.7 also shows a trend of the centroid with constant dispersion, which is however more pronounced compared to pattern (C): Both the first and the last sample show an overshoot of the distribution model with respect to the specification limits. Therefore, although the process is controlled when the cause of the trend is known, it is not quality capable. Sample (E) in Fig. 16.8 shows a controlled but not quality capable process. The characteristic distribution model is repeated in centroid and dispersion, it may be assumed that the process is controlled. However, the dispersion is too high in relation to the specification limits, so that a continuously high proportion of overshoot is evident. Pattern (F) in Fig. 16.8 shows a manufacturing process which is neither controlled nor quality capable: The characteristic distribution model changes both in centroid and dispersion. Furthermore, the distribution model also changes (change between normal distribution model and mixed distribution model). A prognosis with regard to a potential reject rate of the subsequent batch is not feasible. The same applies to pattern (G) in Fig. 16.9: distribution model as well as centroid and dispersion change. Although there is no statistically expected reject rate until the last observation point of time. But a prognosis with regard to the following samples (distribution model and parameters) cannot be made either. The process is not controlled.

Fig. 16.8 Possible patterns (E) and (F) within a process capability analysis

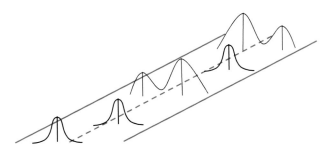

Fig. 16.9 Possible pattern (G) within a process capability analysis

16.3 Machine and Process Analysis: Capability Indices

The process could be judged as quality capable, but no prognosis regarding the quality capability can be made. In this respect, a direct process analysis for the detection and elimination of the process influences is recommended.

Notes: Assessment of capability indices and process distribution model.
A combined consideration of process capability indices and distribution models is mandatory; two examples in the following illustrate this.

Example 1 The calculation of the capability indices results in $C_p = 1.8$ and $C_{pk} = 1.5$. The typical limit value requirement of $C_p \geq 1.67$ and $C_{pk} \geq 1.33$ is thus fulfilled (cf. Table 16.3). If the results of the distribution model analysis can be assigned to pattern (B) over different, successive points in time, then the process may be assessed as controlled and of quality capable if the systematic influences are known.

Example 2 The calculation of the capability indices yield $C_p = 1.71$ and $C_{pk} = 1.55$. The typical threshold requirement of $C_p \geq 1.67$ and $C_{pk} \geq 1.33$ is thus fulfilled (cf. Table 16.3). However, if the analyses of the distribution models based on the successive samples yield a pattern that resembles pattern (G), then it would not be possible to predict the process trajectories going forward; even if the process has not been shown to be controlled to date, but it has been shown to be capable of quality. Ergo, jumps of the process medium position respectively changes of the distribution model should be examined and eliminated if necessary.

16.3.5 Case Study: Capability Analysis of a Machining Process

A machine capability study (MFU; Sect. 16.3.1) is carried out within the engine piston rod case study; cf. Sect. 4.1. 200 piston rods are manufactured within a sample, the focus is on the final machining process honing for the realization of the large connecting rod outer diameter; cf. Table 4.1 (The series of measured values serves as the basis for a machine capability study in this chapter). The representation of the measured values in a histogram is shown in Fig. 16.10. Table 16.4 shows the characteristic description, the determined indicators and parameters of the sample as well as the machine capability indices C_m and C_{mk} calculated from them.

First of all, a normal distribution model can be assumed with regard to the distribution of the measured values (an analysis could be provided, for example, with the Kolmogorov–Smirnov goodness-of-fit test, cf. Chap. 13). The machine capability indices are determined according to Eqs. 16.1 and 16.2; however, both indices do not meet the usual limit requirements according to Table 16.3. The characteristic value C_m already indicates that the process variation is too high in relation to the specification (cf. Fig. 16.10 and Table 16.4). In fact, four rejects are present in relation to the LSL, but the expected reject fraction based on the adjusted normal distribution model is 3112 ppm with respect to the upper and lower specification limits (Table 16.4). The capability indices C_m and C_{mk} are not

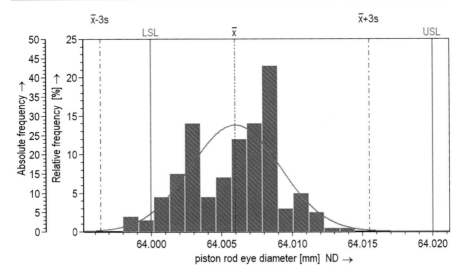

Fig. 16.10 Machine capability study using the example of the honing manufacturing process, feature "diameter connecting rod eye" within the case study engine piston rod

close to each other, which allows the conclusion that there is an offset between the process position and the target value (middle of the tolerance). Accordingly, a machine optimization would focus on both a reduction of the dispersion and a centering (correction of the offset).

16.4 Multivariate Process Evaluation

The assessment of machines and production processes can be carried out via the analysis of capability indices. The state of the art provides for the calculation of univariate machine or or process capability indices: A capability index, e.g. the process capability indices C_p or C_{pk}, is determined in relation to a function-critical product characteristic (Sect. 16.3). In the manufacture of a technically complex product, however, several function-critical characteristics can also be evaluated, i.e. a process capability index can be determined on the basis of a set of characteristics. The explanations of the present chapter regarding the multivariate process evaluation represent a brief overview of the explanations in Bracke (2016). The procedure and the most important approaches were revised and summarized for this chapter.

If a statement regarding a manufacturing process is to be made on the basis of several function-critical characteristics, two paths can be considered (Bracke 2016):

16.3 Machine and Process Analysis: Capability Indices

Table 16.4 Results of a machine capability study, honing machining process. Characteristic diameter of piston rod eye; cf. data set Table 4.1

Item no. No.	Specification	Designation	Case study connecting rod
1	Target value (Set point)	S	64.010
2	Upper specification limit	USL	64.020
3	Lower specification limit	LSL	64.000
4	Tolerance	T	0.020
5	Tolerance center	T_M	64.010
No.	Characteristic value, designation	Characteristic value, formula symbol	Example, Case study piston rod
1	Sample size	n	200
2	Minimum value	x_{min}	63.999 mm
3	Maximum value	x_{max}	64.014 mm
4	Measured values within specification	n_T	196
5	Measured values above USL	$n_{>USL}$	0
	Probability based on normal distribution model: Measured values > USG	$p_{>USL}$	0.0000051
6	Measured values below LSL	$n_{<LSL}$	4
	Probability based on normal distribution model: Measured values < LSL	$p_{<LSL}$	0.0311099
7	Arithmetic mean	$\bar{x}$	64.00594 mm
8	Median	$\tilde{x}$	64.0060 mm
9	Standard deviation	s	0.00318667 mm
10	Variance	s^2	0.00001 mm²
11	Span (Range)	R	0.015 mm
Machine Capability Indices			
1	Scatter/tolerance ratio	C_m	1.05
2	Scatter/tolerance ratio, inclusion of bias	C_{mk}	0.62

1. A univariate capability characteristic value can be calculated for each function-critical product characteristic, so that the statement on the overall process validation is based on a set of capabiliy indices (capabilities of the individual manufacturing processes).
2. An approach is chosen in which a characteristic value represents the process capability in relation to several characteristics. The capabilities of individual manufacturing processes are therefore combined under one characteristic index. This multivariate characteristic capability index then refers to a characteristic set of function-critical product characteristics.

For the calculation of multivariate process capability index there is currently no industry or normative standard. This chapter explains approaches for determining multivariate indices as an analogue to the standardized, univariate process capability index. The aim is the multivariate process capability assessment with a process capability index on the basis of a product characteristic set. The approaches are outlined, a detailed presentation with case studies can be found in Bracke (2016).

In Sect. 16.4.1, the starting point of the approaches presented here for the determination of multivariate process capability indices is presented: The relationship between distribution model of the measurement data and statistically expected scrap probability within the manufacturing process related to the specification. Section 16.4.2 describes the determination of a multivariate process capability index based on the estimation of a reject probability. For the estimation of the reject probability (respectively the reject rate) related to several characteristics of a characteristic set, three approaches are outlined.

16.4.1 Multivariate Process Capability Index

The starting point for determining a process capability on the basis of several important, function-critical characteristics is the correlation between the exceedance proportion (related to the specification) and the process capability indices C_p or C_{pk}, as it results from a classical, univariate analysis. In a univariate capability analysis, a distribution model is first fitted to the measurement data. The determination of the probability with respect to a reject percentage can be done by integrating the distribution model, with the feature specification representing the integration limits. Each capability index is thus assigned a reject probability (cf. Table 16.2). Figure 16.11 shows this relationship between reject probability (with adjusted normal distribution model) and the process capability index: The higher the capability index (abscissa, dimensionless), the lower the reject percentage to be expected (ordinate, [ppm]). Using an adjusted exponential equation, the direct, functional relationship between capability index and reject rate can be represented, cf. Eq. 16.11 with the parameters β_0 (Eq. 16.14) and β_1 (Eq. 16.15). If the probability of rejection (or the proportion of rejects) is known, the associated capability index can be determined directly via the functional relationship with Eq. 16.12, which is obtained after rearranging Eq. 16.11.

16.4 Multivariate Process Evaluation

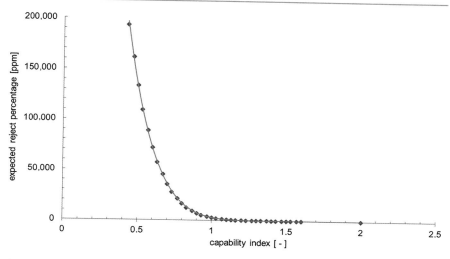

Fig. 16.11 Functional relationship between capability index and statistically expected reject percentage for a centered process

On the basis of this direct correlation between the statistically expected exceedance proportion (y) and the capability index (x), several function-critical characteristics of a product can in principle be assessed as a conglomerate: First, the expected exceedance proportion (y) of several function-critical characteristics is determined. Subsequently, Eq. 16.12 is used to calculate for the analyzed feature set the capability index (x) is determined directly; as an analogy to the univariate approach.

The designation C_{pm} is used for the multivariate capability index and the designation p is used for the associated reject probability (p = process; m = multivariate). Thus, the multivariate process capability index C_{pm} is obtained directly from Eq. 16.12 using the parameters given in Eqs. 16.14 and 16.15. Note that, strictly speaking, this is an approximation, since the assumption is a normal distribution model regarding the measured values of the analysed characteristics. If another distribution model is used to represent the distribution of the measured values (e.g. log-normal distribution), the procedure outlined can be carried out analogously.

$$y = \beta_0 \exp(\beta_1 \exp(x)) \tag{16.11}$$

$$x = \ln\left(\frac{\ln\left(\frac{y}{\beta_0}\right)}{\beta_1}\right) \tag{16.12}$$

$$C_{pm} = \ln\left(\frac{\ln\left(\frac{p}{\beta_0}\right)}{\beta_1}\right) \tag{16.13}$$

Parameters:

$$\beta_0 = 4.61153 \cdot 10^7 \tag{16.14}$$

$$\beta_1 = -3.54913 \tag{16.15}$$

16.4.2 Reject Probability Regarding Characteristic Sets

The starting point for deciding which of the approaches shown is suitable for multivariate process evaluation is an analysis of the dependency of the characteristics within the present set of product characteristics. The dependency can be examined by means of correlation coefficient: For normally distributed measurement data, the Bravais-Pearson correlation coefficient can be used. If the measured values do not follow a specific distribution model, the dependency can be analyzed using the correlation coefficient according to Spearman (cf. Chap. 10).

Approach 1: Combination of rejection probabilities – addition law
If the characteristics of the characteristic set are largely independent, the rejection probability can be determined on the basis of all characteristics of the characteristic set using the addition law of the probability calculation (Chap. 5). The reject probability based on the individual characteristic is obtained from Eq. 16.16. A density function is first fitted to the measured values of each characteristic, if necessary tested with a goodness-of-fit test. The approach is thus suitable for arbitrary distribution models. The integration (with specification limits as integral limits) yields the probability P_i that a component is within specification, the complement yields the reject probability $P_{A,i}$. Afterwards, the reject probability P_{Ages} related to the feature set is determined with the addition law (Eq. 16.17). The advantage of this procedure is that each feature can follow any distribution model, only process control should be present. Furthermore, this approach is mathematically simple compared to the other methods. A disadvantage is that there is no possibility of parameter interpretation, e.g. with respect to a higher-level distribution model, and thus for technical interpretation. It should also be noted that, strictly speaking, the relationship outlined in Sect. 16.4.1 requires that all characteristics of the characteristic set follow one type of distribution model.

$$P_{A,i} = 1 - P_{i=1...n}(x) = 1 - \int_{LSL}^{USL} f(x_i) dx \tag{16.16}$$

$$P_{Ages}(x) = 1 - \prod_{i}^{n} (1 - P_i(x)) \tag{16.17}$$

Approach 2: Multivariate distribution models: Normal distribution or Weibull distribution
Another approach to estimating a reject probability can be done by fitting a multivariate distribution model related to the product characteristic set. Each dimension of the distribution model corresponds to one characteristic. The precondition

16.4 Multivariate Process Evaluation

for this approach is also the independence of the characteristics to be analyzed. The complement of the integration of the adjusted distribution model related to the specification limits results in the reject probability.

Fitting a multivariate distribution model – explicitly the multivariate Weibull distribution model – certainly involves more complications (as will be discussed later) than combining the rejection probabilities via addition law. However, one advantage could be the interpretation of the parameters of the distribution model in terms of the production process. Such an interpretation is not possible with the addition law approach.

If all characteristics of the feature set follow a normal distribution model (Eq. 16.18), then a multivariate normal distribution model can be used to represent the distribution of the product feature set, cf. Eq. 16.19.

$$f(x) = \frac{1}{\sigma\sqrt{2\pi}} e^{-\frac{1}{2}\left(\frac{x-\mu}{\sigma}\right)^2} \tag{16.18}$$

$$f_X(x_1,\ldots,x_k) = \frac{1}{\sqrt{(2\pi)^k |\Sigma|}} exp\left(-\frac{1}{2}(x-\mu)^T \Sigma^{-1}(x-\mu)\right) \tag{16.19}$$

Parameters:

- $\sigma =$ standard deviation
- $\mu =$ mean value (expected value)
- $\Sigma =$ covariance matrix
- $k =$ number of dimensions
- $T =$ index quantity
- $X =$ $(X_1, \ldots, X_k)^T =$ absolutely continuous random vector

If the characteristics do not follow a normal distribution model, a possible starting point could also be the use of a multivariate Weibull distribution model. The advantage of the multivariate Weibull distribution over the normal distribution model is that the analyzed characteristics (when viewed univariately) do not have to be normally distributed, but can also follow other distribution models. Although the complete description of a multivariate Weibull distribution model does not seem to have been achieved yet, this interesting approach is at least outlined here. Following Rinne (2009), the starting point is an exponential approach, Eq. 16.20, here with two components. The exponential function forms the basis for the Weibull distribution and Eq. 16.21 follows for the density function and Eq. 16.22 for the distribution function.

$$f_i(x_i|\lambda_i) = \lambda_i \exp\{-\lambda_i x_i\}, \quad x_i \geq 0, \lambda_i > 0; i = 1,2 \tag{16.20}$$

$$f_i(y_i|\lambda_i, c_i) = c_i \lambda_i (\lambda_i y_i)^{c_i-1} \exp\{-(\lambda_i y_i)^{c_i}\}, \quad y_i \geq 0, \lambda_i, c_i > 0; i = 1,2 \tag{16.21}$$

$$F(y) = 1 - \exp\left\{-\sum_{s \in S} \lambda_s \cdot max(s_i y_i)\right\} \text{ with } y \geq 0 \tag{16.22}$$

Parameters:

F(y) = probability
λ_s = shock rate
S = set of vectors $(s_1, s_2, \ldots, s_p)$, where the vectors can take either the value $s_j = 0$ or $s_j = 1$

The multivariate Weibull distribution model thus arises from the sum of the various univariate Weibull distributions. The limits of the distribution, the shape and position parameters are incorporated and taken into account. The applicability of this approach is also possible for more than two characteristics.

The creation and thus also the fitting of a multivariate Weibull distribution to measurement data are currently not unambiguous. This is partly due to the fact that there seems to be no clear definition of the multivariate Weibull distribution, which is due to a lack of clarity in the definition of a multivariate exponential function.

A summary of the development of the multivariate Weibull distribution and thus also of the existing, different definitions is provided, for example, by Rinne (2009). Rinne states that the first definitions of multivariate Weibull distributions are not absolutely continuous, but have Weibull distributions as marginal distributions. This flaw has been fixed with more recent definitions. A look at other distributions whose marginal distributions are Weibull distributed, and a possible classification of definitions of multivariate Weibull distributions is provided by Lee (1979), among others. In order to solve various practical problems using multivariate Weibull distributions, well-known definitions were often further developed or adapted. However, this often happened only in the area of a few dimensions, such as for forms of the bivariate Weibull distribution, for more in-depth information see Villanueva et al. (2013), Lee (1979), Lu and Bhattacharyya (1990), Lee and Wen (2010) and Rinne (2009).

Approach 3: Monte Carlo simulation using the multivariate normal distribution
The reject probability can also be determined using Monte Carlo simulation (MC simulation). The advantage of this method is that it is irrelevant whether the characteristics of the feature set are dependent or independent of each other. The detailed procedure is described in Bracke (2016: 75 ff.). In the following, the main steps of the approach are outlined and discussed.

The reject rate or overrun rate with respect to the specification limits is determined by means of a multivariate normal distribution, which is fitted to a previously transformed sample. This sample contains information about characteristics of a product which should lie within certain specification limits and, after modeling the distributions of the individual characteristics and transforming these empirical marginal distributions, allows the construction of the estimated marginal distributions of the multivariate normal distribution. The theoretical foundations and underlying definitions can be taken from Rizzo (2008).

In principle, the simulation as well as the determination of the reject probability based on characteristic sets can be carried out as follows:

16.4 Multivariate Process Evaluation

1. Determine distributions that adequately represent the measured data and fit appropriate probability densities f_{xi} to the samples of the marginal distributions.
2. Determine the function $\Phi = (\Phi_1, \ldots \Phi_q)^T$, with $\Phi_i (X_i) \sim N(0,1)$.
3. Transform all samples S_i of the marginal distributions using Φ and determine the estimators $\widehat{\Sigma}$ and $\widehat{\mu}$ of the multivariate normal distribution from the transformed data.
4. Generating n realizations of the Multivariate Normal Distribution $N\_q\left(\widehat{\mu}, \widehat{\Sigma}\right)$. Where q corresponds to the number of marginal distributions and n should be as large as possible.
5. Transform the specification lower and specification upper bounds of the marginal distribution using Φ and determine the empirical rejection probability of the n realizations obtained in step 4.

The empirical reject probability obtained by simulation should be an approximation of the actual, but estimated, reject probability.

Here, the probability densities of the marginal distributions are modeled so that the samples of the marginal distribution can subsequently be transformed into standard normally distributed samples. A multivariate normal distribution is then estimated from the transformed samples, over which the Monte Carlo simulation is ultimately performed. However, there is no guarantee that the transformed marginal distributions are really multivariate normally distributed. This is a model assumption and is approximately certain to be true. From $x_1, \ldots, x_q$ univariate normally distributed, however, it generally does not necessarily follow that $x = (x_1, \ldots, x_q)$ is also multivariate normally distributed, cf. Hamerle and Fahrmeier (1984).

16.5 Process Visualization and Control Technology: Control Charts

Statistical process control deals with the analysis of manufacturing process data as well as the evaluation and control of the manufacturing process with the help of process capability indices as well as the visualization and analysis by means of value patterns. The fundamentals in the industrial application context were first presented in detail in (Shewhart 1931). For the visualization of value patterns, control charts (or sometimes also called quality control charts (QC charts)) are used. Control charts allow the visualization and analysis of measurement/inspection data of function-critical characteristics of components from the current production as well as elementary sample characteristic values. Control charts can thus be used to detect process changes at an early stage; for example due to systematic influences. On the one hand, systematic changes in the process can already be detected through visualization, on the other hand through the exceeding of mathematically determined warning and intervention limits. This makes it possible to intervene in a changing process in a corrective manner before the component specification limits are exceeded with regard to the measured values and thus before defective parts are produced. There are a number of possibilities for the design of a control chart.

In the context of this chapter the most common combinations of data and characteristic values as well as classical calculations of the limit values for continuous (process orientation or specification orientation) and discrete characteristics are presented.

This chapter is structured as follows: The basics of the design and application of quality control charts (QC charts) are shown in Sect. 16.5.1, followed by the design of quality control charts for continuous characteristics according to Shewhart (with and without specification requirements) (Sect. 16.5.2) and illustrated by an example (Sect. 16.5.3). The design of control charts for discrete characteristics according to Shewhart is shown in Sect. 16.5.4; an exemplary illustration is given in Sect. 16.5.5. The last section deals with the analysis of typical process scenarios by means of control charts, Sect. 16.5.6.

With the outline of the Shewhart control charts, the basic production-technical standard of control charts according to the principles of Shewhart (1931) is presented; cf. also the explanations in (Rinne and Mittag 1995). For more in-depth questions and further design possibilities, (Rinne and Mittag 1995) as well as (Wilrich 1987) or (Linß 2005) are recommend.

16.5.1 Basics: Structure of Control Charts

Control charts respectively quality control charts (QC charts) represent a method for the analysis, evaluation and control of manufacturing processes on the basis of measurement and test data. The use of a quality control chart allows both visually and mathematically (quantitatively) the determination of indicators that a process can no longer be assessed as controlled or quality capable.

The development of a control chart generally follows the following steps:

1. Analysis of the product and manufacturing process characteristics: Definition of the function-critical characteristics (continuous/discrete) with regard to monitoring.
2. Process analysis: Process assessment with regard to known or potential disturbances, determination of distribution models with regard to function-critical characteristics and analysis of process capability.
3. Design and layout of the control chart: decision regarding orientation to the specification limits or to the manufacturing process when determining the control chart type. Determination of the parameters and number of tracks of the control chart as well as calculation of the warning and intervention limits.
4. Verification: Use and adaptation of the quality control chart within the manufacturing process.

As a rule, quality control charts are maintained for function-critical, safety-critical or inspection-relevant characteristics (cf. definition in Chap. 3) in order to identify trends, jumps or periodicities at an early stage, for example, and thus avoid the production of potential rejects. Figure 16.12 shows the basic structure of a control

16.5 Process Visualization and Control Technology: Control Charts

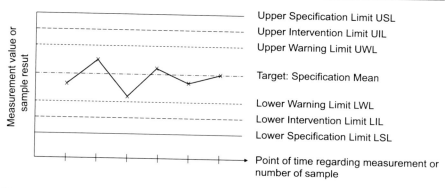

Fig. 16.12 Basic structure of a control chart

chart: The component specification USL/LSL (Upper/Lower Specification Limit), the intervention limits UIL/LIL (upper/lower intervention limit) and the warning limits UWL/LWL (upper/lower warning limit). The structure of the control chart is illustrated using a track (example: original values of the sample). In addition to original values, other tracks, e.g. sample characteristic values (for example, arithmetic mean values or medians of the samples) can also be displayed. If the control chart is to enable control with regard to both process location and process dispersion, two-track or multi-track control charts are used. An overview of the most common control charts is shown in Table 16.5. The original chart is single-track, all other combinations mentioned are double-track control charts. Of course, further combinations are possible with other sampling characteristics.

Table 16.5 Standard control charts in the context of production according to (Shewhart 1931)

Control chart type (continuous characteristics)	Designation
Original value chart	x – card
Arithmetic mean—Standard deviation chart	$(\bar{x} - s)$ – card
Arithmetic mean—Range chart	$(\bar{x} - R)$ – card
Median—Standard deviation	$(\tilde{x} - s)$ – card
Median—range chart	$(\tilde{x} - R)$ – card
Control chart type (discrete characteristics)	Designation
Number of faulty units (process pre-series)	x – card
Proportion of defective units	p – card
Number of nonconforming units (sample)	np – card
Number of failures per unit (several samples)	x – card
Number of failures per unit (sample size constant)	c – card

Determination of warning and intervention limits

The values for the limits UWL/LWL (Upper Warning Limit/Lower Warning Limit) and UIL/LIL (Upper Intervention Limit/Lower Intervention Limit) are determined on the basis of the measurement data with regard to the characteristic to be monitored. The following limits based on the random scatter range are usually used:

Option A) Exact orientation to the process dispersion with $\pm k\sigma$:

1. UWL/LWL map the random scatter range with P=0.9545; this corresponds to $\pm 2\sigma$ for normally distributed measurements.
2. UIL/LIL map the random scatter range with P=0.9973; this corresponds to $\pm 3\sigma$ for normally distributed measurements.

Option B) Approximate orientation to the process dispersion with $\pm k\sigma$:

3. UWL/LWL correspond to the random scatter plot with P=0.95.
4. UIL/LIL correspond to the random scatter range with P=0.99.

Note: The random scatter range is the range in which the value of a random variable lies with a previously defined error probability (Chap. 3). If X_{un} is the lower and X_{ob} the upper limit of the random scattering range and α is the error probability, then the probability P is such that the random variable X lies in the random scattering range. Figure 16.13 shows random scatter ranges $\pm\sigma, \pm 2\sigma$ as well

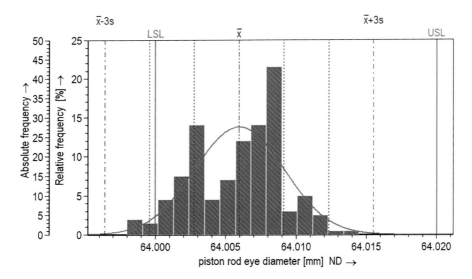

Fig. 16.13 Random scattering ranges $\pm\sigma, \pm 2\sigma$ as well as $\pm 3\sigma$ for normally distributed measured values; data basis engine connecting rod case study, characteristic "diameter of large connecting rod eye" (cf. Sect. 4.1)

as $\pm 3\sigma$ for normally distributed measured values (dashed lines). These limits are added to the value pattern and shown together with the measured values of the current production.

The determination of the control chart limits is carried out according to the following procedure:

1. Definition of the function-critical and/or safety-relevant characteristic.
2. Carrying out a process pre-series: estimation of mean value and standard deviation on the basis of the recorded series of measured values.
3. Determination of the limit values for UWL/LWL/UIL/LIL:
    3.1: Calculation of the limits using probability density or distribution function based on the random scatter range.
    3.2: Graphical determination of the limits from nomograms (e.g.: Binomial distribution/Larson nomogram (Larson 1967), Poisson distribution/Thorndike nomogram (Thorndike 1926), normal distribution model/Wilrich nomogram (Wilrich 1987), which represent the density function and distribution function, respectively.
    3.3: Reading of characteristic values from a corresponding conversion table and use of a simplified set of formulae, cf. (Graebig 2006) and (Graebig 2010).

Note: The options 3.2 and 3.3 (step 3) are no longer up-to-date, since the use of commercially available computers and appropriate software allows exact calculation in a simple way. Nevertheless, the diagrams are mentioned here because statistical process control became the industry standard in the 1970s and 1980s. Accordingly, the literature of these and immediately following years explicitly refers to nomograms and tables, since computers were not available with the same capacity or to the same performance as today and thus a calculation was not easily possible in industrial practice.

16.5.2 Control Charts for Continuous Characteristics

Two common designs of control charts for continuous characteristics are presented in this chapter: Shewhart cards without specification default (hereafter called Shewhart control charts) and Shewhart charts with specification default (hereafter called acceptance control charts); cf. (Shewhart 1931). Both map designs were developed by Shewhart, but the aforementioned names, Shewhart control chart and acceptance control chart, have become accepted. While in the case of the acceptance control chart the intervention and warning limits are defined on the basis of the component specification limits, the definition of the intervention and warning limits in the case of the Shewhart chart is based exclusively on the process sequence. The goal of using a Shewhart chart is thus the process improvement (process centering and reduction of process scatter) and thus a component realization in the range of the specification center (with two-sided specification). The lower the process scatter, the tighter the Shewhart limits. The goal of the acceptance control chart is also a process improvement (process centering and reduction of process scatter), but for the purpose of a wide use of the specification range, which usually results in an optimized use of the specification and thus a positive effect on the process costs.

For more detailed questions and further design possibilities of control charts example: EWMA chart (Exponentially Weighted Moving Average), CUSUM chart (Cumulative SUM) cf. (Rinne and Mittag 1995).

The Acceptance Control Chart
Acceptance control charts are used with the goal of monitoring a process with regard to the specification limits of a characteristic. The basis for determining the intervention and warning limits are the given specification limits, for example component specifications (component geometries such as length, diameter, etc.), assembly specifications (leakage rate, etc.) or process specifications (torque, angle of rotation, etc.). Accordingly, the primary goal is the optimal use of the given specification; exceeding the specification limits is to be prevented by the use of the UIL/LIL. Process improvement by minimizing the spread or centering of the process is also the goal. Figure 16.14 shows the schematic diagram for determining the intervention limits when designing an acceptance control chart: The distribution of the process mean values is set in relation to the upper and lower specification limits. A reduction of the process dispersion leads to a convergence of the intervention limits to the specification limits and thus to the possibility of using a large part of the specification range. The laws for determining the limits relating to mean chart (Eqs. 16.23 and 16.24), median chart (Eqs. 16.25 and 16.26), and original value chart (Eqs. 16.27 and 16.28) are given below. Information is required on the sample size n, current failure proportion p (exceeding proportion), intervention probability $(1 - P_a)$ and the corresponding quantiles of the standard normal distribution u (cf. Appendix Table A.2). The parameter c_n represents the quotient of the standard deviation of the medians of the samples of size n and the standard deviation of the arithmetic mean values of the samples of size n with normally distributed measured values.

Acceptance Control Chart process location: Mean value chart

$$UIL = USL - \left(u_{1-p} + \frac{u_{1-P_a}}{\sqrt{n}}\right) \cdot \widehat{\sigma} \qquad (16.23)$$

$$LIL = LSL + \left(u_{1-p} + \frac{u_{1-P_a}}{\sqrt{n}}\right) \cdot \widehat{\sigma} \qquad (16.24)$$

Acceptance Control Chart process location: Median value chart

$$UIL = USL - \left(u_{1-p} + \frac{c_n \cdot u_{1-P_a}}{\sqrt{n}}\right) \cdot \widehat{\sigma} \qquad (16.25)$$

$$LIL = LSL + \left(u_{1-p} + \frac{c_n \cdot u_{1-P_a}}{\sqrt{n}}\right) \cdot \widehat{\sigma} \qquad (16.26)$$

16.5 Process Visualization and Control Technology: Control Charts

Acceptance Control Chart process location: Original value chart

$$UIL = USL - \left(u_{1-p} - u_{\sqrt[n]{P_a}}\right) \cdot \hat{\sigma} \tag{16.27}$$

$$LIL = LSL + \left(u_{1-p} - u_{\sqrt[n]{P_a}}\right) \cdot \hat{\sigma} \tag{16.28}$$

Shewhart Control Chart

The use of the Shewhart control chart presupposes that the manufacturing process can be assessed as robust and that there is process control or quality capability: The characteristic distribution does not change or changes only in a known manner (ergo: systematic failures are eliminated as far as possible); the statistically expected reject rate is below the required limit values. This condition should be maintained and anomalies should be detected at an early stage. The primary goal

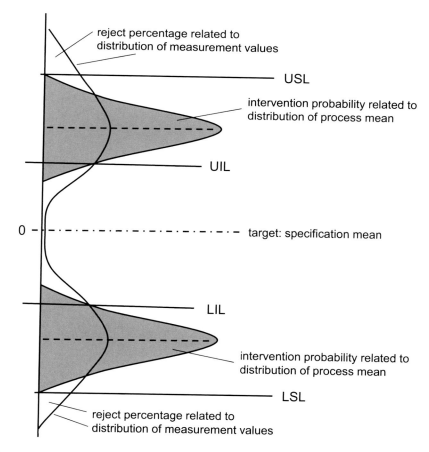

Fig. 16.14 Schematic diagram for determining the intervention limits with orientation to the upper and lower specification limits (USL and LSL) when designing an acceptance control chart

is therefore the continuous improvement of the manufacturing process with optimized centroid and low process dispersion. The determination of the limits UWL/LWL and UIL/LIL is therefore based on the manufacturing process and not on the optimal use of the given specification. Since the manufacturing process has already been judged to be controlled and capable of quality, conformity with the specification limits is assumed.

The basis for determining the intervention or warning limits are the estimated values for the mean value (process centroid) and standard deviation of the measurement series of a function-critical characteristic. Since the intervention or warning limits are dependent on the process location and scatter (cf. random scatter range), the component or specification limits are not part of the determination. Since the quality capability of the process should be a precondition, the calculated intervention or warning limits lie within the component or specification limits.

The control chart usually comprises two tracks for process location and for process dispersion. The process location (as the first track) can be assessed with arithmetic mean, median or directly via the original values: Based on the process location, the warning and intervention limits are determined using Eqs. 16.29–16.40. A normal distribution model is assumed, the two-sided selected random scatter range (Ex: UIL/LIL corresponds to $P=0.99$; UWL/LWL corresponds to $P=0.95$) is taken into account via the failure probability α – which depends on the selected random scatter range – with $u_{1-\alpha/2}$ as a standardized quantity of the normal distribution. Ergo, the equations for the UIL and UWL or LWL and LIL are identical, but are shown separately here for completeness. The parameter c_n refers to the normalized standard deviation related to different sample sizes (cf. acceptance chart).

In contrast to the acceptance control chart, the intervention and warning limits are determined exclusively on the basis of the characteristic values mean value and dispersion of the process; the specification limits are not taken into account in the calculation. Ideally, a process improvement would mean a shift of the intervention/warning limits to the specification midpoint. Thus, the component realization would be oriented to the specification midpoint and the specification range would not be used to its full extent.

Shewhart Control Chart process location: mean value chart

$$UIL = \mu + u_{1-\alpha/2} \cdot \frac{\hat{\sigma}}{\sqrt{n}} \qquad (16.29)$$

$$UWL = \mu + u_{1-\alpha/2} \cdot \frac{\hat{\sigma}}{\sqrt{n}} \qquad (16.30)$$

$$LWL = \mu + u_{1-\alpha/2} \cdot \frac{\hat{\sigma}}{\sqrt{n}} \qquad (16.31)$$

$$LIL = \mu + u_{1-\alpha/2} \cdot \frac{\hat{\sigma}}{\sqrt{n}} \qquad (16.32)$$

16.5 Process Visualization and Control Technology: Control Charts

Shewhart Control Chart process location: Median chart

$$UIL = \mu + u_{1-\alpha/2} \cdot \frac{c_n \cdot \hat{\sigma}}{\sqrt{n}} \tag{16.33}$$

$$UWL = \mu + u_{1-\alpha/2} \cdot \frac{c_n \cdot \hat{\sigma}}{\sqrt{n}} \tag{16.34}$$

$$LWL = \mu + u_{1-\alpha/2} \cdot \frac{c_n \cdot \hat{\sigma}}{\sqrt{n}} \tag{16.35}$$

$$LIL = \mu + u_{1-\alpha/2} \cdot \frac{c_n \cdot \hat{\sigma}}{\sqrt{n}} \tag{16.36}$$

Shewhart Control Chart process location: Original value chart

$$UIL = \mu + u_{(\sqrt[n]{1-\alpha})} \cdot \hat{\sigma} \tag{16.37}$$

$$UWL = \mu + u_{(\sqrt[n]{1-\alpha})} \cdot \hat{\sigma} \tag{16.38}$$

$$LWL = \mu - u_{(\sqrt[n]{1-\alpha})} \cdot \hat{\sigma} \tag{16.39}$$

$$LIL = \mu - u_{(\sqrt[n]{1-\alpha})} \cdot \hat{\sigma} \tag{16.40}$$

The process dispersion is often monitored in the second track of the control chart and can be assessed with standard deviation or range. Based on the process dispersion, estimated on the basis of the measured values from the process run-up, the warning and intervention limits are determined according to Eqs. 16.41–16.48. The dispersion behaviour of the standard deviation is described by the χ^2-distribution (Appendix, Table A.5) with parameter degree of freedom f (f = n − 1). The two-sided random dispersion range (Ex: UIL/LIL corresponds to P = 0.99; UWL/LWL corresponds to P = 0.95) is taken into account via the failure probability α. Since α is not specified here, the equations for the UIL and UWL or LWL and LIL are identical, but are shown separately for completeness. The factor $w_{1-\alpha/2}$ represents the standardized sample span; i.e. the quotient of the span R_n and the standard deviation σ_t of the measured value distribution related to the random variable X.

Shewhart Control Chart process dispersion: standard deviation chart

$$UIL = \sqrt{\frac{\chi^2_{f;1-\alpha/2}}{f}} \cdot \hat{\sigma} \tag{16.41}$$

$$UWL = \sqrt{\frac{\chi^2_{f;1-\alpha/2}}{f}} \cdot \hat{\sigma} \tag{16.42}$$

$$LWL = \sqrt{\frac{\chi^2_{f;1-\alpha/2}}{f}} \cdot \hat{\sigma} \qquad (16.43)$$

$$LIL = \sqrt{\frac{\chi^2_{f;1-\alpha/2}}{f}} \cdot \hat{\sigma} \qquad (16.44)$$

Shewhart Control Chart process dispersion: range chart

$$UIL = w_{n;1-\alpha/2} \cdot \hat{\sigma} \qquad (16.45)$$

$$UWL = w_{n;1-\alpha/2} \cdot \hat{\sigma} \qquad (16.46)$$

$$LWL = w_{n;\alpha/2} \cdot \hat{\sigma} \qquad (16.47)$$

$$LIL = w_{n;\alpha/2} \cdot \hat{\sigma} \qquad (16.48)$$

16.5.3 Case Study Engine Piston Rod: Acceptance and Shewhart Control Charts

Two control charts – Shewhart chart and acceptance chart – are designed on the basis of the data set of the case study engine connecting rod, characteristic "diameter of large piston rod eye" (cf. Sects. 4.1 and Chap. 16.3.5; Table 16.4). This is based on the assumption that the 200 original values originate from a process pre-series and represent forty samples, each with a size of $n_i = 5$. First, a two-track Shewhart control chart is constructed based on the sample characteristics arithmetic mean and standard deviation (Xbar-S chart) is designed, cf. Fig. 16.12; Table 16.5. The warning limits should include ± 2 s (P=0.9545), the intervention limits ± 4 s (P = 0.9973) of the random scatter range of the process pre-series. For the control chart trace of the arithmetic mean, the limits are based on Eqs. 16.29–16.32: UIL=64.01027 mm; UWL=64.00883 mm; M=64.00594 mm; LWL=64.00305 mm; LIL=64.00161 mm. For the second trace standard deviation of the control chart (Eqs. 16.41–16.44) the warning limits are also defined with ± 2 s, the intervention limits with ± 3 s of the random dispersion range of the process lead (related to the scatter). This results in the following for the limits of the control chart trace for the standard deviation and the center position: UIL=0.006815 mm; UWL=0.005446 mm; M=0.003037 mm; LWL=0.001096 mm; LIL=0.000525 mm. Figure 16.15 shows the two-track Xbar-S control chart according to Shewhart. The orientation of the limits at the process pre-series can be seen: Since the process has an offset from the specification center, the calculated limits are close to the lower specification limit LSL.

16.5 Process Visualization and Control Technology: Control Charts

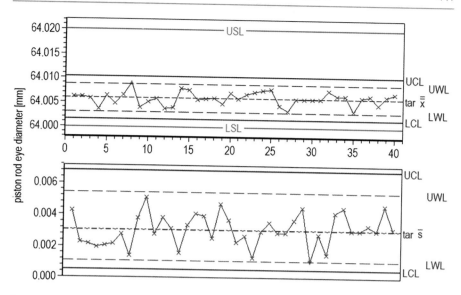

Fig. 16.15 Two-track $\bar{x}$-s control (Shewart) chart within case study engine piston rod, characteristic diameter large piston rod eye

Furthermore, on the basis of the data set (Table 16.4) an acceptance control chart is designed regarding the process location; cf. Fig. 16.16. The limits for the mean chart are based on Eqs. 16.23 and 16.24. Furthermore, an intervention probability of $P=0.9$ is defined. The limits result to: UIL = 64.01804 mm and LIL = 64.00196 mm. The center position results from the specification limits to $M=64.01$ mm. The proximity of the intervention limits UIL/LIL to the specification limits USL/LSL can be seen. The wide range of tolerance use of the acceptance control chart (Fig. 16.16) compared to the Shewhart control chart (Fig. 16.15) becomes visible due to the orientation to the specification limits.

16.5.4 Control Charts for Discrete Characteristics

Control charts for discrete characteristics were also developed by Shewhart (1931). In the context of the present chapter, the following control charts for discrete characteristics are outlined:

1. x-control chart for number of faulty units (process pre-series)
2. p-control chart for the proportion and np-control chart for the number of non-conforming units (samples)
3. x-control chart for number of failures per unit (samples)
4. c-control chart for the number of failures per unit (sample size constant)

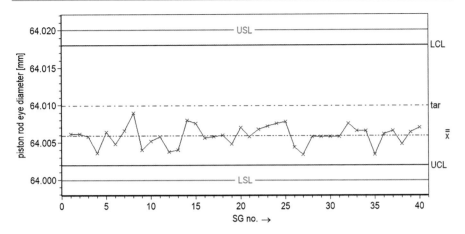

Fig. 16.16 Acceptance control chart process location: mean chart; case study piston rod, characteristic diameter large piston rod eye

Notes:

1. The equations for determination of warning limits UWL/LWL and intervention limits UIL/LIL differ only in the choice of the extent of the random dispersion region γ, with $\alpha = 1 - \gamma$. In this respect, the equations are given in generalized form, α is chosen accordingly to calculate the respective warning or intervention limit. Common values for the random scattering range were presented in Sect. 16.5.1.
2. The calculation of a lower warning or intervention limit is not always meaningful. Often the analyzed characteristics are faulty units, ergo the target value is usually an number of faulty units or the faulty percentage should be zero units. In this respect, a one-sided random dispersion range should be chosen with only an upper warning and/or intervention limit. Accordingly, the failure probability α is not to be halved here, as in the two-sided case.

x-control chart for number of faulty units (process pre-series)
First, the failure proportion is determined with x_i = number of nonconforming units per sample, with n_i = sample size, k = number of samples with Eq. 16.49. Assuming an infinite population with respect to the units to be produced, the intervention and warning limits (Eqs. 16.51 and 16.52) can be determined directly on the basis of the binomial distribution model (Eq. 16.50). Here, α is chosen according to the random dispersion range; e.g. $\alpha = 0.01$ with respect to the intervention limits and $\alpha = 0.05$ with respect to the warning limits. If the binomial distribution model is approximated by a normal distribution model (cf. preconditions in Chap. 7), Eq. 16.53 is used to determine the warning and intervention limits $G_{up/lo}$ (with $u_{1-\alpha/2}$ as the standardized quantity of the normal distribution; Appendix Tab A.2).

16.5 Process Visualization and Control Technology: Control Charts

$$\hat{p} = \frac{\sum_{i=1}^{k} x_i}{\sum_{i=1}^{k} n_i} \tag{16.49}$$

$$F(x) = P(X \leq x) = \sum_{k \leq x} \binom{n}{k} p^k \cdot q^{n-k} \tag{16.50}$$

$$F(x_{up}; n, p) \leq \alpha/2 \tag{16.51}$$

$$F(x_{lo}; n, p) \geq 1 - \alpha/2 \tag{16.52}$$

$$G_{UL/LL} = n\hat{p} \pm u_{1-\alpha/2} \cdot \sqrt{n\hat{p}(1-\hat{p})} \tag{16.53}$$

p-control chart for the proportion and np-control chart for the number of nonconforming units

The focus of a p-control chart refers to the proportion of nonconforming units x_i within a defined sample n_i according to Eq. 16.54. For m samples, the estimator for the mean proportion of nonconforming units is obtained according to Eq. 16.55. To calculate the intervention and warning limits (Eqs. 16.56 and 16.57) of a p-control chart, the normal distribution model is used to approximate the binomial distribution; the k-fold of the standard deviation is used to determine the positions of the limits (e.g. to determine UIL/LIL, k = 3 is chosen for P = 0.9973). Multiple samples form the basis, ergo the average sample size $\bar{n}$ is considered. If the focus of the control is not on the proportion of defective units, but on the number of defective units within an np-control chart, Eqs. 16.58 and 16.59 (approximation via normal distribution; quantiles cf. Appendix Tab A.2) result for calculating the intervention and warning limits G_{LL} and G_{UL}. Here, α is selected in relation to the random dispersion range; e.g. $\alpha = 0.01$ with regard to the intervention limits and $\alpha = 0.05$ with regard to the warning limits.

$$p_i = \frac{x_i}{n_i} \tag{16.54}$$

$$\hat{p} = \frac{x_1 + x_2 + \ldots + x_m}{n_1 + n_2 + \ldots + n_m} = \frac{\sum_{i=1}^{m} x_i}{\sum_{i=1}^{m} n_i} \tag{16.55}$$

$$UIL = \hat{p} + k \cdot \sqrt{\frac{\hat{p}(1-\hat{p})}{\bar{n}}} \tag{16.56}$$

$$LIL = \hat{p} - k \cdot \sqrt{\frac{\hat{p}(1-\hat{p})}{\bar{n}}} \tag{16.57}$$

$$G_{UL} = n\widehat{p} + u_{1-\alpha/2} \cdot \sqrt{n\widehat{p}(1-\widehat{p})} \qquad (16.58)$$

$$G_{LL} = n\widehat{p} - u_{1-\alpha/2} \cdot \sqrt{n\widehat{p}(1-\widehat{p})} \qquad (16.59)$$

x-control chart for number of failures per unit
If the number of defects is related to a unit, the Poisson distribution model forms the basis for determining the warning and intervention limits. Accordingly, there are m samples and x_i failures per unit in each case. A higher number of failures is possible compared to the number of analyzed units (cf. Chap. 7). First, the mean number of failures related to all samples m and the failures per sample x_i are determined. The warning and intervention limits (Eqs. 16.62 and 16.63) can be determined directly from the Poisson distribution model (Eq. 16.61) on the basis of the single parameter mean μ. If the conditions for approximation using the normal distribution model are met (cf. Chap. 7), Eqs. 16.64 and 16.65 can be used to calculate the limits G_{LL} and G_{UL}. Both when using the Poisson distribution model and when approximating by means of the normal distribution model, α is selected in relation to the random dispersion range; e.g. $\alpha = 0.01$ with regard to the intervention limits and $\alpha = 0.05$ with regard to the warning limits.

$$\widehat{\mu} = \frac{\sum_{i=1}^{m} x_i}{m} \qquad (16.60)$$

$$F(x) = P(X \leq x) = e^{-\mu} \cdot \sum_{k \leq x} \frac{\mu^k}{k!} \qquad (16.61)$$

$$F(x_{LL}; \mu) \leq \alpha/2 \qquad (16.62)$$

$$F(x_{UL}; \mu) \geq 1 - \alpha/2 \qquad (16.63)$$

$$G_{UL} = \mu + u_{1-\alpha/2} \cdot \sqrt{\mu} \qquad (16.64)$$

$$G_{LL} = \mu - u_{1-\alpha/2} \cdot \sqrt{\mu} \qquad (16.65)$$

c-control chart for the number of failures per unit
The statement of a c-control chart refers to the number of defective parts per units x_i within a defined sample n_i. Since the number of failures per unit is the focus, the Poisson distribution model forms the basis – as in the case of the x-control chart explained earlier. For m samples, the estimator for the mean defective proportion is obtained according to Eq. 16.66. To calculate the intervention and warning limits (Eqs. 16.67 and 16.68) of a c-control chart, the normal distribution model is used to approximate the Poisson distribution; whereby the positions of the limits

16.5 Process Visualization and Control Technology: Control Charts

G_{UL} and G_{LL} are determined via the k-fold of the standard deviation (e.g. to determine the UIL and LIL, k = 3 is chosen for P = 0.9973).

$$\hat{\mu} = \frac{\sum_{i=1}^{m} x_i}{m} \quad (16.66)$$

$$G_{UL} = \mu + k \cdot \sqrt{\mu} \quad (16.67)$$

$$G_{LL} = \mu - k \cdot \sqrt{\mu} \quad (16.68)$$

16.5.5 Case Study Grey Cast Iron Component Production: Design of a Control Chart

Let there be a primary forming production process in which housing components are manufactured from grey cast iron. A process run with n = 200 parts is carried out, an analysis results in 12 defective parts. A control chart (x-control chart) is to be designed with regard to a subsequent process monitoring.

Interpretation of an x-control chart
An x-control chart for the number of defective units is designed. In terms of the manufacturing process, a reject rate with p = 0 would be ideal. Therefore, only an upper warning limit UWL and upper intervention limit UIL are calculated. The expected defect fraction is estimated using information from the pre-production process according to Eq. 16.69. It is a discrete characteristic, so a binomial distribution model is used as a baseline to determine UWL and LIL, Eq. 16.70. UWL is expected to cover P = 0.95, UIL is expected to cover P = 0.99 of the random dispersion range (one-sided, above). Equation 16.70 cannot be resolved by n, but the cumulative probabilities $F_{UWL}(x)$ can be calculated manually, so that the value P = 0.95 is approximated. The approximation is based on the fact that it is a discrete characteristic and it is not possible to determine exactly the cumulative probability P = 0.95 with respect to an integer set of components. Therefore, for UWL with x = 18, Eq. 16.71 is obtained. For UIL, P = 0.99 related to the one-sided random dispersion range is chosen, Eq. 16.72 is obtained. Accordingly, the limit values sought are obtained according to Eqs. 16.73 and 16.74.

$$\hat{p} = \frac{x}{n} = \frac{12}{200} = 0.06 \quad (16.69)$$

$$F_{UWL}(x) = P(X \leq x) = \sum_{k \leq x} \binom{n}{k} p^k \cdot q^{n-k} = 0.95 \quad (16.70)$$

$$F_{LWL}(x = 18) = P(X \leq x = 18) = \sum_{k \leq x} \binom{200}{k} 0.06^k \cdot (1 - 0.06)^{200-k}$$
$$= 0.9672 \quad (16.71)$$

$$F_{UIL}(x=20) = P(X \leq x = 20) = \sum_{k \leq x} \binom{n}{k} \widehat{p}^k \cdot (1-\widehat{p})^{n-k} = 0.9907$$
(16.72)

$$UWL = 18 \tag{16.73}$$

$$UIL = 20 \tag{16.74}$$

16.5.6 Analysis of Typical Process Scenarios by Means of Control Charts

With the aid of control charts (respectively quality control chart (QCC)) based on measurement and test data, manufacturing processes can be analyzed, evaluated and corrected. By using the control chart, signals can be indicated both visually (qualitatively) and computationally (quantitatively) in the event that a process can no longer be assessed as being under control or quality capable (cf. explanations in Sect. 16.1). Random and/or systematic influences can be detected and interpreted on the basis of the visualization as well as by analyzing the time series behind them. The present chapter shows an overview of schematic diagrams of control charts with common process scenarios and a short interpretation of the same. Furthermore, statistical significance tests are pointed out, which can be used to examine the significance of the observations regarding the process scenarios.

For the sake of simplicity, the intervention limits are drawn as solid lines, the warning limits as dashed lines, and the center line (relative to the specification) as dash-dotted lines on the schematic diagrams. The time of the measurement is plotted on the abscissa and the measured value or the sample characteristic value is plotted on the ordinate.

Figure 16.17 (left) shows a typical scenario of a controlled and quality capable process. The recorded data follow a regular, symmetrical distribution related to the center line. Both, the intervention limits and the warning limits are not exceeded, or are exceeded only at a few points in time. The randomness of the data sequence can be tested, for example, by means of a significance test according to Wallis and Moore (Sect. 13.3.1). On the other hand, Fig. 16.17 (right) shows

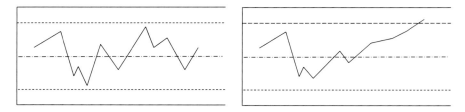

Fig. 16.17 Process scenario of a controlled and quality capable process (left); process scenario trend development (right)

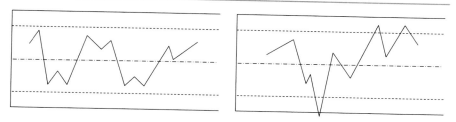

Fig. 16.18 Process scenario Periodicity (left); process scenario Outlier I (right)

a trend behaviour caused by a systematic influence (e.g. tool wear or temperature increase). If the cause of the trend is known, the process can be corrected accordingly when the warning limit is exceeded (e.g. tool change). The analysis of the trend behavior can be done with the trend test according to Cox and Stuart (cf. Sect. 13.3.6). When using an acceptance control chart the objective can be the optimal use of the tool, since the calculation of the intervention limits is based on the specification limits.

Figure 16.18 (left) shows the process scenario of a pattern recurring at regular intervals. Systematic causes for periodicities can be, for example, recurring shift changes – and the associated changing machine/process parameters – or seasonal variations. Indirectly, the application of a significance test for randomness (Wallis and Moore test, cf. Sect. 13.3.1) to investigate periodicity is feasible. Figure 16.18 (right) shows the process scenario Outlier I, in which the intervention limits are exceeded once or several times. The significances of outliers can be determined by appropriate hypothesis tests, e.g. the test for outlier (Sect. 13.3.4). Outliers usually have a (measurement) technical cause (e.g.: incorrect recording of measured values, machine malfunction), so that no unfounded elimination of the outlier should be carried out.

Another process scenario is a conspicuous change of the process position (cf. Fig. 16.9, left), which rarely occurs. Typical causes (systematic influences) are changes of machine parameters or the change of a raw material batch. Proof of the significance of a process location change can be provided by examining an (approximately) equal volume of measurement data before and after the process location change (formation of two random samples) using the Mann–Whitney

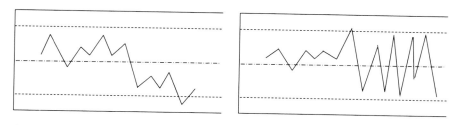

Fig. 16.19 Process scenario location change (left); process scenario dispersion increase (right)

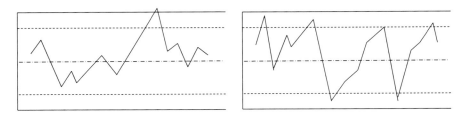

Fig. 16.20 Process scenario Outlier II (left); process scenario Instability (right)

U-test (Sect. 13.4.1). A systematic influence can also lead to a process scenario which is characterized by an increase or decrease in the process scatter. Figure 16.9 (right) shows an increase in process dispersion with no change in process location. The proof of the significance of such a change can be carried out here, for example, by means of the Siegel-Tukey test. The application of the Siegel-Tukey test is based here on the formation of two samples by dividing the measurement data at the point of the scatter widening (Fig. 16.19).

Figure 16.20 (left) shows an inconspicuous process course in the process scenario Outlier II, process position as well as process dispersion do not change, however, a (rarely occurring) outlier can be recognized. In this case – in contrast to the process scenario Outlier I (Fig. 16.18, right – the difficulty of the reproducibility of the measured value and thus the difficulty of a technical cause analysis often becomes apparent. Figure 16.20 (right) shows the process scenario Instability, which is characterized by a high variance of the measured values. In addition to metrological failures, causes can also be, for example, worn tool guides or component fixtures.

Appendix

Table A.1: Standard Normal Distribution

Table A.1 shows the distribution function of the standard normal distribution $\Phi(u)$ with step size: $\Delta u = 0.01$. Accordingly, column 1 of the table refers to the pre-decimal or first decimal place, row 1 to the second decimal place. With negative argument, due to axis symmetry, $\Phi(-u) = 1 - \Phi(u)$; with $u > 0$. For $u \geq 4$, the value of the distribution function $\Phi(u) \approx 1$.

Tables A.2 a, b: Quantiles of the Standard Normal Distribution

Table A.2a contains the associated quantile u_p (one-sided delimitation upwards) for some probabilities P, which are used, for example, when determining confidence intervals or performing significance tests.

P: Probability with $(0 < P < 1)$
u_p: Quantile belonging to the probability P (upper bound)

Furthermore, the following relationships apply:
$u_{1-P} = -u_P$ and $u_P = -u_{1-P}$

Table A.2b contains the associated quantile u_p (one-sided as well as two-sided delimitation) for some probabilities P, which are used, for example, in the determination of confidence intervals or the performance of significance tests.

Table A.3: Gosset ("Student") t-distribution

Table A.3 shows the quantiles of the t-distribution of William Gosset ("Student"). The table contains for special values of P the respective associated quantile $z_{(P;f)}$ depending on the degree of freedom f (one-sided delimitation upwards).

P: Probability; with $0 < P < 1$
f: Number of lines of freedom; with $f = n - 1$
$t_{(P;f)}$: Quantile at f degrees of freedom and probability P (upper bound)

© Springer-Verlag GmbH Germany, part of Springer Nature 2024
S. Bracke, *Reliability Engineering*, https://doi.org/10.1007/978-3-662-67446-8

The following relationships apply:
$t_{(1-P;f)} = -t_{(P;f)}$ and $t_{(P;f)} = -t_{(1-P;f)}$

Tables A.4 a–l: F-distribution

Tables A.4a, A.4b, A.4c, A.4d, A.4e, A.4f, A.4g, A.4h, A.4i, A.4j A.4k–A.4l show the quantiles of Ronald Fisher's F-distribution based on the quotient of two variables following the χ^2-distribution model; depending on the degrees of freedom n_1 and n_2.

Table A.5: χ^2-distribution

Table A.5 contains the quantile $z_{(P;f)}$ of the χ^2-distribution depending on the probability P as well as the degree of freedom f (one-sided upper boundary).
P: Probability with $(0 < P < 1)$
f: Number of degrees of freedom $(f = n - 1)$
$z_{(P;f)}$:   Quantile of the χ^2-distribution (upper bound)

Tables A.6 a–f: U-Test; critical values

Critical values of U for the Mann and Whitney (and Wilcoxon) test for one-sided and two-sided tests and different levels of significance (see Table A.6).

Table A.7: Quantiles of the Kolmogorov–Smirnov test statistic

Table A.7 shows the quantiles k_P (Q k_P) of the Kolmogorov–Smirnov test statistic. The quantile is tabulated depending on the analysed size n of the measurement data as well as the probability P; cf. also (Miller 1956).

Table A.8: Error probabilities for the H-test according to Kruskal and Wallis

Table A.8 shows the error probabilities for the H-test from Kruskal and Wallis (Kruskal and Wallis 1952) and the errata in J. Amer. Statist, Ass. 48 (1953:910), and supplementary values from (Iman et al. 1975); adapted from (Hedderich and Sachs 2020).

Table A.9: Means and Standard Deviations of Reduced Extremes according to Gumbel

Table A.9 shows Means and Standard Deviations of Reduced Extremes according to (Gumbel 1958).

Table A. 10 a, b: Clopper-Pearson

Tables A.10a, A.10b show the confidence intervals (lower bound π_o and upper bound π_o) for the proportion value from a dichotomous population, Clopper-Pearson estimation for confidence level 95%, percentage values with n = sample size and x = number of hits according to Hedderich and Sachs (2020). It holds:
Proportional value: p = x/n
Ex: for p = 20/100 = 0.2 = 20.0%, the upper bound is $\pi_u = 12.67\%$ and the lower bound is $\pi_o = 29.18\%$.

Table A.1 Distribution function of the standard normal distribution $\Phi(u)$ with step size: $\Delta u = 0.01$

u	0	1	2	3	4	5	6	7	8	9
0	0.5000	0.5040	0.5080	0.5120	0.5160	0.5199	0.5239	0.5279	0.5319	0.5359
0.1	0.5398	0.5438	0.5478	0.5517	0.5557	0.5596	0.5636	0.5675	0.5714	0.5754
0.2	0.5793	0.5832	0.5871	0.5910	0.5948	0.5987	0.6026	0.6064	0.6103	0.6141
0.3	0.6179	0.6217	0.6255	0.6293	0.6331	0.6368	0.6406	0.6443	0.6480	0.6517
0.4	0.6554	0.6591	0.6628	0.6664	0.6700	0.6736	0.6772	0.6808	0.6844	0.6879
0.5	0.6915	0.6950	0.6985	0.7019	0.7054	0.7088	0.7123	0.7157	0.7190	0.7224
0.6	0.7258	0.7291	0.7324	0.7357	0.7389	0.7422	0.7454	0.7486	0.7518	0.7549
0.7	0.7580	0.7612	0.7642	0.7673	0.7704	0.7734	0.7764	0.7794	0.7823	0.7852
0.8	0.7881	0.7910	0.7939	0.7967	0.7996	0.8023	0.8051	0.8079	0.8106	0.8133
0.9	0.8159	0.8186	0.8212	0.8238	0.8264	0.8289	0.8315	0.8340	0.8365	0.8389
1	0.8413	0.8438	0.8461	0.8485	0.8508	0.8531	0.8554	0.8577	0.8599	0.8621
1.1	0.8643	0.8665	0.8686	0.8708	0.8729	0.8749	0.8770	0.8790	0.8810	0.8830
1.2	0.8849	0.8869	0.8888	0.8907	0.8925	0.8944	0.8962	0.8980	0.8997	0.9015
1.3	0.9032	0.9049	0.9066	0.9082	0.9099	0.9115	0.9131	0.9147	0.9162	0.9177
1.4	0.9192	0.9207	0.9222	0.9236	0.9251	0.9265	0.9279	0.9292	0.9306	0.9319
1.5	0.9332	0.9345	0.9357	0.9370	0.9382	0.9394	0.9406	0.9418	0.9430	0.9441
1.6	0.9452	0.9463	0.9474	0.9485	0.9495	0.9505	0.9515	0.9525	0.9535	0.9545
1.7	0.9554	0.9564	0.9573	0.9582	0.9591	0.9599	0.9608	0.9616	0.9625	0.9633
1.8	0.9641	0.9649	0.9656	0.9664	0.9671	0.9678	0.9686	0.9693	0.9700	0.9706
1.9	0.9713	0.9719	0.9726	0.9732	0.9738	0.9744	0.9750	0.9756	0.9762	0.9767
2	0.9773	0.9778	0.9783	0.9788	0.9793	0.9798	0.9803	0.9808	0.9812	0.9817
2.1	0.9821	0.9826	0.9830	0.9834	0.9838	0.9842	0.9846	0.9850	0.9854	0.9857
2.2	0.9861	0.9865	0.9868	0.9871	0.9875	0.9878	0.9881	0.9884	0.9887	0.9890
2.3	0.9893	0.9896	0.9898	0.9901	0.9904	0.9906	0.9909	0.9911	0.9913	0.9916
2.4	0.9918	0.9920	0.9922	0.9925	0.9927	0.9929	0.9931	0.9932	0.9934	0.9936
2.5	0.9938	0.9940	0.9941	0.9943	0.9945	0.9946	0.9948	0.9949	0.9951	0.9952
2.6	0.9953	0.9955	0.9956	0.9957	0.9959	0.9960	0.9961	0.9962	0.9963	0.9964
2.7	0.9965	0.9966	0.9967	0.9968	0.9969	0.9970	0.9971	0.9972	0.9973	0.9974
2.8	0.9974	0.9975	0.9976	0.9977	0.9977	0.9978	0.9979	0.9979	0.9980	0.9981
2.9	0.9981	0.9982	0.9982	0.9983	0.9984	0.9984	0.9985	0.9985	0.9986	0.9986
3	0.9987	0.9987	0.9987	0.9988	0.9988	0.9989	0.9989	0.9989	0.9990	0.9990
3.1	0.9990	0.9991	0.9991	0.9991	0.9992	0.9992	0.9992	0.9992	0.9993	0.9993
3.2	0.9993	0.9993	0.9994	0.9994	0.9994	0.9994	0.9994	0.9995	0.9995	0.9995
3.3	0.9995	0.9995	0.9995	0.9996	0.9996	0.9996	0.9996	0.9996	0.9996	0.9997
3.4	0.9997	0.9997	0.9997	0.9997	0.9997	0.9997	0.9997	0.9997	0.9997	0.9998
3.5	0.9998	0.9998	0.9998	0.9998	0.9998	0.9998	0.9998	0.9998	0.9998	0.9998

(continued)

Table A.1 (continued)

u	0	1	2	3	4	5	6	7	8	9
3.6	0.9998	0.9999	0.9999	0.9999	0.9999	0.9999	0.9999	0.9999	0.9999	0.9999
3.7	0.9999	0.9999	0.9999	0.9999	0.9999	0.9999	0.9999	0.9999	0.9999	0.9999
3.8	0.9999	0.9999	0.9999	0.9999	0.9999	0.9999	0.9999	0.9999	0.9999	0.9999
3.9	1.0000	1.0000	1.0000	1.0000	1.0000	1.0000	1.0000	1.0000	1.0000	1.0000

Table A.2a Quantiles for special probabilities (one-sided)

P	u_P	P	u_P
0.900	1.282	**0.100**	−1.282
0.950	1.645	**0.050**	−1.645
0.975	1.960	**0.025**	−1.960
0.990	2.326	**0.010**	−2.326
0.995	2.576	**0.005**	−2.567
0.999	3.090	**0.001**	−3.090

Table A.2b Quantiles for special probabilities (two-sided and one-sided)

1 − P	two-sided	one-sided
0.000001	4.891638	4.753424
0.00001	4.417173	4.264891
0.0001	3.890592	3.720016
0.001	3.290527	3.090232
0.005	2.807034	2.575829
0.01	2.575829	2.326348
0.02	2.326348	2.053749
0.025	2.241400	1.959964
0.03	2.170090	1.880794
0.04	2.053749	1.750686
0.05	1.959964	1.644854
0.06	1.880794	1.554774
0.07	1.811911	1.475791
0.08	1.750686	1.405072
0.09	1.695398	1.340755
0.1	1.644854	1.281552
0.2	1.281552	0.841621
0.3	1.036433	0.524401
0.4	0.841621	0.253347
0.5	0.674490	0.000000

Table A.3 Quantiles of the t-distribution of William Gosset ("Student")

f	P				
	0.9	0.95	0.975	0.99	0.995
1	3.078	6.314	12.707	31.820	63.654
2	1.886	2.920	4.303	6.965	9.925
3	1.638	2,353	3.182	4.541	5.841
4	1.533	2.132	2.776	3.747	4.604
5	1.476	2.015	2.571	3.365	4.032
6	1.440	1.943	2.447	3.143	3.707
7	1.415	1.895	2.365	2.998	3.499
8	1.397	1.860	2.306	2.896	3.355
9	1.383	1.833	2.262	2.821	3.250
10	1.372	1.812	2.228	2.764	3.169
11	1.363	1.796	2.201	2.718	3, 106
12	1.356	1.782	2.179	2.681	3.055
13	1.350	1.771	2.160	2.650	3.012
14	1.345	1.761	2.145	2.624	2.977
15	1.341	1.753	2.131	2.602	2.947
16	1.337	1.746	2.120	2.583	2.921
17	1.333	1.740	2.110	2.567	2.898
18	1.330	1.734	2.101	2.552	2.878
19	1.328	1.729	2.093	2.539	2.861
20	1.325	1.725	2.086	2.528	2.845
22	1.321	1.717	2.074	2.508	2.819
24	1.318	1.711	2.064	2.492	2.797
26	1.315	1.706	2.056	2.479	2.779
28	1.313	1.701	2.048	2.467	2.763
30	1.310	1.697	2.042	2.457	2.750
40	1.303	1.684	2.021	2.423	2.704
50	1.299	1.676	2.009	2.403	2.678
60	1.296	1.671	2.000	2.390	2.660
100	1.290	1.660	1.984	2.364	2.626
200	1.286	1.653	1.972	2.345	2.601
500	1.283	1.648	1.965	2.334	2.586
∞	1.282	1.645	1.960	2.326	2.576

Table A.4a Quantiles of the F-distribution for n_1 within [1,7] and n_2 within [1,8]

n_1 n_2	γ	1	2	3	4	5	6	7
1	0.990	4052.181	4999.500	5403.352	5624.583	5763.650	5858.986	5928.356
	0.975	647.789	799.500	864.163	899.583	921.848	937.111	948.217
	0.950	161.448	199.500	215.707	224.583	230.162	233.986	236.768
	0.900	39.860	49.500	53.590	55.830	57.240	58.200	58.910
2	0.990	98.503	99.000	99.166	99.249	99.299	99.333	99.356
	0.975	38.506	39.000	39.166	39.248	39.298	39.332	39.355
	0.950	18.513	19.000	19.164	19.247	19.296	19.330	19.353
	0.900	8.530	9.000	9.160	9.240	9.290	9.330	9.350
3	0.990	34.116	30.817	29.457	28.710	28.237	27.911	27.672
	0.975	17.443	16.044	15.439	15.101	14.885	14.735	14.624
	0.950	10.128	9.552	9.277	9.117	9.014	8.941	8.887
	0.900	5.540	5.460	5.390	5.340	5.310	5.280	5.270
4	0.990	21.198	18.000	16.694	15.977	15.522	15.207	14.976
	0.975	12.218	10.649	9.979	9.605	9.365	9.197	9.074
	0.950	7.709	6.944	6.591	6.388	6.256	6.163	6.094
	0.900	4.540	4.320	4.190	4.110	4.050	4.010	3.980
5	0.990	16.258	13.274	12.060	11.392	10.967	10.672	10.456
	0.975	10.007	8.434	7.764	7.388	7.146	6.978	6.853
	0.950	6.608	5.786	5.410	5.192	5.050	4.950	4.876
	0.900	4.060	3.780	3.620	3.520	3.450	3.400	3.370
6	0.990	13.745	10.925	9.780	9.148	8.746	8.466	8.260
	0.975	8.813	7.260	6.599	6.227	5.988	5.820	5.696
	0.950	5.987	5.143	4.757	4.534	4.387	4.284	4.207
	0.900	3.780	3.460	3.290	3.180	3.110	3.050	3.010
7	0.990	12.246	9.547	8.451	7.847	7.460	7.191	6.993
	0.975	8.073	6.542	5.890	5.523	5.285	5.119	4.995
	0.950	5.591	4.737	4.347	4.120	3.972	3.866	3.787
	0.900	3.590	3.260	3.070	2.960	2.880	2.830	2.780
8	0.990	11.259	8.649	7.591	7.006	6.632	6.371	6.178
	0.975	7.571	6.060	5.416	5.053	4.817	4.652	4.529
	0.950	5.318	4.459	4.066	3.838	3.688	3.581	3.501
	0.900	3.460	3.110	2.920	2.810	2.730	2.670	2.620

Table A.4b Quantiles of the F-distribution for n_1 within [8,14] and n_2 within [1,8]

n_1	γ	8	9	10	11	12	13	14
n_2								
1	0.990	5981.070	6022.473	6055.847	6083.000	6106.000	6126.000	6143.000
	0.975	956.656	963.285	968.627	973.000	976.700	979.800	982.500
	0.950	238.883	240.543	241.882	243.000	243.900	244.700	245.400
	0.900	59.440	59.860	60.190	60.470	60.710	60.900	61.070
2	0.990	99.374	99.388	99.399	99.410	99.420	99.420	99.430
	0.975	39.373	39.387	39.398	39.410	39.410	39.420	39.430
	0.950	19.371	19.385	19.396	19.400	19.410	19.420	19.420
	0.900	9.370	9.380	9.390	9.401	9.408	9.415	9.420
3	0.990	27.489	27.345	27.229	27.130	27.050	26.980	26.920
	0.975	14.540	14.473	14.419	24.370	14.340	14.300	14.280
	0.950	8.845	8.812	8.786	8.763	8.745	8.729	8.715
	0.900	5.250	5.240	5.230	5.222	5.200	5.180	5.180
4	0.990	14.799	14.659	14.546	14.450	14.370	14.310	14.250
	0.975	8.980	8.905	8.844	8.793	8.751	8.715	8.684
	0.950	6.041	5.999	5.964	5.936	5.912	5.891	5.873
	0.900	3.950	3.940	3.920	3.907	3.896	3.885	3.877
5	0.990	10.289	10.158	10.051	9.962	9.888	9.824	9.770
	0.975	6.757	6.681	6.619	6.568	6.525	6.487	6.455
	0.950	4.818	4.773	4.735	4.704	4.678	4.655	4.636
	0.900	3.340	3.320	3.300	3.282	3.268	3.257	3.247
6	0.990	8.102	7.976	7.874	7.789	7.718	7.657	7.605
	0.975	5.600	5.523	5.461	5.409	5.366	5.329	5.297
	0.950	4.147	4.099	4.060	4.027	4.000	3.976	3.956
	0.900	2.980	2.960	2.940	2.919	2.870	2.840	2.820
7	0.990	6.840	6.719	6.620	6.538	6.469	6.410	6.359
	0.975	4.899	4.823	4.761	4.709	4.666	4.628	4.596
	0.950	3.726	3.677	3.637	3.603	3.575	3.550	3.629
	0.900	2.750	2.720	2.700	2.684	2.668	2.654	2.643
8	0.990	6.029	5.911	5.814	5.734	5.667	5.609	5.558
	0.975	4.433	4.357	4.295	4.243	4.200	4.162	4.129
	0.950	3.438	3.388	3.347	3.313	3.284	3.259	3.237
	0.900	2.590	2.560	2.540	2.500	2.502	2.488	2.475

Table A.4c Quantiles of the F-distribution for n_1 within $[15, \infty)$ and n_2 within $[1, 8]$

n_1		15	20	24	30	40	60	120	∞
n_2	γ								
1	0.990	6157.285	6208.730	6234.631	6260.649	6286.782	6313.030	6339.391	6365.864
	0.975	984.867	993.103	997.249	1001.414	1005.598	1009.800	1014.020	1018.258
	0.950	245.950	248.013	249.052	250.095	251.143	252.196	253.253	254.314
	0.900	61.220	61.740	62.002	62.265	62.529	62.794	63.061	63.328
2	0.990	99.433	99.449	99.458	99.466	99.474	99.482	99.491	99.499
	0.975	39.431	39.448	39.456	39.465	39.473	39.481	39.490	39.498
	0.950	19.429	19.446	19.454	19.462	19.471	19.479	19.487	19.496
	0.900	9.425	9.441	9.450	9.458	9.466	9.475	9.483	9.491
3	0.990	26.872	26.690	26.598	26.505	26.411	26.316	26.221	26.125
	0.975	14.253	14.167	14.124	14.081	14.037	13.992	13.947	13.902
	0.950	8.703	8.660	8.639	8.617	8.594	8.572	8.549	8.526
	0.900	5.170	5.184	5.176	5.168	5.160	5.151	5.143	5.134
4	0.990	14.198	14.020	13.929	13.838	13.745	13.652	13.558	13.463
	0.975	8.657	8.560	8.511	8.461	8.411	8.360	8.309	8.257
	0.950	5.858	5.803	5.774	5.746	5.717	5.688	5.658	5.628
	0.900	3.869	3.844	3.831	3.817	3.804	3.790	3.775	3.761
5	0.990	9.722	9.553	9.466	9.379	9.291	9.202	9.112	9.020
	0.975	6.428	6.329	6.278	6.227	6.175	6.123	6.069	6.015
	0.950	4.619	4.558	4.527	4.496	4.464	4.431	4.399	4.365
	0.900	3.170	3.207	3.191	3.174	3.157	3.140	3.123	3.105

(continued)

Table A.4c (continued)

n_1 n_2	γ	15	20	24	30	40	60	120	∞
6	0.990	7.559	7.396	7.313	7.229	7.143	7.057	6.969	6.880
	0.975	5.269	5.168	5.117	5.065	5.012	4.959	4.904	4.849
	0.950	5.269	5.168	5.117	5.065	5.012	4.959	4.904	4.849
	0.900	2.800	2.836	2.818	2.800	2.781	2.762	2.742	2.722
7	0.990	6.314	6.155	6.074	5.992	5.908	5.824	5.737	5.650
	0.975	4.568	4.467	4.415	4.362	4.309	4.254	4.199	4.142
	0.950	3.511	3.445	3.411	3.376	3.340	3.304	3.267	3.230
	0.900	2.632	2.595	2.575	2.555	2.535	2.514	2.493	2.471
8	0.990	5.515	5.359	5.279	5.198	5.116	5.032	4.946	4.859
	0.975	4.101	4.000	3.947	3.894	3.840	3.784	3.728	3.670
	0.950	3.218	3.150	3.115	3.079	3.043	3.005	2.967	2.928
	0.900	2.464	2.425	2.404	2.383	2.361	2.339	2.316	2.293

Table A.4d Quantiles of the F-distribution for n_1 within [1,7] and n_2 within [9,16]

n_1 n_2	γ	1	2	3	4	5	6	7
9	0.990	10.561	8.022	6.992	6.422	6.057	5.802	5.613
	0.975	7.209	5.715	5.078	4.718	4.484	4.320	4.197
	0.950	5.117	4.257	3.863	3.633	3.482	3.374	3.293
	0.900	3.360	3.010	2.810	2.690	2.610	2.550	2.510
10	0.990	10.044	7.559	6.552	5.994	5.636	5.386	5.200
	0.975	6.937	5.456	4.826	4.468	4.236	4.072	3.950
	0.950	4.965	4.103	3.708	3.478	3.326	3.217	3.136
	0.900	3.290	2.920	2.730	2.610	2.520	2.460	2.410
11	0.990	9.646	7.206	6.217	5.668	5.316	5.069	4.886
	0.975	6.724	5.256	4.630	4.275	4.044	3.881	3.759
	0.950	4.844	3.982	3.587	3.357	3.204	3.095	3.012
	0.900	3.230	2.860	2.660	2.540	2.450	2.390	2.340
12	0.990	9.330	6.927	5.953	5.412	5.064	4.821	4.640
	0.975	6.554	5.096	4.474	4.121	3.891	3.728	3.607
	0.950	4.747	3.885	3.490	3.259	3.106	2.996	2.913
	0.900	3.180	2.810	2.610	2.480	2.390	2.330	2.280
13	0.990	9.074	6.701	5.739	5.205	4.862	4.620	4.441
	0.975	6.414	4.965	4.347	3.996	3.767	3.604	3.483
	0.950	4.667	3.806	3.411	3.179	3.025	2.915	2.832
	0.900	3.140	2.760	2.560	2.430	2.350	2.280	2.230
14	0.990	8.862	6.515	5.564	5.035	4.695	4.456	4.278
	0.975	6.298	4.857	4.242	3.892	3.663	3.501	3.380
	0.950	4.600	3.739	3.344	3.112	2.958	2.848	2.764
	0.900	3.100	2.730	2.520	2.390	2.310	2.240	2.190
15	0.990	8.683	6.359	5.417	4.893	4.556	4.318	4.142
	0.975	6.200	4.765	4.153	3.804	3.576	3.415	3.293
	0.950	4.543	3.682	3.287	3.056	2.901	2.791	2.707
	0.900	3.070	2.700	2.490	2.360	2.270	2.210	2.160
16	0.990	8.531	6.226	5.292	4.773	4.437	4.202	4.026
	0.975	6.115	4.687	4.077	3.729	3.502	3.341	3.219
	0.950	4.494	3.634	3.239	3.007	2.852	2.741	2.657
	0.900	3.050	2.670	2.460	2.330	2.240	2.180	2.130

Table A.4e Quantile F-distribution for n_1 within [8,14] and n_2 within [9,16]

n_1 n_2	γ	8	9	10	11	12	13	14
9	0.990	5.467	5.351	5.257	5.177	5.111	5.054	5.005
	0.975	4.102	4.026	3.964	3.912	3.868	3.830	3.798
	0.950	3.230	3.179	3.137	3.102	3.073	3.047	3.025
	0.900	2.470	2.440	2.420	2.396	2.397	2.364	2.351
10	0.990	5.057	4.942	4.849	4.771	4.706	4.649	4.600
	0.975	3.855	3.779	3.717	3.665	3.621	3.583	3.550
	0.950	3.072	3.020	2.978	2.943	2.913	2.887	2.864
	0.900	2.380	2.350	2.320	2.302	2.284	2.269	2.255
11	0.990	4.744	4.632	4.539	4.462	4.397	4.341	4.293
	0.975	3.664	3.588	3.526	3.470	3.430	3.391	3.358
	0.950	2.948	2.896	2.854	2.818	2.788	2.761	2.738
	0.900	2.300	2.270	2.250	2.227	2.209	2.193	2.179
12	0.990	4.499	4.388	4.296	4.219	4.155	4.099	4.051
	0.975	3.512	3.436	3.374	3.321	3.277	3.239	3.206
	0.950	2.849	2.796	2.753	2.717	2.687	2.660	2.637
	0.900	2.240	2.210	2.190	2.166	2.147	2.131	2.117
13	0.990	4.302	4.191	4.100	4.024	3.960	3.905	3.857
	0.975	3.388	3.312	3.250	3.197	3.153	3.115	3.081
	0.950	2.767	2.714	2.671	2.634	2.604	2.577	2.553
	0.900	2.200	2.160	2.140	2.115	2.097	2.080	2.066
14	0.990	4.140	4.030	3.939	3.863	3.800	3.745	3.697
	0.975	3.285	3.209	3.147	3.094	3.050	3.011	2.978
	0.950	2.699	2.646	2.602	2.565	2.534	2.507	2.483
	0.900	2.150	2.120	2.100	2.073	2.054	2.037	2.022
15	0.990	4.004	3.895	3.805	3.730	3.666	3.611	3.563
	0.975	3.199	3.123	3.060	3.007	2.963	2.924	2.891
	0.950	2.641	2.588	2.544	2.506	2.475	2.448	2.424
	0.900	2.120	2.090	2.060	2.036	2.017	2.000	1.985
16	0.990	3.890	3.780	3.691	3.616	3.553	3.497	3.450
	0.975	3.125	3.049	2.986	2.933	2.889	2.850	2.817
	0.950	2.591	2.538	2.494	2.456	2.425	2.397	2.373
	0.900	2.090	2.060	2.030	2.005	1.985	1.968	1.953

Table A.4f Quantile F-distribution for n_1 within $[15,\infty)$ and n_2 within $[9,16]$

n_1 n_2	γ	15	20	24	30	40	60	120	∞
9	0.990	4.962	4.808	4.729	4.649	4.567	4.483	4.398	4.311
	0.975	3.769	3.667	3.614	3.560	3.505	3.449	3.392	3.333
	0.950	3.006	2.937	2.901	2.864	2.826	2.787	2.748	2.707
	0.900	2.340	2.298	2.277	2.255	2.232	2.208	2.184	2.159
10	0.990	4.558	4.405	4.327	4.247	4.165	4.082	3.996	3.909
	0.975	3.522	3.419	3.365	3.311	3.255	3.198	3.140	3.080
	0.950	2.845	2.774	2.737	2.700	2.661	2.621	2.580	2.538
	0.900	2.244	2.201	2.178	2.155	2.132	2.107	2.082	2.055
11	0.990	4.251	4.099	4.021	3.941	3.860	3.776	3.690	3.602
	0.975	3.330	3.226	3.173	3.118	3.061	3.004	2.944	2.883
	0.950	2.719	2.646	2.609	2.571	2.531	2.490	2.448	2.405
	0.900	2.167	2.123	2.100	2.076	2.052	2.026	2.000	1.972
12	0.990	4.010	3.858	3.780	3.701	3.619	3.535	3.449	3.361
	0.975	3.177	3.073	3.019	2.963	2.906	2.848	2.787	2.725
	0.950	2.617	2.544	2.506	2.466	2.426	2.384	2.341	2.296
	0.900	2.105	2.060	2.036	2.011	1.986	1.960	1.932	1.904
13	0.990	3.815	3.665	3.587	3.507	3.425	3.341	3.255	3.165
	0.975	3.053	2.948	2.893	2.837	2.780	2.720	2.659	2.595
	0.950	2.533	2.459	2.420	2.380	2.339	2.297	2.252	2.206
	0.900	2.053	2.007	1.983	1.958	1.931	1.904	1.876	1.846
14	0.990	3.656	3.505	3.427	3.348	3.266	3.181	3.094	3.004
	0.975	2.949	2.844	2.789	2.732	2.674	2.614	2.552	2.487
	0.950	2.463	2.388	2.349	2.308	2.266	2.223	2.178	2.131
	0.900	2.010	1.962	1.938	1.912	1.885	1.857	1.828	1.797
15	0.990	3.522	3.372	3.294	3.214	3.132	3.047	2.959	2.868
	0.975	2.862	2.756	2.701	2.644	2.585	2.524	2.461	2.395
	0.950	2.403	2.328	2.288	2.247	2.204	2.160	2.114	2.066
	0.900	1.972	1.924	1.899	1.873	1.845	1.817	1.787	1.755
16	0.990	3.409	3.259	3.181	3.101	3.018	2.933	2.845	2.753
	0.975	2.788	2.681	2.625	2.568	2.509	2.447	2.383	2.316
	0.950	2.352	2.276	2.235	2.194	2.151	2.106	2.059	2.010
	0.900	1.940	1.891	1.866	1.839	1.811	1.782	1.751	1.718

Table A.4g Quantile F-distribution for n_1 within [1,7] and n_2 within [17,28]

n_1 n_2	γ	1	2	3	4	5	6	7
17	**0.990**	8.400	6.112	5.185	4.669	4.336	4.102	3.927
	0.975	6.042	4.619	4.011	3.665	3.438	3.277	3.156
	0.950	4.451	3.592	3.197	2.965	2.810	2.699	2.614
	0.900	3.030	2.640	2.440	2.310	2.220	2.150	2.100
18	**0.990**	8.285	6.013	5.092	4.579	4.248	4.015	3.841
	0.975	5.978	4.560	3.954	3.608	3.382	3.221	3.100
	0.950	4.414	3.555	3.160	2.928	2.773	2.661	2.577
	0.900	3.010	2.620	2.420	2.290	2.200	2.130	2.080
19	**0.990**	8.185	5.926	5.010	4.500	4.171	3.939	3.765
	0.975	5.922	4.508	3.903	3.559	3.333	3.172	3.051
	0.950	4.381	3.522	3.127	2.895	2.740	2.628	2.544
	0.900	2.990	2.610	2.400	2.270	2.180	2.110	2.060
20	**0.990**	8.096	5.849	4.938	4.431	4.103	3.871	3.699
	0.975	5.872	4.461	3.859	3.515	3.289	3.128	3.007
	0.950	4.351	3.493	3.098	2.866	2.711	2.599	2.514
	0.900	2.970	2.590	2.380	2.250	2.160	2.090	2.040
22	**0.990**	7.945	5.719	4.817	4.313	3.988	3.758	3.587
	0.975	5.786	4.383	3.783	3.440	3.215	3.055	2.934
	0.950	4.301	3.443	3.049	2.817	2.661	2.549	2.464
	0.900	2.949	2.561	2.351	2.219	2.128	2.061	2.008
24	**0.990**	7.823	5.614	4.718	4.218	3.895	3.667	3.496
	0.975	5.717	4.319	3.721	3.379	3.155	2.995	2.874
	0.950	4.260	3.403	3.009	2.776	2.621	2.508	2.423
	0.900	2.927	2.538	2.327	2.195	2.103	2.035	1.983
26	**0.990**	7.721	5.526	4.637	4.140	3.818	3.591	3.421
	0.975	5.659	4.266	3.670	3.329	3.105	2.945	2.824
	0.950	4.225	3.369	2.975	2.743	2.587	2.474	2.388
	0.900	2.909	2.519	2.307	2.174	2.082	2.014	1.961
28	**0.990**	7.636	5.453	4.568	4.074	3.754	3.528	3.358
	0.975	5.610	4.221	3.626	3.286	3.063	2.903	2.782
	0.950	4.196	3.340	2.947	2.714	2.558	2.445	2.359
	0.900	2.894	2.503	2.291	2.157	2.064	1.996	1.943

Table A.4h Quantile F-distribution for n_1 within [8,14] and n_2 within [17,28]

n_1 n_2	γ	8	9	10	11	12	13	14
17	0.990	3.791	3.682	3.593	3.518	3.455	3.400	3.353
	0.975	3.061	2.985	2.922	2.869	2.825	2.786	2.752
	0.950	2.548	2.494	2.450	2.412	2.381	2.353	2.329
	0.900	2.060	2.030	2.000	1.960	1.958	1.940	1.925
18	0.990	3.705	3.597	3.508	3.433	3.371	3.316	3.268
	0.975	3.005	2.929	2.866	2.813	2.769	2.730	2.696
	0.950	2.510	2.456	2.412	2.374	2.342	2.314	2.290
	0.900	2.040	2.000	1.980	1.953	1.933	1.915	1.900
19	0.990	3.631	3.523	3.434	3.359	3.297	3.241	3.194
	0.975	2.956	2.880	2.817	2.764	2.720	2.680	2.646
	0.950	2.477	2.423	2.378	2.340	2.308	2.280	2.255
	0.900	2.020	1.980	1.960	1.932	1.912	1.894	1.878
20	0.990	3.564	3.457	3.368	3.293	3.231	3.176	3.129
	0.975	2.913	2.837	2.774	2.720	2.676	2.636	2.602
	0.950	2.447	2.393	2.348	2.310	2.278	2.249	2.225
	0.900	2.000	1.960	1.940	1.913	1.892	1.874	1.859
22	0.990	3.453	3.346	3.258	3.183	3.121	3.066	3.019
	0.975	2.839	2.763	2.700	2.646	2.602	2.562	2.528
	0.950	2.397	2.342	2.297	2.258	2.226	2.197	2.172
	0.900	1.967	1.933	1.904	1.880	1.859	1.841	1.825
24	0.990	3.363	3.256	3.168	3.094	3.032	2.977	2.930
	0.975	2.779	2.703	2.640	2.586	2.541	2.501	2.467
	0.950	2.355	2.300	2.255	2.216	2.183	2.154	2.129
	0.900	1.941	1.906	1.877	1.853	1.832	1.813	1.797
26	0.990	3.288	3.182	3.094	3.020	2.958	2.903	2.856
	0.975	2.729	2.653	2.590	2.536	2.491	2.451	2.417
	0.950	2.321	2.266	2.220	2.181	2.148	2.119	2.093
	0.900	1.919	1.884	1.855	1.830	1.809	1.790	1.774
28	0.990	3.226	3.120	3.032	2.958	2.896	2.841	2.794
	0.975	2.687	2.611	2.547	2.493	2.448	2.408	2.374
	0.950	2.291	2.236	2.190	2.151	2.118	2.088	2.063
	0.900	1.900	1.865	1.836	1.811	1.790	1.770	1.754

Table A.4i Quantile F-distribution for n_1 within $[15,\infty)$ and n_2 within $[17,28]$

n_1 n_2	γ	15	20	24	30	40	60	120	∞
17	0.990	3.312	3.162	3.084	3.003	2.920	2.835	2.746	2.653
	0.975	2.723	2.616	2.560	2.502	2.442	2.380	2.315	2.247
	0.950	2.308	2.230	2.190	2.148	2.104	2.058	2.011	1.960
	0.900	1.912	1.862	1.836	1.809	1.781	1.751	1.719	1.686
18	0.990	3.227	3.077	2.999	2.919	2.835	2.749	2.660	2.566
	0.975	2.667	2.559	2.503	2.445	2.384	2.321	2.256	2.187
	0.950	2.269	2.191	2.150	2.107	2.063	2.017	1.968	1.917
	0.900	1.887	1.837	1.810	1.783	1.754	1.723	1.691	1.657
19	0.990	3.153	3.003	2.925	2.844	2.761	2.674	2.584	2.489
	0.975	2.617	2.509	2.452	2.394	2.333	2.270	2.203	2.133
	0.950	2.234	2.156	2.114	2.071	2.026	1.980	1.930	1.878
	0.900	1.865	1.814	1.787	1.759	1.730	1.699	1.666	1.631
20	0.990	3.088	2.938	2.859	2.778	2.695	2.608	2.517	2.421
	0.975	2.573	2.465	2.408	2.349	2.287	2.223	2.156	2.085
	0.950	2.203	2.124	2.083	2.039	1.994	1.946	1.896	1.843
	0.900	1.845	1.794	1.767	1.738	1.708	1.677	1.643	1.607
22	0.990	2.978	2.827	2.749	2.667	2.583	2.495	2.403	2.305
	0.975	2.498	2.389	2.332	2.272	2.210	2.145	2.076	2.003
	0.950	2.151	2.071	2.028	1.984	1.938	1.889	1.838	1.783
	0.900	1.811	1.759	1.731	1.702	1.671	1.639	1.604	1.567
24	0.990	2.889	2.738	2.659	2.577	2.492	2.403	2.310	2.211
	0.975	2.437	2.327	2.269	2.209	2.146	2.080	2.010	1.935
	0.950	2.108	2.027	1.984	1.939	1.892	1.842	1.790	1.733
	0.900	1.783	1.730	1.702	1.672	1.641	1.607	1.571	1.533
26	0.990	2.815	2.664	2.585	2.503	2.417	2.327	2.233	2.131
	0.975	2.387	2.276	2.217	2.157	2.093	2.026	1.954	1.878
	0.950	2.072	1.990	1.946	1.901	1.853	1.803	1.749	1.691
	0.900	1.760	1.706	1.677	1.647	1.615	1.581	1.544	1.504
28	0.990	2.753	2.602	2.522	2.440	2.354	2.263	2.167	2.064
	0.975	2.344	2.232	2.174	2.112	2.048	1.980	1.907	1.829
	0.950	2.041	1.959	1.915	1.869	1.820	1.769	1.714	1.654
	0.900	1.740	1.685	1.656	1.625	1.593	1.558	1.520	1.478

Table A.4j Quantile F-distribution for n_1 within [1,7] and n_2 within [30,∞)

n_1 n_2	γ	1	2	3	4	5	6	7
30	0.990	7.562	5.390	4.510	4.018	3.699	3.473	3.304
	0.975	5.568	4.182	3.589	3.250	3.027	2.867	2.746
	0.950	4.171	3.316	2.922	2.690	2.534	2.421	2.334
	0.900	2.881	2.489	2.276	2.142	2.049	1.980	1.927
40	0.990	7.314	5.179	4.313	3.828	3.514	3.291	3.124
	0.975	5.424	4.051	3.463	3.126	2.904	2.744	2.624
	0.950	4.085	3.232	2.839	2.606	2.450	2.336	2.249
	0.900	2.835	2.440	2.226	2.091	1.997	1.927	1.873
60	0.990	7.077	4.977	4.126	3.649	3.339	3.119	2.953
	0.975	5.286	3.925	3.343	3.008	2.786	2.627	2.507
	0.950	4.001	3.150	2.758	2.525	2.368	2.254	2.167
	0.900	2.791	2.393	2.177	2.041	1.946	1.875	1.819
80	0.990	6.964	4.882	4.036	3.564	3.256	3.037	2.872
	0.975	5.219	3.865	3.285	2.951	2.730	2.571	2.451
	0.950	3.961	3.111	2.719	2.486	2.329	2.214	2.127
	0.900	2.770	2.370	2.154	2.017	1.921	1.849	1.793
120	0.990	6.851	4.787	3.949	3.480	3.174	2.956	2.792
	0.975	5.152	3.805	3.227	2.894	2.674	2.515	2.395
	0.950	3.920	3.072	2.680	2.447	2.290	2.175	2.087
	0.900	2.748	2.347	2.130	1.992	1.896	1.824	1.767
∞	0.990	6.635	4.605	3.782	3.319	3.017	2.802	2.639
	0.975	0.024	3.689	3.116	2.786	2.567	2.408	2.288
	0.950	3.842	2.996	2.605	2.372	2.214	2.099	2.010
	0.900	2.706	2.303	2.084	1.945	1.847	1.774	1.717

Table A.4k Quantile F-distribution for n_1 within [8,14] and n_2 within [30,∞)

n_1 n_2	γ	8	9	10	11	12	13	14
30	**0.990**	3.173	3.067	2.979	2.905	2.843	2.788	2.741
	0.975	2.651	2.575	2.511	2.457	2.412	2.372	2.337
	0.950	2.266	2.211	2.165	2.125	2.092	2.062	2.037
	0.900	1.884	1.849	1.819	1.794	1.773	1.753	1.737
40	**0.990**	2.993	2.888	2.801	2.727	2.665	2.610	2.563
	0.975	2.529	2.452	2.388	2.334	2.288	2.247	2.212
	0.950	2.180	2.124	2.077	2.037	2.003	1.973	1.947
	0.900	1.829	1.793	1.763	1.737	1.715	1.695	1.677
60	**0.990**	2.823	2.718	2.632	2.558	2.496	2.441	2.393
	0.975	2.412	2.334	2.270	2.215	2.169	2.128	2.092
	0.950	2.097	2.040	1.993	1.952	1.917	1.886	1.860
	0.900	1.775	1.738	1.707	1.680	1.657	1.637	1.619
80	**0.990**	2.743	2.639	2.552	2.478	2.416	2.361	2.313
	0.975	2.356	2.278	2.214	2.158	2.112	2.070	2.034
	0.950	2.057	1.999	1.952	1.910	1.876	1.844	1.817
	0.900	1.748	1.711	1.680	1.652	1.629	1.608	1.590
120	**0.990**	2.663	2.559	2.472	2.398	2.336	2.281	2.233
	0.975	2.299	2.222	2.157	2.101	2.055	2.013	1.976
	0.950	2.016	1.959	1.911	1.869	1.834	1.802	1.774
	0.900	1.722	1.684	1.652	1.625	1.601	1.580	1.561
∞	**0.990**	2.511	2.407	2.321	2.247	2.185	2.129	2.080
	0.975	2.192	2.114	2.048	1.992	1.945	1.902	1.865
	0.950	1.938	1.880	1.831	1.788	1.752	1.719	1.691
	0.900	1.670	1.632	1.599	1.570	1.546	1.523	1.504

Table A.4l Quantile F-distribution for n_1 within $[15,\infty)$ and n_2 within $[30,\infty)$

n_1 / n_2	γ	15	20	24	30	40	60	120	∞
30	0.990	2.700	2.549	2.469	2.386	2.299	2.208	2.111	2.006
	0.975	2.307	2.195	2.136	2.074	2.009	1.940	1.866	1.787
	0.950	2.015	1.932	1.887	1.841	1.792	1.740	1.684	1.622
	0.900	1.722	1.667	1.638	1.606	1.573	1.538	1.499	1.456
40	0.990	2.522	2.369	2.288	2.203	2.114	2.019	1.917	1.805
	0.975	2.182	2.068	2.007	1.943	1.875	1.803	1.724	1.637
	0.950	1.925	1.839	1.793	1.744	1.693	1.637	1.577	1.509
	0.900	1.662	1.605	1.574	1.541	1.506	1.467	1.425	1.377
60	0.990	2.352	2.198	2.115	2.028	1.936	1.836	1.726	1.601
	0.975	2.061	1.945	1.882	1.815	1.744	1.667	1.581	1.482
	0.950	1.836	1.748	1.700	1.649	1.594	1.534	1.467	1.389
	0.900	1.603	1.543	1.511	1.476	1.437	1.395	1.348	1.291
80	0.990	2.272	2.116	2.033	1.944	1.849	1.746	1.630	1.491
	0.975	2.003	1.885	1.821	1.753	1.679	1.598	1.507	1.396
	0.950	1.793	1.703	1.654	1.602	1.545	1.482	1.410	1.322
	0.900	1.574	1.513	1.479	1.443	1.403	1.358	1.306	1.242
120	0.990	2.192	2.035	1.950	1.860	1.763	1.656	1.533	1.381
	0.975	1.945	1.825	1.760	1.690	1.614	1.530	1.433	1.310
	0.950	1.750	1.659	1.608	1.554	1.495	1.429	1.352	1.254
	0.900	1.545	1.482	1.447	1.409	1.368	1.320	1.265	1.193
∞	0.990	2.039	1.878	1.791	1.696	1.592	1.473	1.325	1.000
	0.975	1.833	1.708	1.640	1.566	1.484	1.388	1.268	1.000
	0.950	1.666	1.571	1.517	1.459	1.394	1.318	1.221	1.000
	0.900	1.487	1.421	1.383	1.342	1.295	1.240	1.169	1.000

Table A.5 χ^2-distribution

f	P 0.005	0.01	0.025	0.05	0.1	0.9	0.95	0.975	0.99	0.995
1	0.000	0.000	0.001	0.004	0.016	2.71	3.84	5.02	6.63	7.88
2	0.010	0.020	0.051	0.103	0.211	4.61	5.99	7.38	9.21	10.60
3	0.070	0.115	0.216	0.352	0.584	6.25	7.81	9.35	11.35	12.84
4	0.210	0.297	0.484	0.711	1.064	7.78	9.49	11.14	13.28	14.86
5	0.41	0.55	0.83	1.15	1.16	9.24	11.07	12.83	15.09	16.75
6	0.68	0.87	1.24	1.64	2.20	10.64	12.59	14.45	16.81	18.55
7	0.99	1.24	1.69	2.17	2.83	12.02	14.06	16.01	18.48	20.28
8	1.34	1.65	2.18	2.73	3.49	13.36	15.51	17.53	20.09	21.96
9	1.73	2.09	2.70	3.33	4.17	14.68	16.92	19.02	21.67	23.59
10	2.16	2.56	3.25	3.94	4.87	15.99	18.31	20.48	23.21	25.19
11	2.60	3.05	3.82	4.57	5.58	17.28	19.67	21.92	24.73	26.76
12	3.07	3.57	4.40	5.23	6.30	18.55	21.03	23.34	26.22	28.30
13	3.57	4.11	5.01	5.89	7.04	19.81	22.36	24.74	27.69	29.82
14	4.07	4.66	5.63	6.57	7.79	21.06	23.68	26.12	29.14	31.32
15	4.60	5.23	6.26	7.26	8.55	22.31	25.00	27.49	30.58	32.80
16	5.14	5.81	6.91	7.96	9.31	23.54	26.30	28.85	32.00	34.27
17	5.70	6.41	7.56	8.67	10.09	24.77	27.59	30.19	33.41	35.72
18	6.26	7.01	8.23	9.39	10.86	25.99	28.87	31.53	34.81	37.16
19	6.84	7.63	8.91	10.12	11.65	27.20	30.14	32.85	36.19	38.58
20	7.43	8.26	9.59	10.85	12.44	28.41	31.41	34.17	37.57	40.00
22	8.6	9.5	11.0	12.3	14.0	30.8	33.9	36.8	40.3	42.8
24	9.9	10.9	12.4	13.9	15.7	33.2	36.4	39.4	43.0	45.6
26	11.2	12.2	13.8	15.4	17.3	35.6	38.9	41.9	45.6	48.3
28	12.5	13.6	15.3	16.9	18.9	37.9	41.3	44.5	48.3	51.0
30	13.8	15.0	16.8	18.5	20.6	40.3	43.8	47.0	50.9	53.7
40	20.7	22.2	24.4	26.5	29.1	51.8	55.8	59.3	63.7	66.8
50	28.0	29.7	32.4	34.8	37.7	63.2	67.5	71.4	76.2	79.5
60	35.5	37.5	40.5	43.2	46.5	74.4	79.1	83.3	88.4	92.0
70	43.3	45.4	48.8	51.7	55.3	85.5	90.5	95.0	100.4	104.2
80	51.2	53.5	57.2	60.4	64.3	96.6	101.9	106.6	112.3	116.3
90	59.2	61.8	65.6	69.1	73.3	107.6	113.1	118.1	124.1	128.3
100	67.3	70.1	74.2	77.9	82.4	118.5	124.3	129.6	135.8	140.2

Table A.6a One-sided test: $\alpha = 0.10$; Two-sided test: $\alpha = 0.20$

Critical values of U for the Mann and Whitney (and Wilcoxon) test for the one-sided test: $\alpha = 0.10$; two-sided test: $\alpha = 0.20$ (see Milton, R.C.: An extended table of critical values for the Mann–Whitney (Wilcoxon) two sample statistic, J. Amer. Statist. Ass. 59 (1964), pp. 925–934)

m\n	1	2	3	4	5	6	7	8	9	10	11	12	13	14	15	16	17	18	19	20
1																				
2																				
3		0	1																	
4		0	1	3																
5		1	2	4	5															
6		1	3	5	7	9														
7		1	4	6	8	11	13													
8		2	5	7	10	13	16	19												
9		2	5	9	12	15	18	22	25											
10	0	3	6	10	13	17	21	24	28	32										
11	0	3	7	11	15	19	23	27	31	36	40									
12	0	4	8	12	17	21	26	30	35	39	44	49								
13	0	4	9	13	18	23	28	33	38	43	48	53	58							
14	0	5	10	15	20	25	31	36	41	47	52	58	63	69						
15	0	5	10	16	22	27	33	39	45	51	57	63	68	74	80					
16	0	5	11	17	23	29	36	42	48	54	61	67	74	80	86	93				
17	0	6	12	18	25	31	38	45	52	58	65	72	79	85	92	99	106			
18	0	6	13	20	27	34	41	48	55	62	69	77	84	91	98	106	113	120		
19	1	7	14	21	28	36	43	51	58	66	73	81	89	97	104	112	120	128	135	

(continued)

Table A.6a (continued)

m	n=1	2	3	4	5	6	7	8	9	10	11	12	13	14	15	16	17	18	19	20
20	1	7	15	22	30	38	46	54	62	70	78	86	94	102	110	119	127	135	143	151
21	1	8	15	23	32	40	48	56	65	73	82	91	99	108	116	125	134	142	151	160
22	1	8	16	25	33	42	51	59	68	77	86	95	104	113	122	131	141	150	159	168
23	1	9	17	26	35	44	53	62	72	81	90	100	109	119	128	138	147	157	167	176
24	1	9	18	27	36	46	56	65	75	85	95	105	114	124	134	144	154	164	174	184
25	1	9	19	28	38	48	58	68	78	89	99	109	120	130	140	151	161	172	182	193
26	1	10	20	30	40	50	61	71	82	92	103	114	125	136	146	157	168	179	190	201
27	1	10	21	31	41	52	63	74	85	96	107	119	130	141	152	164	175	186	198	209
28	1	11	21	32	43	54	66	77	88	100	112	123	135	147	158	170	182	194	206	217
29	2	11	22	33	45	56	68	80	92	104	116	128	140	152	164	177	189	201	213	226
30	2	12	23	35	46	58	71	83	95	108	120	133	145	158	170	183	196	209	221	234
31	2	12	24	36	48	61	73	86	99	111	124	137	150	163	177	190	203	216	229	242
32	2	13	25	37	50	63	76	89	102	115	129	142	156	169	183	196	210	223	237	251
33	2	13	26	38	51	65	78	92	105	119	133	147	161	175	189	203	217	131	245	259
34	2	13	26	40	53	67	81	95	109	123	137	151	166	180	195	209	224	238	253	267
35	2	14	27	41	55	69	83	98	112	127	141	156	171	186	201	216	230	245	260	275
36	2	14	28	42	56	71	86	100	115	131	146	161	176	191	207	222	237	253	268	284
37	2	15	29	43	58	73	88	103	119	134	150	166	181	197	213	229	244	260	276	292
38	2	15	30	45	60	75	91	106	122	138	154	170	186	203	219	235	251	268	284	301
39	3	16	31	46	61	77	93	109	126	142	158	175	192	208	225	242	258	275	292	309
40	3	16	31	47	63	79	96	112	129	146	163	180	197	214	231	248	265	282	300	317

Table A.6b One-sided test: $\alpha = 0.05$; Two-sided test: $\alpha = 0.10$

Critical values of U for the Mann and Whitney (and Wilcoxon) test for the one-sided test: $\alpha = 0.05$; two-sided test: $\alpha = 0.10$ (see Milton, R.C.: An extended table of critical values for the Mann–Whitney (Wilcoxon) two sample statistic, J. Amer. Statist. Ass. 59 (1964), pp. 925–934)

m\n	1	2	3	4	5	6	7	8	9	10	11	12	13	14	15	16	17	18	19	20
1																				
2																				
3			0																	
4			0	1																
5		0	1	2	4															
6		0	2	3	5	7														
7		0	2	4	6	8	11													
8			3	5	8	10	13	15												
9		1	4	6	9	12	15	18	21											
10		1	4	7	11	14	17	20	24	27										
11		1	5	8	12	16	19	23	27	31	34									
12		2	5	9	13	17	21	26	30	34	38	42								
13		2	6	10	15	19	24	28	33	37	42	47	51							
14		3	7	11	16	21	26	31	36	41	46	51	56	61						
15		3	7	12	18	23	28	33	39	44	50	55	61	66	72					
16		3	8	14	19	25	30	36	42	48	54	60	65	71	77	83				
17		3	9	15	20	26	33	39	45	51	57	64	70	77	83	89	96			
18		4	9	16	22	28	35	41	48	5	61	68	75	82	88	95	102	109		
19	0	4	10	17	23	30	37	44	51	58	65	72	80	87	94	101	109	116	123	
20	0	4	11	18	25	32	39	47	54	62	69	77	84	92	100	107	115	123	130	138

(continued)

Table A.6b (continued)

m	1	2	3	4	5	6	7	8	9	10	11	12	13	14	15	16	17	18	19	20
21	0	5	11	19	26	34	41	49	57	65	73	81	89	97	105	113	121	130	138	146
22	0	5	12	20	28	36	44	52	60	68	77	85	94	102	111	119	128	136	145	154
23	0	5	13	21	29	37	46	54	63	72	81	90	98	107	116	125	134	143	152	161
24	0	6	13	22	30	39	48	57	66	75	85	94	103	113	122	131	141	150	160	169
25	0	6	14	23	32	41	50	60	69	79	89	98	108	118	128	137	147	157	167	177
26	0	6	15	24	33	43	53	62	72	82	92	103	113	123	133	143	154	164	174	185
27	0	7	15	25	35	45	55	65	75	86	96	107	117	128	139	149	160	171	182	192
28	0	7	16	26	36	46	57	68	78	89	100	111	122	133	144	156	167	178	189	200
29	0	7	17	27	38	48	59	70	82	93	104	116	127	138	150	162	173	185	196	208
30	0	7	17	28	39	50	61	73	85	96	108	120	132	144	156	168	180	192	204	216
31	0	8	18	29	40	52	64	76	88	100	112	124	136	149	161	174	186	199	211	224
32	0	8	19	30	42	54	66	78	91	103	116	128	141	154	167	180	193	206	218	231
33	0	8	19	31	43	56	68	81	94	107	120	133	146	159	172	186	199	212	226	239
34	0	9	20	32	45	57	70	84	97	110	124	137	151	164	178	192	206	219	233	247
35	0	9	21	33	46	59	73	86	100	114	128	141	156	170	184	198	212	226	241	255
36	0	9	21	34	48	61	75	89	103	117	131	146	160	175	189	204	219	233	248	263
37	0	10	22	35	49	63	77	91	106	121	135	150	165	180	195	210	225	240	255	271
38	0	10	23	36	50	65	79	94	109	124	139	154	170	185	201	216	232	247	263	278
39	1	10	23	38	52	67	82	97	112	128	143	159	175	190	206	222	238	254	270	286
40	1	11	24	39	53	68	84	99	115	131	147	163	179	196	212	228	245	261	278	294

Table A.6c One-sided test: $\alpha = 0.025$; Two-sided test: $\alpha = 0.05$

Critical values of U for the Mann and Whitney (and Wilcoxon) test for the one-sided test: $\alpha = 0.025$; two-sided test: $\alpha = 0.05$ (see Milton, R.C.: An extended table of critical values for the Mann–Whitney (Wilcoxon) two sample statistic, J. Amer. Statist. Ass. 59 (1964), pp. 925–934)

m	1	2	3	4	5	6	7	8	9	10	11	12	13	14	15	16	17	18	19	20
1																				
2																				
3																				
4				0																
5			0	1	2															
6			1	2	3	5														
7			1	3	5	6	8													
8		0	2	4	6	8	10	13												
9		0	2	4	7	10	12	15	17											
10		0	3	5	8	11	14	17	20	23										
11		0	3	6	9	13	16	19	23	26	30									
12		1	4	7	11	14	18	22	26	29	33	37								
13		1	4	8	12	16	20	24	28	33	37	41	45							
14		1	5	9	13	17	22	26	31	36	40	45	50	55						
15		1	5	10	14	19	24	29	34	39	44	49	54	59	64					
16		1	6	11	15	21	26	31	37	42	47	53	59	64	70	75				
17		2	6	11	17	22	28	34	39	45	51	57	63	69	75	81	87			
18		2	7	12	18	24	30	36	42	48	55	61	67	74	80	86	93	99		
19		2	7	13	19	25	32	38	45	52	58	65	72	78	85	92	99	106	113	
20		2	8	14	20	27	34	41	48	55	62	69	76	83	90	98	105	112	119	127

(continued)

Table A.6c (continued)

m	1	2	3	4	5	6	7	8	9	10	11	12	13	14	15	16	17	18	19	20
21		3	8	15	22	29	36	43	50	58	65	73	80	88	96	103	111	119	126	134
22		3	9	16	23	30	38	45	53	61	69	77	85	93	101	109	117	125	133	141
23		3	9	17	24	32	40	48	56	64	73	81	89	98	106	115	123	132	140	149
24		3	10	17	25	33	42	50	59	67	76	85	94	102	111	120	129	138	147	156
25		3	10	18	27	35	44	53	62	71	80	89	98	107	117	126	135	145	154	163
26		4	11	19	28	37	46	55	64	74	83	93	102	112	122	132	141	151	161	171
27		4	11	20	29	38	48	57	67	77	87	97	107	117	127	137	147	158	168	178
28		4	12	21	30	40	50	60	70	80	90	101	111	122	132	143	154	164	175	186
29		4	13	22	32	42	52	62	73	83	94	105	116	127	138	149	160	171	182	193
30		5	13	23	33	43	54	65	76	87	98	109	120	131	143	154	166	177	189	200
31		5	14	24	34	45	56	67	78	90	101	113	125	136	148	160	172	184	196	208
32		5	14	24	35	46	58	69	81	93	105	117	129	141	153	166	178	190	203	215
33		5	15	25	37	48	60	72	84	96	108	121	133	146	159	171	184	197	210	222
34		5	15	26	38	50	62	74	87	99	112	125	138	151	164	177	190	203	217	230
35		6	16	27	39	51	64	77	89	103	116	129	142	156	169	183	196	210	224	237
36		6	16	28	40	53	66	79	92	106	119	133	147	161	174	188	202	216	231	245
37		6	17	29	41	55	68	81	95	109	123	137	151	165	180	194	209	223	238	252
38		6	17	30	43	56	70	84	98	112	127	141	156	170	185	200	215	230	245	259
39	0	7	18	31	44	58	72	86	101	115	130	145	160	175	190	206	221	236	252	267
40	0	7	18	31	45	59	74	89	103	119	134	149	165	180	196	211	227	243	258	274

Table A.6d One-sided test: $\alpha = 0.01$; Two-sided test: $\alpha = 0.02$

Critical values of U for the Mann and Whitney (and Wilcoxon) test for the one-sided test: $\alpha = 0.01$; two-sided test: $\alpha = 0.02$ (see Milton, R.C.: An extended table of critical values for the Mann–Whitney (Wilcoxon) two sample statistic, J. Amer. Statist. Ass. 59 (1964), pp. 925–934)

m	1	2	3	4	5	6	7	8	9	10	11	12	13	14	15	16	17	18	19	20
1																				
2																				
3																				
4				0																
5				0	1															
6				1	2	3														
7			0	1	3	4	6													
8			0	2	4	6	7	9												
9			1	3	5	7	9	11	14											
10			1	3	6	8	11	13	16	19										
11			1	4	7	9	12	15	18	22	25									
12			2	5	8	11	14	17	21	24	28	31								
13		0	2	5	9	12	16	20	23	27	31	35	39							
14		0	2	6	10	13	17	22	26	30	34	38	43	47						
15		0	3	7	11	15	19	24	28	33	37	42	47	51	56					
16		0	3	7	12	16	21	26	31	36	41	46	51	56	61	66				
17		0	4	8	13	18	23	28	33	38	44	49	55	60	66	71	77			
18		0	4	9	14	19	24	30	36	41	47	53	59	65	70	76	82	88		
19		1	4	9	15	20	26	32	38	44	50	56	63	69	75	82	88	94	101	
20		1	5	10	16	22	28	34	40	47	53	60	67	73	80	87	93	100	107	114

(continued)

Table A.6d (continued)

m	1	2	3	4	5	6	7	8	9	10	11	12	13	14	15	16	17	18	19	20
21		1	5	11	17	23	30	36	43	50	57	64	71	78	85	922	99	106	113	121
22		1	6	11	18	24	31	38	45	53	60	67	75	82	90	97	105	112	120	127
23		1	6	12	19	26	33	40	48	55	63	71	79	87	94	102	110	118	126	134
24		1	6	13	20	27	35	42	50	58	66	75	83	91	99	108	116	124	133	141
25		1	7	13	21	29	36	45	53	61	70	78	87	95	104	113	122	130	139	148
26		1	7	14	22	30	38	47	55	64	73	82	91	100	109	118	127	136	146	155
27		2	7	15	23	31	40	49	58	67	76	85	95	104	114	123	133	142	152	162
28		2	8	16	24	33	42	51	60	70	79	89	99	109	119	129	139	149	159	169
29		2	8	16	25	34	43	53	63	73	83	93	103	113	123	134	144	155	165	176
30		2	9	17	26	35	45	55	65	76	86	96	107	118	128	139	150	161	172	182
31		2	9	18	27	37	47	57	68	75	89	100	111	122	133	144	156	167	178	189
32		2	9	18	28	38	49	59	70	81	92	104	115	127	138	150	161	173	185	196
33		2	10	19	29	40	50	61	73	84	96	107	119	131	143	155	167	179	191	203
34		3	10	20	30	41	52	64	75	87	99	111	123	135	148	160	173	185	198	210
35		3	11	20	31	42	54	66	78	90	102	115	127	140	153	165	178	191	204	217
36		3	11	21	32	44	56	68	80	93	106	118	131	144	158	171	184	197	211	224
37		3	11	22	33	45	57	70	83	96	109	122	135	149	162	176	190	203	217	231
38		3	12	22	34	46	59	72	85	99	112	126	139	153	167	181	195	209	224	238
39		3	12	23	35	48	61	74	88	101	115	129	144	158	172	187	201	216	230	245
40		3	13	24	36	49	63	76	90	104	119	133	148	162	177	192	207	222	237	252

Table A.6e One-sided test: $\alpha = 0.005$; Two-sided test: $\alpha = 0.01$

Critical values of U for the Mann and Whitney (and Wilcoxon) test for the one-sided test: $\alpha = 0.005$; two-sided test: $\alpha = 0.01$ (see Milton, R.C.: An extended table of critical values for the Mann–Whitney (Wilcoxon) two sample statistic, J. Amer. Statist. Ass. 59 (1964), pp. 925–934)

m	1	2	3	4	5	6	7	8	9	10	11	12	13	14	15	16	17	18	19	20
1																				
2																				
3																				
4																				
5					0															
6				0	1	2														
7				0	1	3	4													
8				1	2	4	6	7												
9			0	1	3	5	7	9	11											
10			0	2	4	6	9	11	13	16										
11			0	2	5	7	10	13	16	18	21									
12			1	3	6	9	12	15	18	21	24	27								
13			1	3	7	10	13	17	20	24	27	31	34							
14			1	4	7	11	15	18	22	26	30	34	38	42						
15			2	5	8	12	16	20	24	29	33	37	42	46	51					
16			2	5	9	13	18	22	27	31	36	41	45	50	55	60				
17			2	6	10	15	19	24	29	34	39	44	49	54	60	65	70			
18			2	6	11	16	21	26	31	37	42	47	53	58	64	70	75	81		
19		0	3	7	12	17	22	28	33	39	45	51	57	63	69	74	81	87	93	
20		0	3	8	13	18	24	30	36	42	48	54	60	67	73	79	86	92	99	105

(continued)

Table A.6e (continued)

m	1	2	3	4	5	6	7	8	9	10	11	12	13	14	15	16	17	18	19	20
21		0	3	8	14	19	25	32	38	44	51	58	64	71	78	84	91	98	105	112
22		0	4	9	14	21	27	34	40	47	54	61	68	75	82	89	96	104	111	118
23		0	4	9	15	22	29	35	43	50	57	64	72	79	87	94	102	109	117	125
24		0	4	10	16	23	30	37	45	52	60	68	75	83	91	99	107	115	123	131
25		0	5	10	17	24	32	39	47	55	63	71	79	87	96	104	112	121	129	138
26		0	5	11	18	25	33	41	49	58	66	74	83	92	100	109	118	127	135	144
27		1	5	12	19	27	35	43	52	60	69	78	87	96	105	114	123	132	142	151
28		1	5	12	20	28	36	45	54	63	72	81	91	100	109	119	128	138	148	157
29		1	6	13	21	29	38	47	56	66	75	85	94	104	114	124	134	144	154	164
30		1	6	13	22	30	40	49	58	68	78	88	98	108	119	129	139	150	160	170
31		1	6	14	22	32	41	51	61	71	81	92	102	113	123	134	145	155	166	177
32		1	7	14	23	33	43	53	63	74	84	95	106	117	128	139	150	161	172	184
33		1	7	15	24	34	44	55	65	76	87	98	110	121	132	144	155	167	179	190
34		1	7	16	25	35	46	57	68	79	90	102	113	125	137	149	161	173	185	197
35		1	8	16	26	37	47	59	70	82	93	105	117	129	142	154	166	179	191	203
36		1	8	17	27	38	49	60	72	84	96	109	121	134	146	159	172	184	197	210
37		1	8	17	28	39	51	62	75	87	99	112	125	138	151	164	177	190	203	217
38		1	9	18	29	40	52	64	77	90	102	116	129	142	155	169	182	196	210	223
39		2	9	19	30	41	54	66	79	92	106	119	133	146	160	174	188	202	216	230
40		2	9	19	31	43	55	68	81	95	109	122	136	150	165	179	193	208	222	237

Table A.6f One-sided test: $\alpha = 0.001$; Two-sided test: $\alpha = 0.002$

Critical values of U for the Mann and Whitney (and Wilcoxon) test for the one-sided test: $\alpha = 0.001$; two-sided test: $\alpha = 0.002$ (see Milton, R.C.: An extended table of critical values for the Mann–Whitney (Wilcoxon) two sample statistic, J. Amer. Statist. Ass. 59 (1964), pp. 925–934)

m	1	2	3	4	5	6	7	8	9	10	11	12	13	14	15	16	17	18	19	20
1																				
2																				
3																				
4																				
5																				
6																				
7						0	1													
8					0	1	2	4												
9					1	2	3	5	7											
10				0	1	3	5	6	8	10										
11				0	2	4	6	8	10	12	15									
12				0	2	4	7	9	12	14	17	20								
13				1	3	5	8	11	14	17	20	23	26							
14				1	3	6	9	12	15	19	22	25	29	32						
15				1	4	7	10	14	17	21	24	28	32	36	40					
16				2	5	8	11	15	19	23	27	31	35	39	43	48				
17			0	2	5	9	13	17	21	25	29	34	38	43	47	52	57			
18			0	3	6	10	14	18	23	27	32	37	42	46	51	56	61	66		
19			0	3	7	11	15	20	25	29	34	40	45	50	55	60	66	71	77	
20			0	3	7	12	16	21	26	32	37	42	48	54	59	65	70	76	82	88

(continued)

Table A.6f (continued)

m	1	2	3	4	5	6	7	8	9	10	11	12	13	14	15	16	17	18	19	20
21			1	4	8	12	18	23	28	34	40	45	51	57	63	69	75	81	87	94
22			1	4	8	13	19	24	30	36	42	48	54	61	67	73	80	86	93	99
23			1	4	9	14	20	26	32	38	45	51	58	64	71	78	85	91	98	105
24			1	5	10	15	21	27	34	40	47	54	61	68	75	82	89	96	104	111
25			1	5	10	16	22	29	36	43	50	57	64	72	79	86	94	102	109	117
26			1	6	11	17	24	31	38	45	52	60	68	75	83	91	99	107	115	123
27			2	6	12	18	25	32	40	47	55	63	71	79	87	95	104	112	120	129
28			2	6	12	19	26	34	41	49	57	66	74	83	91	100	108	117	126	135
29			2	7	13	20	27	35	43	52	60	69	77	86	95	104	113	122	131	140
30			2	7	14	21	29	37	45	54	63	72	81	90	99	108	118	127	137	146
31			2	7	14	22	30	38	47	56	65	75	84	94	103	113	123	132	142	152
32			2	8	15	23	31	40	49	58	68	77	87	97	107	117	127	138	148	158
33			3	8	15	24	32	41	51	61	70	80	91	101	111	122	132	143	153	164
34			3	9	16	25	34	43	53	63	73	83	94	105	115	126	137	148	159	170
35			3	9	17	25	35	45	55	65	76	86	97	108	119	131	142	153	165	176
36			3	9	17	26	36	46	57	67	78	89	101	112	123	135	147	158	170	182
37			3	10	18	27	37	48	58	70	81	92	104	116	127	139	151	164	176	188
38			3	10	19	28	39	49	60	72	83	95	107	119	131	144	156	169	181	194
39			4	11	19	29	40	51	62	74	86	98	110	123	136	148	161	174	187	200
40			4	11	20	30	41	52	64	76	89	101	114	127	140	153	166	179	192	206

Table A.7 Quantiles k_P ($Q\ k_P$) of the Kolmogorov–Smirnov test statistic

$Q\ k_P$ \ n	0.8	0.9	0.92	0.95	0.96	0.98	0.99
1	0.900	0.950	0.960	0.975	0.980	0.990	0.995
2	0.684	0.776	0.800	0.842	0.859	0.900	0.929
3	0.565	0.636	0.658	0.708	0.729	0.785	0.829
4	0.493	0.565	0.585	0.624	0.641	0.689	0.734
5	0.447	0.509	0.527	0.563	0.580	0.627	0.669
6	0.410	0.168	0.485	0.519	0.534	0.577	0.617
7	0.381	0.436	0.452	0.483	0.497	0.538	0.576
8	0.358	0.410	0.425	0.454	0.468	0.507	0.542
9	0.339	0.387	0.402	0.430	0.443	0.480	0.513
10	0.323	0.369	0.382	0.409	0.421	0.457	0.489
11	0.308	0.352	0.365	0.391	0.403	0.437	0.468
12	0.296	0.338	0.351	0.375	0.387	0.419	0.449
13	0.285	0.325	0.338	0.361	0.372	0.404	0.432
14	0.275	0.314	0.326	0.349	0.359	0.390	0.418
15	0.266	0.304	0.315	0.338	0.348	0.377	0.404
16	0.258	0.295	0.306	0.327	0.337	0.366	0.392
17	0.250	0.286	0.297	0.318	0.327	0.355	0.381
18	0.244	0.279	0.289	0.309	0.319	0.346	0.371
19	0.237	0.271	0.281	0.301	0.310	0.337	0.361
20	0.232	0.265	0.275	0.294	0.303	0.329	0.352
21	0.226	0.259	0.268	0.287	0.296	0.321	0.344
22	0.221	0.253	0.262	0.281	0.289	0.314	0.337
23	0.216	0.247	0.257	0.275	0.283	0.307	0.330
24	0.212	0.242	0.251	0.269	0.277	0.301	0.323
25	0.208	0.238	0.246	0.264	0.272	0.295	0.317
26	0.204	0.233	0.242	0.259	0.267	0.290	0.311
27	0.200	0.229	0.237	0.254	0.262	0.284	0.305
28	0.197	0.225	0.233	0.250	0.257	0.279	0.300
29	0.193	0.221	0.229	0.246	0.251	0.275	0.295
30	0.190	0.218	0.226	0.242	0.249	0.270	0.290
31	0.187	0.214	0.222	0.238	0.245	0.266	0.285
32	0.184	0.211	0.219	0.234	0.241	0.262	0.281
33	0.182	0.208	0.215	0.231	0.238	0.258	0.277
34	0.179	0.205	0.212	0.227	0.234	0.254	0.273
35	0.177	0.202	0.209	0.224	0.231	0.251	0.269
36	0.174	0.199	0.206	0.221	0.228	0.247	0.265

(continued)

Table A.7 (continued)

Q k_P n	0.8	0.9	0.92	0.95	0.96	0.98	0.99
37	0.172	0.196	0.204	0.218	0.225	0.244	0.262
38	0.170	0.194	0.201	0.215	0.222	0.241	0.258
39	0.168	0.191	0.199	0.213	0.219	0.238	0.255
40	0.165	0.189	0.196	0.210	0.216	0.235	0.252
>40	$1.07/\sqrt{n}$	$1.22/\sqrt{n}$	$1.27/\sqrt{n}$	$1.36/\sqrt{n}$	$1.40/\sqrt{n}$	$1.52/\sqrt{n}$	$1.63/\sqrt{n}$

Table A.8 Error probabilities for the H-test of Kruskal and Wallis (Kruskal and Wallis 1952) and the errata in J. Amer. Statist. Ass. 48 (1953:910), and supplementary values from (Iman et al. 1975); adapted from (Hedderich and Sachs 2020)

n1	n2	n3	H	P	n1	n2	n3	H	P	n1	n2	n3	H	P	n1	n2	n3	H	P
2	1	1	2.7000	0.5000	4	3	2	6.4444	0.0080	5	2	2	6.5333	0.0080	5	4	4	5.6571	0.0490
								6.3000	0.0110				6.1333	0.0130				6.5176	0.0500
2	2	1	3.6000	0.2000				5.4444	0.0460				5.1600	0.0340				4.6187	0.1000
								5.4000	0.0510				5.0400	0.0560				4.5527	0.1020
2	2	2	4.5714	0.0670				4.5111	0.0980				4.3733	0.0900					
			3.7143	0.2000				4.4444	0.1020				4.2933	0.1220	5		1	7.3091	0.0090
			3.2000	0.3000														6.8364	0.0110
3	1	1			4	3	3	6.7455	0.0100	5	3	1	6.4000	0.0120				5.1273	0.0460
3	2	1	4.2857	0.1000				6.7091	0.0130				4.9600	0.0480				4.9091	0.0530
			3.8571	0.1330				5.7909	0.0460				4.8711	0.0520				4.1091	0.0860
3	2	2	5.3572	0.0290				5.7273	0.0500				4.0178	0.0950				4.0364	0.1050
			4.7143	0.0480				4.7091	0.0920				3.8400	0.1230					
			4.5000	0.0670				4.7000	0.1010						5	5	2	7.3385	0.0100
			4.4643	0.1050	4	4	1	6.6667	0.0100	5	3	2	6.9091	0.0090				7.2692	0.0100
3	3	1						6.1667	0.0220				6.8218	0.0100				5.3385	0.0470
			5.1429	0.0430				4.9667	0.0480				5.2509	0.0490				5.2462	0.0510
			4.5714	0.1000				4.8667	0.0540				5.1055	0.0520				4.6231	0.0970
			4.0000	0.1290				4.1667	0.0820				4.6509	0.0910				4.5077	0.1000
								4.0667	0.1020				4.4945	0.1010	5	5	3	7.5780	0.0100

(continued)

Table A.8 (continued)

n1	n2	n3	H	P	n1	n2	n3	H	P	n1	n2	n3	H	P
3	3	2	6.2500	0.0110				7.0364	0.0060	5	3	3	7.0788	0.0090				7.5429	0.0100
			5.3611	0.0320				6.8727	0.0110				6.9818	0.0110				5.7055	0.0460
			5.1389	0.0610	4	4	2	7.0364	0.0060				5.6485	0.0490				5.6264	0.0510
			4.5556	0.1000				6.8727	0.0110				5.5152	0.0510				4.5451	0.1000
			4.2500	0.1210				5.4545	0.0460				4.5333	0.0970				4.5363	0.1020
								5.2364	0.0520				4.4121	0.1090					
3	3	3	7.2000	0.0040				4.5545	0.0980						5	5	4	7.8229	0.0100
			6.4889	0.0110				4.4455	0.1030	5	4	1	6.9545	0.0080				7.7914	0.0100
			5.6889	0.0290									6.8400	0.0110				5.6657	0.0490
			5.6000	0.0500	4	4	3	7.1439	0.0100				4.9855	0.0440				5.6429	0.0500
			5.0667	0.0860				7.1364	0.0110				4.8600	0.0560				4.5229	0.0990
			4.6222	0.1000				5.5985	0.0490				3.9873	0.0980				4.5200	0.1010
								5.5758	0.0510				3.9600	0.1020					
4	1	1	3.5714	0.2000				4.5455	0.0990						5	5	5	8.0000	0.0090
								4.4773	0.1020	5	4	2	7.2045	0.0090				5.7800	0.0490
4	2	1	4.8214	0.0570									7.1182	0.0100				4.5600	0.1000
			4.5000	0.0760	4	4	4	7.6538	0.0080				5.2727	0.0490					
			4.0179	0.1140				7.5385	0.0110				5.2682	0.0500					
4	2	2	6.0000	0.0140				5.6923	0.0490				4.5409	0.0980	6	6	6	8.2222	0.0100
								5.6538	0.0540				4.5182	0.1010					

(continued)

Appendix

Table A.8 (continued)

n1	n2	n3	H	P	n1	n2	n3	H	P	n1	n2	n3	H	P
			5.3333	0.0330				4.6539	0.0970								5.8011	0.0490
			5.1250	0.0520				4.5001	0.1040								4.6430	0.0990
			4.4583	0.1000						5	4	3	7.4449	0.0100					
			4.1667	0.1050	5	1	1	3.8571	0.1430				7.3949	0.0110	7	7	7	8.3780	0.0100
													5.6564	0.0490				5.8190	0.0490
4	3	1	5.8333	0.0210	5	2	1	5.2500	0.0360				5.6308	0.0500				4.5940	0.0990
			5.2083	0.0500				5.0000	0.0480				4.5487	0.0990					
			5.0000	0.0570				4.4500	0.0710				4.5231	0.1030	8	8	8	8.4650	0.0100
			4.0556	0.0930				4.2000	0.0950	5	4	4	7.7604	0.0090				5.8050	0.0500
			3.8889	0.1290				4.0500	0.1190				7.7440	0.0110				4.5950	0.0990

Table A.9 Means and Standard Deviations of Reduced Extremes according to (Gumbel 1958)

N	$\bar{y}_N$	σ_N	N	$\bar{y}_N$	σ_N
8	0.4843	0.9043	49	0.5481	1.1590
9	0.4902	0.9288	50	0.54854	1.16066
10	0.4952	0.9497	51	0.5489	1.1623
11	0.4996	0.9676	52	0.5493	1.1638
12	0.5035	0.9833	53	0.5497	1.1653
13	0.5070	0.9972	54	0.5501	1.1667
14	0.5100	1.0095	55	0.5504	1.1681
15	0.5128	1.0206	56	0.5508	1.1696
16	0.5157	1.0316	57	0.5511	1.1708
17	0.5181	1.0411	58	0.5515	1.1721
18	0.5202	1.0493	59	0.5518	1.1734
19	0.5220	1.0566	60	0.55208	1.17467
20	0.52355	1.06283	62	0.5527	1.1770
21	0.5252	1.0696	64	0.5333	1.1793
22	0.5268	1.0754	66	0.5538	1.1814
23	0.5283	1.0811	68	0.5543	1.1834
24	0.5296	1.0864	70	0.55477	1.18536
25	0.53086	1.09145	72	0.5552	1.1873
26	0.5320	1.0961	74	0.5557	1.1890
27	0.5332	1.1004	76	0.5561	1.1906
28	0.5343	1.1047	78	0.5565	1.1923
29	0.5353	1.1086	80	0.55688	1.19382
30	0.53622	1.11238	82	0.5572	1.1953
31	0.5371	1.1159	84	0.5576	1.1967
32	0.5380	1.1193	86	0.5580	1.1980
33	0.5388	1.1226	88	0.5583	1.1994
34	0.5396	1.1255	90	0.55860	1.20073
35	0.54034	1.12847	92	0.5589	1.2020
36	0.5410	1.1313	94	0.5592	1.2032
37	0.5418	1.1339	96	0.5595	1.2044
38	0.5424	1.1363	98	0.5598	1.2055
39	0.5430	1.1388	100	0.56002	1.20649
40	0.54362	1.14132	150	0.56461	1.22534
41	0.5442	1.1436	200	0.56715	1.23598
42	0.5448	1.1458	150	0.56878	1.24292
43	0.5453	1.1480	300	0.56993	1.24786

Table A.9 Continued

N	$\bar{y}_N$	σ_N	N	$\bar{y}_N$	σ_N
44	0.5458	1.1499	400	0.57144	1.25450
45	0.54630	1.15185	500	0.57240	1.25880
46	0.5468	1.1538	750	0.57377	1.26506
47	0.5473	1.1557	1000	0.57450	1.26851
48	0.5477	1.1574	∞	0.57722	1.28255

Table A.10a Clopper-Pearson for x [0; 25] and n [25; 100]

x n	25		50		75		100	
	π_u	π_o	π_u	π_o	π_u	π_o	π_u	π_o
0	0.00	13.72	0.00	7.11	0.00	4.80	0.00	3.62
1	0.10	20.35	0.05	10.65	0.03	7.21	0.03	5.45
2	0.98	26.03	0.49	13.71	0.32	9.30	0.24	7.04
3	2.55	31.22	1.25	16.55	0.83	11.25	0.62	8.52
4	4.54	36.08	2.22	19.23	1.47	13.10	1.10	9.93
5	6.83	40.70	3.33	21.81	2.20	14.88	1.64	11.28
6	9.36	45.13	4.53	24.31	2.99	16.60	2.23	12.60
7	12.07	49.39	5.82	26.74	3.84	18.29	2.86	13.89
8	14.95	53.50	7.17	29.11	4.72	19.94	3.52	15.16
9	17.97	57.48	8.58	31.44	5.64	21.56	4.20	16.40
10	21.13	61.33	10.03	33.72	6.58	23.16	4.90	17.62
11	24.40	65.07	11.53	35.96	7.56	24.73	5.62	18.83
12	27.80	68.69	13.06	38.17	8.55	26.28	6.36	20.02
13	31.31	72.20	14.63	40.34	9.57	27.81	7.11	21.20
14	34.93	75.60	16.23	42.49	10.60	29.33	7.87	22.37
15	38.67	78.87	17.86	44.61	11.65	30.83	8.65	23.53
16	42.52	82.03	19.52	46.70	12.71	32.32	9.43	24.68
17	46.50	85.05	21.21	48.77	13.79	33.79	10.23	25.82
18	50.61	87.93	22.92	50.81	14.89	35.25	11.03	26.95
19	54.87	90.64	24.65	52.83	15.99	36.70	11.84	28.07
20	59.30	93.17	26.41	54.82	17.11	38.14	12.67	29.18
21	63.92	95.46	28.19	56.79	18.24	39.56	13.49	30.29
22	68.78	97.45	29.99	58.75	19.38	40.98	14.33	31.39
23	73.97	99.02	31.81	60.68	20.53	42.38	15.17	32.49
24	79.65	99.90	33.66	62.58	21.69	43.78	16.02	33.57
25	86.28	100.00	35.53	64.47	22.86	45.17	16.88	34.66

Table A.10b Clopper-Pearson for x [0; 25] and n [200; 1000]

x \ n	200		300		400		500		1000	
	π_u	π_o	π_u	π_o	π_u	π_o	π_u	π_o	π_u	π_o
0	0.00	1.83	0.00	1.22	0.00	0.92	0.00	0.74	0.00	0.37
1	0.01	2.75	0.01	1.84	0.01	1.38	0.01	1.11	0.00	0.56
2	0.12	3.57	0.08	2.39	0.06	1.79	0.05	1.44	0.02	0.72
3	0.31	4.32	0.21	2.89	0.15	2.18	0.12	1.74	0.06	0.87
4	0.55	5.04	0.36	3.38	0.27	2.54	0.22	2.04	0.11	1.02
5	0.82	5.74	0.54	3.85	0.41	2.89	0.33	2.32	0.16	1.16
6	1.11	6.42	0.74	4.30	0.55	3.24	0.44	2.59	0.22	1.30
7	1.42	7.08	0.94	4.75	0.71	3.57	0.56	2.86	0.28	1.44
8	1.74	7.73	1.16	5.19	0.87	3.90	0.69	3.13	0.35	1.57
9	2.08	8.37	1.38	5.62	1.03	4.23	0.83	3.39	0.41	1.70
10	2.42	9.00	1.61	6.04	1.21	4.55	0.96	3.65	0.48	1.83
11	2.78	9.63	1.84	6.47	1.38	4.87	1.10	3.90	0.55	1.96
12	3.14	10.25	2.08	6.88	1.56	5.18	1.25	4.15	0.62	2.09
13	3.51	10.86	2.33	7.30	1.74	5.49	1.39	4.41	0.69	2.21
14	3.88	11.47	2.57	7.71	1.93	5.80	1.54	4.65	0.77	2.34
15	4.26	12.07	2.83	8.11	2.11	6.11	1.69	4.90	0.84	2.46
16	4.64	12.67	3.08	8.52	2.30	6.41	1.84	5.14	0.92	2.59
17	5.03	13.26	3.34	8.92	2.49	6.72	1.99	5.39	0.99	2.71
18	5.42	13.85	3.59	9.32	2.69	7.02	2.15	5.63	1.07	2.83
19	5.82	14.44	3.86	9.71	2.88	7.32	2.30	5.87	1.15	2.95
20	6.22	15.02	4.12	10.11	3.08	7.62	2.46	6.11	1.23	3.07
21	6.62	15.60	4.39	10.50	3.28	7.91	2.62	6.35	1.30	3.19
22	7.02	16.18	4.65	10.89	3.48	8.21	2.78	6.59	1.38	3.31
23	7.43	16.75	4.92	11.28	3.68	8.50	2.94	6.82	1.46	3.43
24	7.84	17.33	5.19	11.67	3.88	8.80	3.10	7.06	1.54	3.55
25	8.26	17.90	5.47	12.06	4.09	9.09	3.26	7.29	1.62	3.67

Table A.11: Evaluation of failure scenarios in an FMEA

The evaluation of potential failure scenarios within the Failure Mode and Effects Analysis (FMEA) is carried out on the basis of the three key indicators B (importance of the failure symptom), A (probability of occurrence of the failure cause) and E (probability of detection of the failure cause). The listed tables are taken from VDA Volume 4 (VDA 2006) and focus on the assessment of risks using the example of automotive engineering. Table A.11.1 and A.11.2 show the tables for a K-FMEA (system FMEA product) and for a P-FMEA (system FMEA process).

K-FMEA (System-FMEA Product)
The allocation of scores for key indicator B (importance of the failure symptom) when carrying out a K-FMEA depending on the potential risk in accordance with VDA Volume 4 (VDA 2006) has already been explained in Sect. 11.3 (Table 11.2), ergo the tables for key indicators A and E can be found in this section.

P-FMEA (System-FMEA Process)

Table FMEA-A.11.1 K-FMEA; allocation of scores for the key indicator A (probability of occurrence) depending on the potential risk in accordance with VDA Volume 4 (VDA 2006)

Points	Probability of occurrence A; qualitative assessment	Failure proportion [ppm]
9, 10	Very high; frequent occurrence of the cause of the fault, unusable, unsuitable design concept	500,000 100,000
7, 8	High; cause of failure occurs repeatedly, problematic, immature design	50,000 10,000
4, 5, 6	Moderate; occasional cause of failure, suitable design advanced in maturity	5000 1000 500
2, 3	Low; occurrence of the cause of the failure is low, proven constructive design	100 50
1	Very low; occurrence of the cause of the failure is unlikely	1

Table FMEA-A.11.2 K-FMEA; allocation of points for the key indicator E (probability of detection) depending on the potential risk in accordance with VDA Volume 4 (VDA 2006)

Points	Probability of detection E; qualitative assessment	certainty of evidence
9, 10	Very low; detection of the failure cause that occurred is unlikely, reliability of the design layout has not been or cannot be demonstrated, detection methods are uncertain	90%
7, 8	Low; detection of the failure cause that occurred is less likely, reliability of the design layout probably cannot be demonstrated, detection methods are uncertain	98%
4, 5, 6	Moderate; discovery of the failure cause that occurred is likely, reliability of design design could perhaps be demonstrated, detection methods are relatively safe	99.7%
2, 3	High; detection of the cause of the failure is highly probable, confirmed by several independent detection methods	99.9%
1	Very high; Cause of failure is reliably detected	99.99%

Table FMEA-A.11.3 P-FMEA; allocation of points for the key indicator B (importance of the failure sequence) depending on the potential risk in accordance with VDA Volume 4 (VDA 2006)

Points	Risk, potential damage scenario or damage symptom
9, 10	Very high; safety risk, noncompliance with legal requirements, break down
7, 8	High; vehicle functionality severely restricted, immediate workshop visit absolutely necessary, function restriction of important subsystems
4, 5, 6	Moderate; vehicle functionality restricted, immediate workshop visit not absolutely necessary, function restriction of important operating and comfort systems
2, 3	Minor; minor functional impairment of the vehicle, removal at the next scheduled workshop visit, functional restriction of operating and comfort systems
1	Very low; very low functional impairment, only detectable by qualified personnel

Table FMEA-A.11.4 P-FMEA; allocation of scores for the key indicator A (probability of occurrence) depending on the potential risk in accordance with VDA Volume 4 (VDA 2006)

Points	Probability of occurrence A; qualitative assessment	Error proportion [ppm]
9, 10	Very high; frequent occurrence of the cause of the failure, unusable, unsuitable process	100,000 50,000
7, 8	High; cause of failure occurs repeatedly, inaccurate process	20,000 10,000
4, 5, 6	Moderate; occasional cause of failure, less accurate process	5000 2000 1000
2, 3	Low; occurrence of the cause of the failure is low, accurate process	100 50
1	Very low; occurrence of the cause of the failure is unlikely	1

Table FMEA-A.11.5 P-FMEA; allocation of points for the key figure E (probability of detection) depending on the potential risk in accordance with VDA Volume 4 (VDA 2006)

Points	probability of detection; qualitative assessment	certainty of evidence
9, 10	Very low; detection of the cause of the failure that occurred is unlikely, the cause of the error will or cannot be checked	90%
7, 8	Low; detection of the failure cause that occurred is less likely, likely failure cause not to be detected, uncertain tests	98%
4, 5, 6	Moderate; detection of the cause of the failure is probable, tests are relatively reliable	99.7%
2, 3	High; detection of the cause of the failure that has occurred is very likely, tests are reliable, e.g. several independent tests	99.9%
1	Very high; Occurred failure cause is reliably detected	99.99%

References

Aiag: Measurement Systems Analysis, MSA, 4th ed. Automotive Industry Action Group (AIAG) (2010)

Anderson, T.W., Darling, D.A.: Asymptotic theory of certain "Goodness of Fit" Criteria based on stochastic processes. Ann. Math. Stat. **23**(2), 193–212 (1952)

Arrhenius, S.: Über die Reaktionsgeschwindigkeit bei der Inversion von Rohrzucker durch Säuren. Z. Phys. Chem. **4**, 226–248 (1889)

Bartlett, M.S.: Properties of sufficiency and statistical tests. Proc. Roy. Stat. Soc. Ser. A. **160**(901), 268–282 (1937)

Basshuysen, R., Schäfer, F.: Lexikon Motorentechnik: Der Verbrennungsmotor von A–Z, 2nd ed. Vieweg, Wiesbaden (2006)

Basshuysen, R., Schäfer, F.: Handbuch Verbrennungsmotor: Grundlagen, Komponenten, Systeme, Perspektiven, 7. ed. Springer, Wiesbaden (2015)

Bender, B., Gericke, K.: Pahl/Beitz Konstruktionslehre: Methoden und Anwendung erfolgreicher Produktentwicklung, 9th ed. Springer Vieweg, Wiesbaden (2021)

Bertsche, B., Lechner, G.: Zuverlässigkeit im Fahrzeug- und Maschinenbau: Ermittlung von Bauteil- und System-Zuverlässigkeiten, 3rd ed. Springer, Berlin (2004)

Bienaymé, J.: Considérations è l'appui de la découverte de Laplace sur la loi de probabilité dans la méthode des moindres carrés. Académie Des Sciences, Paris, Comptes-Rendus Hebdomadaires Des Séances **37**, 309–324 (1853)

Birolini, A.: Quality and reliability of technical systems, 2nd edn. Springer, Berlin (1997)

Bitter, P.: Technische Zuverlässigkeit: Problematik, mathematische Grundlagen, Untersuchungsmethoden, 1st ed. Messerschmidt-Bölkow-Blohm GmbH, ed. Springer, Berlin (1971)

Bitter, P.: Technische Zuverlässigkeit: Problematik, mathematische Grundlagen, Untersuchungsmethoden, Anwendungen, 3rd ed. Messerschmidt-Bölkow-Blohm GmbH, ed. Springer, Berlin (1986)

Bracke, S.: RAPP: A new approach for risk prognosis on technical complex products in automotive engineering. In: Stenbergen, R.D.J.M., Van Gelder, P.H.A.J.M., Miraglia, S., Vrouwenvelder, A.C.W.M. (eds.) Safety, reliability and risk analysis: Beyond the horizon. Proceedings of ESREL, Amsterdam, The Netherlands, 29 September–2 October 2013. European Safety and Reliability Association (ESRA), pp. 1333–1338 (2013)

Bracke, S.: Prozessfähigkeit bei der Herstellung komplexer technischer Produkte: Statistische Mess- und Prüfdatenanalyse, 1st ed. Springer Vieweg, Berlin (2016)

Bracke, S.: Preventive maintenance planning based on Weibull distribution models: The impact of the random scatter behaviour of the threshold parameter. In: Proceedings of AMEST2020, 10–11 September 2020, Virtual Workshop, Cambridge, UK. International Federation of Automatic Control (IFAC). 4th IFAC Workshop on Advanced Maintenance Engineering, Services and Technologies (2020)

Bracke, S., Haller, S.: Analysing multiple damage causes of complex products using DCD algorithm and WCF approach. In: Bris, R., Guedes Soares, C., Martorell, S. (eds.) Reliability, risk and safety: Theory and applications. Proceedings of the European safety and reliability conference, Prague, 07–10 September 2009. European Safety and Reliability Association (ESRA). Taylor & Francis Group, London, pp. 819–826 (2009)

Bracke, S., Neupert, F.-G.: Estimation of failure probabilities for maintenance activities regarding product fleets in use: Censored data handling – An automotive engineering case study. In: Baraldi, P., Di Maio, F., Zio, E. (eds.) Proceedings of ESREL 2020 PSAM 15, 01–06 November 2020, Venice, Italy. European Safety and Reliability Association (ESRA); International Association for Probabilistic Safety Assessment and Management. Research Publishing Services, Singapore (2020)

Bracke, S., Puls, A.: Weibull distribution model: Empirical study with regard to the scatter behaviour of threshold parameter – impact on risk analytics. In: Baraldi, P, DI Maio, F., Zio, E. (eds.) Proceedings of ESREL 2020 PSAM 15, 01–06 November 2020, Venice, Italy. European Safety and Reliability Association (ESRA); International Association for Probabilistic Safety Assessment and Management. Research Publishing Services, Singapore (2020)

Bracke, S., Sochacki, S.: The estimation and prognosis of failure behaviour in product fleets within the usage phase – RAPP method. In: Podofillini, L., Sudret, B., Stojadinovic, B., Zio, E., Kröger, W. (eds.) Safety and reliability of complex engineered Systems. Proceedings of ESREL, Zurich, Switzerland, September 07–10, 2015. European Safety and Reliability Association (ESRA), pp. 1141–1148 (2015)

Bracke, S., Sochacki, S.: Saturation models to assess risk in product fleets during the use phase. In: Beer, M., Zio, E. (eds.) Proceedings of ESREL, Hannover, Germany, September 22–26, 2019. European Safety and Reliability Association (ESRA). Research Publishing, Singapore, pp. 1018–1024 (2019)

Braess, H.-H., Seiffert, U.: Vieweg Handbuch Kraftfahrzeugtechnik, 7th edn. Springer Fachmedien Wiesbaden, Wiesbaden (2013)

Braess, H.-H., Widmann, U., Ehlers, C., Breitling, T., Grawunder, N., Liskowsky, V.: Produktentstehungsprozess. In: Braess, H.H., Seiffert, U., (eds.) Vieweg Handbuch Kraftfahrzeugtechnik. ATZ/MTZ-Fachbuch. Springer Vieweg, Wiesbaden (2013)

British Standards Institution: BS 4778-3.2:1991-10-31.Quality; terms. Reliability. Terms from IEC 60050-191. Beuth Verlag, Berlin (1991)

Brühwiler, B.: Risk-Management als Führungsaufgabe: Methoden und Prozesse der Risikobehältigung für Unternehmen, Organisationen, Produkte und Projekte. Haupt, Bern (2003)

Büning, H., Trenkler, G.: Nichtparametrische statistische Methoden, 2nd edn. De Gruyter, Berlin (1994)

Chan, H.A., Englert, P.J.: Accelerated stress testing handbook: Guide for achieving quality products. IEEE Press, New York (2001)

Clopper, C.J., Pearson, E.S.: The use of confidence or fiducial limits illustrated in the case of the binomial. Biometrika **26**, 404–413 (1934)

Coffin, L.F.: A study of the effects of cyclic thermal stresses on a ductile metal. Trans. ASME. **76**(5), 931–950 (1954)

Coffin, L.F., Jr.: Fatigue at high temperature-prediction and interpretation. Proc. Inst. Mech. Eng. **188**(1), 109–127 (1974)

Coles, S.: An introduction to statistical modeling of extreme values. Springer, London (2001)

Conn, A.R., Gould, N.I.M., Toint, P.L.: Trust-region methods. Society for Industrial and Applied Mathematics, Philadelphia (2000)

Conover, W.J.: Practical nonparametric statistics, 3rd edn. Wiley, New York (1999)

Conover, W.J., Iman, R L.: Multiple-comparisons procedures. Informal report. Tech. Rep. LA-7677-MS, Los Alamos Scientific Laboratory. United States (1979)

References

Cox, D.R., Stuart, A.: Some quick sign tests for trend in location and dispersion. Biometrika **42**(1–2), 80–95 (1955)

Czado, C., Schmidt, T.: Mathematische Statistik. Springer-Publisher, Berlin (2011)

Dietrich, E., Schulze, A.: Statistische Verfahren zur Maschinen- und Prozessqualifikation, 5th edn. Hanser, Munich (2005)

Dietrich, E., Schulze, A.: Prüfprozesseignung: Prüfmittelfähigkeit und Messunsicherheit im aktuellen Normenumfeld. Hanser, Munich (2014)

Deutsches Institut für Normung e. V.: DIN 55302-2:1967-01 Statistische Auswertungsverfahren; Häufigkeitsverteilung, Mittelwert und Streuung, Rechenverfahren in Sonderfällen. Beuth Publisher, Berlin (1967)

Deutsches Institut für Normung e. V.: DIN 55302-1:1970-11. Statistical methods; frequency distribution, mean value and dispersion, fundamental terms and general methods. Beuth Publisher, Berlin (1970)

Deutsches Institut für Normung e. V.: DIN 55350-17:1988-08. Begriffe der Qualitätssicherung und Statistik; Begriffe der Qualitätsprüfungsarten. Beuth Publisher, Berlin (1988)

Deutsches Institut für Normung e. V.: DIN 55350-12:1989-03. Begriffe der Qualitätssicherung und Statistik; Merkmalsbezogene Begriffe. Beuth Publisher, Berlin (1989)

Deutsches Institut für Normung e. V.: DIN 40041. Zuverlässigkeit – Begriffe. Beuth Publisher, Berlin (1990)

Deutsches Institut für Normung e. V.: DIN IEC 60050-191:1994-08. Internationales Elektrotechnisches Wörterbuch – Teil 191: Zuverlässigkeit und Dienstgüte (IEC 50(191):1990). Beuth Publisher, Berlin (1994)

Deutsches Institut für Normung e. V.: DIN ISO 281 Beiblatt 2:1994-09 – Wälzlager; Dynamische Tragzahlen und nominelle Lebensdauer. Beuth Publisher, Berlin (1994)

Deutsches Institut für Normung e. V.: DIN 1319-1:1995-01. Grundlagen der Messtechnik – Teil 1: Grundbegriffe. Beuth Publisher, Berlin (1995)

Deutsches Institut für Normung e. V.: DIN 1319-3:1996-05. Grundlagen der Messtechnik – Teil 3: Auswertung von Messungen einer einzelnen Messgröße, Messunsicherheit. Beuth Publisher, Berlin (1996)

Deutsches Institut für Normung e. V.: DIN 1319-4:1999-02. Grundlagen der Messtechnik – Teil 4: Auswertung von Messungen; Messunsicherheit. Beuth Publisher, Berlin (1999)

Deutsches Institut für Normung e. V.: DIN 1319-2:2005-10. Grundlagen der Messtechnik – Teil 2: Begriffe für Messmittel. Beuth Publisher, Berlin (2005)

Deutsches Institut für Normung e. V.: DIN 55350-11:2008-05. Begriffe zum Qualitätsmanagement – Teil 11: Ergänzung zu DIN EN ISO 9000:2005. Beuth Publisher, Berlin (2008)

Deutsches Institut für Normung e. V.: DIN ISO 281:2010-10: Wälzlager – Dynamische Tragzahlen und nominelle Lebensdauer (ISO 281:2007). Beuth Publisher, Berlin (2010)

Deutsches Institut für Normung e. V.: DIN EN ISO 9000:2015-11. Qualitätsmanagementsysteme – Grundlagen und Begriffe (ISO 9000:2015). Deutsche und Englische Fassung EN ISO 9000:2015. Beuth Publisher, Berlin (2015)

Deutsches Institut für Normung e. V.: DIN 50100:2016-12. Schwingfestigkeitsversuch – Durchführung und Auswertung von zyklischen Versuchen mit konstanter Lastamplitude für metallische Werkstoffproben und Bauteile. Beuth Publisher, Berlin (2016)

Deutsches Institut für Normung e. V.: DIN ISO 15226 1999. Technische Produktdokumentation – Lebenszyklusmodell und Zuordnung von Dokumenten. Beuth Publisher, Berlin (2017)

Deutsches Institut für Normung e. V.: DIN ISO 31000:2018-10. Risikomanagement – Leitlinien (ISO 31000:2018). Beuth Publisher, Berlin (2018)

Deutsches Institut für Normung e. V.: DIN ISO 22514-2:2019-07. Statistische Verfahren im Prozessmanagement – Fähigkeit und Leistung – Teil 2: Prozessleistungs- und Prozessfähigkeitskenngrößen von zeitabhängigen Prozessmodellen. Beuth Publisher, Berlin (2019)

Deutsches Institut für Normung e. V. und Bauformumstahl: Eurocode 3. Bemessung und Konstruktion von Stahlbauten. Band 2: Anschlüsse: DIN EN 1993-1-8 mit Nationalem Anhang, Kommentar und Beispiele. D. Ungermann, R. Puthli, T. Ummenhofer, K. Weynand, E. Preckwinkel, Wilhelm Ernst & Sohn Verlag für Architektur und technische Wissenschaften. Beuth Publisher, Berlin (2015)

Dubey, A.D.: On some permissible estimators of the location parameter of the Weibull and certain other distributions. Technometrics. **9**(2), 293–307. Taylor & Francis, Ltd. on behalf of American Statistical Association and American Society for Quality, London (1967)

Duncan, A.J.: Quality control and industrial statistics, 5th edn. Richard D. Irwin Inc., Homewood (1986)

Eckel, G.: Bestimmung des Anfangsverlaufs der Zuverlässigkeitsfunktion von Automobilteilen. Qualität und Zuverlässigk. **22** (9), 206–208. Hanser, München (1977)

Ericson, C.: Fault tree analysis. Sys. Saf. Conf. Orlando, Florida. **1999**, 1–9 (1999)

Eyring, H.: The activated complex in chemical reactions. J. Chem. Phys. **3**, 107 (1935)

Fisher, R.A.: On an absolute criterion for fitting frequency curves. Messenger Math. **41**, 155–160 (1912)

Fisher, R.A., Tippett, L.H.C.: Limiting forms of the frequency distribution of the largest or smallest member of a sample. Proc. Camb. Philos. Soc. **24**, 180–190 (1928)

Ford Motor Co. EU 1880B: Guideline: Capability of measuring systems and measuring equipment. (in German language: Richtlinie: Fähigkeit von Messsystemen und Messmitteln; Translation of EU 1880A) (1997)

Gauß, J.C.F.: Theoria combinationis observationum erroribus minimis obnoxiae. In: Commentationes Societatis Regiae Scientiarum Gottingensis recentiores 5 (classis mathematicae); Pars prior. (First part; February 15, 1821), pp. 33–62; Pars posterior. (Second part; February 2, 1823), pp. 63–90. Publisher Dieterich, Göttingen (1823)

Graebig, K.: Formelsammlung zu den statistischen Methoden des Qualitätsmanagements. DGQ **11**(5), 3rd edn. Beuth Publisher, Berlin (2006)

Graebig, K.: Tabellen, Auswerteblätter und Nomogramme zu den statistischen Methoden des Qualitätsmanagements. DGQ **18**(105), Frankfurt (2010)

Gumbel, E.J.: Statistics of extremes. Columbia Univ, Press (1958)

Haibach, E.: Betriebsfestigkeit: Verfahren und Daten zur Bauteilberechnung, 3rd edn. Springer, Berlin (2006)

Hamerle, A., Fahrmeir, L.: Multivariate statistische Verfahren. de Gruyter, Berlin (1984)

Hanley, J.A., Lippman-Hand, A.: If nothing goes wrong, is everything all right? Interpreting Zero Numerators. J. Am. Med. Assoc. **249**(13), 1743–1745 (1983)

Hartung, J., Epelt, B., Klösener, K.-H.: Statistik: Lehr- und Handbuch der angewandten Statistik. 15th edn. Oldenburg (2009)

Hedderich, J., Sachs, L.: Angewandte Statistik: Methodensammlung mit R, 17th ed. Springer, Springer Spektrum, Berlin (2020)

Henderson, B.: Gospel of the flying spaghetti monster. Harper Collins, New York (2006)

Hinz, M., Temminghoff, P., Bracke, S.: Contribution to multivariate analysis of quality and reliability data in the product testing phase using the example of automotive engineering. In: Gröger, S., Eiselt, T., Schuldt, J. (eds.) Reports on quality management. Gesellschaft für Qualitätswissenschaft e. V., vol. 16, pp. 93–113, Shaker Publisher, Düren (2014)

von Hippel, P.T.: Mean, median, and skew: Correcting a textbook rule. J. Stat. Educ. **13**(2) (2005)

Hoischen, H., Fritz, A.: Technisches Zeichnen: Grundlagen, Normen, Beispiele, Darstellende Geometrie: Lehr-, Übungs- und Nachschlagewerk für Schule, Fortbildung, Studium und Praxis, mit mehr als 100 Tabellen und weit über 1000 Zeichnungen. 35th, revised and expanded edition. Cornelsen Publisher, Berlin (2016)

Hornbogen, E.: Werkstoffe: Aufbau und Eigenschaften von Keramik-, Metall-, Polymer- und Verbundwerkstoffen, 6th edn. Springer, Berlin (1994)

Iman, R.L., Quade, D., Alexander, D.A.: Exact probability levels for the Kruskal-Wallis test. In: Harter, H.L., Owen, D.B. (eds.) Selected tables in mathematical statistics, pp. 329–384. Institute of Mathematical Statistics and American Mathematical Society, New York (1975)

International Organization for Standardization: ISO/IEC Guide 98-3:2008. (GUM, 1995). Uncertainty of measurement – Part 3: Guide to the expression of uncertainty in measurement. Beuth Publisher, Berlin (2008)

International Organization for Standardization: ISO 14253-1, 2017-10. Geometrical product specifications (GPS) – Inspection by measurement of workpieces and measuring equipment – Part 1: Decision rules for verifying conformity or nonconformity with specifications. Beuth Publisher, Berlin (2017)

International Organization for Standardization: ISO 31000-2018. Risk Management – Guidelines. Beuth Publisher, Berlin (2018)

International Organization for Standardization: ISO 22514-7, 2021. Statistical methods in process management. In: Capability and performance Part 7: Capability of measurement processes. Beuth Publisher, Berlin (2021)

Johnson, L.G.: The statistical treatment of fatigue experiments. Elsevier, Amsterdam (1964)

Jonckheere, A.R.: A distribution-free k-sample test against ordered alternatives. Biometrika **41**(1/2), 133–145 (1954)

Kamiske, G.F., Brauer, J.-P.: Qualitätsmanagement von A bis Z: Erläuterungen moderner Begriffe des Qualitätsmanagements, 2nd edn. Hanser, Munich (1995)

Kaplan, E., Meier, P.: Nonparametric estimation from incomplete observations. J. Am. Stat. Assoc. **53**(282), 457–481 (1958)

Keferstein, C.P., Marxer, M., Bach, C.: Fertigungsmesstechnik: Alles zu Messunsicherheit, konventioneller Messtechnik und Multisensorik, 9th ed.: Springer Fachmedien, Springer Vieweg, Wiesbaden (2018)

Kendall, M.G., Gibbons, J.D.: Rank correlation methods, 5th edn. Edward Arnold, London (1990)

Kendall, M.G., Stuart, A.: The advanced theory of statistics., Vol. 1, 2nd edn. Hafner Publishing, New York (1967)

Khamis, I.H., Higgins, J.J.: An alternative to the Weibull step-stress model. Int. J. Qual. Reliab. Manage. **16**(2), 158–165 (1999)

Kolmogorov, A.N.: Sulla determinazione empirica di una legge di distribuzione. Giornale Dell'istituto Italiano Degli Attuari **4**, 83–91 (1933)

Kolmogorov, A.N.: Foundations of the theory of probability, 2nd ed. in English, Chelsea Publications, New York (1956)

Kruskal, W.H.: A nonparametric test for the several sampling problem. Ann. Math. Stat. **23**(4), 525–540 (1952)

Kruskal, W.H., Wallis, W.A.: Use of ranks in one-criterion variance analysis. J. Am. Stat. Assoc. **47**(260), 583–621 (1952)

Kruskal, W.H., Wallis, W.A.: Use of ranks in one criterion variance analysis. J. Am. Stat. Assoc. **48**(264), 907–911 (1953)

Kühlmeyer, M.: Statistische Auswertungsmethoden für Ingenieure mit Praxisbeispielen. Springer, Berlin (2001)

Labisch, S., Weber, C.: Technisches Zeichnen: Selbstständig lernen und effektiv üben, 4th edn. Springer Fachmedien, Wiesbaden (2014)

Lai, C.D.: Generalized Weibull distributions. Springer, Berlin (2014)

Larson, H.R.: Un nomogramme de la distribution cumulative binomiale. Revue de Statistique Appliquée. **15**(2), 39–58 (1967)

Lee, L.: Multivariate distributions having Weibull properties. J. Multivar. Anal. **9**(2), 267–277. Elsevier Inc, Amsterdam (1979)

Lee, C.K., Wen, M.-J.: A multivariate Weibull distribution. Pak. J. Stat. Oper. Res. **5**, 55–66 (2010)

Levene, H.: Robust tests for equality of variances. In: Olkin, I., Hotelling, H. et al. (eds.) Contributions to probability and statistics: Essays in honor of harold hotelling. Stanford University Press, pp. 278–292 (1960)

Leveson, N.G.: White paper on approaches to safety engineering. Massachusetts (2003)

Linß, G.: Qualitätsmanagement für Ingenieure, 2nd edn. Fachbuchverlag, Leipzig (2005)

Linß, G.: Qualitätssicherung – Technische Zuverlässigkeit. Lehr- und Arbeitsbuch. Hanser, Munich (2016)

Liu, J., Zenner, H.: Berechnung von Bauteilwöhlerlinien unter Berücksichtigung der statistischen und spannungsmechanischen Stützziffer. Mater. Sci. Eng. **26**, 14–21 (1995)

Lu, J.-C., Bhattacharyya, G.K.: Some new constructions of bivariate Weibull models. Ann. Inst. Stat. Math. **42**, 543–559 (1990)

Lundberg, G., Palmgren, A.: Dynamic capacity of rolling bearings. Acta Polytech. Mech. Eng. Ser. **1**(3). The Royal Swedish Academy of Engineering Sciences (1947)

Lundberg, G., Palmgren, A.: Dynamische Tragfähigkeit von Rollenlagern. Skf Vereinigte Kugellagerfabriken, Schweinfurt (1953)

Mandel, J.: The statistical analysis of experimental data. InterScience Publisher, New York (1964)

Mann, N.R., Fertig, K.N.: A goodness-of-fit test for the two parameter vs. three parameter weibull; confidence bounds for threshold. Technometrics **17**(2), 237–245 (1975)

Mann, N.R., Fertig, K. N., Scheuer, E.M.: Confidence and tolerance bounds and a new goodness-of-fit test for two-parameter Weibull or extreme -value distributions. Aerospace Research Laboratories, Ohio (1971)

Mann, H., Whitney, D.: On a test of whether one of two random variables is stochastically larger than the other. Ann. Math. Stat. **18**(1), 50–60 (1947)

Manson, S.S.: Behaviour of materials under conditions of thermal stress. National Advisory Committee for Aeronautics, Tech. Note 2933 (1953)

Manson, S.S.: Fatigue: A complex subject – Some simple approximations. Exp. Mech. **5**(7), 193–226 (1965)

Martensen, A.L., Butler, R.W.: NASA Technical Memorandum 89098: The fault-tree compiler. NASA (National Aeronautics and Space Administration), Langley Research Center, Hampton, Virginia (1987)

Matthews, R.: Der Storch bringt die Babys zur Welt ($p = 0.008$). Stochastik in der Schule **21**(2), 21–23 (2001)

Meyer-Bahlburg, H.F.L.: A nonparametric test for relative spread in k unpaired samples. Metrika **15**(1), 23–29 (1970)

Military Handbook 338: Electronic reliability design. US Military (1998)

Military Procedure MIL-P-1629: Procedures for performing a failure mode, effects and criticality analysis. US Military (1949)

Miller, L.H.: Table of percentage points of Kolmogorov statistics. J. Amer. Statist. Assoc. **51**(273), 111–121 (1956)

Milton, R.C.: An extended table of critical values for the Mann-Whitney (Wilcoxon) two-sample statistic. J. Am. Stat. Assoc. **59**(307), 925–934 (1964)

Miner, M.A.: Cumulative damage in fatigue. J. Appl. Mech. **12**(3), 159–164 (1945)

Moré, J.J.: The Levenberg-Marquardt algorithm: Implementation and theory. In: Watson, G.A. (ed.) Numerical analysis. Lecture notes in mathematics. **630**, Springer, Berlin (1978)

Nelson, W.: Applied life data analysis. Wiley, New York (1982)

Nelson, W.: Accelerated testing: Statistical models, test plans and data analyses. Wiley, New York (1990)

Palmgren, A.: Die Lebensdauer von Kugellagern. VDI-Z. **68**(14), 339–341 (1924)

Papula, L.: Mathematik für Ingenieure und Naturwissenschaftler Band 3: Vektoranalysis, Wahrscheinlichkeitsrechnung, mathematische Statistik, Fehler- und Ausgleichsrechnung, 5th ed. Springer Fachmedien, Springer Vieweg, Wiesbaden (2008)

Pfeifer, T.: Qualitätsmanagement: Strategien, Methoden, Techniken, 3rd ed. Hanser, Munich. (English: 2002. Quality Management: Strategies, Methods, Techniques. 3rd ed. Munich: Hanser) (2001)

Ramos, P., Nascimento, D., Louzada, F.: The long term Fréchet distribution: Estimation, properties and its application. Biometrics & Biostatistics Int. J. **6**, 1–7 (2017)

Ramos, P.L., Louzada, F., Ramos, E., Dey, S.: The Fréchet distribution: Estimation and application – An overview. J. Stat. Manage. Sys. **23**(3), 549–578 (2020)

Rao, C.V., Swarupchand, U.: Multiple comparison procedures – A note and a bibliography. J. Stat. **16**(1), 66–109 (2009)

Rayleigh, J., Strutt, V.: On the problem of random vibrations and of random flights in one, two, and three dimensions. Philos. Mag. **37**(220), 321–347 (1919)

Rhodes, E.C.: Population mathematics-III. J. R. Stat. Soc. **103**(3), 362–387 (1940)

Rinne, H.: The Weibull distribution: A handbook. CRC Press, Boca Raton (2009)

Rinne, H., Mittag, H.-J.: Statistische Methoden der Qualitätssicherung, 3rd ed. Hanser, Munich (English: 2003. *Statistical Methods of Quality Assurance*. 1st ed. London: Chapman & Hall) (1995)

Rizzo, M.L.: Statistical computing with R. Chapman & Hall/CRC (2008)

Rogers, W.P.: Introduction to system safety engineering. Wiley, New York (1971)

Schulze, A.: Statistical Process Control (SPC). In: Pfeifer, T., Schmitt, R. (eds.) Masing Handbuch Qualitätsmanagement, 6th ed. Hanser, Munich (2014)

Shewhart, W.A.: Economic control of quality of manufactured product. Orig: American Society for Quality Control. Post Print: Martino Fine Books, 2015 (1931)

Shewhart, W.A.: Statistical method from the viewpoint of quality control. [N. p.], orig: Washington, The Graduate School, The Department of Agriculture; Post Print: Dover Publications. Dover Books on Mathematics, 2012 (1939)

Siegel, S., Tukey, J.W.: A nonparametric sum of rank procedure for relative spread in unpaired samples. J. Am. Stat. Assoc. **55**(291), 429–445 (1960)

Singh, V.P.: Entropy-based parameter estimation in hydrology. Kluwer, Alphen aan den Rijn (1998)

SKF (Svenska Kugellagerfabriken): Wälzlager. Leaflet PUB BU/P1 10000/2 EN (2014)

Smirnov, N.: Table for estimating the goodness of fit of empirical distributions. Ann. Math. Stat. **19**(2), 279–281 (1948)

Sochacki, S., Bracke, S.: The comparison of the estimation and prognosis of failure behaviour in product fleets by the RAPP method with state-of-the-art risk prognosis models within the usage phase. In: Cepin, M., Bris, R. (eds.) Safety and reliability – Theory and applications. Proceedings of the ESREL, Portorož, Slovenia, 18–22 June 2017. European Safety and Reliability Association (ESRA). CRC Press, pp. 3481–3490 (2017)

Spearman, C.: The proof and measurement of association between two things. Am. J. Psychol. **15**, 72–101 (1904)

Student: The probable error of a mean. Biometrika **6**(1), 1–25 (1908)

Tait, N.R.S.: Robert Lusser and Lusser's Law. Saf. Reliab. **15**(2), 15–18. Taylor and Francis, London (1995)

Thorndike, F.: Applications of poisson's probability summation. Bell Sys. Tech. J. **5**(4), 604–624 (1926)

Tintner, G.: Eine neue Methode für die Schätzung der logistischen Funktion. Metrika-Zeitschrift Für Theoretische Und Angewandte Statistik **1**, 154–157 (1958)

Chebyshev, P.L.: Sur les fonctions qui différent le moins possible de zéro. Journal De Mathématiques Pures Et Appliquées **19**, 319–346 (1874)

Tukey, J.W.: Exploratory data analysis. Addison-Wesley, Boston (1977)

Van't Hoff, J.H.: Études de dynamique chimique. Recl. Trav. Chim. Pays-Bas. **3**, 333–336 (1884)

Verband der Automobilindustrie e.V.: VDA Volume 3.2. Zuverlässigkeitssicherung bei Automobilherstellern und Lieferanten: Verfahren und Beispiele. Verband der Automobilindustrie e.V. (VDA). / Qualitätskontrolle in der Automobilindustrie, Berlin (1976)

Verband der Automobilindustrie e.V.: VDA Volume 4.2. Qualitätsmanagement in der Automobilindustrie, Sicherung der Qualität vor Serieneinsatz, 1st ed. Verband der Automobilindustrie e.V. (VDA), Berlin (1996)

Verband der Automobilindustrie e.V.: VDA Volume 4: Quality assurance before series production – Product and process FMEA, 2nd ed. Verband der Automobilindustrie e.V. (VDA), Berlin (2006)

Verband der Automobilindustrie e.V.: VDA Volume 5. Capability of measurement processes, 2nd ed. Verband der Automobilindustrie e.V. (VDA), Berlin (2010)

Verband der Automobilindustrie e.V.: Reliability assurance of car manufacturers and suppliers – Reliability management, 4th ed. Verband der Automobilindustrie e.V. (VDA), Berlin (2019a)

Association of the Automotive Industry e.v.: AIAG & VDA FMEA Handbook. Design FMEA, Process FMEA, Supplemental FMEA for Monitoring & System Response, 1st ed. Verband der Automobilindustrie e.V. (VDA), Berlin (2019b)

Verein deutscher Ingenieure e.V.: VDI/VDE/DGQ 2619:1985-06. Prüfplanung. Beuth Verlag (successor to VDI/VDE 2600 Sheet 3; planned publication in 2022-10), Berlin (1985)

Verein deutscher Ingenieure e.V.: VDI 4001 Blatt 1:1998. Allgemeine Hinweise zum VDI-Handbuch Technische Zuverlässigkeit. Beuth Publisher, Berlin (1998)

Verhulst, P.-F.: Notice sur la loi que la population poursuit dans son accroissement. Correspondance Mathématique Et Physique **10**, 113–121 (1838)

Verhulst, P.-F.: Recherches mathematiques sur la loi d'accroissement de la population. Nouveaux Mémoires De L'académie Royale Des Sciences Et Belles-Lettres De Bruxelles **18**, 14–54 (1845)

Villanueva, D., Feijóo, A., Pazos, J.L.: Multivariate Weibull distribution for wind speed and wind power behavior assessment. Resources **2**(3), 370–384 (2013)

Wallis, W.A., Moore, G.H.: A significance test for time series analysis. J. Am. Stat. Assoc. **36**(215), 401–409 (1941)

Watson, H.A.: Launch control safety study. Bell Telephone Laboratories, Murray Hill (1961)

Weibull, W.: A statistical distribution function of wide applicability. ASME J. Appl. Mech. **18**(3), 293–297 (1951)

Westkämper, E., Warnecke, H.-J.: Einführung in die Fertigungstechnik, 8th, updated and extended ed. Vieweg+Teubner / Springer Fachmedien Wiesbaden, Wiesbaden (2010)

Wilcoxon, F.: Individual comparisons by ranking methods. Biometrics Bulletin **1**(6), 80–83 (1945)

Wilrich, P.-T., Graf, U., Henning, H.-J.: Formeln und Tabellen der angewandten mathematischen Statistik, 3rd edn. Springer, Berlin (1987)

Wirtz, M.A., Dorsch, F., Strohmer, J.: Dorsch – Lexikon der Psychologie, 18th edn. Hogrefe, Bern (2017)

Yang, G., Zaghati, Z.: Accelerated life tests at higher usage rates: a case study. Proceedings of the annual reliability and maintainability symposium (RAMS '06), pp. 313–317. (2006)

Index

100% inspection, 408

A
Absolute frequencies, 79
Absolute frequency, 46, 47
Accelerated Binary Test, 343
Accelerated Destructive Degradation Test, 343
Accelerated Life Test, 343
Accelerated Repeated Measures Degradation Test, 343
Accelerated testing, 4, 12, 342, 366, 367
Acceleration factor, 366, 367
Acceptance control chart, 435, 436
Addition law, 428
Addition theorem, 68
Adjustment, 38
α-error, 281
Alternative hypothesis, 276, 278
Alternative model, 141
Amstetten locomotive, 348
Analysis of variance, 333, 389
AND-linkage, 62
ANOVA, 333
Arithmetic mean, 89, 157
Arrhenius law, 344
Average Range Method, 389
Axioms of Kolmogorov, 66

B
B_3 parameter, 236
Bar diagram, 46
Bartlett-test, 337
Behrens-Fisher problem, 324
Beta binomial distribution, 177
Beta distribution, 173
β-error, 281
Bias, 385
Binomial coefficient, 130
Binomial distribution, 128, 133
Binomial test, 296
Block diagram, 208, 215
Body, 10
Boolean algebra, 207
Boole Theorems and axioms, 207
Boxplot, 52
Brake system, 213
Bravais-Pearson, 183
Bravais-Pearson correlation coefficient, 185
Bridge circuit, 216
B_Z-parameters, 236

C
Calibration, 38
Candidate prognosis, 251
Candidates, 242, 262
Capability characteristics, 395
Capability indices, 420
Capability of test equipment, 402
Carry-over-parts (COP), 15
Cause-effect chains, 203
c-control chart, 444
Censored data, 251
Censoring, 34
Central limit theorem, 85
Centroid, 88, 311
C_g-/C_{gk}-Study, 382, 384
C_g capability index, 385
Characteristic life span, 175
Characteristic lifetime, 111
Characteristics, 25
Chassis, 10
X^2-distribution, 123, 336, 338, 340
Classification, 81
Clopper-Pearson approach, 172, 174

CNOMO standard, 82
Coffin-Manson model, 357
Commanded failure, 202
Comment, 203
Competing model, 143
Component specification, 27
Component test, 252
Concept phase, 11
Condition-based maintenance, 35
Condition monitoring, 36
Confidence interval, 3, 155–157, 167, 177, 237, 239
Confidence level, 155
Confounding variable, 197
Conjunction, 203
Connecting rod eye, 43
Conover, 336
Consistency, 282
Consistent, 84
Continuous, 25
Continuous characteristics, 94
Continuous distribution models, 94
Continuously distributed random variable, 95
Control chart, 5, 431, 432
Control chart limits, 435
Control charts for continuous characteristics, 435
Control charts for discrete characteristics, 441
Controlled process, 404
Coolant pump, 243
Correct value, 37
Correlation, 3, 181
Correlation analysis, 181
Correlation coefficient, 183
Cox and Stuart trend test, 307
Cumulative diagram, 48

D

Damage appearance, 262
Damage data, 33
Damage scenarios, 262
Data classification, 81
Data review, 232
Data Visualization, 421
Decision certainty, 18
Decision-making, 16
Definition of Technical Reliability, 31
De Morgan's law, 208
Density function, 95, 121, 127
Dependence of characteristics, 181
Dependent samples, 282
Design, 12
Design FMEA, 226

Destruction limit, 372
Deviation, 24
Device under test, 361, 362
Dice roll, 62
Disc brake, 202
Discrete, 25
Discrete distribution, 127
Discretely distributed random variable, 127
Disjunction, 203
Dispersion, 23, 89, 311, 318
Distribution-free significance tests, 283
Distribution function, 22, 94, 95, 121, 127
Distribution models, 3
Double logarithmic, 49
Double logarithmizing, 146
Downward trend, 307
Dual-circuit braking system, 214
Dubey, 150

E

Early failure behaviour, 32, 238
Eckel method, 257
Efficiency, 84
Electronics, 10
Elementary events, 61
End of Production (EOP), 14
Endurance range, 350
Erlang, 115
Erlang distribution, 115
Error of the 1st kind, 280
Error of the 2nd kind, 280
Estimation of parameters, 145
Estimator, 84
Euler, 61, 64
Event, 62
Event Tree Analysis, 208
Event trees, 69
Expanded measurement uncertainty, 396
Expectation-true, 84
Expected value, 23, 37
Exponential distribution, 111
Extreme value distributions, 117
Extreme value theory, 117
Eyring law, 345
Eyring model, 360

F

Failure, 24
Failure behaviour, 233, 243, 367
Failure behaviour due to runtime, 32
Failure-free time, 150
Failure Mode and Effects Analysis, 201, 225

Index

Failure probability, 32, 121, 209, 361
Failure rate, 32, 121, 237
Fatigue strength range, 350
Fault tree analysis, 201, 202
F-distribution, 125, 173, 179
Fiedler, 5
Field support phase, 14
Fisher distribution, 125
Fisher-Tippett distribution, 117
Fisher-Tippett-Gnedenko theorem, 117
Fisher's F-distribution, 331
FMEA, 3, 225
Folded normal distribution, 104
Fractional importance, 213
Fréchet distribution, 118
Frontloading approach, 19
FTA, 3, 202
F-Test, 331
Fuel consumption, 190
Full inspection, 408
Functional characteristics, 409
Function-critical characteristics, 27, 378

G

Galton board, 129
Gamma distribution, 114
Gamma function, 122
Gauging, 38
Gauss, 96, 189
Gaussian normal distribution, 104
Global average temperature, 199
Goat Problem, 75
Goodness, 282
Gosset, 121
Graph paper, 49
Grauburgunder, VIII
Growth process, 137
%GRR indicator, 388
%GRR study, 382, 387
GUM, 396
Gumbel, 152, 234
Gumbel distribution, 118

H

HALT, 372
HASS, 373
Highly Accelerated Life Test, 372
Highly Accelerated Stress Screening, 373
Histogram, 47
H-test, 334
Hypergeometric distribution, 130, 135

Hypothesis, 275
Hypothesis test, 276

I

Idempotent law, 207
Independent samples, 282
Inspection characteristic, 28
Interval censored data, 34
Inverse Power Law, 356
Inverse Power Weibull Model, 357
Isochrone diagram, 54

J

Johnson ranking, 253

K

Kaplan-Meier estimator, 257
Key indicators, 235
Kolmogorov, 66
Kolmogorov-Smirnov goodness-of-fit test, 291
Kolmogorov's axioms, 66
Kruskal and Wallis, 334
KS goodness-of-fit test, 291
KS test statistic, 293

L

Laplace experiment, 64
Larson nomogram, 49
Left censored data, 34
Let's Make a Deal, 75
Lifetime distribution, 32
Life time distributions, 119
Lifetime variable, 32
Likelihood function, 147
Linear damage accumulation, 350
Linearization, 146
Location parameter, 110
Logarithmic normal distribution, 107
Logarithmic paper, 49
Logarithmization, 146
Logistic function, 137
Logistic model, 137
Long-term process capability, 406
Long-term process capability analysis, 414
Lord of Cork, 59
Lundberg-Palmgren equation, 358
Lusser, 5

M

Machine capability, 406
Machine capability analysis, 413
Machine capability indices, 407, 413
Machine capability study, 423
Maintenance, 35
Mann-Whitney U-Test, 311
Marginal importance, 213
Mathematical papers, 49
Maximum likelihood estimator, 147, 234
Me 262, 6
Mean, 85
Mean Time Between Failures, 35
Mean Time To Failure, 35
Mean Time To Repair, 35
Mean value, 88
Measured value, 37
Measured variable, 37
Measurement, 36, 37
Measurement deviation, 37
Measurement object, 37
Measurement process uncertainty, 392
Measurement result, 37, 379
Measurement standard, 38
Measurement system analysis, 4
Measurement system uncertainty, 392
Measurement uncertainty, 36, 379, 381
Measuring, 36
Measuring system, 38
Median, 89
Mercury, 192
Messerschmitt-Bölkow-Blohm, 6
Metric, 25
Meyer-Bahlburg test, 339
Miner Eurocode, 355
Miner-Haibach, 355
Miner-Liu-Zenner, 355
Miner-Palmgren, 354
Miner rule, 351
Minimal cut sets, 216
Minimum success paths, 216
Mixed distribution model, 140
Modal value, 89
Monte Carlo simulation, 430
Monty Hall Problem, 75
Multiple sample case, 333
Multiplication theorem, 68
Multi-sample case, 286
Multivariate characteristic capability index, 426
Multivariate distribution model, 429
Multivariate process capability index, 427
Multivariate process evaluation, 424

N

Negation, 203
Nelson procedure, 254
Nominal, 26
Nomograms, 49
Non-failed units, 242
Non-parametric significance tests, 283
Normal distribution, 96
Null hypothesis, 276, 278

O

One-sample case, 286
One-sided confidence interval, 157
Operating load limit, 372
Ordinal, 25
Original value, 22
OR-linkage, 62
Outlier, 37

P

Parallel arrangement, 209
Parameter-free significance tests, 283
Parametric significance tests, 283
p-control chart, 443
Pirates on earth, 198
Piston rod, 43
Poisson distribution, 132
Poisson process, 114
Population, 22
Power, 282
Power accelerating factor, 360
Powertrain, 10
Pre-development, 12
Predictive maintenance, 35
Preliminary process capability, 406
Preliminary process capability indices, 407, 414
Preliminary process capability study, 414
Primary failure, 202
Probabilistics, 2
Probability, 22, 61
Probability density function, 22
Probability of error, 155
Probability of failure, 209
Probability of survival, 209
Probability paper, 49
Procedure X, 402
Process capability index, 415
Process capability indices, 407, 415
Process characteristics, 24
Process FMEA, 226

Index 505

Process scenarios, 446
Process suitability analyses, 406
Product characteristics, 24
Product design, 12
Product emergence process, 2, 9
Product fleet, 33, 232, 242, 251, 259
Production phase, 13
Product life cycle, 9
Proportion value, 86, 167
Prototype, 12
Prototype testing, 341, 360
P-value, 279

Q

Q_{MS}-/Q_{MP}-Study, 382, 392
Qualitative Accelerated Test, 342
Qualitative characteristics, 25
Quality, 36
Quality capable process, 404
Quality control charts, 431, 432
Quantile approach, 417
Quantitative Accelerated Test, 342
Quantitative characteristics, 25

R

Ramp-up, 13
Random experiment, 21, 61
Random failure behaviour, 32
Random failures, 239
Random influences, 405
Random measurement deviation, 37
Random result, 21
Random variable X, 22
Range, 89
Rank assignment, 313
RAPP procedure, 259
Ratio of damaged units, 34
Rayleigh distribution, 106
Recycling phase, 10
Regression, 3
Regression analysis, 146, 182, 186
Reject probability, 428, 430
Relative frequencies, 79
Relative frequency, 46, 47
Reliability, 28, 30
Reliability Analysis, 232
Reliability planning, 16
Repeatability and reproducibility, 387
Resolution, 38
Resolution analysis, 384
Right-censored data, 34

Risk, 39
Risk analysis, 40, 41
Risk Analysis and Prognosis of complex
  Products (RAPP), 258
Risk avoidance, 19
Risk elimination, 18
Risk handling, 41
Risk identification, 41
Risk matrix, 40
Risk monitoring, 41
Risk prevention, 18
Risk priority number, 228
Robust manufacturing, 405
Robustness, 282
Ronald Aylmer Fisher, 331
Runtime-related failure behaviour, 239

S

Safety-critical characteristics, 27, 378, 409
Safety-critical features, 13, 408
Sample, 22, 79
Sample space, 61
Sampling inspection, 71
Sampling inspection strategy, 409
Saturation function, 137, 153
Saturation limit, 137
Saturation model, 262
Scatter, 23, 89
Schematic diagrams of control charts, 446
Secondary failure, 202
Separation, 216
Series development, 12
Series preparation, 12
Series production, 13
Series structure, 209
Service life ratio, 363
Shape parameter, 110, 175
Shewhart control chart, 432, 435, 437
Short-term strength range, 350
Short-term test equipment capability, 407
Siegel-Tukey test, 318
Significance level, 278
Significance test, 4, 275, 276
Sign test, 296
Single logarithmic, 49
Single sample case, 285
Sinusoidal stress curve, 349
Skewness, 91
Software, 10
SOP, 13
SPC *see* statistical process control
SPC control loop, 410

Spearman, 183
Spearman correlation coefficient, 185
Specification, 23
Specification Limit, 24
Standard deviation, 90
Standard input, 203
Standard normal distribution, 99, 101
Start of production, 13
Statistical process control (SPC), 4, 403, 409
Statistical significance test, 278
Statistical test, 276
Steiner's law, 234
Storks, 197
Structural importance, 212
Student's t-distribution, 122
Sudden death test, 251
Sudden death testing, 252
Suitability of test processes, 402
Surface topography, 194
Survival probability, 32, 121, 209, 237, 361
Systematic measurement deviation, 37
System FMEA, 226
Systemic influences, 405

T
Target value, 23
Taylor equation, 359
Taylor series expansion, 396
T-distribution, 121, 307
Technical drawing, 26
Technical reliability, 2, 28
Test/measuring device, 37
Test distributions, 121
Test equipment capability, 381, 384, 406
Test for outliers, 302
Testing, 36
Test process suitability, 4, 39, 377, 381, 406, 407
Test Process Suitability Study, 392
Test statistic, 278
Thorndike nomogram, 49
Threshold parameter, 110, 150, 176
Timelkam railway station, 348
Tintner and Rhodes, 153
Tolerance, 23
Total probability, 70
Transmission input/output, 203
Trend, 307
True value, 37
T-test, 306, 324

Two sample case, 285
Two-sample case, 311
Two-sided confidence interval, 157

U
Unbiasedness, 282
Uncertainty, 38
Uncertainty components, 393
Uniform distribution, 112
Upward trend, 307
Urn, 72
Usage profile, 255
U-test, 311

V
Value added, 17
Value beam diagram, 46
Value pattern, 45
Van't Hoff, 344
Van't Hoff rule, 346
Variance, 23, 85, 90, 160, 161
Vehicle, 10
Vehicle fleet, 256
Venn, 61, 64
Verfahren X, 402
Verhulst, 137
Verhulst saturation model, 139, 269

W
Wallis-Moore Test, 287
Wallis-Moore test for randomness, 286
Weibull, 109
Weibull distribution, 118, 146, 148, 175, 233, 240
Weibull distribution model, 269, 362, 366
Weibull probability paper, 50, 109
Whisker, 54
Wilcoxon test, 317
Wöhler lines, 349
Wöhler test, 348

X
x-control chart, 442, 444
X-Y plot, 59

Printed in the United States
by Baker & Taylor Publisher Services